PERCOLATION PROCESSES: THEORY AND APPLICATIONS

NATO ADVANCED STUDY INSTITUTES SERIES

Proceedings of the Advanced Study Institute Programme, which aims at the dissemination of advanced knowledge and the formation of contacts among scientists from different countries.

The series is published by an international board of publishers in conjunction with NATO Scientific Affairs Division

| A | Life Sciences | Plenum Publishing Corporation |
| B | Physics | London and New York |

| C | Mathematical and Physical Sciences | D. Reidel Publishing Company Dordrecht and Boston |

| D | Behavioural and Social Sciences | Sijthoff & Noordhoff International Publishers B.V. |
| E | Applied Sciences | Alphen aan den Rijn, The Netherlands and Rockville, Maryland, USA |

Series E: Applied Sciences - No. 33

PERCOLATION PROCESSES: THEORY AND APPLICATIONS

edited by

ALIRIO E. RODRIGUES
Professor of Chemical Engineering, University of Oporto,
Oporto, Portugal

and

DANIEL TONDEUR
Laboratoire des Sciences du Génie Chimique du CNRS, Nancy, France

SIJTHOFF & NOORDHOFF   1981
Alphen aan den Rijn, The Netherlands
Rockville, Maryland, USA

Proceedings of the NATO Advanced Study Institute on
Percolation Processes: Theory and Applications
Espinho, Portugal, July 17-29, 1978

ISBN-13: 978-94-009-8581-0   e-ISBN-13:978-94-009-8579-7
DOI: 10.1007/978-94-009-8579-7

PREFACE

In a general way, "percolation" might be defined as a
process in which at least two phases, one continuous and the
other dispersed, are in bulk relative movement and exchange heat
and/or mass through their interface.

In practical terms this covers operations such as adsorption,
ion exchange, leaching, washing, fluid-fluid displacement in porous media, deep bed filtration, chromatography.

Apart from the fact that the design of these processes is
almost empirical we have concluded to the necessity of organizing
a Summer School on this topic because

1. Powerful new concepts have emerged in this area in the last
    decade, centered around unsteady-state, non-linearly coupled,
    multi component systems, and low energy processes

2. There is, potentially, a phenomenology that is common to all
    percolation operations, and which even extends to other mi-
    gration phenomena such as electrophoresis, sedimentation,
    traffic flow

3. There is a need for a synthetic and didatic approach to these
    problems and its spreading will be most fruitful for the de-
    velopment of separation science.

We have tried to choose lecturers who have, themselves, made significant contributions toward the development of such an
approach.

The lectures given at the Summer School held at Espinho,
Portugal in July 17-29, 1978 were compiled in this volume. It is

divided in three parts. In the first part some general and intro-
ductory notions common to all operations are presented and chroma
tography is analised in detail.

The second part is concerned with multi component problems;
starting with the equilibrium model and then introducing deviations (nonisothermal, nonequilibrium cases); the analysis of ion
exchange with chemical reaction closes this section illustrating
the use of theoretical concepts.

In part three cyclic operations are studied, namely parametric pumping, pressure swing adsorption, etc. and a novel applica
tion -the chromatographic reactor- is treated in some detail.

I want to acknowledge my coworkers who run the social programme and the secretary of this course, Maria Idalina who pati-
ently typed some of the manuscripts.

This Summer School was made possible because of the finan-
cial support of the NATO Advanced Study Institute Programme. The
participants and ourselves are very indebted to this organization.

A.E. Rodrigues

D. Tondeur

TABLE OF CONTENTS

VIII

Part 1

GENERAL PRINCIPLES OF PERCOLATION PROCESSES
AND CHROMATOGRAPHIC APPLICATIONS

POPULATION MIGRATION AND WAVE PHENOMENA IN PERCOLATION OPERATIONS

Daniel TONDEUR

Maître de Recherche at the Centre National de la Recher-
che Scientifique.
Ecole Nationale Supérieure des Industries Chimiques -
Laboratoire des Sciences du Génie Chimique - 1, rue
Grandville - 54042 NANCY-FRANCE.

1. INTRODUCTION

This text gathers a variety of concepts and representations
that are used in different areas of physics and other fields such
as traffic flow and chemical engineering. Its aim is not to present
a unified, universal and self-sufficient approach to all problems,
but rather to incite the reader to look occasionnaly outside his
own field for the solution of his problems.

The first concept we ought to introduce is certainly that of
percolation. The chemical engineer might define it as a process in
which one phase flows through another phase, exchanging with it
energy and/or matter across their common interface.

Although this definition may seem simple and common sense, its
accuracy relies entirely on the term "flows through", the exact
meaning of which is not obvious : First, it implies some topologi-
cal properties of the phases : they are "non-compact", that is,
they have "holes" filled by the other phase. Which phase flows
through the other is not always obvious unless one phase is immobile. For instance, in dense-bed liquid-liquid extraction, one phase
flows as a bed of adjacent drops, while the other flows in the intervals between the drops. But the distinction is probably unimportant for the purpose of this course; we shall restrict the scope to
cases where one phase has a permanent open (connected) pore structure, in other words to a fluid flowing through a granular or porous
material.

The terms "flows through" also implies an orientation : the
flow is not random, the phases are not mixed. For the purpose of

4

this course, we restrict ourselves to the case of one preferential direction, and possibly two.

Many of the concepts we shall use to describe percolation apply to processes not described by the above definition, even in its widest meaning. They apply even to processes where no physical phases can be recognised: the example of traffic flow used in this text illustrates this.

Finally, we should keep aware that the term "percolation" or "percolation transition" is being used for a different but neighbouring concept : that of the connectedness of a network. This notion has been introduced by the British mathematicien Hammersley around 1956, to describe the formation of infinite, connected clusters of one metal component in an alloy. An excellent review of the wide application of the concept is made by De Gennes (1), for such areas as the properties of semi-conductors, macroscopic magnetism, the formation of large macromolecules, or the propagation of an epidemy. From a chemical engineer's point of view, it would apply also to the drainage, or to immiscible fluid displacement in a solid. In the language we use here, this "transition percolation" is somewhat analogous to the "breakthrough".

2. PERCOLATION AS A POPULATION MIGRATION

2.1 Cars on a highway

Car traffic is a percolation process among others, with the difference that all of us have probably had some practice, but little theory. Choosing this process to introduce the basic concepts of percolation will allow us to illustrate the generality of these concepts.

Our purpose is to determine what type of information can be obtained and how it can be obtained. More precisely, we shall be concerned with quantitative information, thus with measurable quantities. The questions we propose to examine are :

- where, how, what to observe ?

- What information is contained in the observations ?

Our purpose is not to modelize traffic, a task approached in a very clear fashion in a book by Prigogine and Herman (2).

2.2 There are many different ways of observing

The table below summarizes the essential concepts. The reader may find out which are transposable to usual percolation.

OBSERVATION

WHERE ?	HOW ?	WHAT ?
Standing Still Flying high above Moving along	Follow individual cars or not Impose a perturbation or not Observe at a point or on a fi- nite interval	Count numbers Measure velocities Measure times

The concepts introduced briefly here are discussed in more detail below, where we examine specific combinations of where, how and what.

2.3 "Point" Observation : non-individual, non-perturbing measurement

By "point" observation, we mean an observation which concerns the local, or the instantaneous behavior of cars, and not their behavior over a finite interval of time or distance.

The standing observer counts anonymous cars as a function of time at a given location z, and can draw the two graphs of Figure 1, $m_z(t)$ and $M_z(t)$ such that :

. $m_z(t)$ dt is the number of cars that pass at location z in a time interval (t, t + dt)

. $M_z(t)$ is the total number of cars that have passed at location z from t = 0 to t.

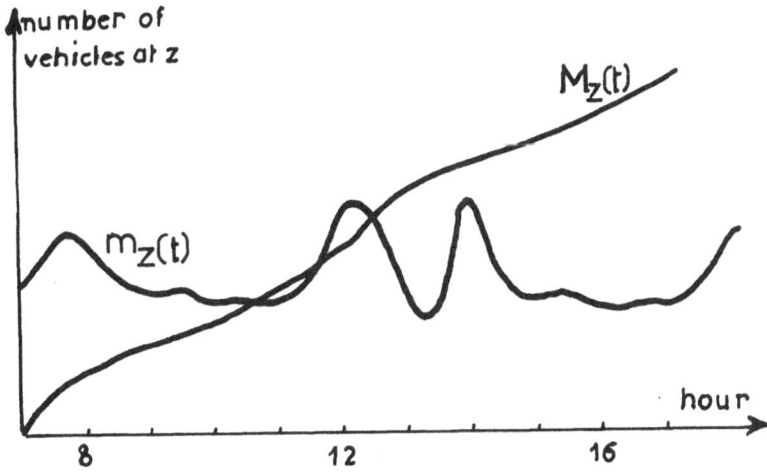

Figure 1 : Example of local density history, and integral

Clearly,

$$M_z(t) = \int_o^t m_z(t) \, dt \qquad (1)$$

The first curve bears several names : <u>time-distribution</u>, time-density curve, time frequency curve, local density (or concentration) history, isoplane, flow curve. Names for the second are obtained by adding "cumulative", or "integral" in front of the previous names.

With a suitable instrument, the standing observer can also measure the instantaneous velocities of the cars, and draw the two graphs of Figure 2, of $m_z(u)$ and $M_z(u)$ versus velocity u, such that

. $m_z(u)$ du is the number of cars that pass at location z with a velocity in the interval (u, u + du) :

. $M_z(u)$ is the total number of cars that have passed at location z with a velocity smaller than u. Clearly,

$$M_z(u) = \int_o^u m_z(u) \, du \qquad (2)$$

The first curve is called the <u>local velocity distribution</u> and the second the integral local velocity distribution.

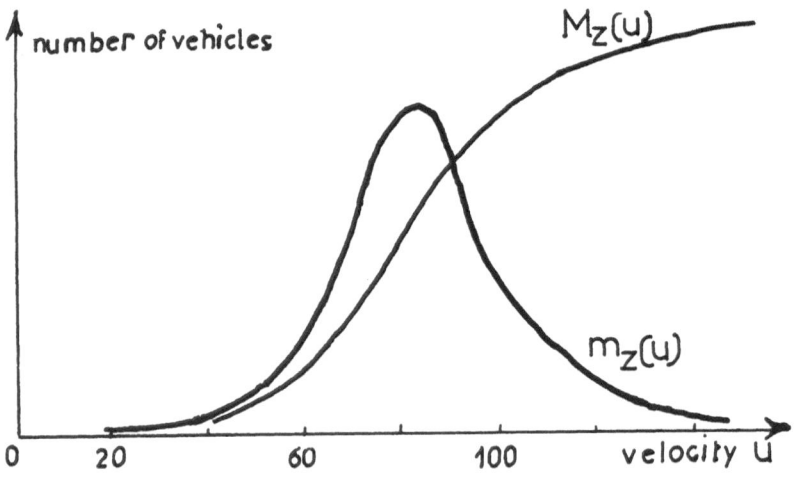

Figure 2 : Local velocity distribution and integral

Finally, if he is near a rest area or a gasoline station, the standing observer may count the number of cars that stop in a time interval (t, t + dt), the number of cars that start moving in this interval (stopping and starting distributions) and the total number of cars that <u>are</u> stopped at any time (local holdup distribution). He thus establishes three distributions $n_s(t)$, $n_m(t)$ and $N_z(t)$ respectively, such that

$$N_z(t) = N_z(o) + \int_o^t (n_s - n_m)\, dt \tag{3}$$

The standing observer is probably the most frequent situation in percolation. However the velocity distribution, and the fixed phase concentration, (non-moving cars) cannot always be measured. We see that here, the observer has not perturbed the system, and he has not considered cars individually.

The <u>flying (spatial) observer</u> can see at one glance a long stretch of highway, and take photographs of it. On one picture, he will be able to count the cars along the stretch, at the time of the picture. If he has taken long exposures, or two photographs at a short time interval, he will be able to determine the velocities of the cars from their displacement in that time interval. The following graphs of $m_t(z)$, $M_t(z)$, $m_t(u)$, $M_t(u)$ may then be drawn, which are qualitatively similar to that of Figures 1 and 2.

. $m_t(z)$ dz is the number of cars which at time t are in a distance interval (z, z + dz).

. $M_t(z)$ is the total number of cars in the stretch from the begining to some distance z, and clearly

$$M_t(z) = \int_o^z m_t(z)\, dz$$

. $m_t(u)$ du is the number of cars with a velocity in the interval (u, u + du) at time t.

. $M_t(u)$ is the total number of cars which at time t, have a velocity smaller than u, and $M_t(u) = \int_o^u m_t(u)\, du$. $m_t(z)$ is called <u>space-distribution</u>, space density or space frequency curve, instantaneous concentration or density profile, isochrone. $m_t(u)$ is the <u>instantaneous velocity distribution</u>. $M_t(z)$ and $M_t(u)$ are called integral or cumulative space or velocity distribution respectively.

The flying observer may also determine the distribution $n_t(z)$ of non-moving cars along the highway stretch and its integral $N_t(z)$.

<u>The driving (Lagrangian) observer</u> can count the cars that

pass him, cars that he passes, non-moving cars, and measure the velocities. If he drives with a known schedule, say a known velocity u_0 (not necessarily constant), its position at any time is known; time and position are thus interchangeable for him. He may define the graphs $m_+(z)$, $m_-(z)$, $M_+(z)$, $M_-(z)$, $m_L(u)$, $M_L(u_r)$, $M_L(u)$, $M_L(u_r)$, so that:

. m_+dz, is the number of cars that pass the observer, m_-dz is the number passed by him, in a distance interval $(z, z + dz)$. $M+$ and $M-$ are the distance integrals (from o to z) of these Lagrangian passing distributions.

. $m_L(u)$ is the <u>absolute Lagrangian velocity distribution</u>, seen by the observer and $M_L(u)$ its integral, defined as for the standing or spatial observers.

. $m_L(u_r)$ is the <u>relative Lagrangian velocity distribution</u>, where the relative velocity u_r is defined by

$$u_r = u - u_0 \qquad (4)$$

$M_L(u)$ and $M_L(u_r)$ are the integral (over u on u_r) of these distributions. Note that the two velocity distributions are not superposable, except when u_0 is constant.

The techniques envisaged in this section do not rely on following individual cars, nor on imposing a perturbation to the system, but they still hold in either of this situation.

2.4 Finite interval, or two-point observation

In order to obtain information on the behaviour of cars over a finite distance or time interval, we must make at least two observations on each car and preferably one at each end of the interval. In addition, we must make sure to match properly the two observations. This implies either that each car can be recognised, marked in some way, or that something is known on the situation of each car at one end of the interval. There are three classical situation, considered below : individual follow up experiments, tracer experiments and perturbation methods.

<u>Individual follow-up experiments</u>. They consist in marking certain cars so as to recognise them individually. They do not perturb the flow, in principle. For instance, several observers standing at different locations may measure the passage time of individual cars, and possibly their local velocity, or a flying observer may follow them continuously. It is then possible to draw the <u>space-time trajectory</u> of the sample studied (figure 3), that is the position as a function of time of each car. The slope of this

curve at any time is the instantaneous velocity of the car at this
time. Observations may be limited to two for each car ; for instance the passage time may be determined at the two ends of a stretch
of road, the difference of these times being the residence time τ
on the stretch. An overall velocity v may then be defined as the
ratio of the length of the stretch to the residence time. The residence time distribution m (τ) and the overall velocity distribution m (v) may be drawn (figure 4) as well as their integral.
Several such distributions can be defined, depending on the sample
observed ; for instance the cars counted may be those leaving the
freeway in a given time interval, or the cars entering in a given
interval, or those that were present at time t = o, or some selection, or random sampling of either of the preceding. All these distributions are in general different and it is important that the
sampling be carefully specified. Only at steady state are these
distributions identical.

Individual follow up is hardly practical in chemical engineering,except may be in the area of particle transport.

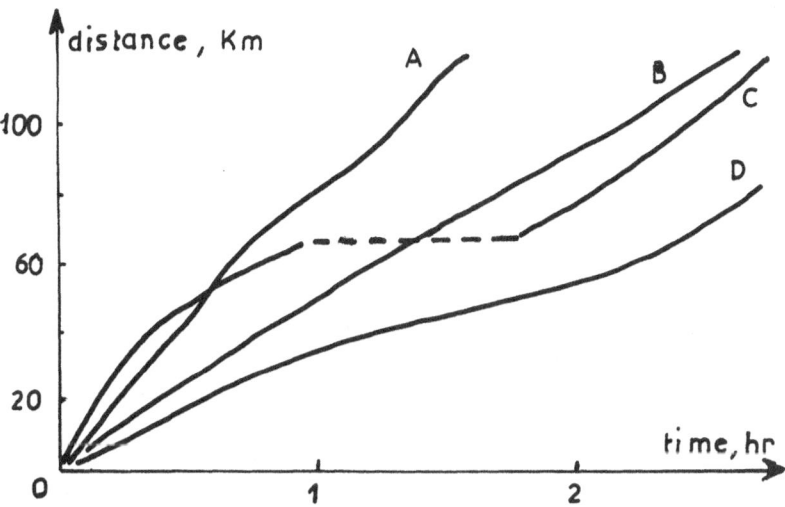

Figure 3 : Examples of space-time trajectories

Tracer experiments. They consist in marking a limited and
well defined sample of cars, and observing the behaviour of the
sample as a whole, not of the individuals. There are again several
ways of doing this, depending on the sample definition and the measurement. Basically, four types of sample definition are used :
random, cyclic, impulse, step. We leave the first two aside ; they

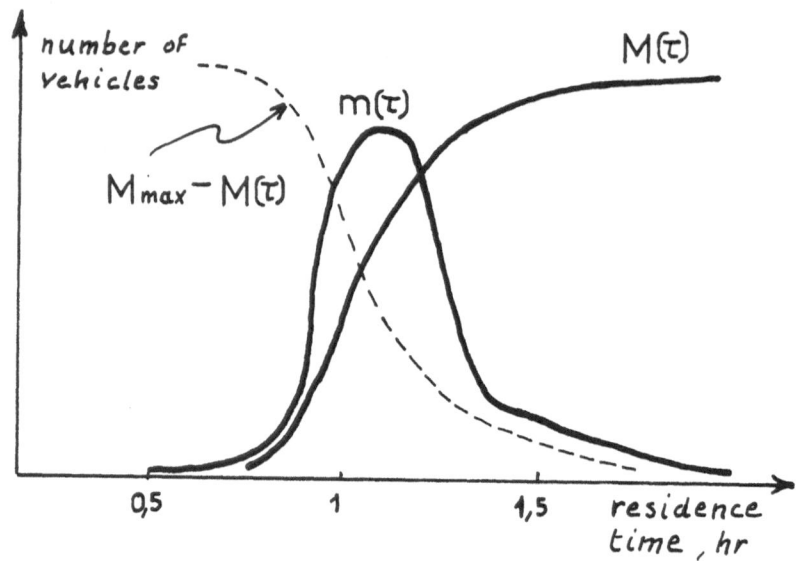

Figure 4 : Example of residence time distribution, integral, and complementary integral

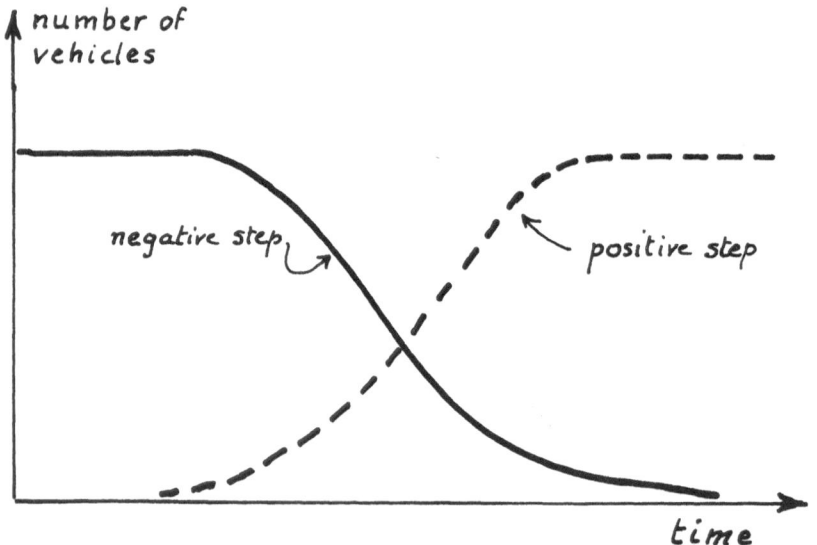

Figure 5 : Examples of response to step changes

may give information on the overall kinetics, but not on the behaviour of the cars except under restrictive conditions. In an impulse experiment, the sample consists in the cars that enter a
stretch of highway within a time interval short enough with respect to the residence time, so that it may be assumed that they
all start at the same time. Their passage time is then measured at
a given location, and the residence time distribution m (τ) deduced
as well as its integral M (τ). At steady state, such an experiment
thus gives the same information as the individual follow up experiment, as far as residence time is concerned. The residence time
distribution is sometimes called the "exit age distribution function".

In a step experiment, the sample consists in all cars that
are present in the stretch at time t = o (negative step) or all
cars that enter after time t = o, and up to a sufficient time (positive step). In the same way as above, passage times are measured,
and the time distribution is plotted (figure 5). It is not immediately obvious what information these step responses contain about
car behaviour and this will be considered later. The response to a
negative step is sometimes called the "washout" curve, or the elution curve. The response to a positive step is sometimes called the
F-curve, or breakthrough curve. More detailed analysis of tracer
kinetics may be found for instances in Levenspiel (3), Buffham
(4,5), Giddings (6).

Flow perturbations. Such perturbations consist for example in
interrupting the traffic for a while. They do not require to mark
individual cars, but give information only on some overall kinetics,
not on the behaviour of cars.

2.5. Defining averages

The arithmetic average of a collection of observations is the
sum of the values observed, divided by the number of observations.
Applying this definition to the distributions defined earlier, considered as continuous, we obtain for example :

. the average flow of cars at z, $\overline{m}_z = \dfrac{\int_o^t m_z(t)\ dt}{\int_o^t dt} = \dfrac{M_z(t)}{t}$ 	(5)

. the average local velocity $\overline{u}_z = \dfrac{\int_o^\infty m_z(u)\ u\ du}{\int_o^\infty m_z(u)\ du = M_z(\infty)}$ 	(6)

The average spatial density $\overline{m}_t$ is defined in a way similar to
$\overline{m}_z$, the average instantaneous velocity $\overline{u}_t$, the average residence

time $\bar{\tau}$, the average overall velocity $\bar{v}$, are defined in the same way as $\bar{u}_z$. We notice that for distributions with a finite support, (bounded integral) such as u_z, u_r, τ , v, the average is the first moment of the distribution, normalized by the total area of the distribution curve. This property does not apply to the unbounded distributions such as $m_z(t)$.

With the above definition of $\bar{u}_z$, we can write the following relation between $m_z(t)$ and $m_t(z)$, which is a first-order approximation, for a small section dz and time interval dt :

$$m_z(t) \; dt = \bar{u}_z(t) \; m_t(z) \; dz \tag{7}$$

flow into the section = average velocity $\times$ density in section.

Let us now consider a given sample of marked cars. We wish to define an average moving velocity u on a stretch of road which is somehow related to the average residence time on one hand, and to the average velocities defined above. If the local velocities u_z of the cars of the sample were measured at various positions on the stretch (o,z), we can define a global velocity distribution $M(u)$ such that $M(u) \; du$ is the number of observations (not of cars) giving a velocity between u and $u + du$. In the same way as above, an average velocity $\bar{U}$ can be defined as the first normalized moment of this distribution. But the same $\bar{U}$ can be obtained as averages of $\bar{u}_z$ or $\bar{u}_t$ defined on the same sample and the same stretch. Then we have

$$\left. \begin{array}{l} \text{average moving velocity} \\ \text{of sample on stretch } (o,z) \end{array} \right\} \; \bar{U} = \frac{\displaystyle\int_o^\infty M(u) \; u \; du}{\displaystyle\int_o^\infty M(u) \; du} = \frac{1}{z} \int_o^z \bar{u}_z \; d_z$$

$$= \frac{1}{t} \int_o^t \bar{u}_r \; dt \tag{8}$$

and $\bar{U} = z/\overline{td} \tag{9}$

where $\overline{td}$ is the average time necessary to drive from o to z.

We now relate the overall velocities v, and $\bar{v}$, to u and $\bar{U}$. (recall that v is related to the residence time by $v = z/\tau$, and the same relation holds between $\bar{v}$ and $\bar{\tau}$). For doing this, we need to introduce the stopping time distribution $n(ts)$ and the average stopping time $\overline{ts}$. $n(ts)$ is defined in such way that $n(ts) \; dt$ is the number of cars that stop for a duration in the interval $(ts, ts + dt)$. The average $\overline{ts}$ of this distribution is defined as above, and we have

$$\overline{td} + \overline{ts} = \bar{\tau} \tag{10}$$

$$g = \frac{\overline{ts}}{\overline{\tau}} \tag{11}$$

$$\text{average fraction of} \atop \text{time cars are stopped} \quad = \quad \frac{\text{average stopping time}}{\text{average residence time}}$$

and $f = 1 - g = \overline{td}/\overline{\tau}$ (12)

where f is called the fractional moving time.

With these definitions, the average overall velocity $\overline{v}$ is related to the average "moving" velocity $\overline{U}$ by

$$\overline{v} = f\,\overline{U} \tag{13}$$

These relations hold when all the averages are defined on a given sample of cars and on a given stretch of road.

2.6 Unsteady flow regime

The unsteady regime is characterized by the fact that at any location, the flow of cars and their velocities change with time. In other words, the local density distribution is not uniform, the local velocity distribution depends on the interval of observation chosen. In this situation, no accessible relations exist between the various observations described so far, (except in the cases considered below), and these observations are thus independent, each bringing additional information on the unsteady regime. Exceptions are :

. when a marked sample can be observed at many times and locations, then the relations established in the previous section 2.5 hold between average velocities, stopping times and residence times.

. when the regime can be considered as fluctuations (that is relatively small deviations) around a steady regime.

. when the transition from a steady regime to another is considered (step changes), or the momentaneous deviation from a steady regime (impulse). In what follows, we shall consider only steady regimes, steps and impulses.

2.7 The steady flow regime

It is defined in such way that the observations made at a given location are independent of time. The first consequences are that the various distributions are time independent : the time distribution is uniform, that is, $m_z(t) = $ constant, and $M_z(t)$ is a straight line ; the flows of stopping and starting cars n_s and n_m are constant and equal ; the number of non-moving cars at any location N_z is constant ; the space distributions $n_t(z)$, $m_t(z)$ are independent of time ; the local velocity distribution $m_z(u)$ is independent of the interval of observation (if it is large enough) ; the instantaneous velocity distribution u_t is independent of the time ; for the Lagrangian observer, with a given driving schedule u_o, his observations are independent of the time chosen to do the trip ; the residence time distributions obtained by individual follow up are independent of the sample chosen, and are thus all identical to the impulse response.

An interesting consequence of this is that the average values can be obtained in a less restrictive manner : there is no need to follow a marked sample, and the average values are independent of sampling providing the sample is large enough. In particular, the average moving velocity $\overline{U}$ is equal to the average instantaneous velocity $\overline{u}_t$ since the latter is independent of time, and we have

$$\overline{v} = z/\overline{\tau} = f\,\overline{u}_t \tag{14}$$

overall average velocity = fractional driving time x average instantaneous velocity.

Probably the most important consequence of the steady regime is the relation between average times and velocities and the "holdup" of cars. At steady regime, we may write

mo	x	$\overline{td}$	=	$M_t(z)$ (15)
flow of cars at entrance of stretch per unit time		average driving time necessary to arrive to z		number of cars moving on the stretch (o,z)

mo	x	$\overline{\tau}$	=	$M_t(z) + N_t(z)$ (16)
		average time cars spend on the stretch = residence time		number of cars moving + number of cars stopped.

$M_t(z)$ and $N_t(z)$ are the integrals of the space distributions defined in section 2.3. Using the relations between times and velocities, and Equations 15 and 16, we can write:

$$f \quad = \quad \frac{\overline{td}}{\overline{\tau}} \quad = \quad \frac{M}{M+N} \quad = \quad \frac{\overline{v}}{\overline{U}} \tag{17}$$

fractional driving time $= \dfrac{\text{driving time}}{\text{total time}} = \dfrac{\text{number of moving cars}}{\text{total number of cars}} \quad \dfrac{\text{overall velocity}}{\text{moving velocity}}$

2.8 Linear processes

A linear process is a process in which any perturbation entails a proportional change in response. Important consequences of
this definition are that if several linear processes are occuring
simultaneously in a system, their overall effect is also a linear
process, furthermore, the response of a linear system to several
perturbations occuring simultaneously or not, is the sum of its responses to each perturbation considered individually. We consider
below two applications of this concept.

Relation between step and impulse response in steady flow. In
the sense of distributions, the "Heaviside step function" is the
integral of the "Dirac delta function", so that a positive unit
step at time o may be written

$$H(o) = \int_o^\infty \delta(t)\ dt$$

with $\delta(t) = \infty$ at $t = 0$
$\delta(t) = o$ at any other time.

If the system is linear, the response to a positive unit step
is the integral of the responses to unit impulses. The single impulse response was seen to be the steady-state residence time distribution $m(\tau)$. Its integral $M(\tau)$ (figure 4) is thus essentially
the response to a positive step (fig. 5), multiplied by some adjusting constant (we have not considered, in the traffic example, unit
step or impulses). Symmetrically, the "complement" of $M(\tau)$, that
is $M(\infty) - M(\tau)$, is essentially the response to a negative step.
The value of $M(\infty) - M(\tau)$ for any τ measures the number of cars
that have a residence time larger than τ, thus the response to a
a negative step, or the $M(\infty) - M(\tau)$ curve, is sometimes called
the "internal age distribution function".

Dispersion. The spreading of the distributions we have considered can be measured by the second moment of the distribution.
The second moment centered around the average is the square of the
variance σ^2. For instance, for the residence time distribution,
we write :

$$\bar{\tau} = \frac{\int_o^\infty m(\tau)\tau\ d\tau}{\int_o^\infty m(\tau)\ d\tau} = \text{first moment} \tag{18}$$

$$\sigma_\tau^2 = \frac{\int_o^\infty m(\tau)\ (\tau-\bar{\tau})^2\ d\tau}{\int_o^\infty m(\tau)\ d\tau} = \text{second moment = variance} \tag{19}$$

The use of moments for studying impulse response in linear chromatography is considered in detail in another conference of this
course. Let us merely mention that in linear processes, the varian-

ces can be shown to be additive. For example, if the spreading
of the residence time distribution is due to the spreading of the
velocity distribution and of the stopping time distribution, letting σ_u^2 and σ_s^2 be the variances of these two distributions we may
write

$$\sigma_\tau^2 = \sigma_u^2 + \sigma_s^2 \tag{20}$$

Open and closed section. In all the preceding discussions, we
have implicitly assumed that no vehicle entering a stretch of road
turns around and goes back out, and that no car going out of the
stretch observed comes back into it. In chemical reactor engineering, we call this a "closed vessel", that is, closed to diffusion
or backmixing at both ends. Accounting for "open" conditions amounts
to introducing a flux at the boundaries which is not due simply to
bulk flow. "Open boundary conditions are considered in the course on
linear chromatography".

3 - PERCOLATION AS A WAVE PHENOMENON

A wave is the propagation of a perturbation of a flow. Describing percolation as a wave phenomenon thus amounts to following perturbations rather than individuals. Since perturbations are usually applied volontarily in chemical engineering system, to effect separation, or chemical reaction, the wave approach is the most commonly used in this field.

3.1 The material balance

We look at a small stretch of road, of length dz without exit or entrance, during a time interval dt. We write that the flow of cars coming in, m_z dt, plus the initial holdup of moving and standing cars $(m_t + n_t)$ dz, equals the flow out m_{z+dz} dt, plus the final holdup $(m_{t+dt} + n_{t+dt})$ dz. Expressing the small variations as differentials, we write

$$m_{z+dz} = m_z + \left(\frac{d\,m_z}{dz}\right)_t \; ;$$

and similarly for m_{t+dt} and n_{t+dt}. Thus the material balance is :

$$m_z dt + m_t dz + n_t d_z = \left(m_z + \frac{\partial m_z}{\partial z}\right) dt + \left(m_t + \frac{\partial m_t}{\partial t}\right) dz$$

$$+ \left(n_t + \frac{\partial n_t}{\partial t}\right) dz$$

$$\text{or} \quad \left(\frac{\partial m_z}{\partial z}\right)_t dt + \left(\frac{\partial m_t}{\partial t}\right)_z dz + \left(\frac{\partial n_t}{\partial t}\right)_z dz = o$$

recalling from section 25, Equation 7, that

$$m_z(t)\ dt = \bar{\bar{U}}_z(t)\ m_t(z)\ dz$$

we can eliminate m_z, to obtain finally

$$\boxed{\frac{\partial}{\partial z}\,(\bar{U}_z m_t) + \frac{\partial}{\partial t}\,(m_t + n_t) = o} \qquad (21)$$

We recognize here the classical form of a conservation equation in a system with no source, no sink, no diffusion and one directional flow. The general form of such a conservation is

$$\frac{\partial \rho}{\partial t} \quad = \quad - \text{div J} \tag{22}$$

accumulation,
time derivative   =   divergence of a flux
of a concentration
or density

The divergence operator : $\text{div J} = \frac{\partial J_x}{\partial x} + \frac{\partial J_y}{\partial y} + \frac{\partial J_z}{\partial z}$
is here reduced to one coordinate derivative. The flux is purely convective, it is the product of a velocity (cm sec^{-1}) by a density or concentration (cars cm^{-1}) on unit cross-section. In a more general situation the flux would be the sum of this convective term and of a diffusion or dispersion term, of the form :

$$J = u\, c - D \frac{\partial c}{\partial z}$$

We see also that m_t and n_t play the role of concentration ; we shall consider them as such from here on, and to use notations more in harmony with the rest of this course, we let : $m_t = c =$ concentration (density) of mobile vehicles (mobile phase) $n =$ concentration (density) of immobile vehicles (stationnary phase).

3.2. Boundary and initial conditions

To the differential conservation equation, we must adjoin boundary conditions which express the flow, thus the concentration at the inlet of the system, and occasionally at the outlet. Although perturbations will usually be applied at the inlet, constraints may still occur at the outlet. This is the case in particular when the system is said "open to diffusion or backmixing". In terms of traffic, diffusion or backmixing implies that a vehicle that has entered a stretch of road can turn around and go back. Boundary conditions of this type are considered in the text on linear chromatography. We shall restrict the present study to systems closed to diffusion, and to step perturbations, and we write for example

Inlet Boundary condition $z = 0$, $t < 0$ $c = c_0$ (23)
                                   $t > 0$ $c = c_i$

Initial condition $z > 0$ $t = 0$ $\begin{cases} c = c_0 \\ n = n_0 \end{cases}$ (24)

The inlet B.C. expresses a step change in concentration at time 0. The initial condition expresses that the system is initially at uniform concentration (mobile and immobile).

3.3. The differentiel balance as a kinematic wave equation.

We want to rewrite the conservation equation in a form which is close to the kinematic wave equations as studied for example by Whitham (8) :

$$\frac{\partial \rho}{\partial t} + v \frac{\partial \rho}{\partial z} = 0 \tag{25}$$

For doing this, we do the following operations on equation

- assume $\bar{u}$ is constant : the derivatives of $\bar{u}$ disappear
- assume a relation exists between mobile and immobile concentrations of the form $n = n \, (c,z)$. $\tag{26}$

With the assumption of constant $\bar{u}$, the material balance is written

$$u \frac{\partial c}{\partial z} + \frac{\partial c}{\partial t} + \frac{\partial n}{\partial t} = 0 \tag{27}$$

using 26 to eliminate n :

$$\left(\frac{\partial n}{\partial t}\right)_z = \left(\frac{\partial n}{\partial c}\right)_z \left(\frac{\partial c}{\partial t}\right)_z$$

and substituting into 27 :

$$\frac{\partial c}{\partial t} + \frac{u}{1 + \dfrac{\partial n}{\partial c}} \frac{\partial c}{\partial z} = 0 \tag{28}$$

which is the form of the kinematic wave equation 25 , with

$$v_c = \frac{u}{1 + \dfrac{\partial n}{\partial c}} = v \, (c,z) \tag{29}$$

This equation expresses that the concentration c propagates with velocity v_c. This can be seen by writing a chain relation between the derivatives

$$\left(\frac{\partial c}{\partial t}\right)_z \left(\frac{\partial t}{\partial z}\right)_c \left(\frac{\partial z}{\partial c}\right)_t = -1$$

giving, together with equation 28 :

$$-\frac{\partial c/\partial t}{\partial c/\partial z} = v_c = \frac{u}{1 + \dfrac{\partial n}{\partial c}} = \left(\frac{\partial z}{\partial t}\right)_c = \text{velocity of propaga} \atop \text{tion of } c \tag{30}$$

This relation is very important, as it relates the velocity of a concentration to the average velocity of flow and to the slope of the n (c) curve.

Note 1 : Some authors introduce the change of variable

$$\theta = ut - z \tag{31}$$

which amounts to taking an origin of times moving at velocity u. It is thus a Lagrangian description. The change of variable leads to

$$\frac{\partial c}{\partial \theta} + v_L \frac{\partial c}{\partial z} = o \tag{32}$$

with

$$v_L = \frac{1}{\partial n/\partial c} \tag{33}$$

and the relation between the velocities is

$$\frac{v_c}{u} = \frac{v_L}{1 + v_L} \tag{34}$$

Note 2 : we have established, in section 2. 7 , a relation at steady regime between the velocities $\bar{v}$, $\bar{u}$, and the mobile and immobile holdup M and N. This was Equation 17 :

$$\frac{\text{driving time}}{\text{residence time}} = \frac{\bar{v}}{\bar{u}} = \frac{M}{M + N}$$

Let M = c and N = n, and we have exactly the overall, integral version of equation 30 .

3.4 Geometrical analysis of the kinematic wave equation. The quasi-linear equation

Let us remind some definitions, to allow the reader to make his way through the mathematical litterature. Consider the differential equation

$$P \frac{\partial c}{\partial z} + Q \frac{\partial c}{\partial t} = R \tag{35}$$

This equation is said to be

   . first order, because there are only first derivatives
   . linear with constant coefficients, if P, Q and R are constant
   . linear, with variable coefficients, if P, Q, R are functions
                                    of z and t only.
   . semi-linear, if P and Q are functions of z and t,
                      and R function of z,t and c
   . quasi-linear, if P, Q, R are functions of z, t and c
                   but not of their derivatives.
   . homogeneous, if R = o.

    In the kinematic wave equation 28 , the unique coefficient
v_c is a function of c and z, since we have postulated a relation
n = n (c,z). The equation is thus quasi-linear, first order, homogeneous. Books like Courant and Hilbert (11) treat in detail the
geometrical properties of such equations and in particular the
notion of characteristics.

    The characteristics. The solution of the differential equa-
tions , or , can be represented in the three-dimensional space
(c,z, t). A priori, it is likely to be a family of surfaces in this
space. Let M (c, z,t) be a point of such a surface. It can be shown
(7) that the vector of directing coefficients, P, Q, R is tangent
to the surface at point M. Along this "characteristic" vector, a
coordinate may be defined, so that one can write

$$\frac{dz}{ds} = P \quad ; \quad \frac{dt}{ds} = Q \quad ; \quad \frac{dc}{ds} = R \tag{36}$$

which is sometimes written

$$\frac{dz}{P} = \frac{dt}{Q} = \frac{dc}{R} = ds \tag{37}$$

in the present case, we have P = v_c, Q = 1 and R = o.

    Since dc/ds = o, c is constant along a characteristic. If in
addition, we assume n is a function of c only, not of z, then v_c
depends only on c, and is thus constant along a characteristic. As
Q = 1 is constant, all the directing coefficients are constant, and
the characteristic is a straight line, parallel to the (z,t) plane.

    For constructing a solution corresponding to given initial and
boundary conditions, one draws a network of characteristics starting from the "initial curve".Figure 6 shows the network of
straight characteristics corresponding to a step from c_o to c_i at
z = o, t = o. The initial curve here is the segment of the

c-axis comprised between c_0 and c_i.

Figure 6 : Characteristics in (c,z, t) space, corresponding to a step perturbation.

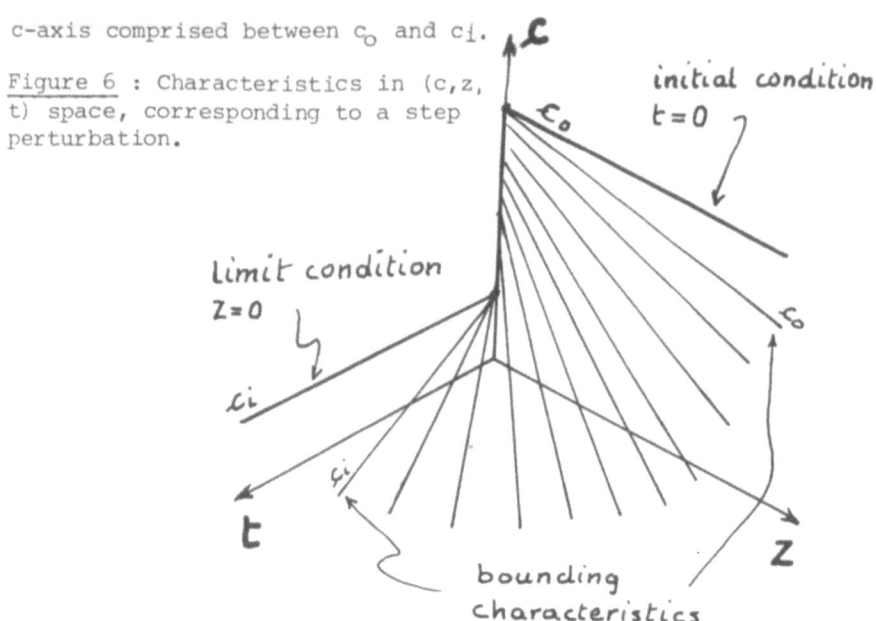

Two-dimensional representations. If we cut the network of characteristics by planes at t = constant (parallel to the c, z, plane), the intersections represent succesive concentrations profiles or space distributions. (see figure 7). Similarly, cutting by planes at z = constant, we obtain time distributions or concentration histories at different abscissae; only one is shown on Figure 7, figure 8 shows the projection of the characteristics on the (z,t) plane. The slope of each characteristic in this plane is v_c. These representations resemble the ones we have introduced in section 2. However, keep in mind that the (z,t) space-time representation concerns here concentrations, not individuals.

The complete solution. It can be seen by simple inspection of the differential equation that a trivial solution is c = constant. This trivial solution which represents the boundary and initial conditions has to be combined with the non-trivial curve to give the complete solution. Geometrically, this means that plateaus corresponding to the initial condition c_0 and inlet condition c_i bound the non-trivial c - curve as is visible on Figure 5 . The length of the plateaus vary as the front moves ahead. This is also expressed by the fact that the initial and inlet concentrations both propagate forward.

3.5 Properties of the step response

In the example considered above, the initial curve is on the c-axis, and describes a step input. Figure 8 shows that in the

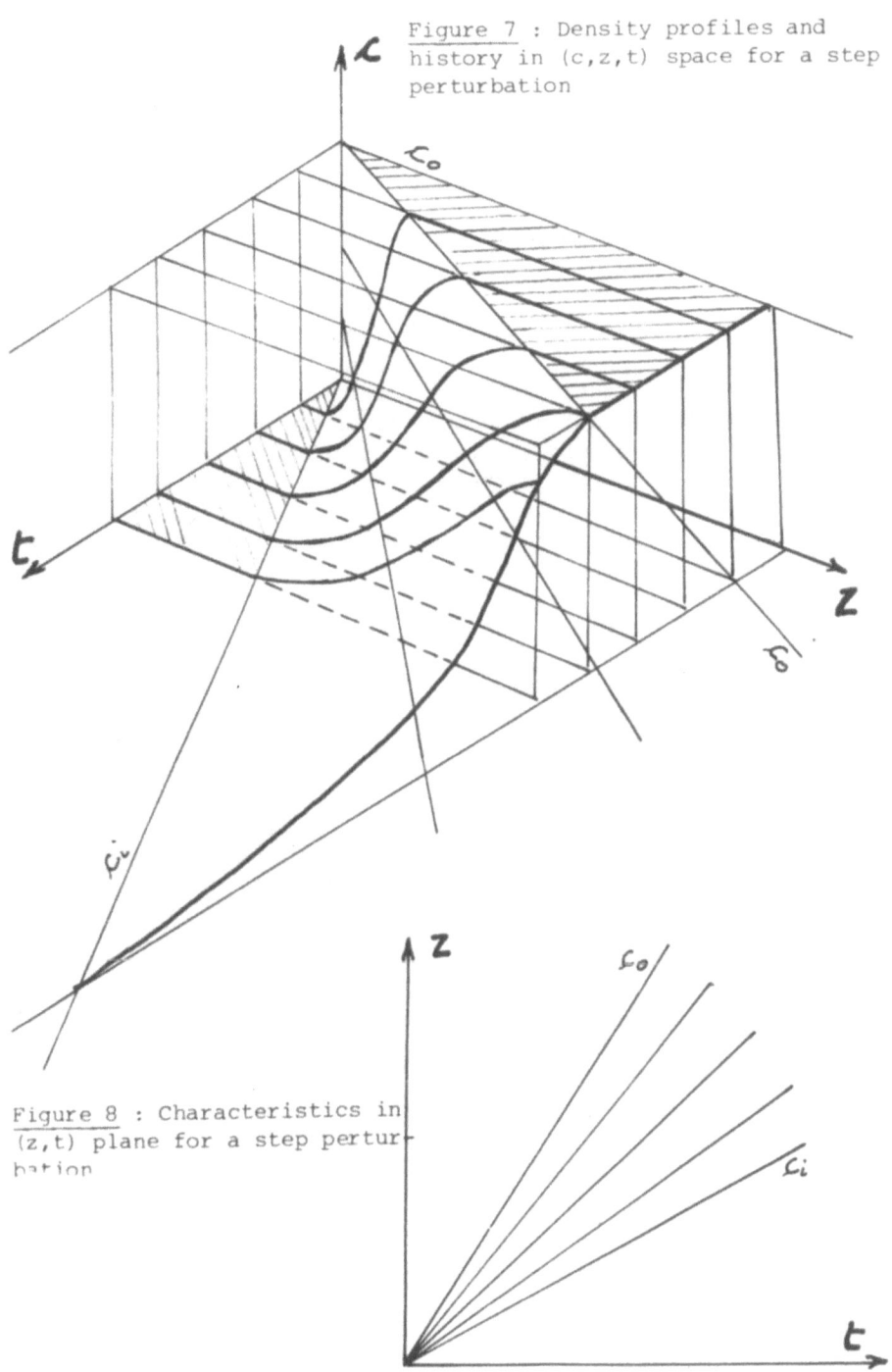

Figure 7 : Density profiles and history in (c,z,t) space for a step perturbation

Figure 8 : Characteristics in (z,t) plane for a step perturbation

(z,t) plane, all characteristics converge at the origin. The wave is called a centered wave (7). The slope of the characteristics is here

$$V_c = \left(\frac{\partial z}{\partial t}\right)_c = \left(\frac{z}{t}\right)_c = \frac{u}{1 + \dfrac{dn}{dc}} \tag{38}$$

from the last equality, we obtain, using equations 31 and 33:

$$n' = \frac{dn}{dc} = \frac{ut - z}{z} = \frac{\theta}{z} = \frac{1}{v_L} \tag{39}$$

This can be considered as a relatively explicit solution , since it is a relation between c (through n'), θ and z. The quantity $\theta/z = 1/v_L$ is closely related to the throughput parameter T used for adsorption and ion-exchange later in this course. The use of the Lagrangian velocity v_L allows velocity profiles and time distributions to be constructed that are geometrically related to the n (c) curve. This relation is illustrated on Figure 9 The c versus $1/v_L$ curve can be considered as a generalized concentration history, and the c versus v_L curve, as a generalized concentration profile. These curves contain implicitly all the profiles or histories of Figure 7

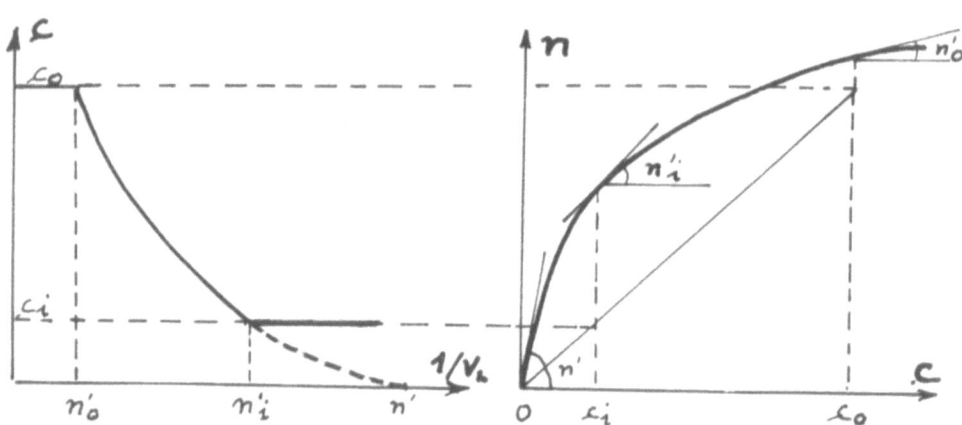

Figure 9 :   Relation between n (c) curve and inverse velocity
                profile

Negative versus positive step. The manipulations done so far are independent of the fact that the step input, between c_o and c_i is negative or positive. The network of characteristics is identical, but the bounds (initial and inlet concentration) are interchanged. The example we chose, was such that the successive concentration profiles spread out while propagating (wash-out curve). Clearly if we interchange initial and inlet concentration, we will obtain "overhanging" profiles, such as shown on Figure 10 Such a situation is in most cases impossible (exception : sea waves). Our solution, although it satisfies the differential equation and boundary conditions, is physically impossible. This is due to the fact that, in writing a differential material balance, we have implicitly assumed that the solution was continuous. In fact, the impossible overhanging profile should be replaced by a discontinuity, for which the differential material balance does not hold.

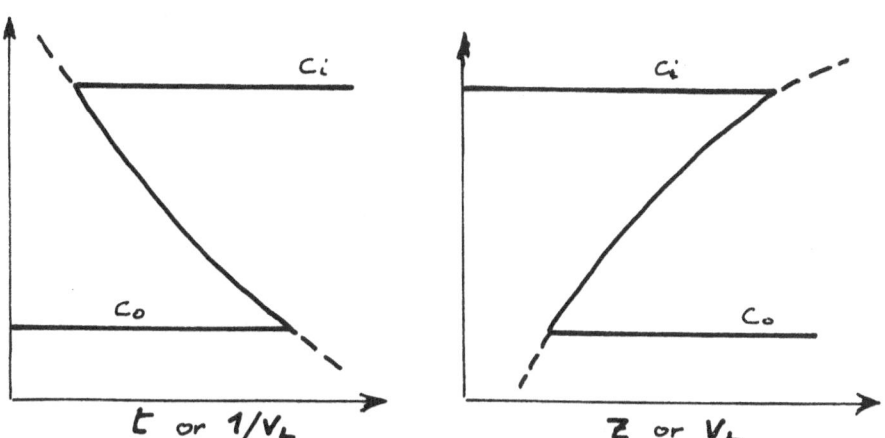

Figure 10 : Overhanging history or profile, when differential approach does not apply.

Discontinuous material balance. A material balance established under the same condition as equation 21 but with a finite change in the concentration yields

$$v_\Delta = \left(\frac{z}{t}\right)_\Delta = \frac{u}{1 + \dfrac{\Delta n}{\Delta c}} \tag{40}$$

with $\Delta c = c_o - c_i$ $\Delta n = n_o - n_i$

This has the same form as the differential balance, and shows that the discontinuity propagates with a unique velocity v_Δ. In the same way as the concentration velocities v_c were related to the local slope of the c (n) curve, v_Δ is related to the slope of the chords to the c (n) curve, as shown on Figure 11

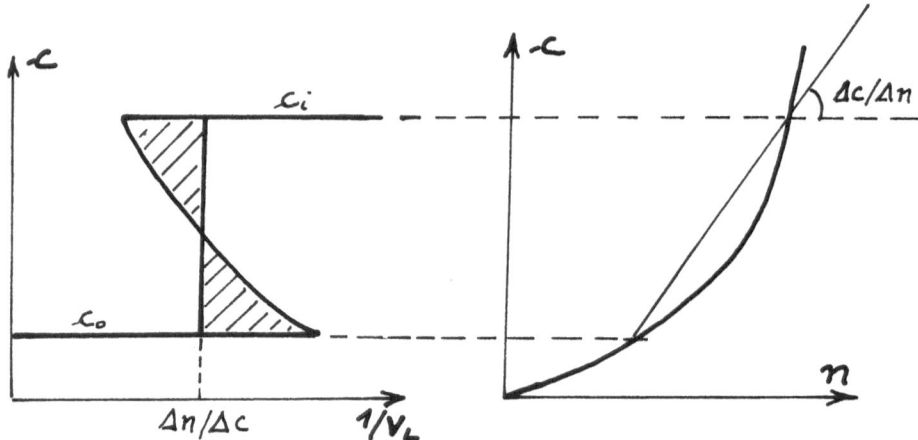

Figure 11 : Relation between n (c) curve and discontinuous jump.

Obviously, discontinuities are hardly more realistic, physically, than overhanging profiles. In practice, dispersion effects will tend to smoothen out the profile, and eventually, a balance will be attained between this smoothening effect and the overhanging tendency.

3.6 Compression - Dispersion

Inspection of Figure 9   shows that the velocity profile c versus v_L is connected to the local slope of the c (n) curve. To large slopes correspond large velocities. In the case considered, the largest slopes correspond to the highest concentrations. Thus, the high concentrations move faster than the smaller. If the step change is positive (increase from c_o to ci), the incoming concentrations will tend to override the initial concentrations. The tendency is toward an overhanging profile, but will result in a discontinuity. We call this a compressive behaviour.

On the contrary if the step is negative, the profile will tend to spread out. We call this a dispersive or diffusive behaviour. The table below summarises the   possible situations, according to the sign of the step and the curvature of the c (n) curve.

step change concavity of $c(n)$	positive $c_i > c_o$	negative $c_i < c_o$
toward the c-axis $c''(n) > 0$	compressive	dispersive
linear	indifferent	indifferent
toward the n-axis $c''(n) < 0$	dispersive	compressive

The compression/dispersion is a phenomenon characteristic of non-linear processes. In chromatography, where the $c(n)$ curve can be taken to be the equilibrium isotherm, it entails peak dissymetry (one flank being compressive, the other dispersive) as soon as one departs from small concentrations. It is a "first-order" phenomenon, independent of the mass-transfer.

4. ANALOGOUS PHYSICAL PROCESSES

Many physical situations lead to first-order, quasi linear equations, and thus to wave behaviour, particularly in fluid-flow, where the compressive phenomenon gives rise to the shock waves. The classical Rankine-Hugoniot relation for shocks in compressible flow is analogous to equation 40. The book by Landau and Lifschitz (12) gives a good treatment of this aspect. Kinematic waves in hydrodynamic and vehicular traffic have been reviewed by Whitham (8) and Prigogine (2). In these problems, the curve relating mobile and immobile concentrations, $c(n)$, is replaced by a relation between a flow and a "concentration" which is for instance the density in gas compression, and the height of water layer in water waves.

The wave approach has been widely used in sedimentation, (see the book by Wallis, 10) which closely resembles vehicular traffic, insofar as the flow-concentration relation presents a maximum, and thus has a slope of changing sign. This entails wave-propagation in both positive and negative directions relative to flow (e. g. the "countercurrent" propagation of a traffic jam). In the area of electricity, the book by Moore (Travelling wave engineering ,9) gives many related examples. The most interesting synthesis in the realm of chemical engineering is to be found in the book of Aris and Amundson (7).

Many problems expressed by second-order differential equations lead to single wave-phenomena : second order operators may sometimes be decomposed into products of two first order "wave operators"

$$\left(\frac{\partial}{\partial t} + a \frac{\partial}{\partial z} \right)$$

see for instance Whitham (8).

The examples below illustrate cases where systems of two or several first-order equations arise. Such coupled systems will be studied in detail in   other conferences of this session, in relation to chromatography. Let us stress here these analogies which are probably far from having been exploited.

<u>Compressible fluid flow</u> (isentropic, mono dimensional)

Material balance : $\frac{\partial \rho}{\partial f} + u \frac{\partial \rho}{\partial z} + \rho \frac{\partial u}{\partial z} = 0$

Momentum balance : $\rho \frac{\partial u}{\partial t} + \rho u \frac{\partial u}{\partial z} + c^2 \frac{\partial \rho}{\partial z} = 0$

ρ : density ; p : pressure $= k\rho^\gamma$; $c^2 = \partial p / \partial \gamma$: sound velocity

<u>High flux heat propagation</u>

Heat balance : $\frac{\partial q}{\partial t} + c\rho \frac{\partial T}{\partial t} = 0$

Conduction law : $\tau_R \frac{\partial q}{\partial t} + \lambda \frac{\partial T}{\partial z} = 0$

λ : conductivity ; c = specific heat ; ρ = density ; τ_R : relaxation time

<u>Electricity propagation</u>

Potentiel drop : $\frac{\partial v}{\partial z} + L \frac{\partial i}{\partial t} = - R_i$

Intensity drop : $\frac{\partial i}{\partial z} + C \frac{\partial v}{\partial t} = - G v$

v : potential, volts ; i = intensity, amps ; R : resistivity, ohm/cm
L : inductance, henrys/cm ; c : capacitance, farads/cm ; G = leakage, mhos/cm

Electrophoresis (one cation c and one anion A)

Material balances :

$$\frac{\partial c_c}{\partial t} + \frac{I}{FS}\ \frac{\partial T_c}{\partial z} = 0$$

$$\frac{\partial c_A}{\partial t} + \frac{I}{FS}\ \frac{\partial T_A}{\partial z} = 0$$

c_c, c_A : cation and anion concentrations, equiv. g./cm^3

I : current intensity, amps ; S : cross section, cm^2 ; F : Faraday's constant

T_c, T_A : transport numbers of A and c, non-dimensional.

Two component-chromatography

Materiel balances

$$\frac{\partial c_A}{\partial t} + \frac{1 - \varepsilon}{\varepsilon}\ \frac{\partial n_A}{\partial t} + u\ \frac{\partial c_A}{\partial z} = 0$$

$$\frac{\partial c_B}{\partial t} + \frac{1 - \varepsilon}{\varepsilon}\ \frac{\partial n_B}{\partial t} + u\ \frac{\partial c_B}{\partial z} = 0$$

c_A, c_B, n_A, n_B ; mobile phase and fixed phase concentrations of A and B, ε = void fraction of bed, u = interstitial velocity.

REFERENCES

1   DE GENNES P.G., La percolation, un concept unificateur
     La Recherche, 7, n° 72, 919, 1976

2   PRIGOGINE I., HERMAN R., Kinetic Theory of Vehicular Traffic,
     Elsevier, New-York, 1971

3.   LEVENSPIEL O., Chemical Reactor Engineering, chap. 9, Wiley,
     1964

4.   BUFFHAM B.A., KROPHOLLER H.V., Tracer kinetics : some general
     properties, the mean residence time, and applications to pha-
     se and chemical equilibria
     Chem. Eng. Science, 28, 1081, 1973

5.   BUFFHAM B.A., Model-independent aspects of tracer chromato-
     graphy theory
     Proc. Royal Soc. London, A 333, 89, 1973

6.   GIDDINGS, C., Dynamics of chromatography , Part 1
     Marcel Dekker, New York, 1965

7.   ARIS.R., AMUNDSON N.R., Mathematical methods in chemical
        Engineering, volume 2, First-order Partial Differential
        Equations. Prentice Hall, 1973

8.   WHITHAM G.B., Linear and Non-linear Waves
        Wiley, 1974

9.   MOORE R.K., Travelling-wave Engineering
        Mc Graw-Hill, 1960

10.   WALLIS G.B., One-dimensional two-phase flow
        Mc Graw Hill, 1969

11.   COURANT R., HILBERT D., Methods of Mathematical Physics
        Vol. II, Partial Differential Equations
        Interscience, 1962

12.   LANDAU L.D., LIFSCHITZ E.M., Course of Theoretical Physics,
        vol. 6 : Fluid Mechanics,
        Pergamon,1959

MODELING OF PERCOLATION PROCESSES

Alírio E. Rodrigues

Department of Chemical Engineering
University of Porto, Porto, Portugal

1. INTRODUCTION

A percolation process occurs when a fluid passes through a fixed
bed of a "support". This simple definition covers some classic
unit operations such as ion exchange, adsorption, chromatography,
drying, washing, etc.

These operations are performed in order to obtain:

a) - separation of products

b) - purification of diluents (e.g. air cleaning by removal of
        pollutants)

c) - recovery of solutes (recovery of metals from a solution)

A percolation process is a mass transfer operation between a fluid
phase and a solid phase and thus it is possible to identify a solu
te, a solvent and a diluent. In Table I we present some examples
of percolation processes |1|.

In the case of fixed bed operations the solvent is contained in
the column and the feed is a mixture of solute and diluent.

This analogy can be extended to the mathematical formulation of
the process.

TABLE I - Examples pf percolation operations

Nature of the unit operation	Diluent	Solute	Solvent
Liquid-solid adsorption (removal of phenol by DUOLITE ES-861)	water	phenol	adsorbent resin: DUOLITE ES-861
Gas-solid adsorption (removal of SO$_2$ by alumina)	air	SO$_2$	alumina
Ion exchange (water softening)	water	Ca^{++}	ion exchange resin, RNa$^+$
Large scale chromatography (Parex Process |2,3|)	other C$_8$ a romatics (xylene i somers and ethyl benzene)	p-xylene	molecular sieve
Leaching of copper sulpha te ore	water	copper sulphate	roasted ore (10% soluble CuSO$_4$, 85% gangue and 5% water)

The methodology used in the mathematical treatment of a percolation operation follows the general approach to chemical enginee-
ring problems |4|, which involves writting:

a) conservation equations

b) equilibrium laws at the solid/fluid interface

c) kinetic laws of transport and/or chemical reaction

d) boundary and initial conditions

1.1 Conservation equations

These equations describe the conservation of an extensive property: mass, energy, momentum.

Let us consider the mass balance equation over a volume element
dV= Ω dz of a fixed bed with cross section Ω; assuming that axial
dispersion is important, the flow of -a component in the z-direction is the sum of a bulk convective term and a diffusional term
relatively to the mass center

$$N_z = u_i \, c - D_{ax} \frac{\partial c}{\partial z} \qquad |1|$$

where u_i is the interstitial velocity in cm/sec, D_{ax} is the axial dispersion coefficient (cm²/sec) and c the solute concentration in the fluid phase.

The mass balance over dV is then

$$\varepsilon \, \Omega \, N_z = \varepsilon \, \Omega \, N_{z+dz} + \varepsilon \, \Omega \, \frac{\partial c}{\partial t} \, dz + (1 - \varepsilon) \, \Omega \, \frac{\partial q}{\partial t} \, dz \qquad |2|$$

flow in   flow out   accumula-   accumulation in
tion in the   the solid phase
fluid phase

which leads to

$$D_{ax} \frac{\partial^2 c}{\partial z^2} = u_i \frac{\partial c}{\partial z} + \frac{\partial c}{\partial t} + \frac{1-\varepsilon}{\varepsilon} \frac{\partial q}{\partial t} \qquad |3|$$

where q is the solute concentration in the solid phase referred to the volume of solid, t is the time variable, z is the space variable and ε the bed porosity.

Equation $|3|$ is a parabolic partial differential equation; when $D_{ax}=0$ we get the mass balance for plug flow.

1.2 Equilibrium laws

As Eq. $|3|$ has two dependent variables, c and q, we have to add an equation relating these two variables in order to solve the problem. This equation is the equilibrium isotherm, which relates the fluid and solid concentrations at the interface

$$q^* = f(c^*) \qquad |4|$$

Obviously, if there is no resistance to mass transfer, then $c=c^*$ and $q=q^*$.

Several cases should be considered, as shown in Figure 1:

1 - linear isotherm, $f''(c^*) = 0$

2 - favorable isotherm, $f''(c^*) < 0$

3 - unfavorable isotherm, $f''(c^*) > 0$

4 - isotherm with inflection point

5 - irreversible isotherm

34

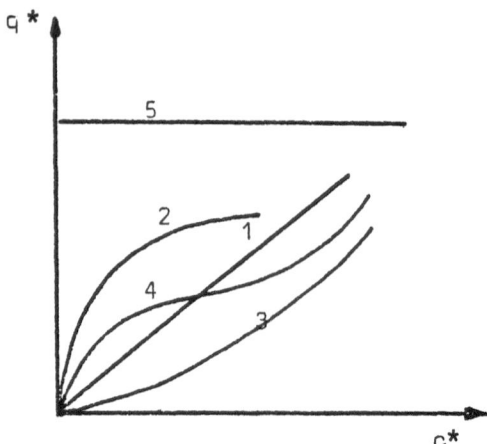

Figure 1 - Equilibrium isotherms

We are dealing with equilibrium isotherms as mathematical relations without taking care of their physical meaning; however we
recognize that the Langmuir isotherm is classified above in class
2 and the B.E.T. equation belongs to class 4.

1.3 Kinetic laws

In general the kinetic law of transport and/or chemical reaction
can be expressed as

$$\frac{\partial q}{\partial t} = g(c, q, c^*, q^*)$$
|5|

According to the controlling rate of transport, Equation |5| takes
different forms.

When film mass transfer resistance is the only resistance to mass
transfer, the kinetic law can be written as

$$\frac{\partial q}{\partial t} = K_f a_p (c - c^*)$$
|6|

which means that the mass flow is the product of a driving force
(c-c*) and a conductance. In Equation |6|, K_f is the film mass
transfer coefficient expressed in cm/sec and a_p the specific area
of the particle (in cm^{-1}). The concentration profile in such a

case is sketched in Figure 2.
When the resistance to mass transfer is concentrated inside the
particle,two cases should be considered according to the nature
of the solid phase.
If the solid phase is a,or can be considered as,a homogeneous medium (e.g. a gel type ion exchange resin),then the equation governing diffusion in a particle is

$$\frac{\partial q}{\partial t} = D \left(\frac{\partial^2 q}{\partial \rho^2} + \frac{\alpha}{\rho} \frac{\partial q}{\partial \rho} \right) \qquad |7|$$

where D is a constant diffusion coefficient of the solute inside
the particle, α=2,1 or 0 for spherical,cylindrical or slab simmetry,respectively and ρ the radial position in the particle.

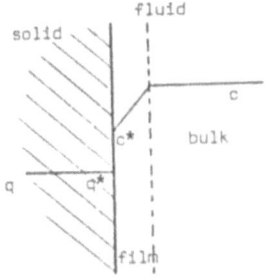

Figure 2 -Concentration profile
for the case of film mass transfer resistance

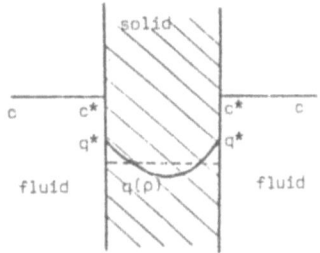

Figure 3 -Concentration profile for
the case of particle mass transfer
resistance

For porous particles the differential mass balance leads to

$$\varepsilon_p \frac{\partial c_p}{\partial t} + \frac{\partial q}{\partial t} = \varepsilon_p D_{pf} \left(\frac{\partial^2 c_p}{\partial \rho^2} + \frac{\alpha}{\rho} \frac{\partial c_p}{\partial \rho} \right) + D_s \left(\frac{\partial^2 q}{\partial \rho^2} + \frac{\alpha}{\rho} \frac{\partial q}{\partial \rho} \right) \qquad |8|$$

where c_p is the fluid concentration in the pore, ε_p the particle
porosity, D_{pf} the solute diffusivity in the pore and D_s the solute diffusivity in the solid phase.
Figure 3 shows a sketch of the concentration profile when particle mass transfer resistance is the only mechanism to be taken
into account.

(z,t) plane, all characteristics converge at the origin. The wave is called a centered wave (7). The slope of the characteristics is here

$$v_c = \left(\frac{\partial z}{\partial t}\right)_c = \left(\frac{z}{t}\right)_c = \frac{u}{1 + \dfrac{dn}{dc}} \tag{38}$$

from the last equality, we obtain, using equations 31 and 33:

$$n' = \frac{dn}{dc} = \frac{ut - z}{z} = \frac{\theta}{z} = \frac{1}{v_L} \tag{39}$$

This can be considered as a relatively explicit solution , since it is a relation between c (through n'), θ and z. The quantity $\theta/z = 1/v_L$ is closely related to the throughput parameter T used for adsorption and ion-exchange later in this course. The use of the Lagrangian velocity v_L allows velocity profiles and time distributions to be constructed that are geometrically related to the n (c) curve. This relation is illustrated on Figure 9 The c versus $1/v_L$ curve can be considered as a generalized concentration history, and the c versus v_L curve, as a generalized concentration profile. These curves contain implicitly all the profiles or histories of Figure 7

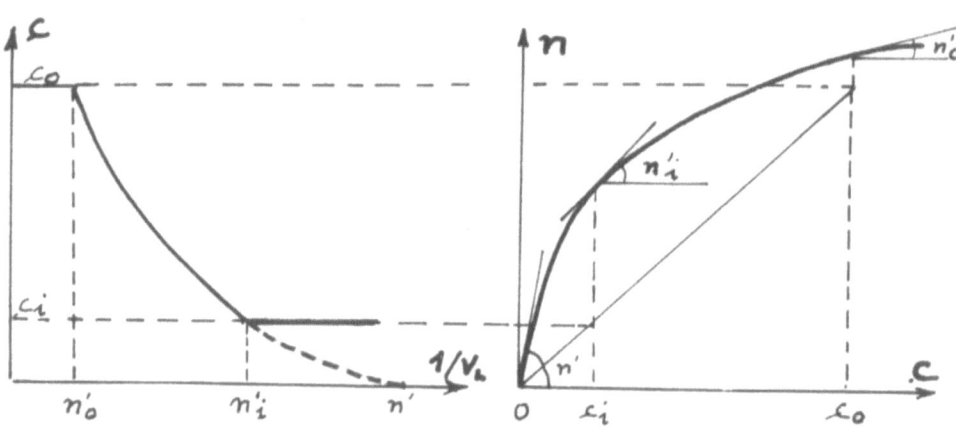

Figure 9 :   Relation between n (c) curve and inverse velocity profile

equilibrium model are the basic features of percolation processes.
The equilibrium model is based on the following assumptions:

- plug flow of the fluid phase

- isothermal operation

- instantaneous equilibrium between fluid and solid at every
  point in the column

- negligible pressure drop

The model equations are then:

$$u_i \left(\frac{\partial c}{\partial z}\right)_t + \left(\frac{\partial c}{\partial t}\right)_z + \frac{1-\varepsilon}{\varepsilon}\left(\frac{\partial q}{\partial t}\right)_z = 0 \qquad |9a|$$

$$q = f(c) \qquad |9b|$$

Combining these two equations and taking into account the chain
rule for partial derivatives, we obtain:

$$u_c = \left(\frac{\partial z}{\partial t}\right)_c = \frac{u_i}{1 + \dfrac{1-\varepsilon}{\varepsilon} f'(c)} \qquad |10|$$

This equation gives the velocity of propagation of the concentration c and shows its dependence on $f'(c)$, the slope of the e-
quilibrium isotherm.

In other terms Equation 10 is the "characteristic direction" of
the system

$$\begin{bmatrix} u_i & 1 + \dfrac{1-\varepsilon}{\varepsilon} f'(c) \\ \\ dz & dt \end{bmatrix} \begin{bmatrix} \left(\dfrac{\partial c}{\partial z}\right)_t \\ \\ \left(\dfrac{\partial c}{\partial t}\right)_z \end{bmatrix} = \begin{bmatrix} 0 \\ \\ dc \end{bmatrix} \qquad |11|$$

Equation 10 was first derived by Don de Vault |7| and it is the
basis for the notions of compressive and dispersive fronts or
waves.

In fact when the isotherm is unfavourable, higher concentrations
travel at lower velocities than smaller concentrations and the
wave tends to be more and more "dispersive", as shown in Figure 4.

In the case of a favorable isotherm higher concentrations travel
faster than lower concentrations and the front tends to be more
and more sharp. The wave is called a "compressive wave". Physically the limit of this shock wave is a discontituity (Figure 5).

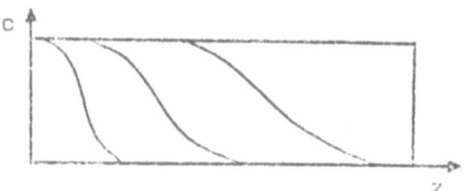

Figure 4 - Dispersive wave

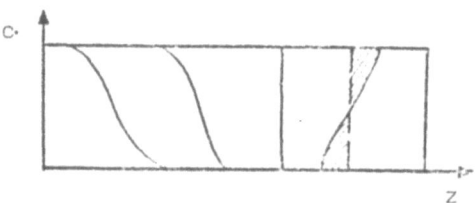

Figure 5 - Compressive wave

A first parameter useful for the design of a percolation column is the time that this discontinuity takes to travel through the column, which is called the stoichiometric time, t_{st}.

The stoichiometric time, t_{st}, can be calculated considering that the state of the column when the discontinuity of concentration appears at the outlet, $z = L$, is:

$\varepsilon \, v \, c_0$ - solute in the void bed volume

$(1-\varepsilon)v \, Q$ - solute retained in the solid phase, Q being the capacity of the solid phase corresponding to the feed concentration c_0

The quantity of solute passed through the column until the stoichiometric time is $U c_0 t_{st}$; then

$$\varepsilon \, v \, c_0 + (1-\varepsilon) \, v \, Q = U \, c_0 \, t_{st} \qquad\qquad |12|$$

and we get*

$$t_{st} = \tau \, (1+\xi) \qquad\qquad |13|$$

where $\xi=\dfrac{1-\varepsilon}{\varepsilon}\dfrac{Q}{c_0}$ is the capacitance parameter. The discontinuity travels through the column with a velocity $\bar{u}=L/t_{st}=u_i/(1+\xi)$.

3. CONCENTRATION HISTORIES OR BREAKTHROUGH CURVES DERIVED FROM THE EQUILIBRIUM MODEL

The notions of propagation of concentration waves on percolation columns enable us to derive the effluent concentration as a function of time, i.e. the concentration histories or breakthrough curves, for any shape of equilibrium isotherm.

We assume that the equilibrium model is valid and that the input is a step function of concentration, $c_E=c_0H(t)$.

a) Favorable isotherm

We have shown before that the shock wave or discontinuity takes a time t_{st} to appear at the outlet; then the concentration history is simply

$$c = c_0 \, H(t - t_{st}) \qquad\qquad |14|$$

b) Linear isotherm

Combining the mass balance, Equation $|9a|$ with the equilibrium isotherm, Equation $|15|$

$$q = \frac{Q}{c_0} \, c \qquad\qquad |15|$$

---

* In fact the stoichiometric time (zero-order moment of the step response of the column, F(t)) is known as retention time in chromatography (first order moment of the Dirac response, E(t), since $|8|$

$$\mu_n(F) = \frac{\mu_{n+1}(E)}{n+1}$$

we get

$$u_i \frac{\partial c}{\partial z} + (1+\xi) \frac{\partial c}{\partial t} = 0 \qquad |16|$$

The transfer function of the column is then

$$G(s) = \frac{\bar{c}\ (L,\ s)}{\bar{c}\ (0,\ s)} = \exp\ (\ -\ t_{st}\ s) \qquad |17|$$

and the breackthrough curve is again

$$c = c_0\ H(t - t_{st}) \qquad |18|$$

which means that the concentration inlet signal travels through the column without deformation and appears at the outlet at $t = t_{st}$ (Figure 6).

c) Unfavorable isotherm

Let us consider that equilibrium isotherms $y = f(x)$ can be described by

$$K = \frac{y\ (1-x)}{x\ (1-y)} \qquad |19|$$

where $x = c/c_0$, $y = q/Q$ are reduced concentrations and K is a constant separation factor. These isotherms, represented in Figure 7, are unfavorable for $K<1$ or $r = \frac{1}{K} > 1$ (r - equilibrium parameter).

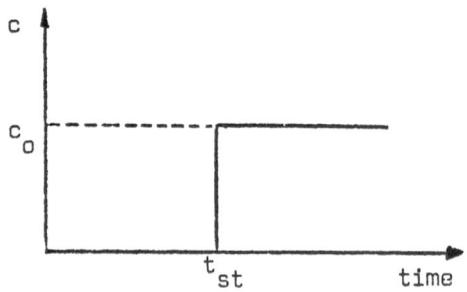

Figure 6- Breakthrough curve : equilibrium model and linear isotherm.

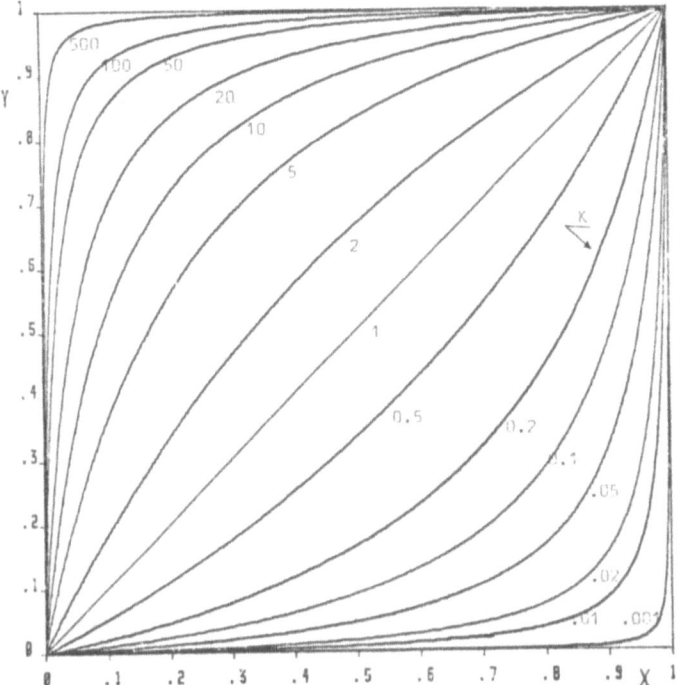

FIGURE 7 - EQUILIBRIUM ISOTHERMS OF "CONSTANT SEPARATION FACTOR"
         TYPE

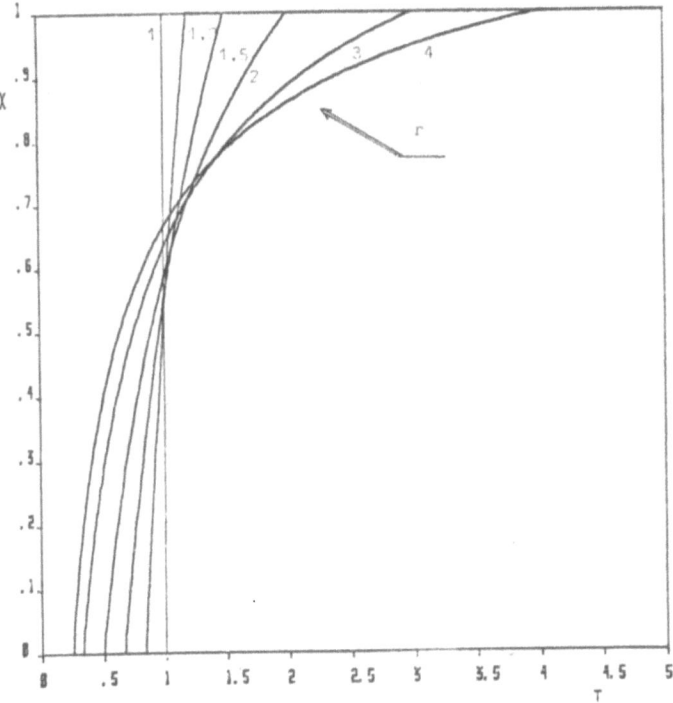

FIGURE 8 - BREAKTHROUGH CURVES:EQUILIBRIUM MODEL;UNFAVORABLE
         ISOTHERMS.

In this case the breakthrough curve can be derived from the fundamental relationship, Eq. $|10|$. In fact, in terms of dimensionless variables $z^*=z/L$ and $\theta=t/t_{st}$, Eq. $|10|$ can be written as

$$u_x = (\frac{\partial z^*}{\partial \theta})_x = \frac{1+\xi}{1+\xi\ y'(x)} \qquad |20|$$

At $z=L$ or $z^*=1$, Eq. $|20|$ leads, by integration, to

$$\theta = \frac{1+\xi\ y'(x)}{1+\xi} \qquad |21|$$

For the particular isotherm represented by Eq. $|19|$ we get

$$x = \frac{\sqrt{\frac{r}{(1+\xi)\theta-1}} - r}{1-r} \qquad |22|$$

If we introduce now the percolation parameter, or throughput parameter T $|9|$, defined as the mass of product per unit mass of solid phase

$$T = \frac{c_0\ (V-\varepsilon\ v)}{(1-\varepsilon)\ Q\ v} \qquad |23|$$

where V is the volume of fluid phase passed through the column of volume v, we obtain

$$x = \frac{\sqrt{\frac{r}{T}} - r}{1-r}, \quad \frac{1}{r} \leqslant T \leqslant r \qquad |24|$$

Figure 8 shows breakthrough curves for the case of unfavorable isotherms.

Equation $|24|$ is easily obtained since $\theta=\frac{1+\xi\ T}{1+\xi}$; for high values of ξ (or $\varepsilon c_0 << (1-\varepsilon)Q$), $\theta=T$.

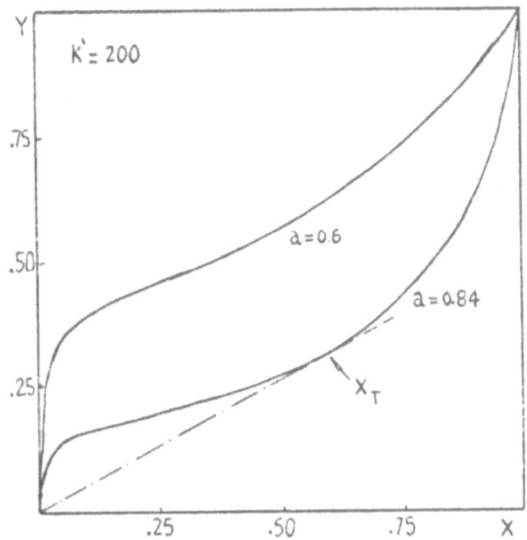

FIGURE 9 -B.E.T.ISOTHERMS

d) Isotherms with inflection point

This is an important class of equilibrium isotherms which includes
B.E.T. isotherms, and can be represented by the following equation
in the case of liquid/solid systems:

$$q = \frac{K' \, Q_m \, c_s \, c}{(c_s - c) \, (c_s - c + K' c)} \qquad |25|$$

In the above equation c_s is the maximum concentration of the solu̲
te in the liquid (solubility) and Q_m is the solid concentration
corresponding to a monolayer.

Introducing the reduced variables $x = c/c_1$, $y = q/q_1$ and $a = c_1/c_s$,
where c_1 is the feed concentration, we get:

$$y = \frac{x(1-a) \, (1-a+K' \, a)}{(1-a \, x) \, (1-a \, x + K' \, a \, x)} \qquad |26|$$

The equilibrium isotherm is now described by two parameters: a
and K'; Figure 9 shows the influence of a in the shape of the
isotherm.

The breakthrough curve $x = x(\theta)$ can be drawn by using Eq. $|21|$. For
K'=200 and a=0.84 the concentration history is shown in Figure 10
by the dashed line. However, this history has no physical meaning;
then we have to locate a discontinuity at some point (θ_D, x_D) in

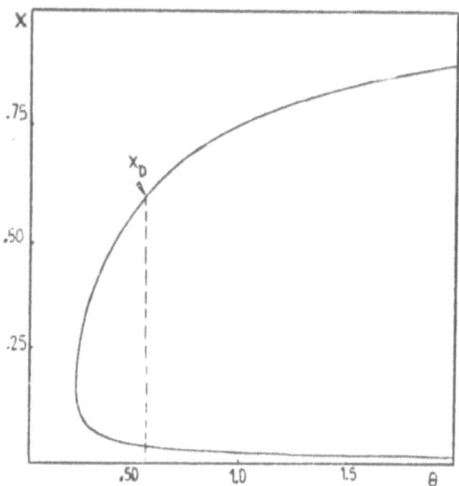

Figure 10- Breakthrough curve :equilibrium
model;B.E.T.isotherm(a=.84,K'=200)

such a way that areas I and II are equal.

This is true if, according to the mean-theorem of calculus,

$$\theta_D \, x_D = \int_0^{x_D} \theta(x) d \, x \qquad\qquad |27|$$

which leads to

$$x_D = \frac{K'-2}{2 \, a(K'-1)} \qquad\qquad |28|$$

On the other hand, the tangency point (x_T, y_T) of the isotherm is given by the intersection of a straight line $y_1 = bx$ and Eq. $|26|$; so

$$\left\{ \begin{array}{l} y(x)\big|_{x=x_T} = b \, x_T \\[2ex] y'(x)\big|_{x=x_T} = b \end{array} \right. \qquad\qquad |29|$$

The solution is

$$x_T = \frac{K'-2}{2\ a(K'-1)} \qquad |30|$$

and it results that $x_D = x_T$.

We can conclude by stating that in the case of a B.E.T. isotherm the equilibrium model predicts that the concentration history will be a discontinuity followed by a dispersive front. The separation between the two parts of the "composite" wave is given by the composition corresponding to the tangency point x_T |10, 11, 12|. This can be extended for any feed point F, and presaturation state of the bed P, and the result is the so-called "Golden rule" |13| (see Klein's lecture in this course).

4. DESIGN IMPLICATIONS OF THE NATURE OF THE EQUILIBRIUM ISOTHERM AND KINETIC FACTORS

Let us suppose we want to design an adsorption column in order to remove a pollutant from a feed with a flowrate U. The pollutant concentration in the feed is c_0. Our goal is to get a good effluent quality (say $c < 0.05\ c_0$) during a time of operation, t_{Bp} by using an adsorbent with capacity Q (corresponding to c_0).

In order to know what volume of adsorbent, $(v_s)_1$ is needed, some people would simply consider that

$$(v_s)_1 = \frac{c_0}{Q}\ \frac{\xi}{1+\xi}\ U\ t_{st} \qquad |31|$$

which results from an overall mass balance.

This would only be true if the equilibrium model were valid and the isotherm were favorable, bscause then $t_{Bp} = t_{st}$. However, in the case of unfavorable equilibrium, the time of operation, using the same volume of adsorbent, is reduced as shown in Figure 11.

In fact, the history of concentrations is then given by Eq. |24| and for x=0 we get

$$t_{Bp} = \frac{1+K\ \xi}{1+\xi}\ t_{st} \qquad |32|$$

and the corresponding volume of adsorbent is

$$(v_s)_2 = \frac{c_0}{Q}\ \frac{\xi}{1+K\ \xi}\ U\ t_{Bp} \qquad |33|$$

For the same t_{Bp} and large values of ξ we need a larger volume of adsorbent in the case of unfavorable isotherms than in the case of favorable isotherms, i.e.

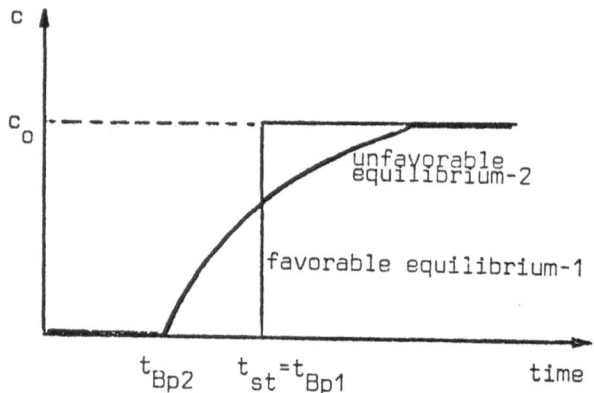

Figure 11- Breakthrough time for favorable
and unfavorable isotherms :equilibrium
model

$$(v_s)_2 = (v_s)_1 /K \qquad\qquad |34|$$

with $K<1$.

The calculation above shows that when designing a percolation column one has to take into account, not only the adsorbent capacity at the feed concentration c_0, but the whole equilibrium isotherm between the feed (F) and the presaturation P (initial state
of the bed) points.

The useful time of operation can be reduced, relatively to the
stoichiometric time by the unfavorable nature of the equilibrium
isotherm and also, in practice, by hydrodynamic and kinetic
factors (axial dispersion, mass transfer resistances, ...)

Some people include these influences in a safety factor, S_f and
calculate

$$v'_s = S_f (v_s)_1 \qquad\qquad |35|$$

which leads, obviously, to a poor design due to uncertainties on
S_f.

Thus, the development of models for the prediction of breakthrough
curves and their dependency on the operating variables appears to
be necessary. From these models we can get useful design parameters: the breakthrough time t_{Bp}, the width of the mass transfer
zone δz, and the leakage L, as shown in Figure 12.

The shape of the breakthrough curve is a result of the influences
of the kinetic factors and of the nature of the equilibrium isotherm. It is usually accepted that in the case of favorable equilibrium, the dispersive effects of kinetic phenomena are compen-

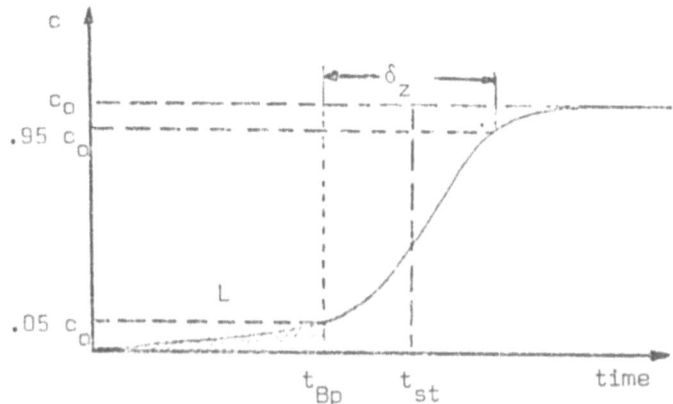

Figure 12- Design parameters from a breakthrough
curve

sated by the compressive effects due to the favorable nature of
the equilibrium isotherm; the result is that after a certain time
of front formation this should propagate at constant velocity,
maintaining its form while passing through the column. The resulting wave is called a stationary front or constant-pattern break-
through or a shock layer $|14|$.

The definition of a stationary front given above implies that
$c(z+dz, t+dt)=c(z, t)$; then from the mass balance, Eq. $|9a|$ we
get

$$\frac{\partial x}{\partial t} = \frac{\xi}{\frac{u_i}{u_c} - 1} \frac{\partial y}{\partial t} \qquad |36|$$

where $u_c=(\frac{\partial z}{\partial t})_c$. In a stationary front, all concentrations travel
at the same velocity and thus $u_c=\text{constant}=\frac{u_i}{1+\xi}$; integration of
Eq. $|36|$ leads to $|15|$:

$$x = y \qquad |37|$$

This relation is sometimes improperly used in situations where a stationary front can not be obtained. This is, for instance, the case of the equilibrium model for linear and unfavorable isotherms when axial dispersion is important.

For linear isotherms and axial dispersion we can show that the asymptotic expression is

$$x = \frac{1}{2} \left[1 - erf \frac{z - \lambda t}{2 \sqrt{\tilde{D} \, t}} \right]$$ |38|

where $\tilde{D} = \frac{D_{ax}}{1+\xi}$ and $\lambda = \frac{u_i}{1+\xi}$. Equation |38| shows that the width of the mass transfer zone is then proportional to $\sqrt{t}$.

For unfavorable isotherms the spreading of the front due to the equilibrium isotherm is proportional to t while the spreading due to axial dispersion is proportional to $\sqrt{t}$. This means that the effect of the isotherm curvature is stronger than the effect of the axial dispersion and to a first approximation we can disre gard the effect of axial dispersion. This is a crude justification of a dynamic method for the determination of an equilibrium iso- therm from a single desorption experiment in column |16|.

Let us close this section with the analysis of the equilibrium model with axial dispersion for the case of a favorable isotherm. If there is a stationary front the solution of Eqs. |3| and |4| can be written in terms of a coordinate moving along with the wave $c = \phi(z - \lambda t)$, where λ is the velocity of the stationary front or shock wave |14|. In this way we pass from a system (z, t) to a one coordinate system, $\chi = z - \lambda t$. The mass balance in terms of χ is then

$$D_{ax} \frac{d^2 x}{d\chi^2} = (u_i - \lambda) \frac{dx}{d\chi} - \xi \lambda \frac{dy}{d\chi}$$ |39|

and the boundary conditions

$$\chi = -\infty \quad , \quad x = 1 \quad , \quad y = 1 \quad , \quad \frac{dx}{d\chi} = 0$$ |40a|

$$\chi = +\infty \quad , \quad x = 0 \quad , \quad y = 0 \quad , \quad \frac{dx}{d\chi} = 0$$ |40b|

Integrating Eq. |39| we get

$$D_{ax} \frac{dx}{d\chi} = (u_i - \lambda) x - \xi \lambda y$$ |41|

and as Eq. |41| should satisfy the boundary conditions we get

$$\lambda = \frac{u_i}{1+\xi} = \bar{u} \qquad \qquad |42|$$

Equation $|42|$ means that the stationary front velocity when $D_{ax} \neq 0$ is the same as the discontinuity obtained when $D_{ax}=0$.

5. MODELING OF PERCOLATION COLUMNS

The design of a percolation column is based on the knowledge of breakthrough curves. These curves are affected by such factors as flowrate, particle diameter, temperature, feed concentration, etc. The effect of these variables can be predicted by modeling the column. The goals of modeling are:

a) prediction of the dynamic behavior of the column

b) optimization of the operating conditions

c) scale up from laboratory to large scale

These goals can be achieved by computer simulation of the column under a wide range of working conditions. Modeling and simulation saves time because then we need only a limited number of experiments.

The philosophy of modeling which we follow below is based in some principles:

    a) start with simple models; obtain from these models all the information which is still valid for more complex models

As an example we can notice that from the equilibrium model we can get some important quantities such as the stoichiometric time, t_{st} velocity of a shock wave, $\bar{u}$, etc.

    b) the validity of a good model is not only the result of a good fitting

The important feature of a good model is its ability to predict the behavior of the column under operating conditions different from those used to obtain the model parameters.

    c) use models to get useful design perameters and their dependency on operating conditions

For instance, we can obtain from a breakthrough curve design parameters such as the breakthrough time, t_{Bp} (time at which the effluent concentration is $c_1 = 0.05 c_0$), the width of the mass transfer zone, δz, and the leakage, L (quantity of solute which leaves the bed during the saturation step).

5.1 Classification of models for percolation columns

There are many models available in the literature but the anarchy
of notations and the diversity of mathematical techniques involved
might disapoint the reader.

Usually the models are divided into staged and differential models
following the approach used to write the conservation equation
which leads to ODE and PDE, respectively.

However, if we classify the models according to the nature of the
kinetic law of transport we get two main classes:

> CLASS I - Models of "Chemical kinetic law" type

> CLASS II - Models of "Physical kinetic law" type

In class I the kinetic law is described by an equation similar to
a rate of chemical reaction.

In class II the kinetic law describes the different mechanisms of
mass transfer (film diffusion, pore diffusion, surface diffusion).

The advantage of this classification is that the models are grou-
ped around "nuclei-models" |17|. These are the Thomas model |18|
for class I and the Rosen model |19| for class II.

From these models we can derive a certain number of other models
as simplifications or extensions.

5.1.1 The Thomas model

This model was developed in 1948 by Thomas; we present in Table
III for historical reasons, the corresponding model equations and
the mathematical solution for the case of the saturation of a bed
initially clean.

The solution presented below uses the notation of Vermeulen & al.
|20|, which reduces the number of model parameters from 4 (in the
original Thomas version) to 3 (r, N and NT).

Some results are presented in Figures 13 a) and b) for the cases
of $r=0$ and $r=1$, in prob-log coordinates |21|.

---

* The complete solution of the Thomas model is clearly presented
in Aris and Amundson's book |22|.

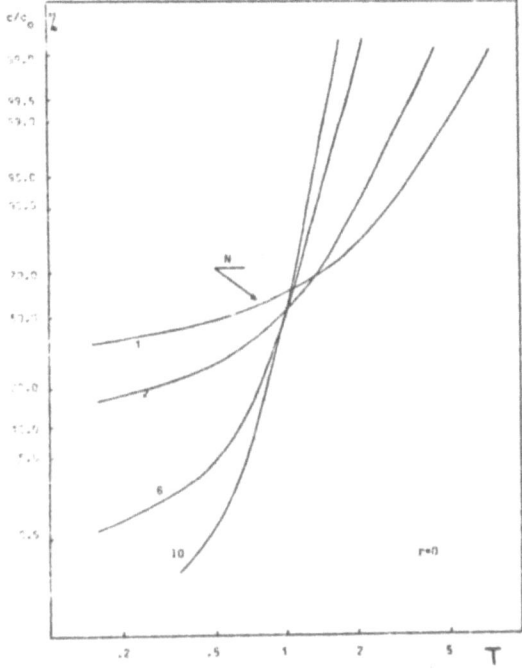

FIGURE 13 a - THOMAS MODEL (r=0)

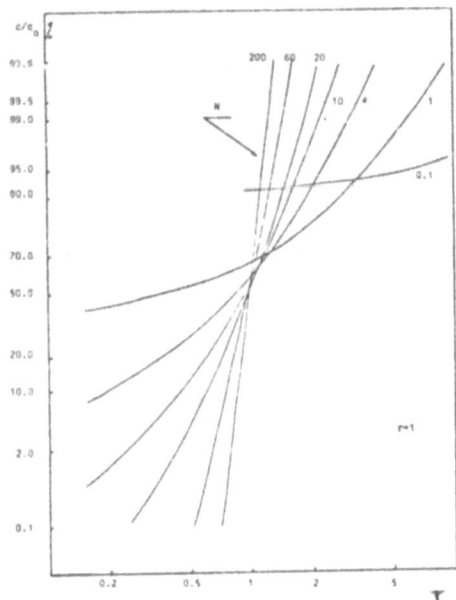

FIGURE 13 b- THOMAS MODEL (r=1)

TABLE III - Thomas model

Mass balance	$u_i \dfrac{\partial c}{\partial z} + \dfrac{\partial c}{\partial t} + \dfrac{1-\varepsilon}{\varepsilon} \dfrac{\partial q}{\partial t} = 0$

Kinetic law $\quad \dfrac{\partial q}{\partial t} = K_1 \left[c(Q-q) - r\, q(c_0 - c) \right]$

Boundary and initial conditions

$z = 0 \qquad c = c_0\, H(t) \qquad \forall\, t$

$t = z/u_i \quad q = 0 \qquad\qquad \forall\, z$

Solution

$$\dfrac{c}{c_0} = \dfrac{J(rN,NT)}{J(rN,NT) + \left[1 - J(N,rNT)\right]\, \exp\left[(r-1)N(T-1)\right]}{}^*$$

$*\quad J(\alpha,\beta) = 1 - \displaystyle\int_0^\alpha \exp(-\beta - \xi) I_0(2\sqrt{\beta\xi})\, d\xi \quad$ is the J Function

$I_0(2\sqrt{\beta\xi}) = \displaystyle\sum_{i=0}^\infty \dfrac{(\beta\xi)^i}{(i!)^2} \quad$ is the modified Bessel function of zero order and first kind

Model parameters:

$r = \dfrac{1}{K} \quad$, equilibrium parameter

$T = \dfrac{c_0(V - \varepsilon\, v)}{(1-\varepsilon)\, Q\, v} \quad$, percolation parameter or throughput parameter

$N = \dfrac{K_1\, Q}{\varepsilon} \dfrac{z}{u_i} \quad$, number of reaction units

---

In Table IV we present some solutions of classic models which can be derived from the Thomas model.

TABLE IV - Simplifications of the Thomas model

Model	Simplification	Solution
Bohart [23]	$r = 0$	$\dfrac{c}{c_0} = \dfrac{\exp(NT)}{\exp(NT)+\exp(N)-1}$
Walter [24]	$r = 1$	$\dfrac{c}{c_0} = J(N,NT)$
Walter [25]	high r, dispersive front	$\dfrac{c}{c_0} = \dfrac{\sqrt{\dfrac{r}{T}} - r}{1-r}$
Klinkenberg [26]	$r = 1$ N and NT high	$\dfrac{c}{c_0} = \dfrac{1}{2}\left[1+\mathrm{erf}(\sqrt{NT}-\sqrt{N})\right]$ *

$$*\quad \mathrm{erf}\ x = \frac{2}{\sqrt{\pi}}\int_0^x e^{-\xi^2}\,d\xi \quad \text{is the "error function"}$$

5.1.2 The Rosen model

The Rosen model is presented here as the nuclei-model for the models of "Physical kinetic law type". However we will recall a simpler model in class II, that of Anzelius and Schumann [27, 28], developed independently by these authors in 1926 and 1929, respectively.

The model is based on the following assumptions: plug flow of the fluid phase, linear equilibrium isotherm and isothermal operation. Furthermore, the only resistance to mass transfer is due to diffusion in a stagnant film around the particles.

The model equations are then

mass balance

$$u_i \frac{\partial c}{\partial z} + \frac{\partial c}{\partial t} + \frac{1-\varepsilon}{\varepsilon}\frac{\partial q}{\partial t} = 0 \qquad |43a|$$

equilibrium law

$$q = \frac{Q}{c_0}\, c_i^* \qquad |43b|$$

kinetic law

$$\frac{\partial q}{\partial t} = (K_f\, a_p)\, (c - c_i{}^*) \qquad\qquad |43c|$$

B.C. and I.C.

$$z = 0\ ,\ c = c_0\, H(t) \qquad\qquad \forall\, t \qquad\qquad |43d|$$

$$t = z/u_i,\ c = q = 0 \qquad\qquad \forall\, z \qquad\qquad |43e|$$

The Anzelius treatment starts with the introduction of the new variables $N_f = \dfrac{K_f\, a}{\varepsilon}\, \dfrac{z}{u_i}$, $N_f T = \dfrac{K_f\, a}{1-\varepsilon}\, \dfrac{c_0}{Q}\, (t-\dfrac{z}{u_i})$, $x = c/c_0$ and $y = \dfrac{q}{Q}$ with $a = a_p(1-\varepsilon)$. The system of Eqs. $|43|$ is then reduced to

$$\frac{\partial\, x}{\partial\, N_f} = x - y \qquad\qquad |44|$$

$$\frac{\partial\, y}{\partial (N_f\, T)} = y - x \qquad\qquad |45|$$

With a new change of variables

$$x = (S - W)\, \exp\!\left[- (N_f + N_f\, T)\right] \qquad\qquad |46|$$

$$y = (S + W)\, \exp\!\left[- (N_f + N_f\, T)\right] \qquad\qquad |47|$$

we get an hyperbolic PDE

$$\frac{\partial^2\, W}{\partial (N_f)\, \partial (N_f\, T)} = W \qquad\qquad |48|$$

Putting $\phi^2 = -4(N_f)(N_f T)$ we obtain a Bessel equation of zero order

$$\frac{d^2\, W}{d\phi^2} + \frac{1}{\phi}\frac{d\, W}{d\phi} + W = 0 \qquad\qquad |49|$$

which has the solution

$$W = \frac{1}{2}\, I_0\, (2\sqrt{(N_f)\, (N_f\, T)}) \qquad\qquad |50|$$

and finally

$$\frac{c}{c_0} = J(N_f,\ N_f\, T) \qquad\qquad |51|$$

However this solution can be easily obtained by using the Laplace transform method. In fact, if we transform Eqs. 44 and 45 relatively to $N_f T$ and we insert $\bar{y}(N_f, s)$ from Eq. 45 in Eq. 44 we get

$$\frac{d\bar{x}(N_f, s)}{d N_f} = \frac{s}{s+1} \bar{x}(N_f, s) \qquad |52|$$

Taking into account that for $N_f=0$, $x=1$ $\forall$ t and then $\bar{x}(0, s)=\frac{1}{s}$
we get

$$\bar{x}(N_f, s) = \frac{1}{s} \exp(-\frac{N_f\ s}{s+1})\ (1 + \frac{1}{s})\ \frac{1}{s+1}\ \exp(-N_f)\ \exp(\frac{N_f}{s+1}) \qquad |53|$$

and finally

$$x(N_f, N_f\ T) = \exp(-N_f)\ \exp(-N_f\ T)\ I_0\ (2\sqrt{(N_f)\ (N_f\ T)} +$$

$$+ \int_0^{N_f\ T} \exp(-N_f - N_f\ T) I_0 (2\sqrt{(N_f)(N_f T)} d(N_f T) \qquad |54|$$

Using the following properties of the J function

$$J(\alpha, \beta) = 1 - \int_0^\alpha e^{-\beta-\xi}\ I_0(2\sqrt{\beta\ \xi})\ d\ \xi$$

$$1 - J(\alpha, \beta) = J(\beta, \alpha) - e^{-\beta-\alpha}\ I_0(2\sqrt{\alpha\ \beta}$$

we get

$$x = \frac{c}{c_0} = J(N_f, N_f\ T) \qquad |55|$$

The transfer function of the column, defined as the ratio between
the output and input signals in the Laplace domain (transforms
taken relatively to $t-z/u_i$) is

$$G(s) = \exp\left[-\frac{N_{f0}\ \tau\ s}{\frac{N_{f0}}{\xi} + \tau\ s}\right] \qquad |56|$$

where $N_{f0}=K_f a_p \frac{1-\varepsilon}{\varepsilon} \tau$ is the number of   film mass transfer
units based on the lenght of the column, z=L. The Van der Laan
theorem $|29|$ enables us to get the moments of the Dirac response;
they are

$$\mu_1 = \xi\ \tau \qquad |57|$$

$$\mu_2 = (\xi\ \tau)^2\ \left[1 + \frac{2}{N_{f0}}\right] \qquad |58|$$

The Rosen model [19] developed in 1954 takes into account film diffusion and also particle diffusion.

The model equations are presented in Table V.

TABLE V - Rosen model

Mass balance	$u_i \dfrac{\partial c}{\partial z} + \dfrac{\partial c}{\partial t} + \dfrac{1-\varepsilon}{\varepsilon} \dfrac{\partial \bar{q}}{\partial t} = 0$
Kinetic law for film mass transfer	$\left(\dfrac{\partial q}{\partial t}\right)_{\rho=R} = (K_f a_p)(c-c_{i*})$
Solid diffusion	$\dfrac{\partial q(\rho,z,t)}{\partial t} = D_{eff}\left[\dfrac{\partial^2 q(\rho,z,t)}{\partial \rho^2} + \dfrac{2}{\rho}\dfrac{\partial q(\rho,z,t)}{\partial \rho}\right]$
Mean concentration in the particle	$\bar{q}(z,t) = \dfrac{3}{R^3}\displaystyle\int_0^R q(\rho,z,t)\rho^2\, d\rho$
Equilibrium isotherm	$q = \dfrac{Q}{c_0} c_i$
B.C. and I.C.	$q(\rho,z,0)=0 \qquad 0 \leqslant \rho \leqslant R \qquad z>0$ $c(0,t)=c_0 H(t), \qquad z=0 \qquad \forall\, t$
Solution	$\dfrac{c}{c_0} = \dfrac{1}{2} + \dfrac{2}{\pi}\displaystyle\int_0^\infty \exp(A)\sin(B)\dfrac{d\lambda}{\lambda}$

A,B functions of the model parameters

The transfer function of the system with respect to $t\dfrac{z}{u_i}$ is

$$G(s) = \exp\left[-\dfrac{1-\varepsilon}{\varepsilon}\tau\, Y_T(s)\right]$$ [59]

where $Y_T(s)=Y_D(s)/(1+R_f Y_D)$ is the overall admittance and $R_f=\dfrac{1}{K_f a_p}$

and $Y_D(s)=\dfrac{6D_{eff}\xi}{R^2}\displaystyle\sum_{n=1}^{\infty}\dfrac{s}{s+\dfrac{D_{eff}\pi^2}{R^2}n^2}$ are the resistances due to the

film and particle diffusion,which are supposed to be in series.
The moments of the Dirac response are

$$\mu_1 = \xi \, \tau \qquad \qquad |60|$$

$$\mu_2 = (\xi \, \tau)^2 \left[1 + \frac{2}{N_{f0}} (1 + \frac{Bi_m}{5}) \right] \qquad \qquad |61|$$

where $Bi_m = \dfrac{K_f R}{D_{eff}}$ is the mass Biot number.

If we compare Eqs. $|58|$ and $|61|$ we can conclude that the Anzelius
and Rosen models are equivalent, by putting simply:

$$(N_{f0})_{Anzelius} = \left[\frac{N_{f0}}{1 + \dfrac{Bi_m}{5 \, \xi}} \right]_{Rosen} \qquad \qquad |62|$$

It is obvious that when the effective diffusivity, $D_{eff} \to \infty$ then
$Bi_m \to 0$ and the Rosen model becomes the Anzelius model.

Using these models we can draw plots of a design parameter,
$\theta_{Bp} = t_{Bp}/t_{st}$ as a function of the model parameters (ξ, N_{f0} and
Bi_m for the Rosen model) and use them for the design of percola-
tion columns.

A detailed review of models for percolation columns is presented
elsewhere $|30|$.

5.2 Staged approach of percolation processes

In the following section we consider the column as a series of
perfectly mixed cells. This approach is well known in reaction
engineering and different from that of Martin and Synge $|31|$. We
shall discuss first the equilibrium model and then introduce mass
transfer resistances.

5.2.1 Equilibrium model

Let us consider the ith cell of the column (Figure 14). The model
equations are:

mass balance

$$U \, c_{i-1} = U \, c_i + V_f \frac{dc_i}{dt} + V_s \frac{dq_i}{dt} \qquad \qquad |63a|$$

equilibrium isotherm

$$q_i = \frac{K\,Q\,c_i^*}{c_0 + c_i^*\,(K - 1)} \qquad |63b|$$

kinetic law

$$c_i = c_i^*$$

$$q_i = q_i^* \qquad |63c|$$

B.C. and I.C.

$$c_E = c_0\,H(t)$$

$$t = 0\,, \quad q_i = c_i = 0 \qquad |63d|$$

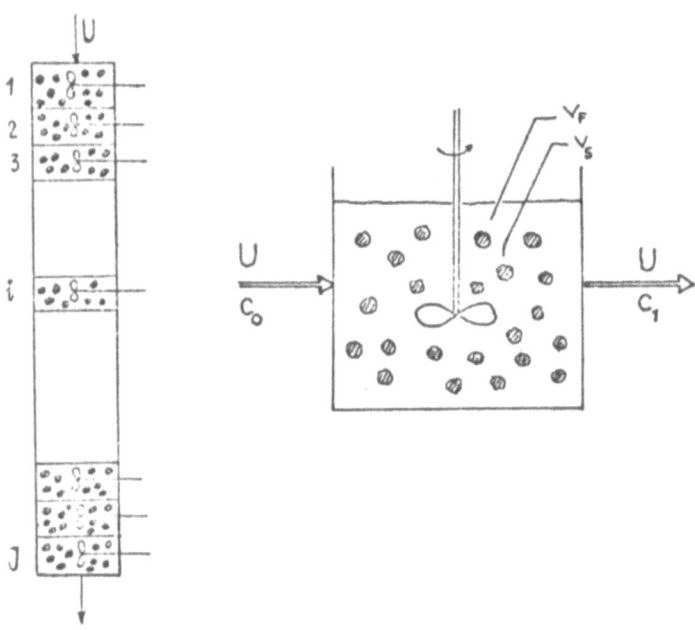

Figure 14- Staged model for a percolation column

In the above equations U is the flowrate, v_f and v_s the fluid and solid volumes in each cell c_{i-1} and c_i the fluid phase con centrations at he inlet and outlet of the cell i , respectively,

Q the maximum capacity of the solid, q_i the solid phase concentration in cell i and K the separation factor, which characterizes (by its deviation from unity) the nonlinearity of the isotherm.

Introducing dimensionless variables

$$x_{i-1} = \frac{c_{i-1}}{c_0} \ , \quad x_i = \frac{c_i}{c_0} \ , \quad y_i = \frac{q_i}{Q} \quad \text{and} \quad \theta = \frac{t}{t_{st}}$$

we get

$$\frac{dx_i}{d\theta} = \frac{x_{i-1} - x_i}{f(x_i)} \tag{64}$$

where $f(x_i) = \dfrac{1}{J(1+\xi)} \left[1 + \dfrac{\xi K}{[1+x_i(K-1)]^2} \right]$.

Let us illustrate the methodology of modeling for the case of only one cell or perfectly mixed "sorber" (J=1). The resulting equation is then a nonlinear ODE which can be solved analitically [30].

The influence of the model parameters K and ξ on the concentration histories is shown is Figures 15 and 16, respectively [32].

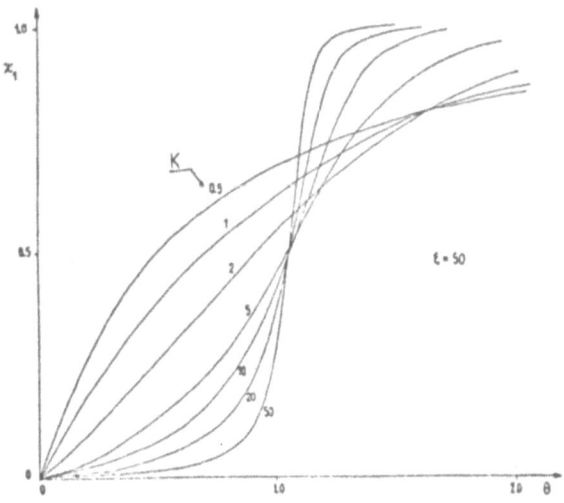

Figure 15- Effect of K on the breakthrough curves

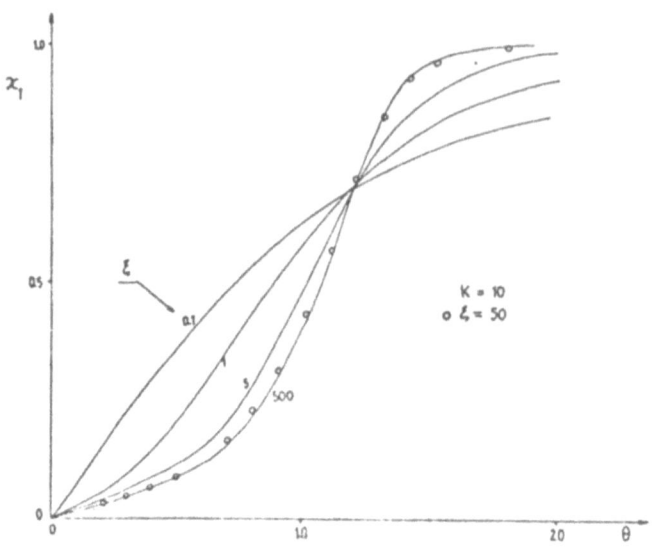

Figure 16- Effect of ξ on the breakthrough curves

For design purposes we can plot a design parameter, θ_{Bp} as a function of K and ξ as shown in Figure 17.

A simple design procedure can then be developed as follows |33|:

- from a laboratory experiment determine the equilibrium isotherm

- choose ϵ, so ξ is fixed

  (as a crude example ξ is about 100 in the saturation step of a water softening operation and about 1 in the regeneration step).

- using Figure 17, read $(\theta_{Bp})_{calc}$ for a given pair of ξ and K values

- the time of operation being fixed (t_1) we get from the value $(\theta_{Bp})_{calc}$

$$\tau = \frac{t_1}{(\theta_{Bp})_{calc}\,(1+\xi)} \quad \text{and then} \quad v_s = \frac{1-\epsilon}{\epsilon}\, U\tau$$

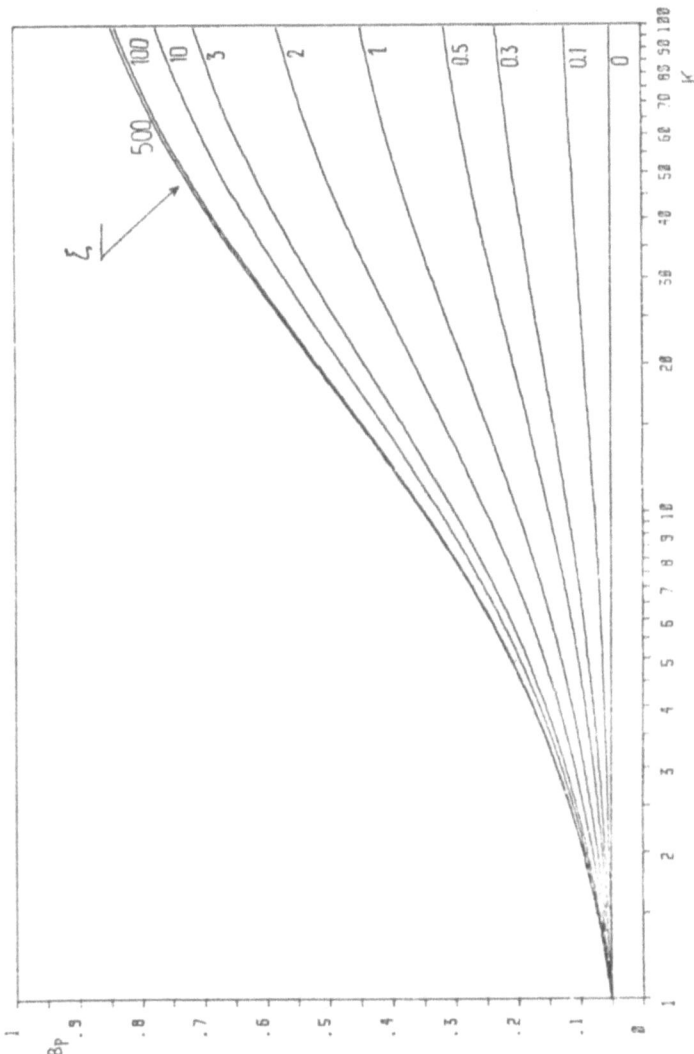

FIGURE 17- $\theta_{Bp} = f(\xi, K)$ FROM THE EQUILIBRIUM MODEL FOR A PERFECTLY MIXED SORBER

When we consider the column as a series of J cells, J being a measure of the axial dispersion of the fluid in the porous medium, we can get some typical breakthrough curves as a function of J. This is shown in Figures 18a, 18b and 18c for favorable, unfavorable and B.E.T. isotherms respectively.

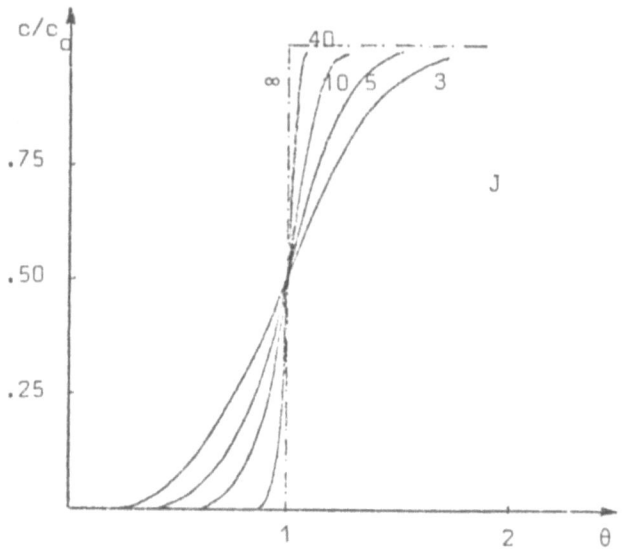

Figure 18a- Equilibrium model for favorable isotherms: $x=f(\theta,J)$ at fixed K and ξ'

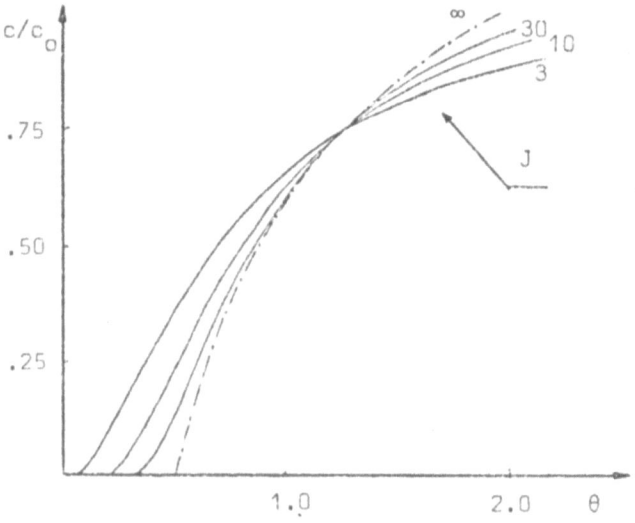

Figure 18b- Equilibrium model for unfavorable isotherms: $x=f(\theta,J)$ at fixed K and ξ.

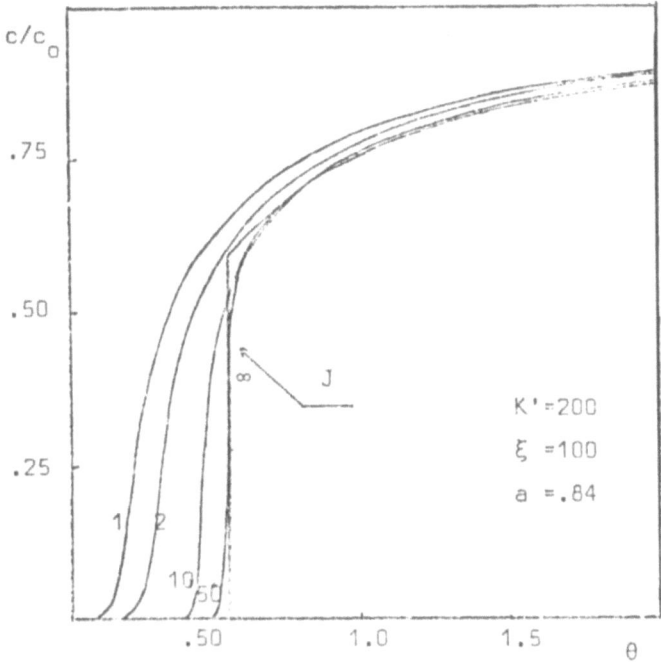

Figure 18c- Breakthrough curves x=f(θ,J) for
B.E.T. isotherms.

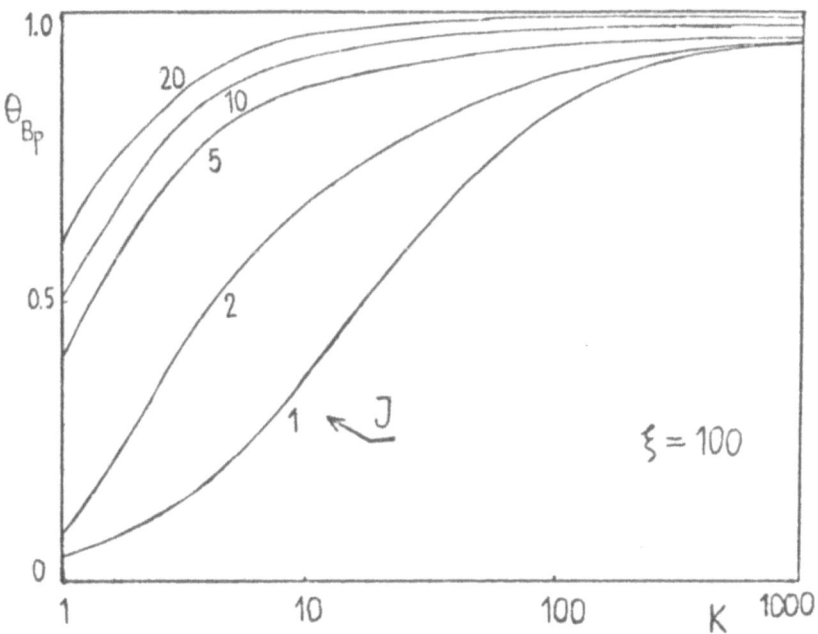

FIGURE 19- THE BREAKTHROUGH TIME AS A FUNCTION OF THE MODEL
PARAMETERS FOR "CONSTANT SEPARATION FACTOR TYPE"ISOTHERMS.

For this model, we also show in Figure 19 the breakthrough time, θ_{Bp} as a function of J and K (for fixed ξ).

For B.E.T. isotherms, which are characterized by two parameters, a and K', a similar graph is drawn is Figure 20.

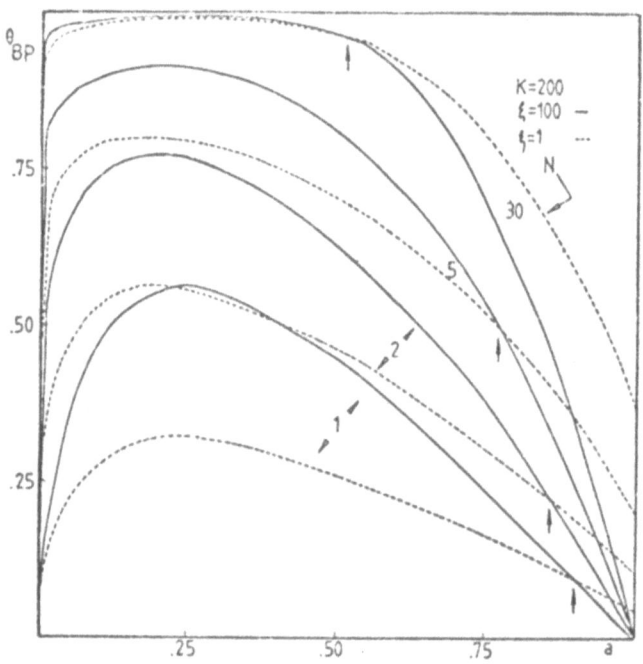

Figure 20- The breakthrough time as a function of the model parameters for B.E.T. isotherms.

A design procedure can be developed, as indicated for one cell, including at the begining one more experiment in order to determine J (measure of the axial dispersion by "tracer" methods).

5.2.2 Non equilibrium model: film diffusion

The complexity of the model can be increased in a realistic way by introducing first a resistance to mass transfer due to film diffusion.*

The model equations for cell i are again: mass balance, Eq. $|63a|$, equilibrium isotherm, Eq. $|63b|$, boundary and initial conditions, Eq. $|63d|$ coupled to the kinetic law

$$\frac{dq_i}{dt} = (K_f a_p) (c_i - c_i*) \qquad |65|$$

The system of Equations 63a, 63b, 63d, and 65, can be reduced to

$$\frac{dx_i}{d\theta} = f_1(x_i, y_i) \qquad |66a|$$

$$\frac{dy_i}{d\theta} = f_2(x_i, y_i) \qquad |66b|$$

where $f_1(x_i,y_i)=(1+\xi)\left[Jx_{i-1}-(J+N_f)x_i+\dfrac{N_f y_i}{y_i+K(1-y_i)}\right]$

$f_2(x_i,y_i)=\dfrac{(1+\xi)N_f}{\xi}\left[x_i-\dfrac{y_i}{y_i+K(1-y_i)}\right]$

and $N_f=\dfrac{1-\varepsilon}{\varepsilon} K_f a_p \tau$ is the number of film mass transfer units.

Let us restrict the analysis of the above system to the case of $J=1$. Due to the stiff nature of this system, the use of explicit methods such as Runge-Kutta is unappropiate.

Lapidus $|34|$ defines the stiffness ratio as the ratio between the two extreme values of the eigenvalues, that is,**

---

* In Annex 1 we present some correlations for the prediction of film mass transfer coefficients

** In this problem the Jacobian matrix is

$$\underline{J} = \begin{bmatrix} \dfrac{\partial f_1}{\partial x_1} & \dfrac{\partial f_1}{\partial y_1} \\[2mm] \dfrac{\partial f_2}{\partial x_1} & \dfrac{\partial f_2}{\partial y_1} \end{bmatrix} = \begin{bmatrix} -(1+\xi)(1+N_f) & (1+\xi)N_f \dfrac{K}{|y_1+K(1-y_1)|^2} \\[3mm] N_f\dfrac{1+\xi}{\xi} & -N_f\dfrac{1+\xi}{\xi} \dfrac{K}{|y_1+K(1-y_1)|^2} \end{bmatrix}$$

The eigenvalues λ_1 and λ_2 are determined by $\det|\underline{J}-\lambda\underline{I}|=0$ where $\underline{I}$ is the identity matrix.

$$S(x) = \frac{\text{Max Re} \ (-\lambda_i)}{\text{Min Re} \ (-\lambda_i)}$$

|67|

It is usually accepted that the degree of stiffness is measured by the deviation of $S(x)$ from unity. However it has been shown that the stiffness ratio is an imprecise criterium of stiffness |35, 36|.

A comparison of different numerical methods |36, 37, 38| of integration of the system of Eqs. 66 a, b is presented elsewhere |36|.

Some typical results are presented in Figure 21, which shows that below a certain value of N_f a "plateau" appears in the breakthrough curve.

Again we can use figures like Figure 21 in order to plot the breakthrough time as a function of the model parameters: K, ξ and N_f. This is shown in Figure 22.

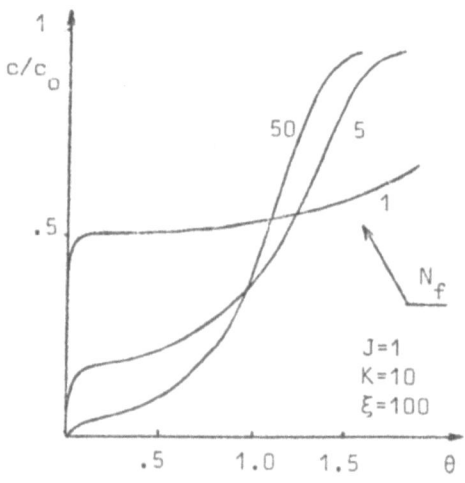

Figure 21- Concentration histories: film diffusion model for one cell

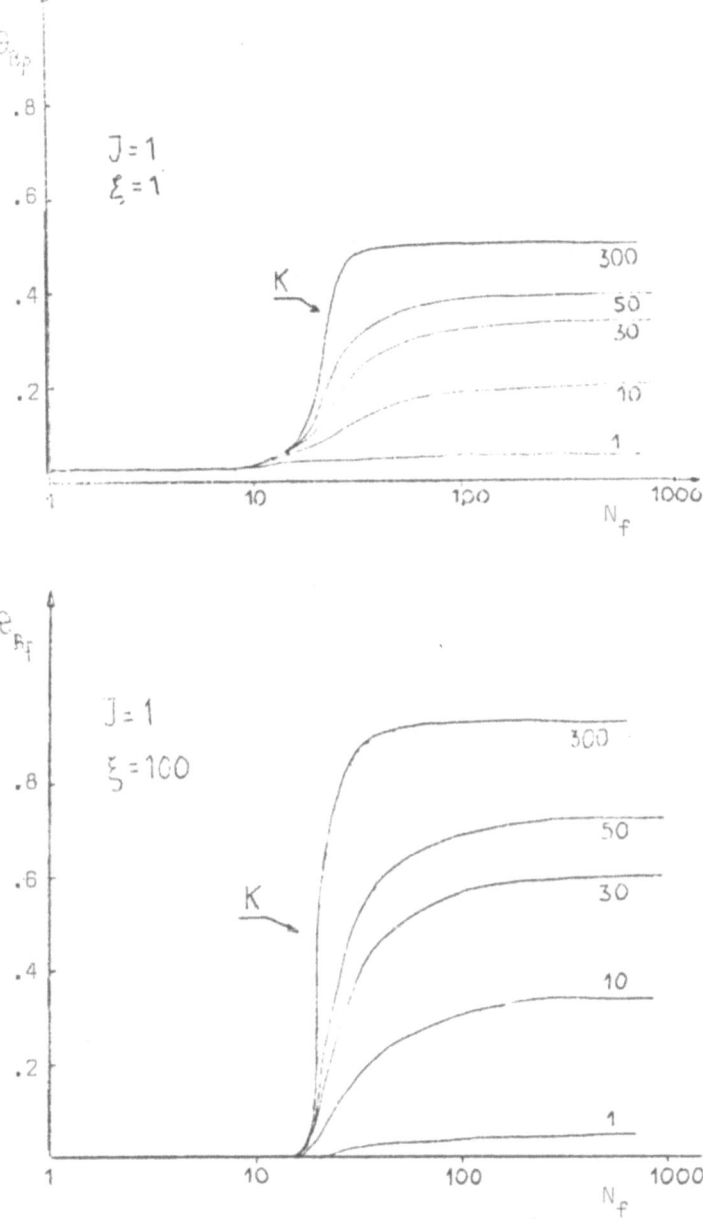

Figure 22-   Breakthrough time as a function of K and
N_f at fixed ξ :film diffusion model (J=1)

A procedure for design can be drawn:

- from a laboratory experiment we get   K
- choose ε, so   ξ   is fixed
- choose τ
- with this   τ   value we can calculate $(\theta_{Bp})_1$ since we know the time of operation t_{Bp}; we can also estimate N_f for that   τ   value
- using the predicted value of N_f we read in Figure 22 a $(\theta_{Bp})_c$ value; if $(\theta_{Bp})_c \neq (\theta_{Bp})_1$ we choose another   τ   value; the procedure is   repeated until we get $(\theta_{Bp})_c = (\theta_{Bp})_1$ for a τ_c value
- finally we get

$$v_r = \frac{1-\varepsilon}{\varepsilon} \frac{U}{J} \tau_c \qquad\qquad |68|$$

which is generalized to   J   cells in series

A convenient plot for   J=1,2   and   3   in which we compare the performance of one, two and three cells in series is shown in Figure 23 |37|.

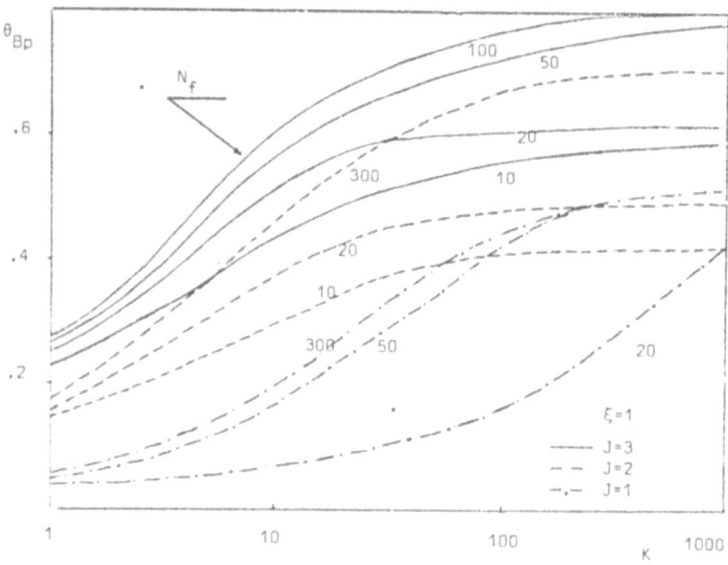

Figure 23- Comparison of 1,2 and 3 cells in series: film diffusion model.

5.2.3 Nonequilibrium model: film diffusion+particle diffusion

Let us consider the solid phase as a gel medium; then we can write
the model equations in terms of dimensionless variables:

mass balance

$$\frac{\partial x_j}{\partial \theta} = J(1+\xi)(x_{j-1} - x_j) - N_f(1+\xi)(x_j - x_j^*) \qquad |69a|$$

particle diffusion in a sphere

$$\frac{\partial y_j}{\partial \theta} = N_D(1+\xi)\left[\frac{\partial^2 y_j}{\partial \rho^2} + \frac{2}{\rho}\frac{\partial y_j}{\partial \rho}\right] \qquad |69b|$$

equilibrium law

$$x_j^* = \frac{y_j(1)}{K + (1-K)y_j(1)} \qquad |69c|$$
$$j = 1, 2, \ldots, J$$

boundary and initial conditions

$$\text{B.C.1.} \qquad \left.\frac{\partial y_j}{\partial \rho}\right|_{\rho=0} = 0 \qquad |69d|$$

$$\text{B.C.2.} \qquad \left.\frac{\partial y_j}{\partial \rho}\right|_{\rho=1} = \frac{N_f}{3N_D \xi}(x_j - x_j^*) \qquad |69e|$$

$$\text{I.C.} \qquad x_0 = H(\theta) \qquad |69f|$$

$$\left.x_j\right|_{\theta=0} = \left.y_j\right|_{\theta=0} = 0 \qquad j>0 \qquad |69g|$$

where $N_D = \dfrac{D_{eff}}{R_p}$ is the number of mass transfer units due to internal diffusion and $\rho = \dfrac{r}{R_p}$ is the dimensionless radius of the
particle.

The system of PDEs was solved by using Crank-Nicholson's finite
difference scheme of integration |38|. Defining

$$x_n^j = \left.x^j\right|_{\theta=n\delta\theta}$$
$$y_{in}^j = \left.y^j\right|_{\rho=(i-1/2)\delta\rho}$$

$$j = 1, 2, \ldots, J$$
$$n = 0, 1, \ldots, \infty$$
$$i = 0, 1, \ldots, N+1$$

where $\delta\theta$ and $\delta\rho$ are the time and radius steps respectively, and then $\delta\rho = 1/N$.
We have then

$$\frac{1}{N_D(1+\xi)} \frac{y^j_{i,n+1} - y^j_n}{\delta\theta} = \frac{y^j_{i+1,n+1} - 2y^j_{i,n+1} + y^j_{i-1,n+1} + y^j_{i+1,n} - 2y^j_{i,n} + y^j_{i-1,n}}{2(\delta\rho)^2} +$$

$$+ \frac{2}{(i-1/2)\delta\rho} \frac{y^j_{i+1,n+1} - y^j_{i-1,n+1} + y^j_{i+1,n} - y^j_{i-1,n}}{4\,\delta\rho} \qquad |70|$$

For $i=1$, as $\left.\frac{\partial y^j}{\partial\rho}\right|_{\rho=0} = 0$ then $y^j_{0,n} = y^j_{1,n} \; \forall\, n$ and Eq. $|70|$ can be written as

$$\frac{1}{N_D(1+\xi)} \frac{y^j_{1,n+1} - y^j_{1,n}}{\delta\theta} \approx \frac{y^j_{2,n+1} - y^j_{1,n+1} + y^j_{2,n} - y^j_{1,n}}{2(\delta\rho)^2} +$$

$$+ \frac{y^j_{2,n+1} - y^j_{1,n+1} + y^j_{2,n} - y^j_{1,n}}{(\delta\rho)^2} \qquad |71|$$

Solving these equations for $y^j_{i,n+1}$ a tridiagonal system of N equations is obtained, which can be transformed into a bidiagonal system using gaussian elimination.
The Nth equation will be written as

$$B_N y^0_{N,n+1} + C_N y^j_{N+1,n+1} = D_N \qquad |72|$$

As there are N+1 unknowns $y^j_{i,n+1}$ we need to introduce the boundary conditions

$$\frac{1}{1+\xi} \frac{x^j_{n+1} - x^j_n}{\delta\theta} = -(N_f + J)\frac{x^j_{n+1} + x^j_n}{2} + J\frac{x^{j-1}_n + x^{j-1}_{n+1}}{2} + N_f \frac{x^{*j}_{n+1} + x^{*j}_n}{2} \qquad |73|$$

which can be solved for x^j_{n+1} .
From the equal flux boundary condition

$$\frac{y^j_{N+1,n+1} - y^j_{N,n+1}}{\delta\rho} = \frac{N_f}{3N_D\xi} (x^j_{n+1} - x^{*j}_{n+1}) \qquad |74|$$

The equilibrium relationship is

$$x*_{n+1}^j = \frac{1/2 \; (y_{N+1,n+1}^j + y_{N,n+1}^j)}{K + (1-K) \; (y_{N+1,n+1}^j + y_{N,n+1}^j)/2} \qquad |75|$$

Solving simultaneously Eqs $|72|$, $|73|$, $|74|$ and $|75|$ we obtain
the breakthrough curves and the profiles inside the particles
at different times.
Figure 24 shows the influence of N_D on the breakthrough curves
for J=20, which is a case of relatively important axial dispersion since J-1=Pe/2 (Pe=$u_i L/D_{ax}$ is the Peclet number).

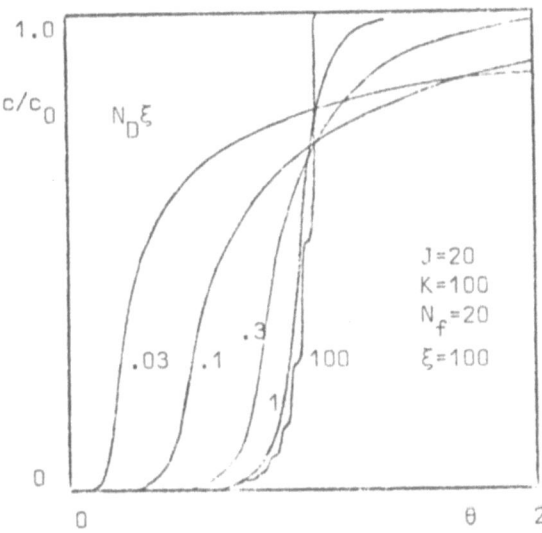

Figure 24-Nonequilibrium model:film diffusion+particle
diffusion.Effect of N_D on the concentration histories.

Other authors use the differential approach for modeling. Svedberg
$|39|$ uses a similar scheme for the case where both particle and pore diffusion are present. This case should be investigated in more
detail, in order to take into account the fact that these resistan
ces can be associated either in series or in paralel. The resulting
equations are not the same for these cases although most of the
authors overlook this question.

Liapis$|40|$ used the collocation method which transforms the origi-
nal system of equations into a system of ODEs; then an implicit

method of integration was applied leading to good results.

Complex profiles will, however, require more collocations points; the result is that the degree of stiffness of the system of ODEs will therefore increase.

Computer calculations suggest that, when axial dispersion is impor tant, the staged approach is a good way of tackling the problem, which can be combined with collocation methods for high Pe.

ANNEX 1.   Mass Transfer Between Fluid and Particles in Percolation
           Columns

Numerous studies deal with the measure of mass transfer rates between a fluid phase and a bed of particles.

Experimental techniques can be divided in the following catego- ries: dissolution of solid particles (e.g. benzoic acid), sublima tion of particles (e.g. naphtalene), evaporation of a pure liquid from the surface of a porous particle or fast catalytic reaction on a non porous solid.

An interesting method is due to Myauchi |41|; he used an ion ex- change reaction accompanied by neutralization (fast and irrever- sible reaction). A pulse of area S_0 is introduced into the column and the output signal of area S is recorded.

It can be shown* that $\frac{S}{S_0} = \frac{c}{c_0}$ where c is the outlet steady state concentration when the column is fed with a constant inlet concen tration, $c_E = c_0 H(t)$.

For the plug-diffusional model we get

$$\frac{c}{c_0} = \frac{4 \, \beta \, \exp\frac{Pe}{2}}{(1+\beta)^2 \, \exp(\frac{\beta \, Pe}{2}) - (1-\beta)^2 \, \exp(-\frac{\beta \, Pe}{2})} = f(Pe, K) \qquad |76|$$

where $\beta = \sqrt{1 + \frac{4N_r}{Pe}}$, $N_r = \frac{KL}{u_i}$ and $Pe = \frac{u_i L}{D_{ax}}$.

---

* Putting $S = \int_0^\infty y(t)dt = Y(s)\Big|_{s=0}$ and $S_0 = \int_0^\infty x(t)dt = X(s)\Big|_{s=0}$ we get $\frac{S}{S_0} = G(s)$; applying the final value theorem $\frac{c}{c_0} = \lim_{s \to 0} G(s) = \frac{S}{S_0}$

Then $\dfrac{S}{S_0}=f(Pe,K)$; if Pe is known we obtain K, which is similar to a film mass transfer coefficient*.

Mass transfer between a fluid and a sphere

If the fluid is stagnant, mass transfer will occur by molecular diffusion. For a sphere of radius R inside a concentrical spherical shell of radius r_s the equation for steady state diffusion from the surface of the sphere to the fluid is

$$4 \pi R^2 N_R = 4 \pi \rho^2 N_\rho = -4 \pi \rho^2 D \frac{dc}{d\rho} \qquad |77|$$

where N_r is the molal flux through the spherical shell at radius ρ $(\rho > R)$.

Integrating between $c=c_i$ at $\rho=R$ and $c=c_s$ at $\rho=r_s$ and then introducing the Sherwood number, $Sh=K_f d_p/D$ (with $d_p=2R$), we get

$$Sh = \frac{2r_s}{r_s - R} \qquad |78|$$

In an infinite medium, $r_s \to \infty$ and then the minimum Sherwood number is $Sh_m=2$.

Any fluid motion will contribute to increase K_f (and Sh); the results can be represented by the Froessling equation |42|

$$Sh = 2 + 0.6 \; Sc^{1/3} \; Re^{1/2} \qquad |79|$$

where $Sc=\dfrac{\nu}{D}$ and $Re=\dfrac{u_0 d_p}{\nu}$ (u_0 is the superficial velocity).

---

* For an irreversible reaction $r=Kc$; the rate of film mass transfer is $K_f(c-c^*)$. In the case of ion exchange accompanied by neutralization $c^*=0$ and then $K=K_f$.

74

Mass transfer in fixed beds

Several correlations have been presented in the litterature and reviewed elsewhere |44, 45|. One of the more widely used is that of Ranz |43|

$$Sh = 2.0 + 1.8 \; Sc^{1/3} \; Re^{1/2} \quad \text{for} \quad Re > 100 \qquad\qquad |80|$$

Experimental data are shown in Figure 25.

Several authors state that in fixed beds the minimum Sherwood number is 2. This is in collision with the existing experimental data, for small particles (dp<<1mm).

However Sh_m lower than 2 can be explained by arguments due to Cornish |46| and reported by Skelland |47|. As shown by Eq. 78, Sh_m is different whether a sphere is isolated or located within an array of spheres.

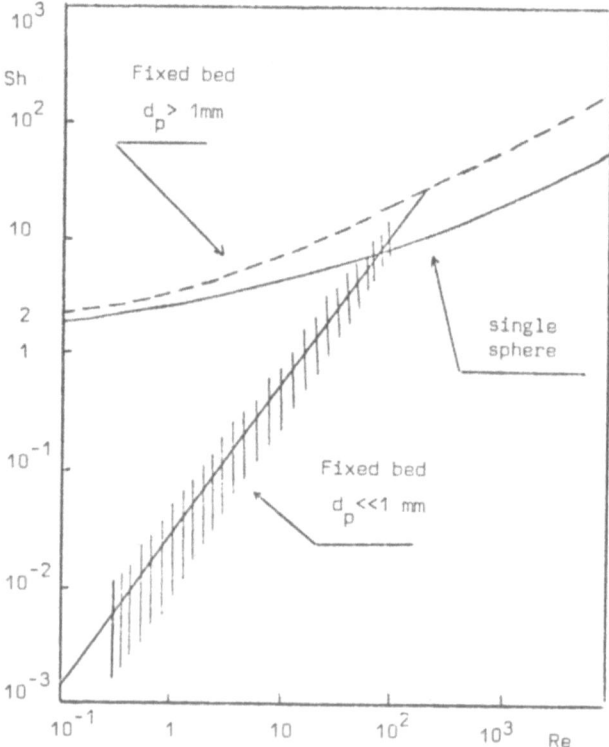

Figure 25- Experimental data for mass transfer in fixed beds.

Only in the case of an infinite medium, that is, when the distance between the centers of spheres ℓ is infinite, we get $Sh_m=2$. Otherwise, we obtain $Sh_m=1.98$ for a distance $\ell=100R$, $Sh_m=1.6$ for $\ell=4R$ and $Sh_m=1.386$ when the spheres are touching , $\ell=2R$.

Recently, Schlunder $|48|$ was able to explain these observations by studying mass transfer in capillary bundles.

When tackling the problem of mass transfer in fixed beds it is interesting to know some theoretical correlations which can be used to predict mass transfer in different Re range. We will mention Carberry $|49|$, Kataoka $|50|$ and Pfeffer $|51|$ correlations.

Carberry's formulation is based on the unsteady state diffusion in a film which can be destroyed and renewed while the fluid passes through the bed.

This correlation is valid for $0.1<Re<1000$ and its form is

$$Sh = \frac{1.15}{\sqrt{\varepsilon}} \, Re^{1/2} \, Sc^{1/3} \qquad\qquad |81|$$

The Kataoka model starts with the analogy of the intraparticular spaces with a series of tubes; the mass transfer rate is analysed as that between a tube surface and a fluid in laminar regime. The starting equation is

$$u_z \frac{\partial c}{\partial z} = D \left(\frac{\partial^2 c}{\partial \rho^2} - \frac{1}{R-\rho} \frac{\partial c}{\partial \rho} \right) \qquad\qquad |82|$$

with the boundary conditions $c=c_s$ at $\rho=0$, $c=c_1$ at $z=0$ and $\frac{\partial c}{\partial \rho}=0$ at $\rho=R$; the result is

$$Sh = \frac{1.85 \, \varepsilon^{-2/3}}{(1-\varepsilon)^{-1/3}} \, Re^{1/3} \, Sc^{1/3} \quad \text{for} \quad Re<100 \qquad\qquad |83|$$

Pfeffer analysis is based on the Happel's free surface model, which assumes no friction at the external layer; combining this model with the Levich solution for a thin boundary layer we get

$$Sh = 1.26 \left(\frac{1-\gamma^5}{w} \right)^{1/3} Pe^{1/3} \quad \text{for} \quad 0.1<Re<50 \qquad\qquad |84|$$

where $\gamma=\frac{a}{b}$ (a, b radius of sphere and external layer, respectively), $w=2-3\gamma+3\gamma^5-2\gamma^6$ and $Pe=Re \cdot Sc = \frac{u_0 d \rho}{D}$. For a sphere in an infinite medium, $\gamma=0$ and $\varepsilon=1$, we get

$$Sh = 0.997 \ Pe^{1/3} \qquad\qquad |85|$$

which is the Levich-Lightill equation.

In summary we can say that mass transfer in fixed beds is control‌led by four groups: Sh, Re, Sc and Gr ($Gr = \dfrac{d_p^3 \rho \Delta \rho g}{\mu^2}$ is the Grashof number, with ρ the fluid density).

Some asymptotic expressions can be derived. For the pure forced convection case, Sh is independent of Gr, that is, for small values of Re we can disregard the inertial effects.

Then

$$Sh = constant \quad , \quad Pe \rightarrow 0 \ and \ small \ Re \qquad\qquad |86|$$

and

$$Sh = b \ Pe^{1/3} \quad , \quad Pe \rightarrow \infty \ and \ small \ Re \qquad\qquad |87|$$

which is similar to the Graetz-Nusselt equation $|41|$.

For high Re we get

$$Sh = c \ Re^n \ Sc^{1/3}, \frac{1}{2} < n < 2/3 \quad , \quad Sc > 10 \qquad\qquad |88|$$

If natural convection is the controlling mechanism then for high Sc we obtain

$$Sh = a(Gr \ Sc)^{1/4} \qquad Laminar \ convection \qquad\qquad |89|$$

and

$$Sh = d \ (Gr \ Sc)^{1/3} \qquad Turbulent \ convection \qquad\qquad |90|$$

Finally it must be pointed out that we have to be careful when using these predicted K_f values for ion exchange operations. In fact, in this case we have to take into account the influence of the electric field on the film mass transfer coefficients. Very little work has been done in this area; however several articles have been published based on Van Brocklin's thesis $|52|$. He deter‌mines a base line for fixed bed mass transfer in the absence of ionic migration; then he uses a R_i factor to correct the previous mass transfer coefficient. The R_i factor is defined as the ratio between molar fluxes in the presence, and in the absence of electric field , J and J_d respectively.

In fact the diffusion flux of a component i is $J_d = -D_i \frac{\partial c_i}{\partial z}$

while in the presence of an electric field

$J = J_d + J_e = -D_i \left[\frac{\partial c_i}{\partial z} + c_i \frac{F}{RT} \frac{\partial V}{\partial z} \right]$ where V is the potential and F the Faraday constant.

The R_i factor can be used either to correct the traditional Chilton-Colburn factor, $j_D (j_D = \frac{K_f}{u_0} Sc^{2/3})$ by calculating a new $j^* = R_i j_D$ or to define a variable mass transfer coefficient $K_f^* = K_f R_i$ (with K_f constant for given hydrodynamic conditions and R_i variable).

Acknowledgements

This work was performed under NATO Research Grant no. 1424.

The author is grateful to his students C.A.Costa, J.C.Lopes, M.Q.Dias and M.R.Costa for their contributions.

NOMENCLATURE

a parameter of the B.E.T. isotherm

a_p specific area of the particle

Bi_m mass Biot number

c fluid phase concentration

c^* fluid concentration at the interface

c_p fluid concentration in the pore

c_s solubility

D_{ax} axial dispersion coefficient

D_{pf} pore diffusivity

D_s solid diffusivity

F Faraday constant

$G(s)$ transfer function

Gr Grashof number

$H(t)$ Heaviside function

J number of cells in series

J_d diffusion flux

J_e flux due to the electric field

k_1 rate constant

K_f film mass transfer coefficient

K constant separation factor

K' parameter of the B.E.T. isotherm

N number of reaction units

N_f number of film mass transfer units

N_z specific mass flux

N_D number of particle mass transfer units

Pe Peclet number

q solid phase concentration

q^* solid concentration at the interface

Q capacity

Q_m solid concentration corresponding to a monolayer

r equilibrium parameter

R ideal gas constant

Re Reynolds number

Sh Sherwood number

Sc Schmidt number

$S(x)$ stiffness ratio

S_f safety factor

t_{st} stoichiometric time

t   time
t_{Bp} breakthrough time
T   absolute temperature;percolation parameter
u_i interstitial velocity
u_c velocity of propagation of a concentration c
U   flowrate
v   bed volume
v_f fluid volume in a cell
v_s solid volume in a cell
V   volume of fluid passed through the column
x   dimensionless fluid concentration
y   dimensionless solid concentration
z   axial coordinate

α parameter which characterizes particle simmetry
δ_z width of the mass transfer zone
ϵ porosity
$\delta(t)$ Dirac delta function
ξ capacity parameter
ρ radial or reduced radial position in a particle or in a tube
λ velocity of a shock layer
τ space time
θ_{Bp} reduced breakthrough time
Ω cross sectional area of the bed
μ_n n^{th} order moments of the Dirac response

REFERENCES

(1)   Rodrigues,A.E.,"Percolation theory I-Basic principles" in
       Stagewise and mass transfer operations,ed.E.Henley and
       J.Calo,CHEMI Project (in press)

(2)   Broughton,D.,Neuzil,R.,Pharis,J. and Brearley,C.,
       Chem.Eng.Prog.,66(9),70(1970)

(3)   de Rosset,A.,Neuzil,R.and Korous,J.,Ind.Eng.Chem.Proc.Des.Dev.
       15(2),261(1976)

(4)   Le Goff,P. Cours de génie chimique,ENSIC,Nancy(1972)

(5)   Danckwerts,P.V.,Chem.Eng.Sci.,2(1),1(1953)

(6)   Langmuir,I.,J.Am.Chem.Soc.,30,1742(1908)

(7) De Vault,D.,J.Am.Chem.Soc.,65,532(1943)

(8) Jeffreson,C.,Chem.Eng.Sci.,25,1319(1970)

(9) Vermeulen,T.,"Separation by adsorption methods" in
     Adv.Chem.Eng.,vol.2,147(1958),Acad.Press

(10) Tudge,A.,Can.J.Phys.,39,1611(1961)

(11) Roberts,J.,Ph.D.Thesis,Georgia Institute of Technology(1951)

(12) Rodrigues,A.E. and Costa,C.A.,6thCHISA Congress,Prague(1978)

(13) Golden,F.M.,Ph.D.Thesis,University of California-Berkeley(1972)

(14) Rhee,H.,Bodin,B.and Amundson,N.,Chem.Eng.Sci.,26,1571(1971)

(15) Glueckauf,E.,J.Chem.Soc.(London),1302(1947)

(16) Glueckauf,E.,"Principles of operation of ion exchange columns"
     in Ion exchange and its applications,Soc.Chem.Ind.,London,
     (1955)

(17) Rodrigues,A.E. and Tondeur,D.,5thCHISA Congress,Prague(1975)

(18) Thomas,H.C.,Ann.N.Y.Acad.Sci.,49,161(1948)

(19) Rosen,J.B.,Ind.Eng.Chem.,46(8),1590(1954)

(20) Vermeulen,T. and Hiester,N.,Chem.Eng.Prog.,48(10)505(1952)

(21) Gladel,Y.,Revue de l'IFP,12(7),864(1957)

(22) Aris,R. and Amundson,N., Mathematical methods in chemical
     engineering,vol.2 First order partial differential equations
     with applications,Prentice Hall (1973)

(23) Bohart,G. and Adams,E.,J.Am.Chem.Soc.,42,523(1920)

(24) Walter,J.E.,J.of Chem.Phys.,13(8),332(1945)

(25) Walter,J.E.,J.of Chem.Phys.,13(6),229(1945)

(26) Klinkenberg,A.,Ind.Eng.Chem.,40(10),1992(1948)

(27) Anzelius,A.,Z.Angew.Math.Mech.,6,291(1926)

(28) Schumann,T.,J.Franklin Inst.,208,405(1929)

(29) Van der Laan,E.,Chem.Eng.Sci.,7,187(1958)

(30) Rodrigues,A.E.,Docteur-Ingenieur Thesis,Univ.Nancy(1973)

(31) Martin,A. and Synge,R.,Biochem.J.,35,1358(1941)

(32) Rodrigues,A.E. and Tondeur,D.,J.Chim.Phys.,72(6),785(1975)

(33) Rodrigues,A.E. and Beira,E.,accepted for publication in
     AIChE J.

(34) Lapidus,1. and Seinfeld,J.,Numerical solution of ordinary
     differential equations,Academic Press(1971)

(35) Birnbaum,I. and Lapidus,L.,Chem.Eng.Sci.,33,415(1978)

(36) Rodrigues,A.,Lopes,J.,Dias,M.,Costa,M. and Costa,C.,
      12th Symp.Computer Applications in Chem.Eng.,Montreux(1979)

(37) Rodrigues,A.,Dias,M.,Lopes,J.and Tondeur,D.,CHEMPOR'78,
      Braga(1978)

(38) Rosenberg,D.,Methods for the numerical solution of PDEs,
      Elsevier(1969)

(39) Svedberg,G.,Ph.D.Thesis,Royal Institute of Technology,
      Stockholm(1975)

(40) Liapis,A.,Ph.D.Thesis,E.T.H.Zurich(1976)

(41) Myauchi,T.and Nomura,T.,Int.Chem.Eng.,12(2),360(1972)

(42) Froessling,W.,Gerland Beitr.Geophys.,52,170(1938)

(43) Ranz,W.E.,Chem.Eng.Prog.,48,247(1952)

(44) Dwivedl,P.and Upadhyay,S.,Ind.Eng.Chem.Proc.Des.Dev.,
      16(2),157(1977)

(45) Karabelas,A.,Wegner,T.and Hanratty,T.,Chem.Eng.Sci.,
      26,1581(1971)

(46) Cornish,A.,Trans.Inst. Chem.Eng.,43,T332(1965)

(47) Skelland,A.,Diffusional mass transfer,J.Wiley (1974)

(48) Schlunder,E.,Chem.Eng.Sci.,32,845(1977)

(49) Carberry,J.,AIChEJ.,6(3),460(1960)

(50) Kataoka,T.,J.of Chem.Eng.Japan,5(2),132(1972)

(51) Pfeffer,R.,Ind.Eng.Chem.Fund.,3(4),380(1964)

(52) Van Brocklin,P.,Ph.D.Thesis,Univ.of Washington(1968)

THEORY OF LINEAR CHROMATOGRAPHY

Jacques VILLERMAUX

Professor of Chemical Engineering at the Ecole Nationale Supérieure des Industries Chimiques.
Director of the Laboratoire des Sciences du Génie Chimique. CNRS-ENSIC. NANCY-FRANCE.

1. INTRODUCTION - BASIC ASSUMPTIONS

Chromatographic processes are frequently encountered in practice,
each time a solute or a mixture of solutes present in a flowing
fluid, undergo a transient and (at least partially) reversible exchange between a mobile phase and a stationary or stagnant phase.
If the solutes have not the same affinity for the stationary phase
this process causes a certain degree of separation of the mixture.
The term "chromatography" may appear to be too specific and restricted to the area of analytical separations. A more general terminology would perhaps be : "transient exchange between a convective flowing fluid and a stationary phase" or "dynamics of fixed
bed absorption".

However, for convenience's sake, we will use the word "chromatography" throughout this paper, being clearly specified that
it must be understood in its general sense. Examples of chromatographic systems are found in the following areas :
- ab/adsorption in porous beds
- analytical separations on sorbent columns
- mass transfer in fixed beds
- hydrodynamics of gas-liquid packed columns or more generally convective fluid/fluid contacting
- dispersion of pollutants and/or fertilizers in soils
- use of tracers in hydrology and medicine
- transit of drugs in organs of living beings.
etc. .

The aim of this paper is to give a general treatment of chromatographic processes relying on the methods of chemical engineering and system dynamics.

Two kinds of problems may be encountered :

1) Given fundamental data of the interaction process : equilibrium properties, kinetic and transport data, hydrodynamic characteristics of the flow pattern, how to predict the response of the chromatographic system to input concentration signals ? What is the influence of interaction parameters on the shape of the output signals ? This is a simulation problem (fig.1-1).

2) Having recorded experimentally a set of input/output signals, how to deduce from this data information on the value of intrinsic interaction parameters ? This is a problem of identification or exploitation (fig. 1-1).

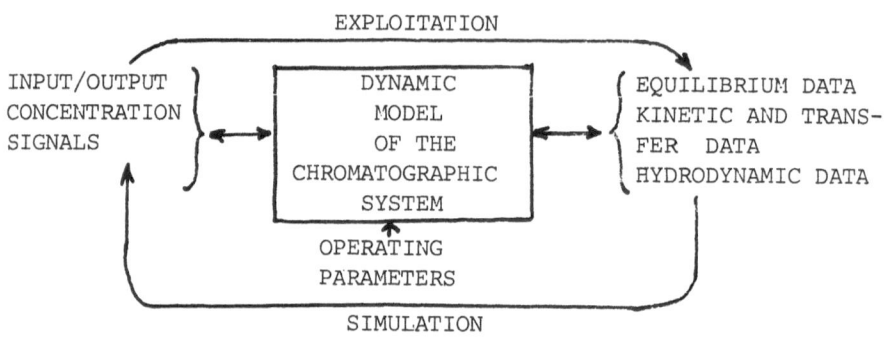

Fig. 1-1

The fundamental assumptions and basic processes involved in the models we will consider here are the following :

- mobile phase flow at steady state characterized for instance by its residence time distribution.

- transient and reversible solute exchange between the mobile phase and stagnant zones or stationary phases, for instance a porous fixed bed.

- no perturbation of the flowrate consecutive to mass exchange (dilute solutions).

- linear partition isotherms.

- external mass transfer, internal diffusion, ab/adsorption into the solid may be involved.

We shall limit ourselves to these processes as a first step by excluding non linear or irreversible processes and chemical reactions in both phases.

2. THE CHROMATOGRAPHIC SYSTEM AS A DYNAMIC SYSTEM

We begin this section by a short refresher on system dynamics. Detailed treatments of these methods can be found in many standard textbooks [1] [2] [3] .

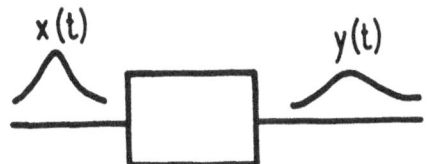

Fig. 2-1

The central problem in linear system dynamics is to predict the relationship existing between the output signal $y(t)$ corresponding to an input signal $x(t)$. Let us define the Laplace transforms of the signals :

$$\overline{x}(s) = \mathcal{L}\left[x(t)\right] = \int_0^\infty x(t)\, e^{-st}\, dt \tag{2.1}$$

$$\overline{y}(s) = \mathcal{L}\left[y(t)\right] = \int_0^\infty y(t)\, e^{-st}\, dt \tag{2.2}$$

The <u>transfer function</u> $G(s)$ of the system is defined as the ratio

$$G(s) = \overline{y}(s)/\overline{x}(s) \tag{2.3}$$

$G(o)$ represents the y/x ratio at steady state in the real domain; $G(s)$ is independant of the shape of the input signal and entirely characterizes the transient behaviour of the system. This behaviour can also be characterized in the real domain by the <u>impulse response</u> $E(t)$ of the system to a Dirac delta function input $x(t) = \delta(t)$. The transfer function is simply the Laplace transform of $E(t)$:

$$G(s) = \mathcal{L}\left[E(t)\right] = \int_0^\infty E(t)\, e^{-st}\, dt \tag{2.4}$$

$E(t)$ is the equation of the so-called "chromatographic peak", which is the response to a pulse injection of solute. The step response $F(t)$ is also frequently considered. $F(t)$ is the response to a step disturbance $x(t) = H(t)$ where $H(t)$ is the unit step function of Heaviside :

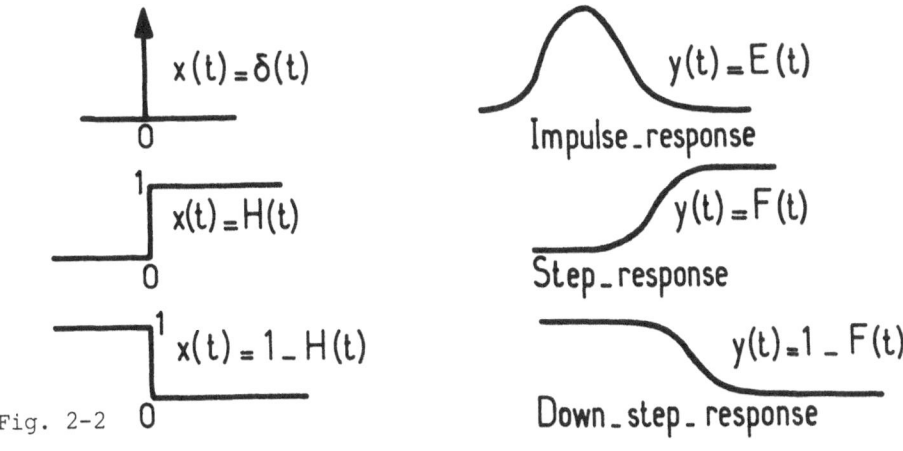

Fig. 2-2

F(t) and E(t) are simply related by

$$F(t) = \int_0^t E(u)\,du \tag{2.5}$$

So that

$$\mathcal{L}\left[F(t)\right] = G(s)/s \tag{2.6}$$

The analysis and modeling of the system consists in finding a mathematical representation of the physicochemical processes tying
the output and input variables together. If the variables undergo
discrete variation in space, the system is described by a set of
ordinary differential equations and G(s) is a function of s alone:
the model is said to be a "lumped parameter" model. An example
will be presented below with the famous "plate model" of a chromatographic column. Conversely, if continuous spatial variations are
taken into account, the system is described by partial differential equation and the model is said to be of the "distributed parameter" kind.

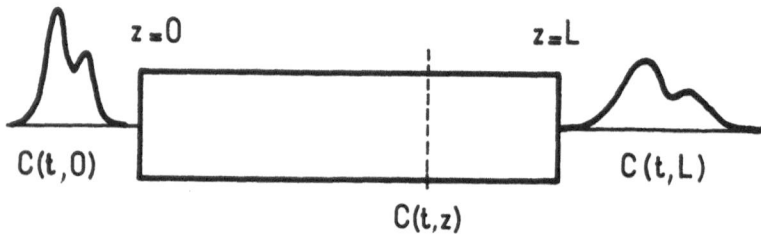

Fig. 2-3

It is also possible to define a transfer function by transforming the equations with respect to time and in integrating over
spatial coordinates

$$\bar{c}(s,z) = \mathcal{L}\left[c(t,z)\right] = \int_0^\infty c(t,z)e^{-st}\,dt \tag{2-7}$$

$$G(s,z) = \bar{c}(s,z)/\bar{c}(s,o) \tag{2-8}$$

The transfer function is a function of spatial position and
can in particular be derived at the outlet of the system.

In many cases, the model equations can easily be solved in
the Laplace domain and the transfer function thus derived while it
would be impossible to obtain a solution in the real domain. Except for very simple systems, the inversion of a Laplace transform
is a difficult problem. In principle, the time domain solution is
given by the Mellin-Fourier inverse transform :

$$f(t) = \mathcal{L}^{-1}\left[\bar{f}(s)\right] = \frac{1}{2\pi i}\int_{a-i\infty}^{a+i\infty} \bar{f}(s)e^{st}\,ds \tag{2.9}$$

Setting a = o and s = $2\pi i\nu$, this integral can be calculated in the Fourier domain

$$f(t) = \int_{-\infty}^{+\infty} \overline{f}(2\pi i\nu) e^{2\pi i\nu t} d\nu \tag{2-10}$$

N points of f(t) at times t =kT/N over the time interval (O,T) are thus given by the discrete form of (2-10) :

$$f(\frac{kT}{N}) = \sum_{n=0}^{N-1} \overline{f}(2\pi i\frac{n}{T}) e^{2\pi ink/N} \tag{2-11}$$

corresponding to frequencies $\nu = n/T$. The powerful algorithm of the Fast Fourier Transform (FFT) $[4]_2$ allows (2-11) to be calculated in $2NLog_2N$ operations instead of N^2 operations provided N is chosen as a power of 2. Notice also that the practical calculation of (2-11) requires analytical separation of the real part and the imaginary part of $\overline{f}(2\pi i\nu)$. Most simulations presented in this paper have been obtained in this way.

The impulse response E(t) may also be characterized by its moments about the origin :

$$\mu_k = \int_0^\infty t^k E(t) dt \tag{2-12}$$

Or its central moments

$$\mu'_k = \int_0^\infty (t - \mu_1)^k E(t) dt \tag{2-13}$$

E(t) is assumed to be normalized so that μ_o = 1. The first moment μ_1 = $\overline{t}$ or mean is a measure of the location of the center of gravity of the output signal while the second central moment, or varian-ce μ'_2 = σ^2 is a measure of the amount of dispersion of the data from the mean. Higher moments measure "skewness"

$$\gamma_1 = \mu'_3/(\mu'_2)^{3/2} \tag{2-14}$$

and "peakedness"

$$\gamma_2 = (\mu'_4/\mu'_2{}^2) -3 \tag{2-15}$$

From (2-12) and (2-13), the following relations are easily esta-blished :

$$\mu'_2 = \sigma^2 = \mu_2 - \mu_1{}^2$$
$$\mu'_3 = \mu_3 - 3\mu_1 \mu_2 + 2\mu_1{}^3 \tag{2-16}$$
$$\mu'_4 = \mu_4 - 4 \mu_1 \mu_3 + 6\mu_1{}^2 \mu_2 - 3\mu_1{}^4$$

An interesting feature is that the moments can be deduced from the sole knowledge of the expression of the transfer function. Comparing (2-12) and (2-4) one obtains by derivation with respect to s the VAN DER LAAN relationship :

$$\mu_k = (-1)^k \left(\frac{\partial^k G}{\partial s^k} \right)_{s=0}$$

(2-17)

which is very useful in evaluating the moments from the model equations. Generally, the impulse response is asymmetric so that the time of the maximum t_{max}, the mean $\bar{t}$ and the median $t_{1/2}$ dividing the curve into two parts of equal surface are distinct.

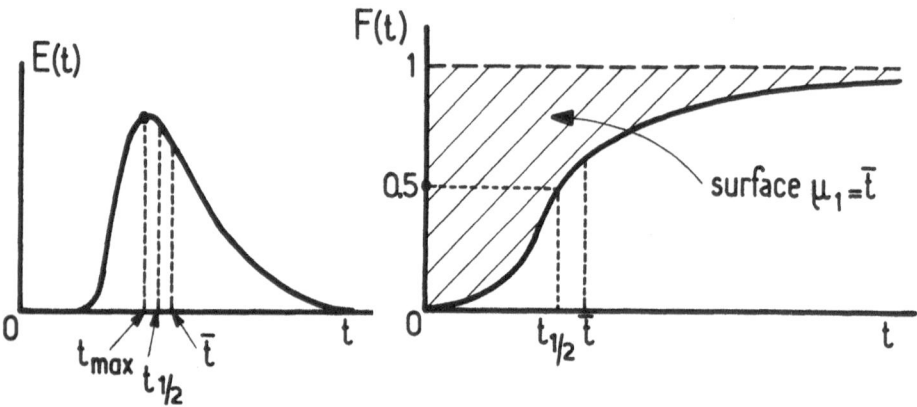

Fig. 2-4

In the same way, the abcissa of the point F = 0.5 is $t_{1/2}$ which may be distinct from the mean $\bar{t}$ in case of asymetry. Notice that $\bar{t}$ appears as the surface between the F curve and the line F = 1 (fig. 2-4)

$$\bar{t} = \int_0^\infty t E(t) \, dt = \int_0^1 t \, dF$$

(2-18)

The numerical evaluation of moments from experimental response curves is difficult and inaccurate, especially for higher order moments, owing to the statistical weight of curve tails and to the noise affecting experimental data. VERGNES [5] has proposed an interesting method relying on successive integrations, especially easy to implement when the data are obtained by sampling at equal time intervals (for instance in a multichannel analyzer). Let E', F', G', H' be respectively the impulse response E'(t) and its successive integrals, measured in arbitrary units

$$F' = \int_0^t E'(u) \, du, \quad G' = \int_0^t F'(u) \, du, \quad H' = \int_0^t G'(u) \, du$$

Unnormalized moments may still be defined

$$m_k = \int_0^\infty t^k E'(t)\,dt \tag{2-19}$$

$$m'_k = \int_0^\infty (t - \bar{t})^k E'(t)\,dt \tag{2-20}$$

But $m_0 \neq 1$ and $E(t) = E'(t)/m_0$. Comparing (2-19) (2-20) to (2-12) (2-13), one finds :

$$\mu_1 = \bar{t} = m_1/m_0 \tag{2-21}$$

$$\mu'_2 = \sigma^2 = m'_2/m_0 = m_2/m_0 - m_1^2/m_0^2 \tag{2-22}$$

Choosing a value of t so large that $E'(t)$ has come back to zero, it can be shown from the definitions that :

$$m_0 = F'(t) \tag{2-23}$$

$$\mu_1 = t - G'(t)/F'(t) \tag{2-24}$$

$$\sigma^2 = 2H'(t)/F'(t) - G'^2(t)/F'^2(t) \tag{2-25}$$

Having recalled these basic concepts, we may now pass on to the derivation of chromatographic models.

3. THE PLATE THEORY OF MARTIN AND SYNGE: A FIRST INTRODUCTORY EXAMPLE

The following treatment is not that which was formally given by MARTIN AND SYNGE in their famous pioneer paper [6]. It is a more modern presentation which is however based on the same physical assumptions.

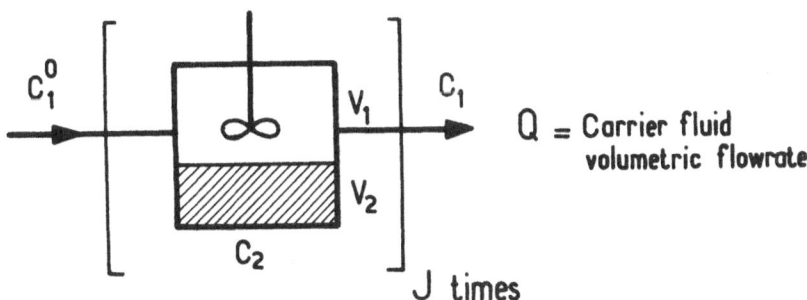

Fig. 3-1

The chromatographic column is regarded as an association of J mixing-cells in series (the "plates") containing a mobile phase (volume V_1) and a stationary phase (volume V_2). A solute (concentration c_1 in the mobile phase) undergoes a reversible exchange with the stationary phase (concentration c_2).

3.1 Instantaneous mass transfer

MARTIN AND SYNGE assume that partition equilibrium is instantaneously achieved on each plate

$$c_2 = \alpha c_1 \tag{3-1}$$

The mass balance in a plate is then written

$$Qc_1^{\circ} = Qc_1 + V_1 \frac{dc_1}{dt} + V_2 \frac{dc_2}{dt} \tag{3-2}$$

Let $t_1 = V_1/Q$ be the mean residence time in the mobile phase for one plate and $K' = \alpha V_2/V_1$ be the capacity factor, defined as the ratio of the amounts of solute present in each phase at equilibrium :

$$K' = (n_2/n_1)_{eq} \tag{3-3}$$

If t_o denotes the mean residence time of an unretained solute (of the mobile phase) in the whole column, then $t_1 = t_o/J$ and (3-2) becomes :

$$c_1^{\circ} = c_1 + \frac{t_o}{J}(1 + K')\frac{dc_1}{dt} \tag{3-4}$$

Each plate behaves like a first-order dynamic system, with the transfer function

$$g(s) = \overline{c_1}/\overline{c_1^{\circ}} = \left[1 + \frac{t_o}{J}(1 + K')s\right]^{-1} \tag{3-5}$$

The overall transfer function of the column is obtained by the multiplication of the elementary transfer function, as J plates are connected in series

$$G(s) = \left[1 + \frac{t_o}{J}(1 + K')s\right]^{-J} \tag{3-6}$$

Set $t_R = t_o(1 + K')$, the transfer function is finally written as :

$$\boxed{G(s) = (1 + t_R s/J)^{-J}} \tag{3-7}$$

If the number of plates J is very high $J \to \infty$, then

$$G(s) \simeq \exp(-st_R) \tag{3-8}$$

This is the transfer function a pure time-delay

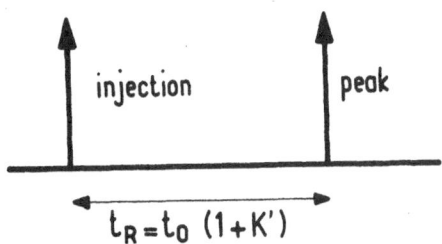

Fig. 3-2

The solutes are separated according to their partition coefficient. The injection peak is undistorted : these are the conditions of ideal chromatography (no axial dispersion, no mass-transfer resistance).

If J is finite, a straightforward inversion of the Laplace transform (3-7) using for instance tables and the translation theorem yields the impulse response (peak equation) :

$$E(t) = \frac{1}{t_R} \cdot \frac{J^J}{(J-1)!} \exp\left(-\frac{Jt}{t_R}\right) \left(\frac{t}{t_R}\right)^{J-1} \tag{3-9}$$

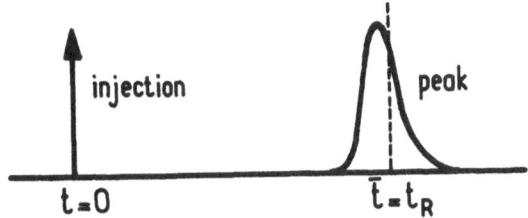

Fig. 3-3

This is a gamma-distribution, identical to that of a cascade of pure mixing-cells in series with a space time t_R.
The moments are easily obtained using Van der Laan's equation (2-17). One finds :

$$\mu_1 = \bar{t} = t_R = t_o(1 + K') \tag{3-10}$$

The center of gravity of the peak is located at the mean retention time t_R.

$$\sigma^2 = t_R^2/J \tag{3-11}$$

When J is large, the width of the peak at half-height is nearly $2\sigma = 2 t_R/ \sqrt{J}$. For instance, with a 10 000 plate column, $2\sigma = t_R/50$. The time of the maximum is such that $dE/dt = 0$. From (3-9) :

$$t_{max} = t_R(1 - \frac{1}{J})$$ (3-12)

very close to t_R when J is large. However, it must be pointed out that the gamma distribution is asymmetric. The peak is not gaussian contrary to the statements that can be found in too many textbooks on chromatography : the gaussian distribution is irrelevant, as it has values for t < 0, before the time of injection ! Nevertheless, it is true that the gaussian distribution is an asymptotic form of (3-9) as $J \to \infty$.

If the column is lengthened by addition of plates in series, it is interesting to notice that t_o (and thence t_R) is proportional to J. Thus, the separation between two adjacent peaks varies as J whereas their width (2σ) varies as $\sqrt{J}$. Consequently, the separation is improved as the number of plates is increased : this is one of the basic properties of chromatographic separations. The main criticism of the simple theory of Martin and Synge is that it does not take into account mass transfer effects. They are accounted for in the following way.

3.2 Finite mass-transfer rate: The MCE model (Mixing cells in Cascade with mass Exchange).

Let us go back to the mass balance on a "plate", but assume a finite mass-transfer rate between mobile and stationary phases expressed by the flux $kS(c_1 - c_2/\alpha)$ where k is the overall mass transfer coefficient and S the interfacial area.

In the mobile phase :

$$Qc_1^o = Qc_1 + V_1 \frac{dc_1}{dt} + kS(c_1 - \frac{c_2}{\alpha})$$ (3-13)

In the stationary phase :

$$kS(c_1 - \frac{c_2}{\alpha}) = V_2 \frac{dc_2}{dt}$$ (3-14)

Define the mass transfer time constant t_m by

$$t_m = \frac{\alpha V_2}{kS}$$ (3-15)

the stationary phase mass balance is written

$$Kc_1 = c_2 + t_m \frac{dc_2}{dt}$$ (3-16)

which is characteristic of a new first order system with the transfer function

$$\frac{\overline{c_2}}{\overline{c_1}} = \frac{K}{1 + t_m s}$$ (3-17)

The overall mass transfer balance in the plate has the same expression as (3-2). In the Laplace domain

$$Q\overline{c_1^o} = Q\overline{c_1} + V_1 \ s\overline{c_1} + V_2 \ s\overline{c_1} \ \frac{\alpha}{1 + t_m s} \tag{3-18}$$

Setting as before $K' = \alpha V_2/V_1$ and $t_o/J = V_1/Q$ we obtain the plate transfer function :

$$g(s) = \frac{\overline{c_1}}{\overline{c_1^o}} = \left[1 + \frac{st_o}{J} \ (1 + \frac{K'}{1 + t_m s}) \right]^{-1} \tag{3-19}$$

and the J-plate column transfer function

$$\boxed{G(s) = \left[1 + \frac{st_o}{J} \ (1 + \frac{K'}{1 + t_m s}) \right]^{-J}} \tag{3-20}$$

If we compare (3-20) to (3-6), we see that K' has simply been replaced by $\frac{K'}{1 + t_m s}$.

Unfortunately, there is no simple analytical expression for $E(t)$, the original of $G(s)$[9]. The impulse response $E(t)$ has to be calculated by numerical inversion of (3-20) as explained above. Later we will study the influence of the parameters on the peak shape (section 7.2). The moments of $E(t)$ can be obtained using Van der Laan's theorem as usual :

$$\boxed{\overline{t} = t_R = t_o \ (1 + K')} \tag{3-21}$$

In spite of the presence of mass-transfer resistance, it is confirmed that the position of the peak, marked by its center of gravity, depends only on equilibrium properties (the partition coefficient). Conversely, finite mass-transfer rate and axial dispersion cause the broadening of the peak, as can be seen on the variance expression

$$\sigma^2 = \frac{t_R^2}{J} + \frac{2K'}{1 + K'} \ t_m t_R \tag{3-22}$$

In reduced form :

$$\boxed{\frac{\sigma^2}{t_R^2} = \frac{1}{J} + \frac{2K'}{1 + K'} \cdot \frac{t_m}{t_R}} \tag{3-23}$$

The first term $(1/J)$ relates to axial dispersion : the larger the number of mixing cells in series, the sharper the peak. The second term relates to mass-transfer kinetics : the important feature is the ratio of the exchange time t_m to the retention time t_R. The larger this ratio, the larger the mass-transfer broadening.

4. THE CONTINUOUS THEORY OF VAN DEEMTER, ZUIDERWEG AND KLINKENBERG: A SECOND INTRODUCTORY EXAMPLE.

The Martin and Synge model and its improved MCE version are lumped parameter models. An alternative approach consists in considering the chromatographic column as a continuous medium with respect to the axial dimension. This leads to a distributed parameter model. The mobile phase occupies the fraction ε of the volume and the stationary phase the fraction $1-\varepsilon$. The intersticial velocity of the mobile phase is u. The specific interfacial area is a (interfacial area per unit column volume)

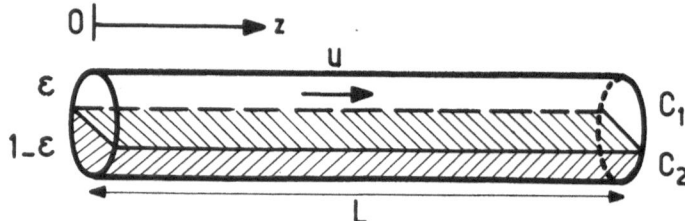

Fig. 4-1

4.1 Instantaneous mass-transfer and plug flow

The mass balance in an element of column of unit cross-section area is written

$$\varepsilon u \frac{\partial c_1}{\partial z} + \varepsilon \frac{\partial c_1}{\partial t} + (1 - \varepsilon) \frac{\partial c_2}{\partial t} = 0 \qquad (4-1)$$

c_1 and c_2 are at equilibrium : $c_2 = \alpha c_1$
Set $K' = \alpha \frac{1-\varepsilon}{\varepsilon}$, this definition of the capacity factor is exactly the same as above, since $V_2/V_1 = (1 - \varepsilon)/\varepsilon$, the mass-balance equation is

$$u \frac{\partial c_1}{\partial z} + (1 + K') \frac{\partial c_1}{\partial t} = 0 \qquad (4-2)$$

Applying the Laplace transform :

$$u \frac{d\overline{c_1}}{dz} + (1 + K') s\overline{c_1} = 0 \qquad (4-3)$$

The general solution of this equation in the Laplace domain is :

$$\overline{c_1} = A(s) \exp\left[-s(1 + K') \frac{z}{u}\right] \qquad (4-4)$$

From which we derive the transfer function $G(s) = \overline{c_1}(z=L)/\overline{c_1}(z=0)$, where L is the column length. The residence time of unretained solutes (of the mobile phase) is here $t_o = L/u$. The transfer function is thus :

$$G(s) = \exp\left[-st_o(1 + K')\right] \qquad (4-5)$$

which is identical to (3-8) : the column behaves as a pure time-delay $t_R = t_o(1 + K')$.

4.2 General case : finite mass-transfer rate and axially dispersed plug flow. The PDE model (Plug flow with axial Dispersion and Exchange)

We assume the mass-transfer flux between mobile and stationary phases in a column element of unit cross section area and lenght dz to be : $kadz(c_1 - c_2/\alpha)$. Moreover, the axial dispersion flux $- D(\partial c_1/\partial z)$ is superimposed on the purely convective flow uc_1. The mass balance in each phase is then :

mobile phase :

$$\varepsilon u \frac{\partial c_1}{\partial z} + \varepsilon \frac{\partial c_1}{\partial t} + ka(c_1 - \frac{c_2}{\alpha}) = \varepsilon D \frac{\partial^2 c_1}{\partial z^2} \qquad (4\text{-}6)$$

stationary phase :

$$(1 - \varepsilon) \frac{\partial c_2}{\partial t} - ka (c_1 - \frac{c_2}{\alpha}) = 0 \qquad (4\text{-}7)$$

This latter equation is that of a firs order dynamic system having the mass-transfer time constant

$$t_m = \frac{\alpha(1 - \varepsilon)}{ka} \qquad (4\text{-}8)$$

Again, (4-8) is identical to (3-15) as $a = S/(V_1 + V_2)$ and $\varepsilon = V_1/(V_1 + V_2)$.

$$\alpha c_1 = c_2 + t_m \frac{\partial c_2}{\partial t} \qquad (4\text{-}9)$$

Substituting $\overline{c}_2 = \dfrac{\alpha c_1}{1 + t_m s}$ in (4-6) after transforming (4-6) by Laplace, yields

$$u \frac{d\overline{c}_1}{dz} + s\overline{c}_1 (1 + \frac{K'}{1 + t_m s}) = D \frac{d^2 \overline{c}_1}{dz^2} \qquad (4\text{-}10)$$

Set $M = \dfrac{K'}{1 + t_m s}$, we obtain the second order linear differential equation,

$$D \frac{d^2 \overline{c}_1}{dz^2} - u \frac{d\overline{c}_1}{dz} - s (1 + M) \overline{c}_1 = 0 \qquad (4\text{-}11)$$

The roots of the characteristic equation are

$$r_1 = \frac{u + \sqrt{\Delta}}{2D} > 0 \text{ and } r_2 = \frac{u - \sqrt{\Delta}}{2D} < 0 \qquad (4\text{-}12)$$

Where $\Delta = u^2 + 4sD(1 + M)$ $\qquad (4\text{-}13)$

The general solution of equation (4-11) is then

$$\overline{c}_1 = A_1 \exp(r_1 z) + A_2 \exp(r_2 z) \qquad (4\text{-}14)$$

To go further in the solution, we require an assumption on the boundary conditions at the inlet and outlet of the column. We shall discuss the influence of the boundary conditions below, when

we deal with axial dispersion models. At the moment, let us only
assume that the column is infinite and that $\overline{c_1}$ should remain
bounded as $z \to \infty$. As r_1 is positive, this requires that $A_1 = 0$.
The transfer function $G(s) = \overline{c_1}(z + L)/\overline{c_1}(z = 0)$ from (4-12)
(4-13) and (4-14) is written :

$$G(s) = \exp\left[\frac{uL}{2D} - \frac{uL}{2D}\sqrt{1 + \frac{4sD(1 + M)}{u^2}}\right] \qquad (4-15)$$

Let us introduce a new parameter to characterize axial dispersion,
the Peclet number $P = uL/D$ and recall that $t_o = L/u$. We finally
obtain :

$$G(s) = \exp\left[\frac{P}{2} - \frac{P}{2}\sqrt{1 + \frac{4st_o(1 + M)}{P}}\right] \qquad (4-16)$$

$$M = \frac{K'}{1 + t_m s}$$

This is one of the PDE models transfer functions. This kind
of model was derived for the first time by Lapidus and Amundson
[7], and applied to chromatography problems by Van Deemter, Zuiderweg and Klinkenberg [8]. Of course, there is no simple analytical solution to (4-16) in the real domain, even if cumbersome
and almost untractable expressions have been derived after many
mathematical manipulations [9] [10]. Here again, impulse responses
will be obtained numerically using the Fast Fourier Transform. We
will comment later on the equivalence between PDE and MCE models.
Let us only say at this point that (3-20) and (4-16) yield almost
the same responses if we put $P = 2J$ provided P and J are large
enough. From Van der Laan's theorem, the moments of the impulse
response are :

$$\overline{t} = t_o (1 + K') = t_R \qquad (4-17)$$

as above (see (3-21))

$$\frac{\sigma^2}{t_R^2} = \frac{2}{P} + \frac{2K'}{1 + K'} \cdot \frac{t_m}{t_R} \qquad (4-18)$$

It is remarkable that the contribution of the mass transfer kinetics is exactly the same as for the lumped parameter model. This
will be explained in the next section. The contribution of axial
dispersion is $2/P = 2D/uL$, to be compared with $1/J$ in the previous
model. Matching these two contributions provides the basis for the
equivalence between the models.

4.3 Negligible axial dispersion : the PE model (Plug flow with
      Exchange)

As $D \rightarrow 0$, $P \rightarrow \infty$. A first order expansion of (4-16) in $1/P$ yields
the transfer function of the PE model

$$G(s) = \exp \left[- st_o (1 + \frac{K'}{1 + t_m s}) \right] \qquad (4-19)$$

One recognizes the kernel of a pure time-delay, combined with the
first order system of mass exchange. We will give examples of
response curves below (section 7-2) The broadening here is purely
due to mass-transfer kinetics

$$\frac{\sigma^2}{t_R^2} = \frac{2K'}{1 + K'} \cdot \frac{t_m}{t_R} \qquad (4-20)$$

5. SYNTHETIC AND PROGRESSIVE MODELING OF CHROMATOGRAPHIC INTERAC-
   TIONS : THE ELEMENTARY LEVEL.

     The chromatographic system may be considered as an associa-
tion of "boxes" or subsystems placed in series, in parallel, or
imbedded one in each other. Each box has its own internal structure. The succession of boxes imbedded in each other define a
hierarchy of levels of increasing complexity. The accuracy of the
description may thus be adjusted to the available information and
to the goal which is aimed at. When a particular "box" is considered, its internal structure is forgotten and one is interested
only in the relationships between input and output parameters.
The art of modeling consists in defining the boxes, choosing the
levels, writing down the input/output relationships and associating the boxes one to another in a proper way to obtain a tracta-
ble representation and to give the best fit to the experimental
behaviour of the system. The transfer function approach is well
adapted to such a description.

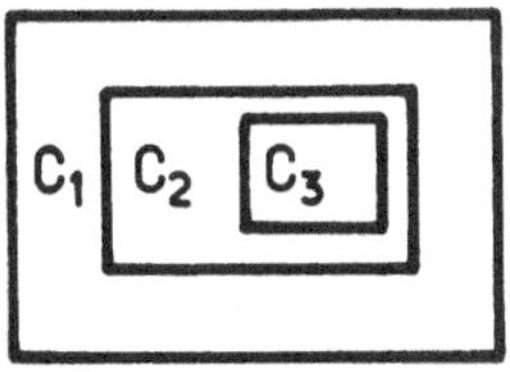

Fig. 5-1

     Imagine for instance a structure of imbedded boxes describing
one part of a porous medium. Box 1 represents the intersticial
fluid, box 2 the pores in a porous particle and box 3 the solid
of the particle. The mean concentrations in each box are c_1, c_2,
c_3 respectively. At equilibrium, partition coefficients may be
defined :

$$\alpha_{21} = c_2/c_1 \qquad \alpha_{32} = c_3/c_2 \qquad etc... \qquad (5-1)$$

Now, if c_1 undergoes a transient perturbation, the transfer functions generalize the partition coefficient and allow the resulting

perturbations on c_2 and c_3 to be calculated in the Laplace domain :

$$G_{21}(s) = \overline{c_2}/\overline{c_1} \quad , \quad G_{32}(s) = \overline{c_3}/\overline{c_2} \quad \text{etc...} \tag{5-2}$$

Following this approach, we shall consider first the exchange of solute between the mobile and the stationary phase at the elementary volume level, we will consider in a second step the whole column level. A priori, the elementary retention mechanisms may appear to be complex. Finally, they can be reduced to a combination of three ideal situations :

- the stationary phase is made up of porous particles into which the solute diffuses before being adsorbed.
- the stationary phase is made up of a pile of homogeneous retention layers in series.
- the stationary phase is made up of a collection of retention sites acting in parallel and exhibiting different "activities" or ease of access.

Let n' be the amount of solute retained in the stationary phase and n the corresponding amount in the mobile phase. At this elementary level the dynamics of interaction is characterized by the transfer function $M(s) = \overline{n'}/\overline{n}$. At equilibrium, the ratio $K' = (n'/n)_{eq} = M(o)$ is the capacity factor. Let us examine the above three situations and establish in each case the expression for $M(s)$.

5.1 Transient exchange in porous particles

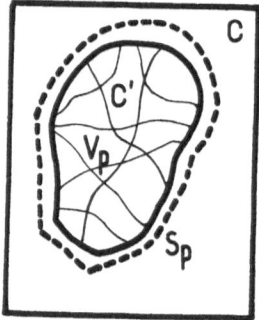

Fig. 5-2

The notation is the following :

c = concentration in the outside fluid (per unit fluid volume)
c" = concentration in the fluid contained in the pores of the particle (per unit fluid volume)
c' = concentration in the particle (overall, per unit particle volume)
c"'= concentration of adsorbed solute (per unit mass of the solid particle)
ρ_p = apparent density of the particle
β = internal porosity = pore volume/particle volume
K_A = adsorption equilibrium constant = $(c'''/c'')_{eq}$

α = partition coefficient = $(c'/c'')_{eq}$ = $(c'/c)_{eq}$ = $\beta + \rho_p K_A$
k_e = external mass transfer coefficient (resistance accross the laminar film)
k_A = adsorption kinetic rate constant (time^{-1})
d^A = V_p/S_p = volume/external surface area ratio
indices = s : at the particle surface, m : mean value averaged over the particle volume

Our goal is to establish the expression for the transfer function $L(s) = \overline{c'}_m/c$ from which $M(s) = \frac{1-\epsilon}{\epsilon}L(s)$ will immediatly result (ϵ = external packing porosity)

1- Instantaneous adsorption
The transient mass balance is written

$$k_e S_p (c - c_s) = V_p \frac{d\overline{c'}_m}{dt} \qquad (5-3)$$

with $c'_s = \alpha c''_s = \alpha c_s$ $\qquad (5-4)$

$$c = \frac{c'_s}{\alpha} + \frac{d}{k_e} \cdot \frac{d\overline{c'}_m}{dt} \qquad (5-5)$$

To go farther, we must know the relation between $\overline{c'}_m$ and c'_s, the mean internal concentration and the superficial concentration. This is a standard diffusion problem, the solution of which can be found in textbooks [10], at least for simple particle shapes. The solutions are shown on table below, in the form of the transfer function $H(s) = \overline{c'}_m/c'_s$

Geometrical shape	$d= V_p/S_p$	$H(s)$	shape factor μ
Slab. Thickness 2e	e	$\tanh \lambda/\lambda$	1/3
Cylinder-Radius r	r/2	$\frac{1}{\lambda}\frac{I_1(2\lambda)}{I_0(2\lambda)}$	1/2
Sphere-Radius r	r/3	$\frac{1}{\lambda}\coth(3\lambda) - \frac{1}{3\lambda^2}$	3/5
Any shape : first order system approximation	d	$\frac{1}{1 + \mu\lambda^2}$	μ

$$\lambda = d(s/D_e)^{1/2}$$

I_0 and I_1 are modified Bessel functions of the first kind and D_e is the effective diffusivity of the solute into the porous particle.

At first sight, it might appear that H(s) is a complicated function depending on the particle shape and on the boundary conditions. This is true in a strict mathematical sense. However, closer study shows that H(s) can be approximated with a good accuracy by the transfer function of a simple first-order system.

$$H(s) \neq \frac{1}{1 + \mu d^2 s/D_e} \qquad (5-6)$$

In matching the exact expression of H(s) and the approximate expression (5-6), for instance in the Fourier domain (s = iω, real
and imaginary part), the shape factor μ can be determined [11].
It is found to be very close to that which is obtained in writing
that both expressions yield the same first order moment of the
impulse response (fig.5-3)

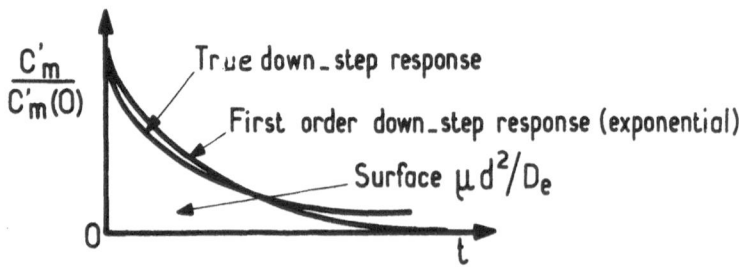

Fig. 5-3

Notice that if we replace s by the kinetic rate constant k of a
first order chemical reaction, H(k) is nothing but the efficiency
factor of the particle and λ the Thiele modulus.
In current practice, it would probably be difficult to devise an
experiment accurate enough to show a difference between the exact
and the approximate solution. Thus, we will now consider that internal diffusion in the particle is like a first-order process
characterized by the diffusion time constant $t_d = \mu d^2/D_e$:

$$H(s) = \frac{1}{1 + t_d s} \qquad (5-7)$$

We may now come back to equation (5-5). In the Laplace domain :

$$\overline{\alpha c} = \overline{c'}_s + \frac{\alpha d}{k_e} s\, c'_m \qquad (5-8)$$

$$\alpha \frac{\overline{c}}{\overline{c'}_m} = \frac{1}{H} + \frac{\alpha d}{k_e} s \qquad (5-9)$$

$t_e = \alpha d/k_e$ is the characteristic external mass-transfer time constant. Finally

$$L(s) = \frac{c'_m}{\overline{c}} = \frac{\alpha}{\frac{1}{H} + t_e s} \neq \frac{\alpha}{1 + (t_d + t_e)\, s} \qquad (5-10)$$

The overall behaviour is also a first-order one.

2- Finite rate adsorption

The transient mass-balance is written here

$$k_e S_p (c - c_s) = V_p (\beta \frac{dc''_m}{dt} + \rho_p \frac{dc'''_m}{dt}) \qquad (5-11)$$

and in the Laplace domain :

$$\bar{c} = \bar{c''}_s + \frac{ds}{k_e} (\beta \bar{c''}_m + \rho_p \bar{c'''}_m) \qquad (5-12)$$

In the solid :

$$\frac{dc'''}{dt} = k_A (K_A c'' - c''') \qquad (5-13)$$

or in the Laplace domain

$$\frac{\bar{c'''}}{\bar{c''}} = \frac{K_A}{1 + s/k_A} \qquad (5-14)$$

As the average over the particle volume is independant of the time, (5-13) and (5-14) also hold for mean values (index m). We assume that the above analysis of the internal diffusion process applies to the solute in the fluid of the pores (concentration c") :

$$\frac{\bar{c''}_m}{\bar{c''}_s} = H(s) \qquad (5-15)$$

Recalling that

$$\bar{c'}_m = \beta \bar{c''}_m + \rho_p \bar{c'''}_m \qquad (5-16)$$

after a few algebraical manipulations, one obtains :

$$\frac{\bar{c}}{\bar{c'}_m} = \frac{1}{\left(\beta + \frac{\rho_p K_A}{1 + s/k_A} \right) H} + \frac{ds}{k_e} \qquad (5-17)$$

Let us define a new adsorption time constant

$$t_a = \frac{\alpha - \beta}{\alpha k_A} = \frac{\rho_p K_A}{\alpha k_A} \qquad (5-18)$$

we find from (5-17) :

$$L(s) = \frac{\bar{c'}_m}{\bar{c}} = \frac{\alpha}{(1 + \frac{st_a}{1 + \frac{\beta t_a}{\rho_p K_A} s}) \frac{1}{H} + t_e s} \qquad (5-19)$$

An approached form of (5-19) is :

$$L(s) = \frac{c'_m}{\overline{\overline{c}}} \ne\ne \frac{\alpha}{1 + (t_a + t_d + t_e)\, s}$$

(5-20)

This expression is obtained by equalizing the first order moments of impulse responses as above. We thus see that the particle in presence of external mass-transfer, internal diffusion and adsorption behaves on the whole as a first order dynamic system characterized by a mass transfer time

$$t_m = t_a + t_d + t_e$$

(5-21)

The mass-transfer time constants

$$t_a = \frac{\rho_p\, K_A}{\alpha\, k_A} \quad ; \quad t_d = \frac{\mu d^2}{D_e} \quad ; \quad t_e = \frac{\alpha d}{k_e}$$

are additive.
From (5-20) and (5-21), it is easy to derive

$$M(s) = \frac{\overline{n'}}{\overline{n}} = \frac{K'}{1 + t_m s}$$

(5-22)

where $K' = \alpha\, \dfrac{1-\varepsilon}{\varepsilon}$

5.2 Retention layers in series [9]

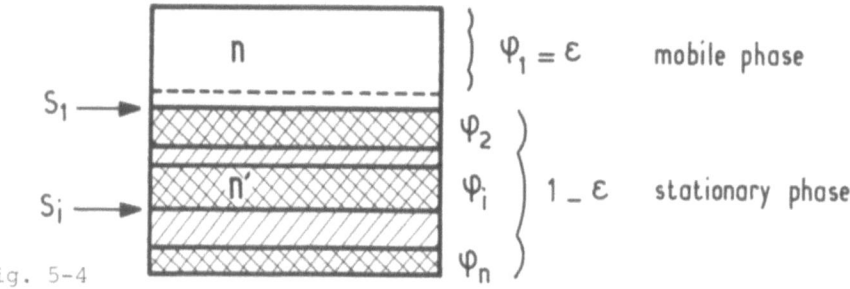

Fig. 5-4

The stationary phase is taken here to consist in a serie of layers occupying volume fractions $\phi_2, \phi_3 \ldots \ldots \phi_n$ ($\phi_i = \varepsilon$ is the volume fraction of the mobile phase). At equilibrium, the solute is distributed among the layers according to the partition coefficients $\alpha_i = (c_{i+1}/c_i)_{eq}$. The transfer flux from layer i to layer i+1 is $k_i(c_i - c_{i+1}/\alpha_i)$. It is convenient to introduce a mass-transfer time constant for each layer $t_i = \alpha_i/(k_i a_i)$ where a_i is the specific interfacial exchange area referred to the total volume $a_i = S_i/V$ and the related time-constant $\tau_i = t_i(1-\varepsilon)$. A straight-forward but tedious derivation relying on the mass balances in each layer leads to the following recurrence relationships :

$$M(s) = \left[st_1 \, \phi_1/\alpha_1 + \Lambda_1(s)\right]^{-1}$$

$$\frac{1}{\Lambda_i(s)} = \frac{\alpha_i \, \phi_{i+1}}{\phi_i} \left[1 + \left(\frac{st_{i+1} \, \phi_{i+1}}{\alpha_i} + \Lambda_{i+1}(s)\right)^{-1}\right] \qquad (5\text{-}23)$$

$$\frac{1}{\Lambda_{n-1}} = \frac{\alpha_{n-1} \, \phi_n}{\phi_{n-1}}$$

Although these recurrent expressions look complicated, a numerical study shows that $M(s)$ can also be approached by an expression of the first order kind:

$$M(s) \not\# K'/ (1 + t_m s)$$

the capacity factor is :

$$K' = \alpha_1 \frac{\phi_2}{\phi_1} + \alpha_1 \, \alpha_2 \, \frac{\phi_3}{\phi_1} + \dots + (\alpha_1 \, \alpha_2 \dots \alpha_{n-1} \frac{\phi_n}{\phi_1}) \qquad (5\text{-}24)$$

The overall time-constant t_m is given by a complicated linear combination of the τ_i. For instance, for 2 stationary layers in series, one finds :

$$t_m = \frac{\phi_2 + \alpha_2 \, \phi_3}{\phi_2 + \phi_3} \, \tau_1 + \frac{\alpha_2 \, \phi_3^2}{(\phi_2 + \alpha_2 \, \phi_3)(\phi_2 + \phi_3)} \, \tau_2 \qquad (5\text{-}25)$$

5.3 Retention sites in parallel [12]

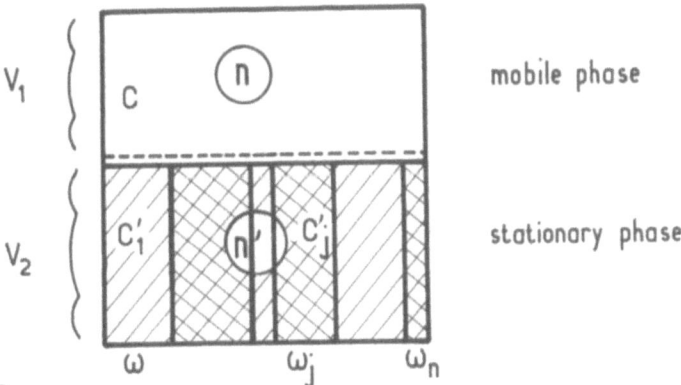

Fig. 5-5

The situation is different for retention sites in parallel. Let ω_j be the fraction of solute which would be absorbed on to the sites j at equilibrium. This means that the corresponding partition coefficient is $\alpha_j = \omega_j \alpha$ with $\Sigma \omega_j = 1$. Let c'_j be the fractional concentration of this fraction of solute referred to the total stationary volume V_2 : $c'_j = n'_j/V_2$. Assume first that there is no external mass transfer resistance. The mass balance for the j-population of sites is written :

$$V_2 \frac{dc'_j}{dt} = k_j \, S_j (c - \frac{c'_j}{\alpha j})$$

(5-26)

where S_j is the exchange surface area of the j-population and k_j the corresponding overall mass-transfer coefficient. Defining the mass-transfer time constant or "transfer time"

$$\tau_j = \frac{\alpha_j V_2}{k_j S_j}$$

(5-27)

(5-26) may be written

$$\tau_j \frac{dc'_j}{dt} + c'_j = \alpha \omega_j c$$

(5-28)

or in terms of the absolute amounts

$$\tau_j \frac{dn'_j}{dt} + n'_j = K' \, \omega_j n$$

(5-29)

where K' has the usual meaning

$$K' = \alpha V_2 / V_1 = \alpha (1 - \varepsilon) / \varepsilon$$

(5-30)

In the Laplace domain, (5-28) yields the transfer function :

$$\frac{\overline{c'_j}}{\overline{c}} = \frac{\alpha \omega_j}{1 + s \tau_j}$$

(5-31)

Summing over all site populations

$$\frac{\overline{c'}}{\overline{c}} = \alpha \sum_j \frac{\omega_j}{1 + s \tau_j} = \alpha H(s)$$

(5-32)

or

$$\frac{\overline{n'}}{\overline{n}} = K' \sum_j \frac{\omega_j}{1 + s \tau_j} = K' H(s)$$

(5-33)

If we now add an external mass transfer resistance, the mass balance becomes

$$k_e S_e (c - c_s) = V_2 \frac{dc'}{dt}$$

(5-34)

with $\overline{c'}/c_s = \alpha H$, from (5-32)

Thus, setting as above $t_e = \frac{\alpha V_2}{k_e S_e}$

$$\frac{\overline{c'}}{\overline{c}} = L(s) = \frac{\alpha}{1/H + t_e s}$$

(5-35)

which is similar to (5-10), or

$$\frac{\overline{n'}}{\overline{n}} = M(s) = \frac{K'}{1/H + t_e s} \qquad (5\text{-}36)$$

which is similar to (5-22) except that

$$H(s) = \sum_j \frac{\omega_j}{1 + s\tau_j} \qquad (5\text{-}37)$$

can no longer be represented by a first order system transfer function.

(5-37) can be extended to the case of a continuous distribution of sites. The sum must then be replaced by an integral

$$H(s) = \int \frac{d\omega}{1 + \tau s} \ . \ \text{Let us define the } \underline{\text{transfer time distribution}}$$

(TTD) $f(\tau)$ as

$$f(\tau) \ d\tau = d\omega \qquad (5\text{-}38)$$

This means that $f(\tau)d\tau$ is the fraction of the total number of sites exhibiting a transfer time between τ and $\tau + d\tau$. Then :

$$H(s) = \frac{\overline{c'}}{\overline{c'}_s} = \int_0^\infty \frac{f(\tau) \ d\tau}{1 + \tau s} \qquad (5\text{-}39)$$

This expression is general and allows any kind of site population to be postulated : monodispersed (1) (one kind of site, leading to a first order lumped model), bidispersed (2) (two kinds of sites having a different ease of access), uniformly (4) or exponentially (3) distributed etc... (fig. 5-6)

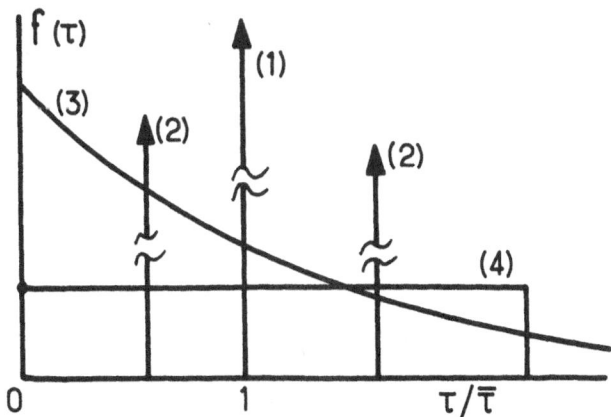

Fig 5-6
The mean internal transfer time is $\bar{\tau} = \int_0^\infty \tau f(\tau)d\tau$ and the mean overall transfer time is $t_m = t_e + \bar{\tau}$ but this is not the time constant of a first order dynamic system.

To sum up, the transient interaction with the stationary phase
may be represented by a first order system (time constant t_m) in
two cases : diffusion and adsorption in porous particles -retention
layers in series. In the third case -retention sites in parallel-
the dynamics is more complex and depends on the transfer time distribution concept. In combining these three ideal situations, more
complicated systems can be dealt with. Consider for instance the
following system :

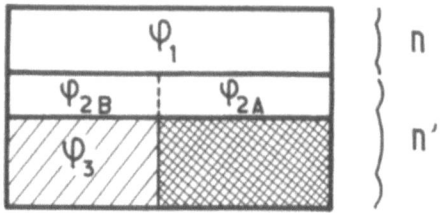

- a mobile phase (fraction ϕ_1)
- a stagnant fluid (fraction $\phi_2 = \phi_{2A} + \phi_{2B}$)
- a porous solid (fraction ϕ_3, partition coefficient ϕ_2) contacting
the ϕ_{2B} fraction.

Applying the methods outlined above, the transfer function $M(s)$ is
written :

$$M(s) = K' \frac{\omega_A}{1 + \tau_A s} + \frac{\omega_B}{1 + \tau_B s} \qquad (5\text{-}40)$$

with $K' = (\alpha_2 \phi_3 + \phi_2)/\phi_1$ \qquad (5-41)

$$\omega_A = \frac{\phi_{2A}}{\phi_2 + \alpha_2 \phi_3} \quad , \quad \omega_B = \frac{\phi_{2B} + \alpha2 \phi_3}{\phi_2 + \alpha_2 \phi_3} \qquad (5\text{-}42)$$

$$\tau_A = \frac{\alpha \omega_A}{k_1 a_A} \quad , \quad \tau_B = \frac{\phi_{2B} + \alpha_2 \phi_3}{\phi_{2B} + \phi_3} \tau_1 + \frac{\alpha_2^2 \phi_3}{(\phi_{2B} + \alpha_2 \phi_3)(\phi_{2B} + \phi_3)} \qquad (5\text{-}43)$$

$$\alpha = \frac{\phi_2 + \alpha_2 \phi_3}{\phi_2 + \phi_3} \qquad (5\text{-}44)$$

6. THE COLUMN LEVEL

After the detailed study of exchange mechanisms at the elementary
level, we may now pass on to the "column" level. We assume that
the porous medium consists of a set of small elementary volumes
connected to each other according to the hydrodynamic flow pattern.
This pattern may be characterized by the residence time distribution (RTD) of the mobile phase $E_0(t)$ and its Laplace transform
$G_0(s)$. This means that the $E_0(t) dt$ fraction of the mobile flowrate
stays between t and $t + dt$ in the column. This function can be
determined experimentally by injecting a tracer undergoing no exchange with the stationary phase (unretained solute). Now if we
inject in the same column and in the same conditions a solute

which can be absorbed into the stationary phase, how can we predict the response of the column to this injection -transfer function G(s)- knowing the RTD of the mobile phase- transfer function
G_o(s)- and the elementary dynamics of exchange - transfer function
M(s) ?

The answer to this question is simply to substitute $s\left(1 + M(s)\right)$
for s in the G_o(s) expression :

$$G(s) = G_o\left[s(1 + M(s))\right]$$ (6-1)

This very useful theorem can be demonstrated as follows : let us
decompose the internal volume of the column into a bundle of elementary parallel streams. Each stream is a small plug flow reactor
with its own residence time. The distribution of these parallel
streams is such that the required RTD of the mobile phase is satisfied. In one of these small columns, the mass balance is expressed as (see 4-1) :

$$u\,\frac{\partial c}{\partial z} + \frac{\partial c}{\partial t} + \frac{1-\varepsilon}{\varepsilon}\,\frac{\partial c'_m}{\partial t} = 0$$ (6-2)

Applying the Laplace transform and taking into account $\overline{c'}_m/\overline{c} = L(s)$

$$u\,\frac{d\overline{c}}{dz} + s\overline{c}\left[1 + \frac{1-\varepsilon}{\varepsilon}\,L(s)\right] = 0$$ (6-3)

Integrating along the column axis for the given residence time t_o,
we obtain the elementary transfer function of the stream :

$$g(s,\,t_o) = \exp\left[-\,st_o(1 + M(s))\right]$$ (6-4)

The whole transfer function of the column is obtained by associating the elementary functions according to the required RTD
$E_o(t_o)$:

$$G(s) = \int_0^\infty g(s,t_o)\,E_o(t_o)\,dt_o$$ (6-5)

From the definition of the Laplace transform :

$$G_o(s') = \int_0^\infty \exp\left[-s't_o\right]E_o(t_o)\,dt_o$$ (6-6)

Comparing (6-5) and (6-6) and taking (6-4) into account, we see
that the two expressions are identical provided we set
$s' = s\left[1 + M(s)\right]$, which demonstrates our theorem.

This result is important, for it shows that the retention dynamics
of solutes and the mobile phase flow hydrodynamics in the column
are <u>uncoupled</u> and can be treated independently. It justifies the
fact that we have first studied the transient exchange at the ele-

mentary level without being interested at this stage in the flow
behaviour of the mobile phase. It also allows any kind of complex
hydrodynamic pattern to be dealt with, provided a model of $G_o(s)$
is available, and provided also that $M(s)$ is the same throughout
the system.

Plug flow with axial dispersion.

In most practical situations, one has to deal with one-dimensional
flow undergoing axial dispersion.
The expression of the RTD (if available) and transfer functions
for axially dispersed plug flow can be found in standard textbooks
[1] [13]. The dispersion parameter is the Peclet number $P = uL/D$
where D is the axial dispersion coefficient. The solution depends
on the boundary conditions at the tube's inlet and outlet, and
also where the input signal is injected. Table 6-1 below summarize
some standard results. In this table $q = (1 + 4st_o/P)^{1/2}$ and
$t_o = L/u = V_1/Q$. I indicates the injection point and M the measurement point. Axial dispersion is assumed to take place throughout
the column.

	Transfer function $G_o(s)$	Mean residence time $\bar{t}$	Variance σ^2
I↓ ▭ ↓M	I. Infinite column : $\frac{1}{q}\exp\left[\frac{P}{2}(1-q)\right]$	$t_o(1 + \frac{2}{P})$	$t_o^2(\frac{2}{P} + \frac{8}{P^2})$
↓I↓M₁ ▭ M₂	II. Infinite column : double measurement $\exp\left[\frac{P}{2}(1-q)\right]$	$\Delta\bar{t} = t_o$	$\Delta\sigma^2 = \frac{2t_o^2}{P}$
I↓ ▭ ↓M	III. Semi-infinite column $\frac{2\exp\left[\frac{P}{2}(1-q)\right]}{1 + q}$	$t_o(1 + \frac{1}{P})$	$t_o^2(\frac{2}{P} + \frac{3}{P^2})$
I↓ ▭ ↓M	IV. Closed column $\frac{4q\ \exp\left[\frac{P}{2}(1-q)\right]}{(1+q)^2 - (1-q)^2\exp(-Pq)}$	t_o	$t_o^2\left[\frac{2}{P} - \frac{2}{P^2}(1-e^{-P})\right]$
J cells	V. Mixing cells in series $\left[\frac{1}{1 + st_o/J}\right]^J$	t_o	$\frac{t_o^2}{J}$

Table 6.1

A simple analytical expression for the impulse response $E_o(t)$ is available only in the case of an infinite column :

$$E_o(t) = \frac{1}{2} \left(\frac{P}{\pi t_o t}\right)^{1/2} \exp\left[-\frac{P(t_o-t)^2}{4t_o t}\right] \tag{6-7}$$

Axial dispersion can also be accounted for by the lumped parameter model of J mixing cells in series :

$$G_o(s) = \left[1 + s\bar{t}_o/\bar{J}\right]^{-J} \tag{6-8}$$

$$E_o(t) = (J/t_o)^J \frac{t^{J-1} \exp(-Jt/t_o)}{(J-1)!} \tag{6-9}$$

When J is large, using Stirling's formula

$$E_o(t) = (J/t_o)^J \frac{t^{J-1} \exp(-Jt/t_o + J - 1)}{(J-1)^{J-1} \sqrt{2\pi(J-1)}} \tag{6-10}$$

Whatever J, the maximum is located at $t_{max} = t_o(1 - 1/J)$ and the inflexion points are symmetrically located with respect to the maximum at a distance $\pm t_o \sqrt{J-1}/J$. When Stirling's approximation holds, the height of the maximum is

$$E_{o,max} = \frac{J}{t_o \sqrt{2\pi(J-1)}} \tag{6-11}$$

At first sight, all theses expression accounting for axial dispersion seem different. This is true when the dispersion coefficient is large (P or J are small). Then, boundary conditions must be carefully taken into account. However, as P or J are large, the response curves are very similar whatever the boundary conditions, with the equivalence $P = 2(J-1)$ or $P \simeq 2J$. All impulse responses go asymptotically to the famous gaussian distribution ($J \to \infty$, $P \to \infty$) :

$$E_o(t) = \frac{1}{2t_o} \sqrt{\frac{P}{\pi}} \exp\left[-\frac{P(1-t)^2}{4t_o^2}\right] \tag{6-12}$$

with $\bar{t} = t_o$ and $\sigma^2 = 2t_o^2/P$

The following figures show the impulse responses (RTD) given by the models of table 6.1 for increasing values of J and P. P has been fixed in each case to $P = 2(J-1)$. The successive values of J are $J = 2$ (fig. 6-1), $J = 5$ (fig. 6-2), $J = 10$ (fig. 6-3), $J = 50$ (fig. 6-4) and $J = 100$ (fig. 6-5). The curves have been obtained by numerical inversion (FFT) of the transfer functions in table 6-1 for the models I = infinite column (P) III = semi-infinite column (P), IV = closed column (P) and V = mixers in series (J).

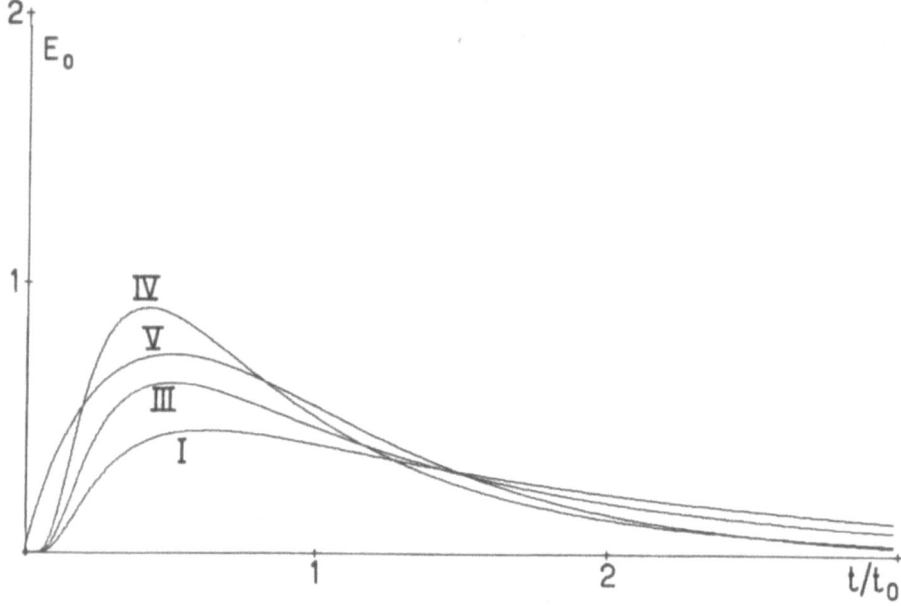

Figure 6-1 .   J = 2   P = 2

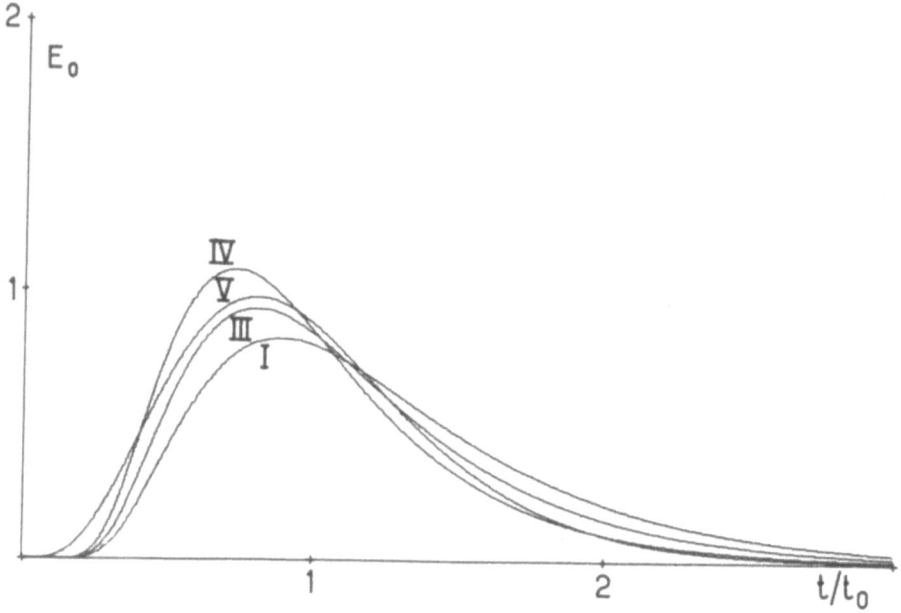

Figure 6-2 .   J = 5   P = 8

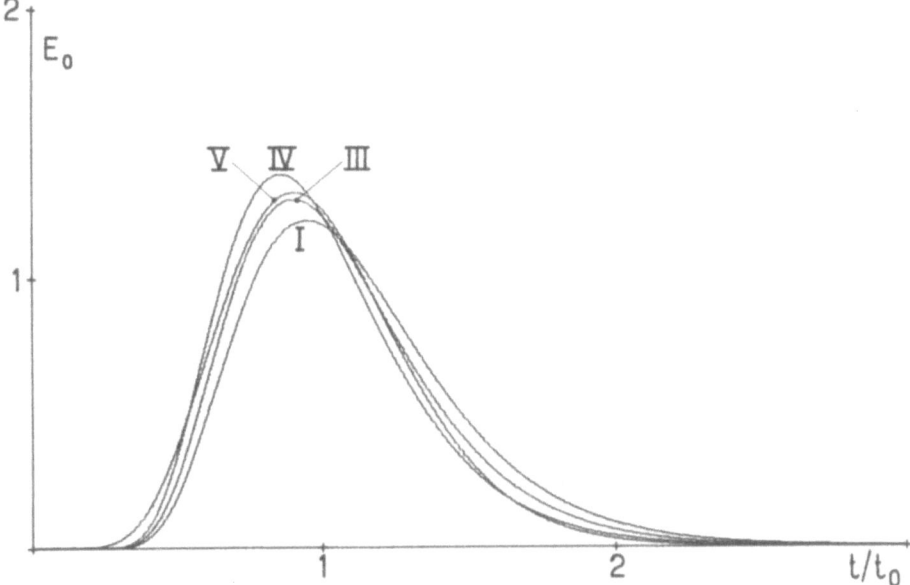

Figure 6-3.   J = 10   P = 18

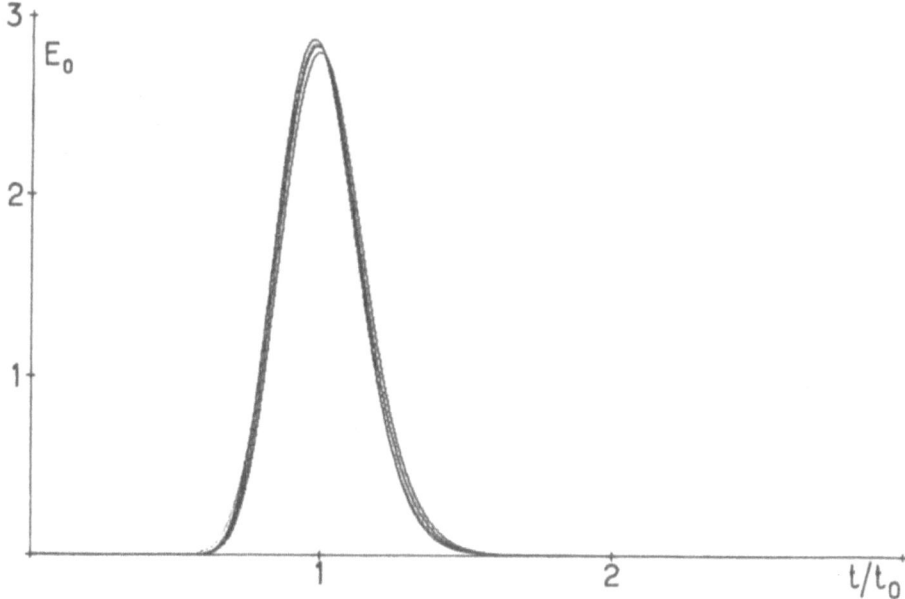

Figure 6-4   J = 50   P = 98

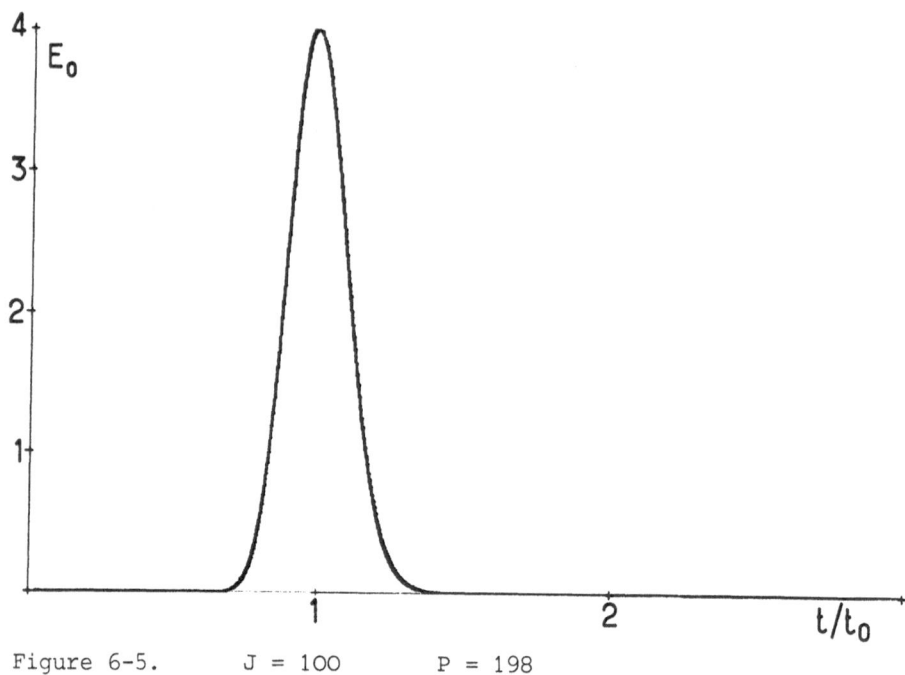

Figure 6-5.   J = 100   P = 198

The influence of boundary conditions and nature of the model (lumped/distributed) is obvious at J = 2 and to a lesser extent at J = 5. The curves are very close to each other at J = 10, nearly co incident at J = 50 and quite undistinguishable at J = 100. If we realize that a column of J = 50 plates is considered as very poor and subjected to a noticeable amount of axial dispersion in most applications of chromatography, we are led to the conclusion that the discussions which have arisen in the literature about the exact boundary conditions to be used are rather academic and without any real importance

For convenience, the study of peak shapes in the following section will be made with the MCE lumped parameter model, which will be shown to give nearly the same results as the distributed parameter PDE model

7. GENERAL MODEL OF A CHROMATOGRAPHIC COLUMN

Taking into account axial dispersion by the mixing cell model, we thus obtain the general transfer function of the column

$$G(s) = \left[1 + \frac{st_o}{J} \left(1 + \frac{K'}{1/H(s) + st_e} \right) \right]^{-J} \tag{7-1}$$

In all cases where mass transfer intervenes only through a single transfer time t_m (see section 5), (7-1) simplifies to

$$G(s) = \left[1 + \frac{st_o}{J} (1 + \frac{K'}{1 + t_m s}) \right]^{-J} \tag{7-2}$$

which is the transfer function of the MCE model. In addition to t_o, the model depends on <u>three</u> parameters :
- the capacity factor K' depending only on equilibrium properties
- the mass transfer time t_m, depending on mass-transfer kinetics, or the dimensionless ratio $\theta = t_m/t_R$.
- the axial dispersion parameter J (number of equivalent mixers in series).

In the literature, related parameters are sometimes used :
- the <u>retention ratio</u> R which is such that the center of gravity of the peak moves at speed Ru within the column. Thus :

$$t_R = t_o/R \text{ and } R = \frac{1}{1 + K'} \, , \, K' = \frac{1 - R}{R} \tag{7-3}$$

- the <u>number of transfer units NTU</u> (or N)

$$NTU = \frac{1 - R}{\theta} = \frac{K't_o}{t_m} = \frac{K'}{(1 + K')\theta} = N \tag{7-4}$$

In terms of NTU, R, J, the transfer function of the MCE model is written

$$G(s') = \left[1 + \frac{s'}{J} (\frac{s'R(1-R) + NTU}{s'(1-R) + NTU}) \right]^{-J} \tag{7-5}$$

In this expression, the retention times are referred to the mean retention time t_R (reduced form) so that $s' = st_R$. The peak equation in the time domain corresponding to (7-5) can be found in reference [9] . As we have already mentioned, it is a very complicated expression which has only an academic interest. Numerical inversion of (7-2) or (7-5) should be preferred in applications.

7.1 HETP concept : the Van Deemter equation

We have already derived the expression of the variance of the chromatographic peak given by the MCE model in section 3 (see 3-22 and 3-23) :

$$\sigma^2 = \frac{t_R^2}{J} + \frac{2K'}{1 + K'} t_m t_R \tag{7-6}$$

If mass transfer is instantaneous, $\sigma^2 = t_R^2/J$ and t_R^2/σ^2 gives the number J of "plates" in the column. When the peak is nearly gaussian, σ^2 can be easily determined in drawing the two inflexion tangents :

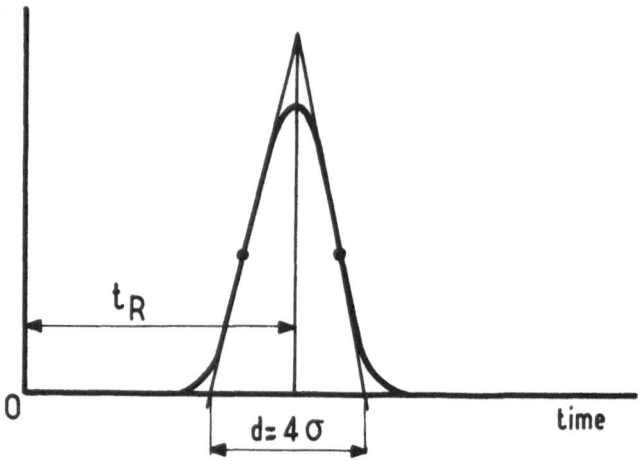

Fig. 7-1

Generalizing this observation, chromatographers use to express the sharpness of their peaks (and thus the resolution achieved by the column) in terms of the number of theoretical plates NTP :

$$NTP = \frac{t_R^2}{\sigma^2} = \frac{16\ t_R^2}{d^2} \tag{7-7}$$

(7-7) is valid even if the time is measured in arbitrary units (length of paper on a strip chart recorder). Another useful concept is that of Height Equivalent to a Theoretical Plate or HETP. L being the column length HETP = L/ NTP. From (7-6) :

$$HETP = \frac{L\ \sigma^2}{t_R^2} = \frac{L}{J} + \frac{2K'}{1 + K'}\ L\ t_m t_R \tag{7-8}$$

Recalling that $J \simeq P/2 = uL/2D$ and $t_R = \frac{L}{u}\ (1 + K')$:

$$HETP = \frac{2D}{u} + \frac{2K'}{(1 + K')^2}\ t_m u \tag{7-9}$$

In a packed bed, axial dispersion D has two main causes : molecular diffusion $\mathcal{D}_m$ (modified by axial tortuosity τ_A) and statistical dispersion created by the presence of the porous medium. The latter is proportional to the dimension of particles d_p and to the intersticial velocity u so that

$$D = \frac{\mathcal{D}_m}{\tau_A} + \frac{ud_p}{Pe_A} \tag{7-10}$$

Pe_A is an axial Peclet number of the order of 0.5 to 2. With this expression for D, the HETP is written

$$HETP = \frac{2\mathcal{D}_m}{u\tau_A} + \frac{2d_p}{Pe_A} + \frac{2K'}{(1 + K')^2}\ t_m u \tag{7-11}$$

In chromatographic language, this equation is known as the VAN
DEEMTER equation. It is generally written :

$$HETP = \frac{A}{u} + B + Cu$$
$$\text{with } A = \frac{2\mathcal{D}_m}{\tau_A} \quad , \quad B = \frac{2d_p}{Pe_A} \quad , \quad C = \frac{2K'}{(1 + K')^2} t_m \tag{7-12}$$

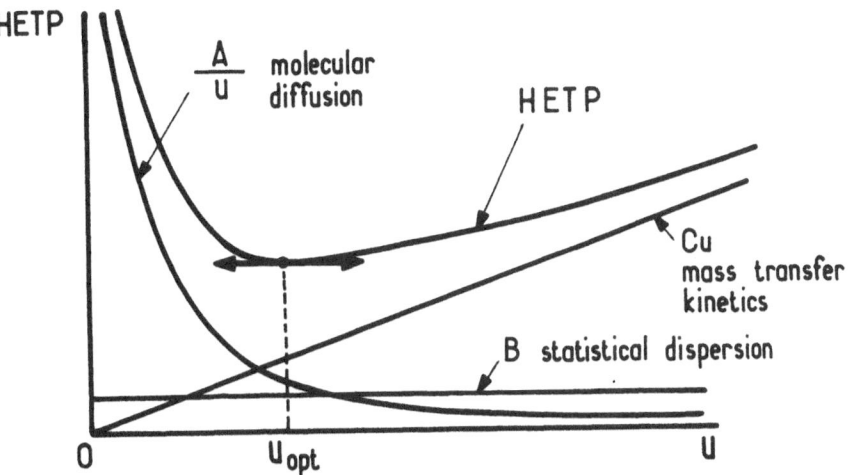

Fig. 7-2

It shows that there exists an optimal intersticial velocity u_{opt}
such that the height equivalent to a theoretical plate is minimum
(the resolution is optimal).
Below u_{opt}, peaks are broadened mainly by molecular diffusion,
above u_{opt}, the main cause of broadening is mass transfer resistance. Equation (7-12) is the simplest form of Van Deemter equation. Many other effects can be taken into account leading to more
sophisticated expressions. For instance if there exists a radial
velocity profile in the column, an additional contribution to
axial dispersion D_{vp} appears in (7-10)

$$D_{vp} = \frac{\times R^2 u^2}{u d_p / Pe_R + \mathcal{D}_m / \tau_R} \tag{7-13}$$

where $\times$ is a constant depending on the shape of the velocity profile ($\times = 1/48$ for a parabolic profile, cf. TAYLOR and ARIS), R
is the column radius, Pe_R a radial Peclet number ($Pe_R \simeq 10$) and
τ_R the radial tortuosity of the packing.
Advanced theories of band spreading in chromatography (especially
in liquid chromatography) can be found in standard texts [15] as
well as in recent publications [14] [16] [17] .

7.2 Influence of chromatographic parameters on peak shapes. Case of one single transfer time.

Let us now study the influence of interaction parameters (retention, mass transfer, dispersion) on peak shapes in the case of the simple MCE model. The simulation below have been obtained by computer inversion of the Laplace transform (7-2) using the Fast Fourier Transform algorithm.
We will first assume axial dispersion to be negligible ($J = P = \infty$, PE model)

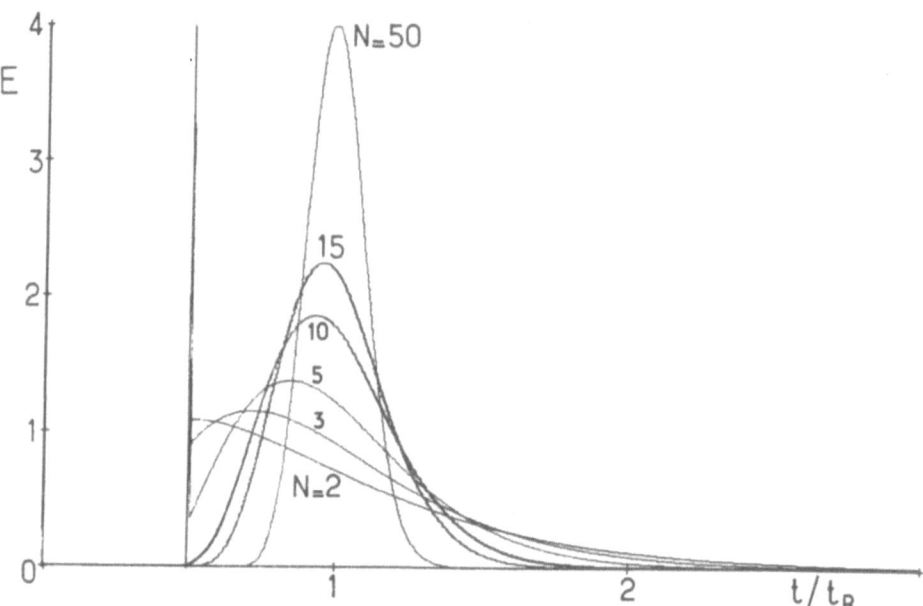

Figure 7-3. No axial dispersion ($J=\infty$), $K'=1$, ($R=0.5$)
Influence of mass-transfer kinetics (N is the number of transfer units defined in 7-4 $N = NTU$)

Fig. 7-3 shows successive peaks for increasing mass-transfer rate. Notice the Dirac delta peak at $t/t_R = R$ (remaining unretained solute) which vanishes almost completely for $N > 10$ as the true chromatographic peak builds itself.

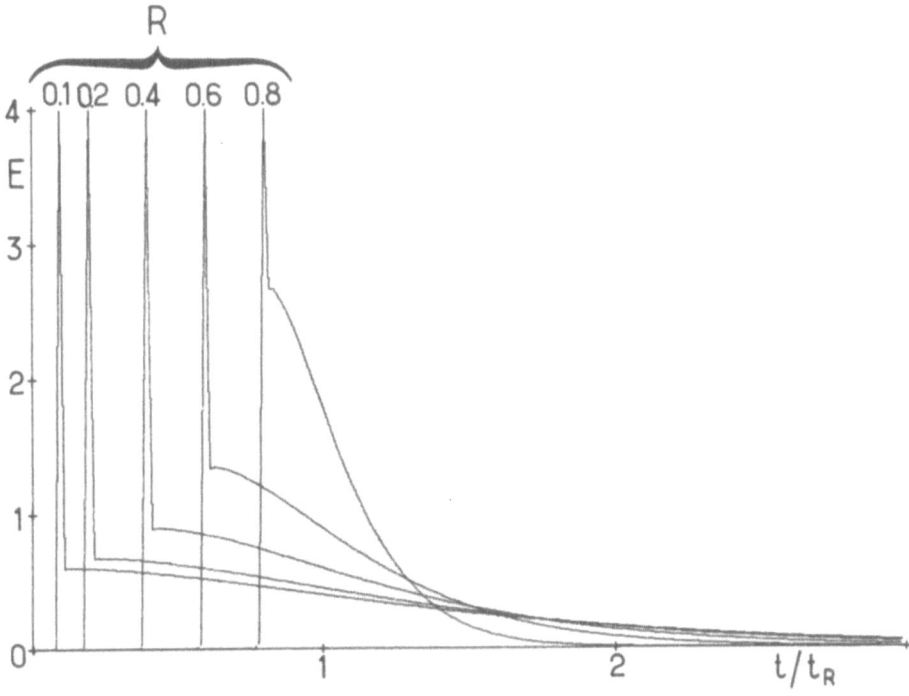

Figure 7-4. No axial dispersion (J=∞), N=2.
Influence of the retention ratio R (defined in 7-3)N = NTU

Fig. 7-4 shows the influence of the retention ratio R at fixed
(and slow) mass transfer kinetics (N = 2). Notice the displacement
of the initial impulse peak and the deformation of its tail. We
may now superimpose axial dispersion and see how the profiles of
figures 7-3 and 7-5 are smoothed out or blurred by diffusion.
Figures 7-5 and 7-6 below correspond to the same values of parameters as 7-7 and 7-4 but now J = 25.

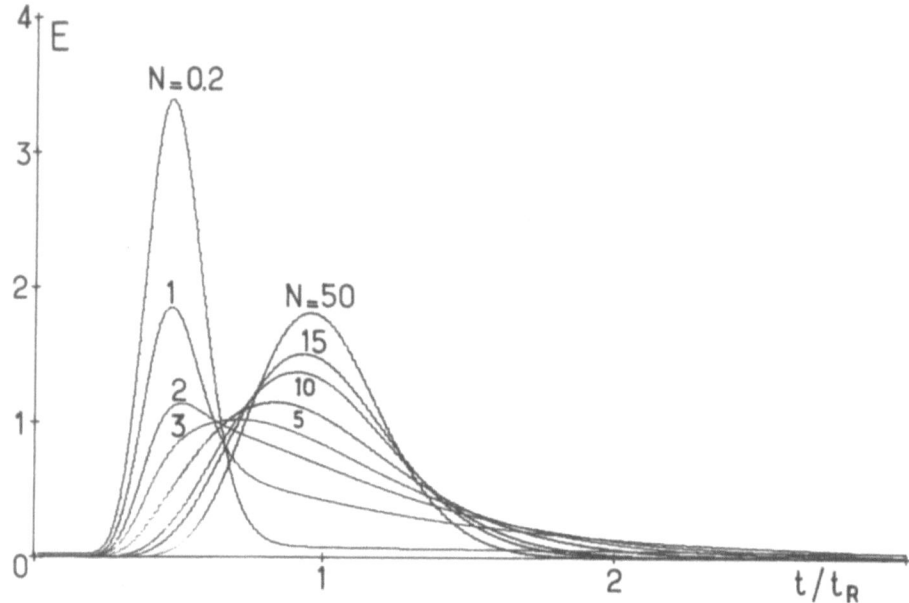

Fig. 7.5. General MCE model J=25, K'=1 (R=0.5)
Influence of mass transfer kinetics (number of transfer units N).

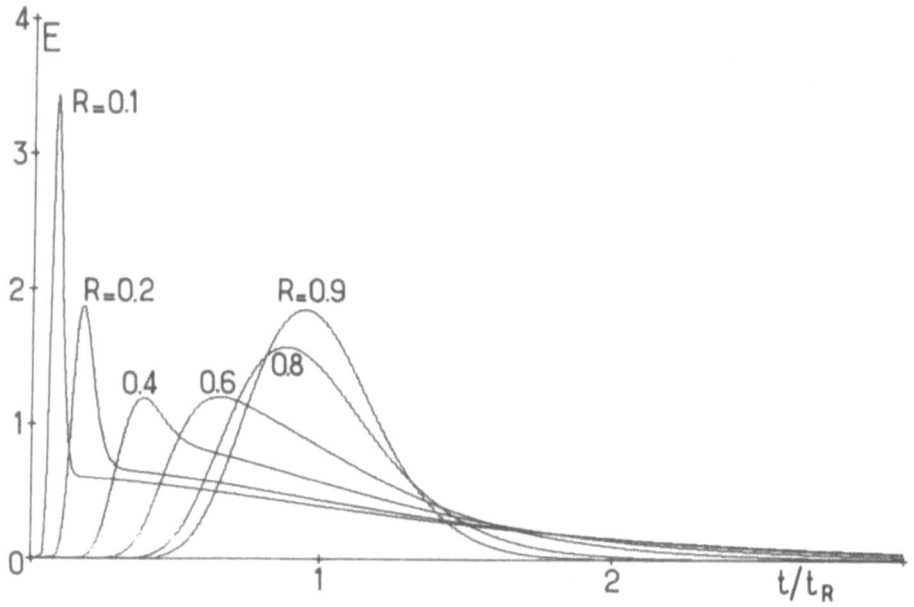

Fig. 7.6. General MCE model J=25, N=2
Influence of the retention ratio R.

Figures 7.7 and 7.8 show more clearly the influence of axial dispersion. As it is increased, it first smoother out accidents and broadens the peak (J = 100 to J = 5), then causes the chromatographic process to completely disappear (J = 5 to J = 1).

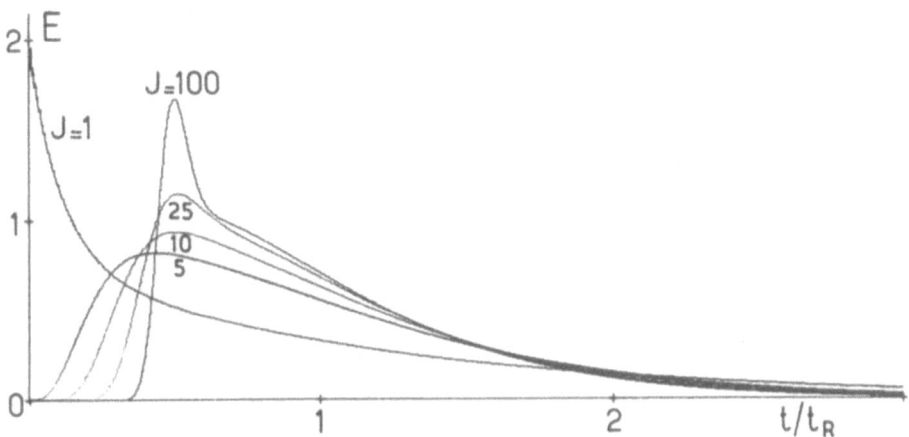

Fig. 7.7. General MCE model. K'=1 (R=0.5), N=2
Influence of axial dispersion

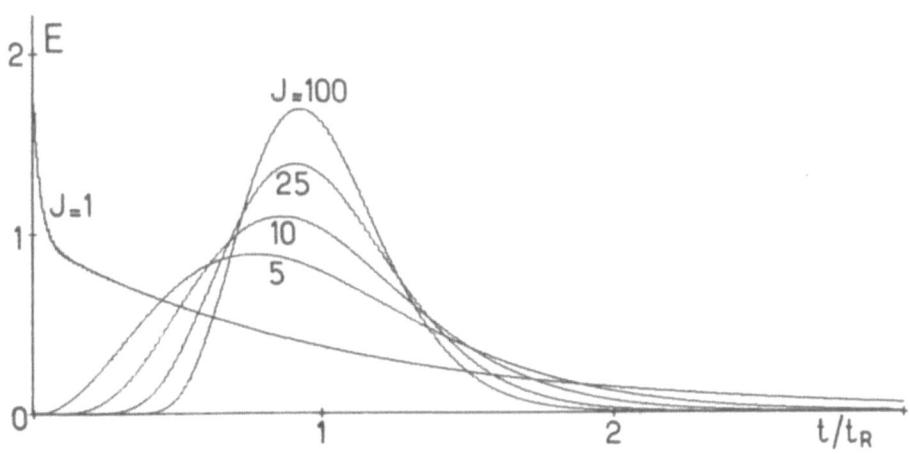

Fig. 7.8. Genral MCE model K'=1 (R=0.5), N≈10
Influence of axial dispersion

120

Finally, figure 7-9 is a more marked version of figure 7-5 to show
the progressive building of the chromatographic peak as mass-transfer rate is increased. At slow mass exchange (N < 5), peaks consist
in a sharp pulse in the region of unretained solutes (t = Rt_R)
followed by a long tail. Then we find an indistinct and large hump
in the intermediate region $Rt_R < t < t_R$ with the possibility of
two maximums (see N = 5). As N passes beyond 7 or 10, the true
chromatographic peak appears and becomes sharper and sharper as
N → ∞. Peaks corresponding to N = 0 (unretained solute) and
N = ∞ (instantaneous mass transfer) are homothetic in the ratio R
and represented by the gamma distribution (3-9). It must be pointed out that all peaks on figure 7-9 have exactly the same mean
$\bar{t} = t_R$ which is obviously distinct from the time of the maximum
in the case of slow mass transfer.
A pattern similar to that of figure 7-9 will be encountered below
when we study two-site chromatography.

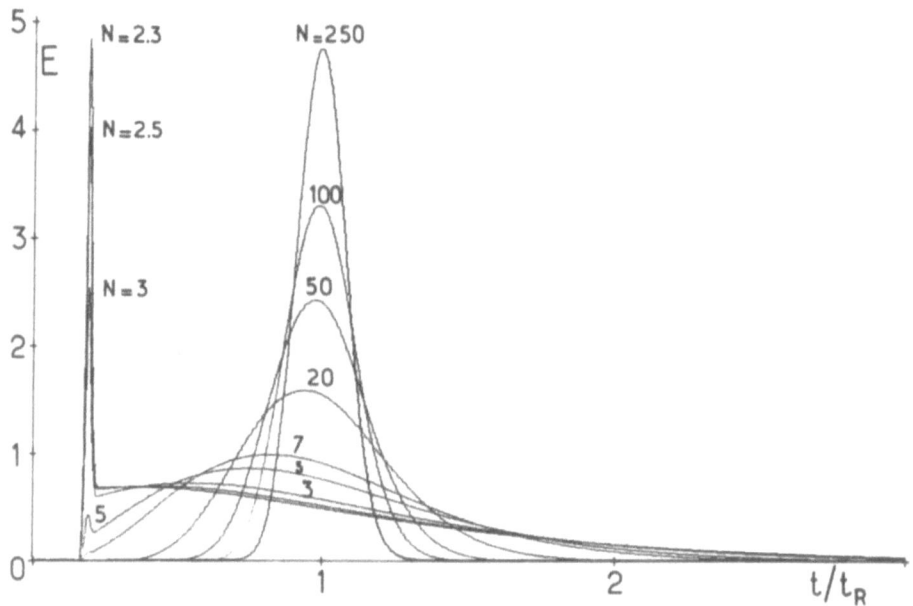

Fig. 7-9. Influence of mass transfer kinetics (MCE model)
R = 0.2 ; J = 500

7.3 Distributed mass transfer. Influence of the transfer time distribution (TTD)

This occurs mainly when several retention sites are exchanging in parallel. Then, equation (7-1) has to be used to represent
chromatographic peaks. The internal mass transfer function H(s) is
either :

$$H(s) = \sum_j \frac{\omega_j}{1 + s\tau_j} \tag{5-37}$$

in the case of a discrete distribution of sites
or

$$H(s) = \int_0^\infty \frac{f(\tau)\,d\tau}{1 + s\tau} \tag{5-39}$$

in the case of a continuous distribution, as shown in section 5-3. Before studying peak shapes, let us examine the expression for the first statistical moment of the peak considered as the retention time distribution. We start from the transfer function :

$$G(s) = G_o \left[s \left(1 + \frac{K'}{1/H(s) + s t_e} \right) \right] \tag{7-14}$$

The moments are calculated using Van der Laan's theorem as explained in section 2 (cf. 2-17) : The mean retention time is always

$$\mu_1 = t_R = t_o (1 + K') \tag{7-15}$$

The variance is derived as

$$\sigma^2 = \sigma_o^2 (1 + K')^2 + 2K' t_o (t_e + \mu_{H1}) \tag{7-16}$$

σ_o^2 is the variance of the RTD of the mobile phase and μ_{H1} is the first moment of $H(s)$ viz. $\mu_{H1} = - (\partial H/\partial s)_o$. From (5-37)

$$\mu_{H1} = \sum_j \omega_j \tau_j \tag{7-17}$$

and from (5-39)

$$\mu_{H1} = \bar{\tau} = \int_0^\infty \tau f(\tau)\,d\tau \tag{7-18}$$

Setting $\bar{t}_m = t_e + \mu_{H1}$, the mass transfer contribution to the variance is thus

$$\sigma_m^2 = 2K' t_o \bar{t}_m \tag{7-19}$$

an expression comparable to (3-22) and (7-6). The peak asymmetry is characterized by its third order central moment μ'_3. A straightforward calculation leads to

$$\mu'_3 = \mu'_3 (1 + K')^3 + 6K' (1 + K') \sigma_o^2 \bar{t}_m$$
$$+ 3K' t_o (\mu_{H2} + 4t_e \bar{t}_m + 2t_e^2) \tag{7-20}$$

where μ'_3 relates to the RTD of the mobile phase and μ_{H2} is the second order moment of $H(s)$ about the origin. Introducing the variance σ_f^2 of the TTD

$$\sigma_f^2 = \int_0^\infty (\tau - \bar{\tau})^2 \, f(\tau) \, d\tau \tag{7-21}$$

The last term of (7-20) is also equal to

$$6K' \, t_o \, (\sigma_f^2 + \bar{t}_m^2) \tag{7-22}$$

From (5-39) it must be noted that the moments of H(s) and those of the TTD, μ_{fn}, are related to each other in a very simple way.

$$\mu_{Hn} = n! \, \mu_{fn} \tag{7-23}$$

This establishes a correspondence between the internal diffusion approach and the TTD description. The presence of σ_f^2 in the expression of μ'_3 proves that the peak shape certainly depends on $f(\tau)$ in a more complicated way than simply through its mean $\bar{\tau}$. This point will become more obvious below.

Let us now pass on to numerical simulations in order to investigate the influence of $f(\tau)$ on peak shapes. The model is that of equation (7-1). Figure 7-10 shows three peaks corresponding to different TTD, namely :

1) monodispersed distribution (classical MCE model)
   $$f(\tau) = \delta(\tau - \bar{\tau}) \tag{7-24}$$

2) uniform distribution
   $$f(\tau) = 1/2\bar{\tau}, \quad 0 < \tau < 2\,\bar{\tau} \tag{7-25}$$

3) exponential distribution
   $$f(\tau) = (1/\bar{\tau}) \exp(-\tau/\bar{\tau}) \tag{7-26}$$

The mass-transfer rate has been assumed to be relatively slow $\bar{\tau} = 0.5 \, t_R$.

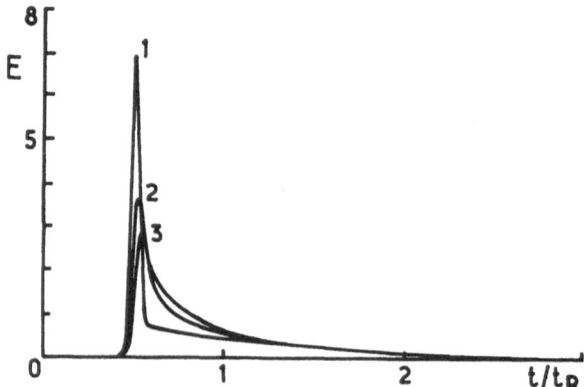

Fig. 7-10. Influence of the TTD on peak shapes. K'=1, J=500 $\bar{\tau}/t_R$ = 0.5. (1) monodispersed TTD ; (2) uniform TTD ; (3) exponential TTD.

As $\bar{\tau}$ is the same, all three peaks have <u>the same mean and the same</u> <u>variance</u> and differ only in their higher order central moments. The spread in $f(\tau)$ obviously reduces the break of the tail and the sharpness of the initial peaking.

Another special case of interest is that of a "two-site" distribution which may be encountered in analytical chromatography when both the support and the active stationary phase exhibit retention properties but do not exchange with the same velocity.

The TTD consists of two Dirac delta functions

$$f(\tau) = \omega_1 \delta(\tau - \tau_1) + \omega_2 \delta(\tau - \tau_2) \qquad (7\text{-}27)$$

Corresponding to

$$H(s) = \frac{\omega_1}{1 + s\tau_1} + \frac{\omega_2}{1 + s\tau_2} \qquad (7\text{-}28)$$

Or, in the real domain, to the system

$$\begin{cases} \tau_1 \dfrac{dn'_1}{dt} + n'_1 = \omega_1 K'n \\[2mm] \tau_2 \dfrac{dn'_2}{dt} + n'_2 = \omega_2 K'n \end{cases} \qquad (7\text{-}29)$$

n'_1 and n'_2 are the amounts of solutes respectively fixed on each population of sites. Defining the two reduced transfer times $\theta_1 = \tau_1/t_R$ and $\theta_2 = \tau_2/t_R$ the reduced variance is

$$\frac{\sigma^2}{t_R^2} = \frac{1}{J} + \frac{2K'}{1 + K'}(\omega_1\theta_1 + \omega_2\theta_2) \qquad (7\text{-}30)$$

The model is characterized by 5 marameters (in addition to t_o) : K', J, ω_1, θ_1 and θ_2.

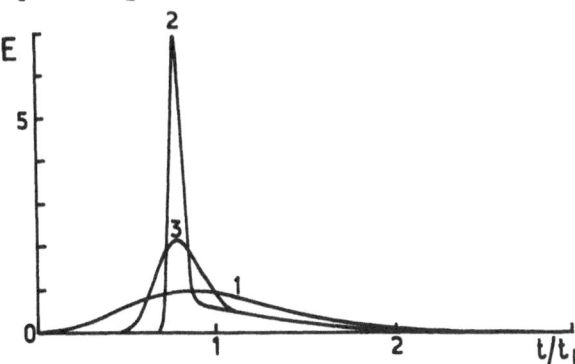

Fig. 7-11. Influence of the TTD on peak shape, "two-site" (bidispersed) distribution, $J=500$, $K'=10$, mean transfer time $\tau/t_R=0.1$, $(t_e=0)$. (1) Monodispersed $\tau_1=\tau_2=\bar{\tau}$; (2) bidispersed $\tau_1=0$, $\omega_1=0.75$, $\tau_2/t_R=0.4$, $\omega_2=0.25$; (3) bidispersed $\tau_1/t_R=0.01$, $\omega_1=0.75$, $\tau_2/t_R=0.37$, $\omega_2=0.25$.

Figure 7-11 again compares peaks having the same mean and variance (same $\bar{\tau} = \omega_1 \tau_1 + \omega_2 \tau_2$) but differing in the (τ_1, τ_2) distribution. As could be expected, the presence of two kinds of site, largely different with respect to their accessibility (rate of mass transfer) strongly affects the peak shape and may lead to important tailing.

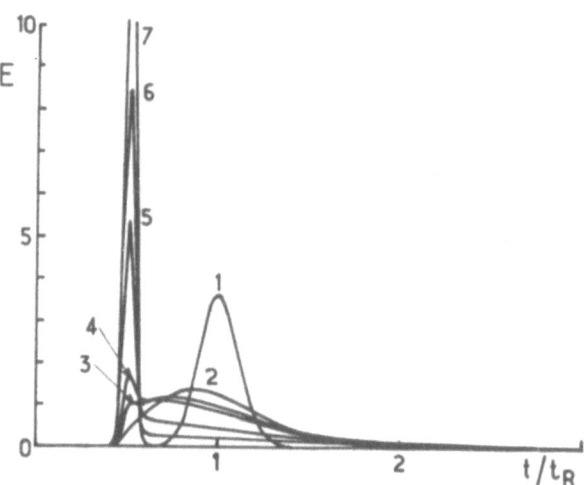

Fig. 7-12. Deformation of peaks under the influence of a gradual increase of the τ_2 transfer time in a "two-site" TTD. J=500, K'=2, t_e=0, ω_1=0.25, $\theta_1=\tau_1/t_R$=0.001, $\theta_2=\tau_2/t_R$ variable ; (1) θ_2=0.01, (2) θ_2=0.10, (3) θ_2=0.15, (4) θ_2=0.20, (5) θ_2=0.50, (6) θ_2=1.00, (7) θ_2=4.00.

Figure 7-12 shows another example of simultation where one of the transfer times θ_1 is short and constant whereas the other one θ_2 is variable and gradually increased. The situation is somewhat analogous to that of figure 7-9 : the peak having its maximum in the region of retention times $t_o (1 + K' \omega_1)$ characterizes only the rapid sites (θ_1) as θ_2 becomes too large.

Furthermore, when axial dispersion is weak, a curious property results from expression (7-28). The transfer function is close to

$$G(s) = \exp\left[-st_o\left(1 + \frac{K'\omega_1}{1+\tau_1 s} + \frac{K'\omega_2}{1+\tau_2 s}\right)\right] = \exp\left[-s\omega_1 t_o\left(1 + \frac{K'}{1+\tau_1 s}\right)\right] .$$

$$\times \exp\left[-s\omega_2 t_o\left(1 + \frac{K'}{1+\tau_2 s}\right)\right] \tag{7-31}$$

This means that the peak shape does not depend on the order in which retention sites are encountered. It would be the same if the fluid flowed first through a column (relative length ω_1) containing all sites 1 then through a column (relative length ω_2) containing all sites 2.

8. FURTHER GENERALIZATIONS OF THE THEORY. DISTRIBUTED RETENTION PROPERTIES AND COMPRESSIBLE MOBILE PHASE

8.1 Distributed retention properties

Up to this point, we have assumed that the solute present in the mobile phase exhibits one single partition coefficient and thus one single value of the capacity factor K'. It may happen however that the injected sample contains a complex solute exhibiting a distribution of retention properties. This is for example the case in gel permeation chromatography or in exclusion chromatography. The fraction of the total amount of solute having a capacity factor between K' and K' + dK'

$$\phi (K')dK'$$

$\phi(K')$ is the capacity factor distribution (CFD). What are the consequences of this distribution on the characteristics of the peak ? Let us consider the MCE model and its parameters K', N (number of transfer units) and J. Let E(t, K',N, J) be the impulse response associated with K'. As consequence of the distribution, the overall impulse response is obtained by superposition

$$\overline{E}(t,N,J) = \int_0^\infty E(t,K',N,J)\, \phi (K')dK' \qquad (8-1)$$

corresponding to the transfer function

$$\overline{G}(s,N,J) = \int_0^\infty G(s,K',N,J)\, \phi (K')dK' \qquad (8-2)$$

The mean retention time is thus

$$t_R = \overline{\mu}_1 = \int_0^\infty t_o (1+K')\, \phi (K')dK' = t_o(1+\overline{K'}) \qquad (8-3)$$

It is the same as that of a monodispersed sample having the mean capacity factor

$$\overline{K'} = \int_0^\infty K'\, \phi (K')dK' \qquad (8-4)$$

In the same way, the variance of the resulting peak is given by

$$\overline{\sigma}^2 = \int_0^\infty (\sigma^2 + \mu_1^2)\, \phi (K')dK' - (\overline{\mu}_1)^2 \qquad (8-5)$$

where $\sigma^2 = t_o^2 \left(\dfrac{2K'^2}{N} + \dfrac{(1+K')^2}{J} \right)$ $\qquad\qquad$ (8-6)

Let $\sigma_{K'}^2$ be the variance of $\phi(K')$

$$\sigma_{K'}^2 = \overline{K'^2} - (\overline{K'})^2 \qquad\qquad (8-7)$$

Let us assume an instantaneous transfer rate ($N = \infty$). A straightforward calculation shows that

$$\frac{\overline{\sigma^2}}{t_R^2} = \frac{\sigma_{K'}^2}{(1+\overline{K'})^2} + \sigma_{K'}^2 \, \sigma_D^2 + \sigma_D^2 \qquad\qquad (8-8)$$

where $\sigma_D^2 = 1/J$ is due to axial dispersion. The important result is that $\overline{\sigma^2}$ is not simply the addition of the variance of the column (σ_D^2) and of the CFD. There is also a coupling term $\sigma_{K'}^2 \; \sigma_D^2$.

In the case of ideal chromatography where $E(t,K',N,J)$ is a Dirac-delta function, then (8-1) becomes simply

$$\overline{E} = \phi \left(\frac{t - t_o}{t_o} \right) \qquad\qquad (8-9)$$

so that the chromatographic peak is in fact a trace of the distribution of the capacity factor. This property is used in the interpretation of gel permeation chromatography data.

8.2 Compressible carrier fluid

At the beginning of this study, we assumed that the carrier fluid velocity u was a constant. If the fluid is a gas, this implies that the pressure drop across the column is negligible. As this is not always true in practice, this assumption appears at first sight as a limitation to our analysis. This difficulty may be overcome in the following way 18 :

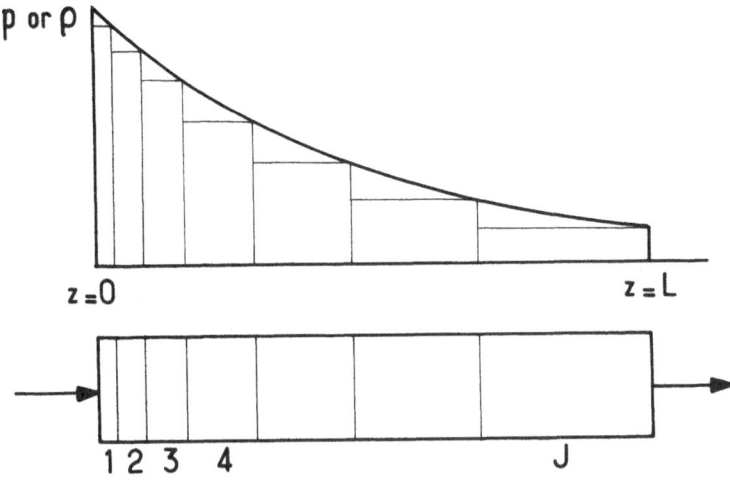

Figure 8-1.

'It is sufficient to assume that the column is divided into J cells containing the same mass of gas. The mean residence time of the carrier fluid is thus the same in each cell i.e. t_o/J and the analysis of preceeding sections applies. t_o is defined from the massflowrate per unit cross section area g and the mass density of the gaz ρ

$$t_o = \frac{1}{g} \int_0^L \rho \, d z \qquad (8\text{-}10)$$

$\rho(z)$ depends on the analysis of the pressure drop problem.

The concentrations of solute are better expressed here as concentrations per unit mass (instead of per unit volume) of the carrier fluid.

The remaining problem is to find the signification of J in terms of the axial dispersion coefficient. It is obviously not the same as in the case of a constant density fluid. However, the meaning of J is always valid as a phenomenological parameter. Further considerations on this problem in the case of the continuous PDE model can be found in reference [18].

9. RELATION BETWEEN CHROMATOGRAPHIC PARAMETERS AND METRIC DATA

OF THE PEAK

Being given an experimental chromatographic peak, useful informations on the values of chromatographic parameters can be obtained from direct measurements made on the peak.

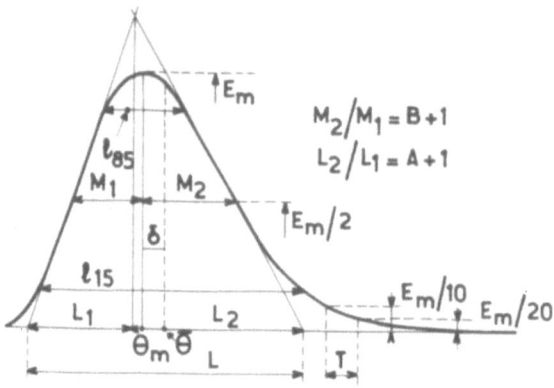

Figure 9-1

Starting from numerical simulations with the MCE model, empirical relationships have been established as an attempt to the solution of this problem [19].

9.1 Estimation of the variance

The peak is assumed to be reduced and normalized.

Let us recall that, in reduced form :

$$\sigma^2 = \frac{2(1-R)^2}{N} + \frac{1}{J} \tag{9-1}$$

From figure 9-1, one finds

$$\sigma = (L/4)(1+\beta) \tag{9-2}$$

β is a correction factor given by

$$\beta^{1.5} = \left[0.63\ N^{-1.15}\right]^{1.5} + \left[\left(\frac{J}{J-1}\right)^{1/2} - 1\right]^{1.5} \tag{9-3}$$

9.2 Peak asymmetry

Two criteria may be proposed. One is deduced from inflexion tangents (see fig. 9.1)

$$A = L_2/L_1 - 1 \tag{9-4}$$

the other one is deduced from the width at half-height

$$B = M_2/M_1 - 1 \tag{9-5}$$

Empirical relationships are the following :

$$A^{1.5} = \left[4.25\ N^{-0.815}\right]^{1.5} + \left[2(J-1)^{-0.5}\right]^{1.5} \tag{9-6}$$

$$B^{1.5} = \left[1.28 \, N^{-0.6}\right]^{1.5} + \left[J^{-0.548}\right]^{1.5} \qquad (9-7)$$

9.3 Peak tailing

T is the difference between abscissae corresponding to $E_m/20$ and $E_m/10$ (see figure 9-1).

$$\text{Then } T^2 = \left[0.856(1-R) \, N^{-0.63}\right]^2 + \left[0.533 \, J^{-0.58}\right]^2 \qquad (9-8)$$

$$\text{If } J \to \infty, \quad \sigma/T = 1.652 \, N^{0.13} \qquad (9-9)$$

9.4 Position and height of the maximum

$$\text{Let } \delta = \frac{t_R - t_{max}}{t_R}$$

For $J \to \infty$ and $N > 2$

$$\delta = \frac{1-R}{2N} \left[3 + (N-1)^{-1.11}\right] \qquad (9-10)$$

$$\text{and } \delta/\sigma^2 = \frac{3 + (N-1)^{-1.11}}{4(1-R)} \qquad (9-11)$$

The height of the maximum of a reduced peak can be calculated by

$$E_m = \frac{1 + \alpha}{\sigma\sqrt{2\pi}} \qquad (9-12)$$

Where α is a corrective factor given by

$$\alpha = \frac{0.357}{N-1} + \sqrt{\frac{J}{J-1}} - 1 \qquad (9-13)$$

9.5 Treatment of an experimental peak

Let us now consider a peak in real coordinates (ordinates in arbitrary units and abscissae in real time).

The surface of the peak may be estimated to better than 1 % by one of the two formulae (see fig. 9-1)(non reduced data are marked by a "dash").

$$S \approx 0.627 \, E'_m \cdot L' \approx 0.5 \, E'_m(l'_{15} + l'_{85}) \qquad (9-14)$$

A first estimate of t_R is

$$t_R \approx t_{max} + L'^2/(16\, t_{max}) \tag{9-15}$$

this assumes that $\delta = \sigma^2$ and $\sigma' = L'/4$.

In the same way, a first estimate of E_m is

$$E_m = E'_m\, t_R/S \tag{9-16}$$

The reader may imagine iterative procedures to improve these first estimates, using the empirical relationships, which are presented in the form of charts in reference [19].

10. EXAMPLES OF APPLICATION

We have selected three examples showing how the notions presented in sections 1-9 can be applied to solve practical problems.

10.1 Measurement of effective diffusivity in porous media 22 .

The first example is concerned with effective diffusivity measurements in packed beds by the method of moments.

A column is packed with spherical beads and a pulse of tracer is injected either into the column or into a gas circuit comprising exactly the same equipment, tubing, flanges etc... as the column circuit. The pressure drops are equalized. The two output signals are recorded in two successive experiments. Then, signal 2 is the convolution product of signal 1 by the impulse response of the packed bed alone. As a result, the mean residence time t_R and the variance σ^2 may be obtained by simple difference between the means and variances of the two signals. Deformations caused by the imperfect input pulse, the external tubing and the detector cancel out in the substraction. This point has often been overlooked by authors who attribute the peak broadening to the chromatographic medium alone. In our opinion, the "two measurement" method is the only correct one.

In a first series of experiments, the column was packed with non porous glass beads. The carrier fluid was helium in laminar flow and the tracer was hydrogen. In these conditions, the peak broadening is controlled by axial dispersion alone. A plot of σ^2/t_R against t_R^2 yields a straight line (see 7-8 and 7-11).

$$\frac{\sigma^2}{t_R} = \frac{\sigma_S^2 - \sigma_E^2}{t_R} = \frac{2 \mathcal{D}_m}{L^2(1+K')\tau_A}\, t_R^2 \tag{10-1}$$

From the slope and the value of $\mathcal{D}_m$, the axial tortuosity can be calculated. One finds $\tau_A = 1.75$, a reasonable value which constitutes a check on the method.

In a second kind of experiment, the column was packed with spherical microporous alumina beads and the tracer was Argon in Helium. In this case, peak broadening was found to be mainly due to internal diffusion into the particles.

A plot of σ^2/t_R against t_R yields a straight line. From (7-8) and (7-11) :

$$\frac{\sigma^2}{t_R} = \frac{\sigma_S^2 - \sigma_E^2}{t_R} = \frac{2 d_p}{L \, Pe_A} t_R + \frac{2K'}{1+K'} t_m \qquad (10-2)$$

The intercept of this line yields $\dfrac{2K'}{1+K'} t_m$ from which t_m is cal-culated. One finds $t_m = \dfrac{0.6 \, d_p^2}{D_e} = 0.13$ seconds. Knowing $d_p = 2.2$ mm, the effective diffusivity of argon in the alumina particles is finally obtained : $D_e = 6 \times 10^{-3}$ cm^2/s at 20°C.

10.2 Chemisorption studies : two site chromatography 12 .

The simulation presented at the end of section 7 showing the deformation of a chromatographic peak in presence of two kinds of sites exchanging with different velocities (fig. 7-12) might appear as a simple mathematical game. This was also my own opi-nion till I was contacted independently by people working in an industrial research center. They were engaged in a chromatographic study of hydrogen chemisorption on some of their catalysts and they were surprised to obtain in some cases unexpected peak sha-pes with two maximums which they could not interpret with classi-cal chromatography theory.

The stationary phase consisted of 10 % nickel on an alumina support. A mixed sample of helium and deuterium was injected into hydrogen as a carrier gas. The first eluted peak was helium, cha-racteristic of an unretained solute. It was followed by a distor-ted HD-D$_2$ peak resulting from isotopic exchange. This second peak may be considered as representative of hydrogen chemisorption on the catalyst. Independent adsorption experiments seemed to show that chemisorption occured on two kinds of adsorption sites :

 - fast exchanging nickel sites (type 1)

 - nicked oxide sites, with a transfer time depending on

temperature (type 2).

The deformation of chromatograms as a function of temperature confirms these assumptions. At low temperature (78°C, 82°C), nickel oxide sites exchange slowly and simply give rise to peak tailing. As temperature is increased, the importance of tailing grows and the initial peaking gets smaller and smaller. At 90°C, a peak with two maximums can be observed. Above 100°C, there remains only one single maximum, close to the mean retention time.

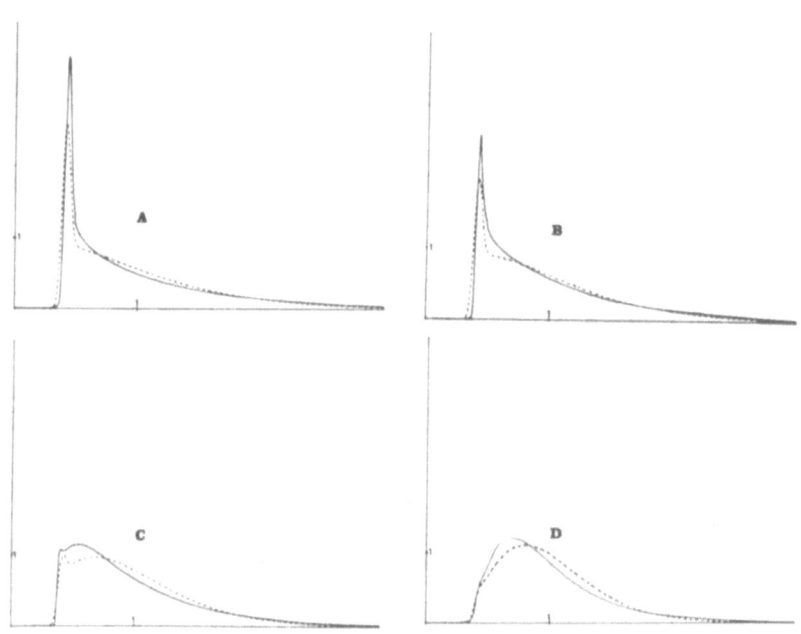

Fig. 10-1. Chemisorption of hydrogen on a nickel/alumina catalyst. Experimental elution peaks (solid lines) compared with simulated peaks (dotted lines). Two site MCE model. J=200, K'=2.40, ω_1=0.16, θ_1=10^{-3}
A : 78°C, θ_2=0.34 . B : 82°C, θ_2=0.28 . C : 90°C, θ_2=0.19 .
D : 100°C, θ_2=0.13

Figure 10-1 shows experimental peaks obtained at four temperatures (A,B,C,D) compared to theoretical curves of a two-site MCE-model where the reduced transfer time θ_2 to nickel oxide sites has been adjusted to give a best fit with experimental shapes. Taking into account the simplicity of the assumption of the model, the agreement may be considered as satisfactory. An Arrhenius plot of the transfer time (a straight line is obtained in plotting $Log_n \theta_2$ against 1/T) confirms the validity of the interpretation and yields the activation energy of the hydrogen chemisorption on nickel oxide sites (12 Kcal). More details can be found in reference [12].

10.3 Applications to hydrology : infiltration of solutes through unsaturated soils.

Leaving adsorption problems, the last example is taken in the field of hydrology and soil science.

When aqueous solutions of various chemical such as herbicides, pesticides, fertilizers, radioactive materials, salts, etc... are spread over a soil surface, or are placed in the soil, they can move downward with the water in the soil and eventually reach the ground water table. On the other hand, sorption and interaction with the soil may delay downward movement. In either case, it is of great importance to be able to predict the time of transit of surface applied chemicals to the aquifer as well as the amount of chemical which will reach and possibly contaminate the aquifer.

In order to study this problem, J.P. GAUDET at the Institute of Fluid Mechanics of Grenoble has made an experimental investigation of the dispersion of solutes in soils columns [21]. The packing was of non porous sand from the alluvial plain of the river Isere. The column was leached with saline water at constant flowrate. At time zero, a rectangular pulse of pure water was injected into the column. The resultant transient changes of concentration were monitored with resistivity probes at several places in the column and with a conductivity cell at the column exit. In some experiments, the column was totally saturated with water filling the whole interparticle porosity. In such cases, the classical axial dispersion plug flow model succeeded in accounting for the observed dispersion. Conversely, other experiments were carried out under unsaturated conditions, which means that there were air pockets in the column between the sand particles. In these experiments, unexpected results were obtained : it was impossible to account for the observed concentration profiles on the basis of single axial dispersion. As no interaction can occur with the sand particles, because of the absence of intraparticle porosity, it must be assumed that a certain fraction of the water in contact with air bubbles and sand particles is stagnant. The other fraction of the intersticial water is supposed to be mobile and to undergo both a reversible exchange with the stagnant fraction and axial dispersion caused by the packing. This is nothing but the MCE model except that both mobile and stationary parts of the intersticial water belong to the same phase.

Let f_o and f_m be the fractions of interparticle porosity occupied respectively by total water and by mobile water, and let R be the ratio $R = f_m/f_o$. The partition coefficient here is $\alpha = 1$ and the capacity factor is $K' = (1/R) - 1$. Application of the MCE model leads to the following Laplace transforms for the relative concentrations of pure water in the column :

In the mobile fraction :

$$Y_1 = \frac{1 - e^{-st_C}}{s} \left(1 + \frac{Wz}{J}\right)^{-J}$$
(10-3)

In the sum of mobile and stagnant fractions (to which probes are sensitive)

$$Y_3 = \frac{1 - e^{-st_C}}{s} \cdot \frac{W}{st_R} \left(1 + \frac{Wz}{J}\right)^{-J}$$
(10-4)

with

$$W = st_R \frac{1 + R\, st_m}{1 + st_m}$$
(10-5)

In these expressions, t_C is the duration of the rectangular pulse of pure water at the inlet, t_R is the mean retention time, which can be expressed as a function of the column length L and Darcy's velocity u_o as $t_R = L\, f_o/u_o$, t_m is the time of transfer between the two water fractions and $z = x/L$ is the relative axial abscissa of probes. The concentration signals are recorded at $x = 7$ cm, 22 cm, 37 cm, 52 cm and 82 cm measured from inlet, and at the outlet $L = 93.5$ cm (signal in mobile water only). The data are $t_C = 0.577$ hr and $u_o = 7$ cm/hr. Parameters to be adjusted are t_R, R, t_m and J. Optimisation is achieved by numerical inversion of Y_1 (internal signals) and Y_3 (outlet signal) and comparison with experimental curves. Adjustement is facilitated by the fortunate fact that the parameters play a relatively independent role on the different characteristics of concentration signals. t_R determines the "time-scale", R the peak height and t_m the "break point" of decay fronts. The influence of J is weak, it produces only some kind of "blurring" of the profiles. Figure 10-2 shows an example of an empirical "adjustement by inspection" so as to obtain an average acceptable fit.

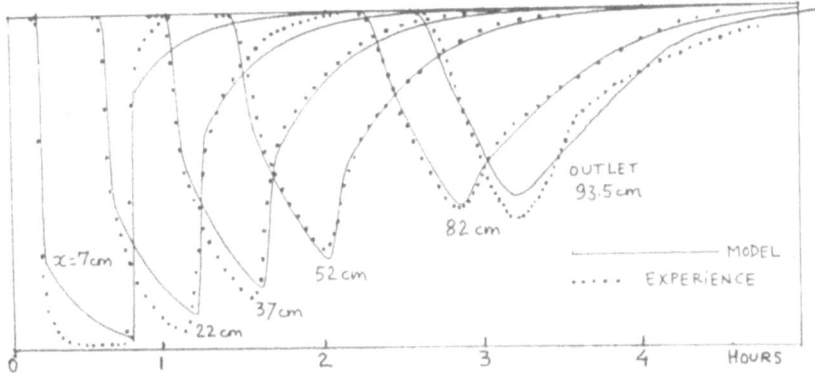

Fig. 10-2. Miscible displacement in unsaturated medium.

The small discrepancies may be explained by a possible variation of the parameters along the column axis. Better results would
likely be obtained by a least square fitting on a particular peak.
Nevertheless, the parameters so determined are t_R = 3.2 hr from
which f_o = 0.23 is deduced (total interparticle porosity is 0.37);
R = 0.83 (There is thus 17 % of stagnant water) ; t_m = 0.22 hr
and J = 3000 (corresponding nearly to an axial diffusion
D = $L^2/(2JRt_R)$ = 10 cm^2/hr). These results are in good agreement
with those obtained by GAUDET and coll. [21] by direct numerical
integration of the mass-balance equations. The conclusion is that
about 17 % of the intersticial water has to be considered as
stagnant and undergoing a very slow exchange with mobile water.
This concept, which seems familiar to chemical engineers was rather new for hydrologists who used to consider axial dispersion
alone. This result will undoubtly provide a better understanding
of solute displacement in unsaturated soils. It also proves the
relevance of the MCE model in this kind of problems.

11. GENERAL CONCLUSION

As shown by the variety of problems which can be dealt with,
the system approach of linear chromatographic interactions proves
to be very effective and fruitful in the treatment, the simulation
and the interpretation of apparently very different phenomena.

The main points established throughout this paper can be
summarized as follows :

1) The response of a chromatographic system to an input concentration signal, and especially the impulse response (chromatographic peak) can be numerically simulated from the expression

of the transfer function G(s).

2) This function is obtained in a simple way from the know-ledge of the hydrodynamic flow pattern of the mobile phase and from the elementary mass transfer mechanism. The two processes are in a certain sense "uncoupled".

3) Mass transfer at the elementary level can be represented by a combination of typical processes : diffusion in a porous solid, retention layers in series, retention sites in parallel, etc... This mass exchange is characterized by the transfer function $M(s) = \bar{n}'/\bar{n}$.

4) In all cases where there is no fixation of the solute on sites in parallel, the dynamics of exchange at the elementary le-vel may be approached by a first order system behaviour of trans-fer function $M(s) = \dfrac{K'}{1 + t_m s}$ where K' is the capacity factor de-fined as the ratio of the amount of bounded solute to that of free solute at equilibrium. t_m is an overall transfer time depending on the kind of retention.

5) Conversely, when there exist sites with different ease of access in parallel, this is no longer true and the "transfer time distribution" has to be taken into account, leading to a possibi-lity of strongly tailing peaks. The case of two-site distributions is very important in practice.

6) Thanks to procedures of numerical inversion of Laplace transforms, transient concentration signals of solute can be simu-lated, starting from transfer function expressions. The three pa-rameter MCE model (dispersion - retention - exchange kinetics) thus applies to many situations encountered in practice.

7) This general treatment can be extended to more complicated cases, for instance a compressible carrier fluid and a distribu-tion of retention properties.

8) Informations on the value of interaction parameters can be obtained directly from metric data measured on chromatographic peaks.

REFERENCES

1. D. Himmelblau and K.B. Bischoff, Process Analysis and Simu-lation, Wiley, 1968.
2. J.M. Douglas, Process dynamics and control, Vol. 1, Analysis of dynamic systems, Prentice Hall, 1972
3. J.C. Friedly, Dynamic behavior of processes, Prentice Hall, 1972

4. J.W. Cooley and J.W. Tukey, Math. Comput., 19, 297-301,
    April 1965
5. F. Vergnes, Nuclear instruments and methods, 96, 421, (1971)
6. M.J.P. Martin and R.L.M. Synge, Biochem. J., 35, 1359 (1941)
7. L. Lapidus and N. Amundson, J. Phys. Chem., 56, 984 (1952)
8. J.J. Van Deemter, F.J. Zuiderweg and A. Klinkenberg, Chem.
    Eng. Sci., 5, 271 (1956)
9. J. Villermaux and W.P.M. Van Swaaij, Chem. Eng. Sci., 24,
    1097 (1969)
10. J. Villermaux, Chem. Eng. Sci., 27, 1231 (1972)
11. R. Aris, "The mathematical theory of diffusion and reaction
    in permeable catalysts", Vol. 1 & 2, Clarendon Press, Oxford
    1975
12. J. Villermaux, J. Chromatog. Sci., 12, 822 (1974)
13. C.Y. Wen and L.T. Fan, "Models for flow systems and chemical
    reactors", Marcel Dekker, New York, 1975
14. R.P.W. Scott, "Contemporary Liquid Chromatography", Techni-
    ques of Chemistry, Vol. XI, Wiley, New York, 1976
15. J.C. Giddings, "Dynamics of Chromatography", Part I "Prin-
    ciples and Theory", Marcel Dekker, New York, 1965
16. C. Horvath and H.J. Lin, J. Chromatog., 126, 401 (1976)
17. C. Horvath and H.J. Lin, Private communication. To be publi-
    shed
18. J. Villermaux, Ve symposium international sur les méthodes
    de séparation : chromatographie sur colonne. Lausanne 1969
    (Supplementum zu Chimia)
19. J. Villermaux, J. of Chromatography, 83, 205 (1973)
20. J. Villermaux, Symposium on Hydrodynamic diffusion and dis-
    persion in porous media, Pavia, April 20-22, 1977
21. J.P. Gaudet, M. Jegat, G. Vachaud and P. Wierenga, Soil Sci.
    Soc. Am. Journal (in press)
22. J. Villermaux and D. Matras, The Canadian Journal of Chemical
    Engineering, 51, 636 (1973)

NOMENCLATURE

a	specific interfacial area	m^2/m^3
A_1, A_2	constants (functions of s)	
A	parameter (Van Demter equation), asymmetry factor	
B	parameter (Van Demter equation), asymmetry factor	
C	parameter (Van Demter equation)	
c	(various subscripts and superscripts, see text) concentration	$mole/m^3$
$d = V_p/S_p$	characteristic particle dimension	m
d_p	particle diameter	m
D	axial diffusivity	
D_e	effective diffusivity porous particle	m^2/s
D_m	molecular diffusivity (fluid)	
e	slab thickness	m
$E(t)$	Impulse response - Peak equation	
$E_0(t)$	RTD of unretained solutes	
E_m	peak maximum	
$f(\tau)$	transfer time distribution (TTD)	
f_0, f_m	porosity fractions	
$F(t)$	step response	
g	mass flowrate per unit cross section area	
$g(s)$	transfer function (plate)	
$G(s)$	transfer function (system, column)	
$G_0(s)$	transfer function (RTD)	
$H(t)$	Heaviside unit step function	
$H(s)$	transfer function (mean/superficial solute concentration)	
HETP	height equivalent to a theoretical plate	m
I_0, I_1	modified Bessel functions of the first kind	
J	number of cells/plates in series	
k	mass transfer coefficient (various subscripts, see text)	m/s
k_A	adsorption rate constant	s^{-1}
K'	capacity factor	
K_A	adsorption equilibrium constant	
L	column length	m
$L(s)$	transfer function (fixed/free solute concentration)	
m_k	unnormalized moment about the origin	
m'_k	unnormalized central moment	
$M(s)$	transfer function (fixed/free solute amount)	
n	(various subscripts and superscripts, see text) amount of solute	moles
N, NTU	number of transfer units	
NTP	number of theoretical plates	

$P = uL/D$	axial Peclet number (overall)	
k_A, k_R	axial radial Peclet numbers (particle)	
$q = (1+4st_o/P)^{1/2}$	parameter	
Q	volumetric flowrate	m^3/s
r	cylinder, sphere radius	m
r_1, r_2	characteristic roots	
R	retention rates $= (1+K')^{-1}$	
S	interfacial area (various subscripts, see text)	m^2
S	peak surface	
s	Laplace parameter	

t	time	t_{max} time of the maximum
		$\bar{t}$ mean
		$t_{1/2}$ median
		t_o mean residence time of unretained solutes
		t_R mean retention time
		t_m mass transfer time constant
		t_d internal diffusion time constant
		t_e external transfer time constant
		t_a adsorption time constant
		t_c rectangular pulse duration

T	time interval, tailing	
u	intersticial velocity	m/s
V	volume (various subscripts, see text)	m^3
W(s)	transfer function	
x(t)	input signal, axial abscissa	(m)
y(t)	output signal	
$Y_1(s), Y_2(s)$	transfer functions	
z	axial abscissa	m or —
α	partition coefficient	
β	intraparticle (internal) porosity	
β	correction factor (σ)	
γ_1	skewness	
γ_2	peakedness	
$\delta(t)$	Dirac impulse delta functions	
Δ	discriminant	
ε	interparticle (external porosity)	
θ_i	reduced transfer time	
	radial velocity profile coefficient	
$\lambda = d\sqrt{s}/D_e$	parameter	
$\Lambda_i(s)$	transfer function (layers in series)	
μ_k	moment about the origin	
μ'_k	central moment	
μ	shape factor	
ν	frequency (Fourier domain)	s^{-1}
ρ, ρ_p	fluid, particle density	kg/m^3

σ^2	variance (various subscripts, see text)	s^2 or —
τ, τ_i	transfer time	s
τ_A	axial tortuosity	
τ_R	radial tortuosity	
ϕ_i	volume fraction (layers in series)	
$\phi(K')$	capacity factor distribution	
ω	pulsation (Fourier domain)	s^{-1}
ω_j	fraction of site occupancy (sites is parallel)	

DESIGN AND OPTIMISATION OF PREPARATIVE CHROMATOGRAPHIC SEPARATIONS

P. VALENTIN

ELF RESEARCH CENTER - SOLAIZE 69 - France
Lecturer at Ecole Polytechnique - PARIS - FRANCE

Introduction

The discovery of chromatography as a separation technique goes back to the beginning of the 20th Century, but its empirical use is certainly much older.

The first instance of "chromatography before the letter" is perhaps to be found, in military art, in the fight of the Horatii and the Curatii (650 BC).

The three Horatius brothers fought for ROME against the three Curatii, the champions of the town of ALBA, in the presence of the two armies - to decide which of the peoples would command the other The third Horatius, uninjured, but the only one surviving, pretended to escape. The three Curatii, unequally injured, raced after him and gradually separated according to their remaining strength. The last of the Horatii, after suddenly stopping, killed them one after the other.

This is chromatography, for we find in this analogy the two characteristic features of this type of process :

1 - Same driving force for all the components to be separated (in this case the flight of the last of the Horatii)

2 - The selective interaction of each of these components with a motionless medium near which they are moving (in the instance, the ground - the interaction is selective because each of the Curatii has more or less difficulty in moving, according to

the seriousness of his injury).

Should we not see in this story that what appealed to Livy and then Corneille was this very empirical use of the chromatography principle ?

Of course such a litterary analogy stands for any chromatographic - or percolation - processes and should not be restricted to gas-liquid chromatography, which is our primary concern here.

Scope of this paper

W'll restrict our discussion to design of gas-liquid preparative chromatographic column and ulterior optimization of a separation. Other parts of the chromatographic unit, which can be calculated by classical methods of Chemical Engineering will be dealt with only in connection with best use of a column.

Gas-solid and liquid-solid chromatography although very similar in principle to gas-liquid chromatography, differ rather strongly in design and optimization of columns as accent has to be set on parameters which can be ignored in GLC., and conversely. So, we shall limit ourselves to GLC., mentionning only which approximations are valid in GSC. and LSC.

Also we consider here a single column, as it is safe first to understand this basic case before going to more intricate systems.

G. L. C. State of the art

Industrial Preparative gas-liquid chromatography uses now routinely from ten centimeters to one meter diameter columns (1,2) Projects can be made for up to 5 meters columns which could produce as much as 10 000 metric tons per year.

(Columns between 1 and 4 centimeters diameter can be labelled laboratory columns).

Topics covered

W'll consider the basic flow-sheet of an industrial preparative G. L. C. unit (IPGLC).

Efficiency of column, probably the best "investment" which can be made in a chromatographic unit depends on many design factors which will be studied.

A feature of IPGLC is its outstanding versatility : the production engineer is faced very often with setting of a new separation : the many parameters of this transient process have to be optimized for lowest overall cost. Theoretical and practical basis for such an optimization will be given.

Interest in Industrial Chromatography with reduced outlet pressure stems from these considerations.

Finally chromatography appears to have advantages of its own faced with other separation processes such as distillation : its very high efficiency factor due to an advanced technology of dispersed medias allows difficult separation to be fulfilled in a short time (a matter of minutes) comparatively to distillation (a matter of hours). New concepts to cope with such a situation have been evolved : these are efficiency factor and number of contacts.

Incidentally introduction of a "tensor extent of separation" allows determination of optimum cut-time in IPGLC, but this definition applies quite generally to the whole field of separation processes.

I - MAIN FEATURES OF A CHROMATOGRAPHIC UNIT

Basic flow-sheet is shown on fig.I. Liquid feed and carrier gas are pumped into the vaporizer where they are mixed. Vapours go through the chromatographic column where they separate. Feed is pumped discontinuously into the vaporizer, and concentration of feed remains constant in the carrier gas at the column inlet for all the injection duration (usually between 10 to 30 seconds). Feed rate and injection duration depend on separation to perform.

At the column inlet we thus have a sequence of pure carrier gas and carrier gas plus mixture flows. Let us follow such a burst of mixture : it will be pushed forward by carrier gas flow, but each of components of the mixture travels along the column at a different velocity according to its affinity for stationary phase (an unvolatile liquid dispersed in a solid packing).

A wide variety of stationary phase can be used and different type of separation can be selected according to vapor pressure and/or polarity.

At the column outlet valves direct gas alternately to the appropriate receivers ("traps") where products are separated, usually by condensation, from carrier gas which goes through dual bed of activated carbon for ultimate purification and is recycled by compression to the vaporizer. Compressor allows a subatmospheric outlet pressure if necessary.

The process is cyclic : injection of feed and opening of the traps for collection of products are operated periodically by an electronic time-based programming device.

Recycling carrier gas (helium or, for better productivity, hydrogen) allows a very low consumption.

Receivers are condensers (in case of sublimable products, gas liquid extractors) but nature of the organic compounds vs nature of the carrier gas tends to promote fog formation and low yield by liquid carry away. To set at its minimum such a detrimental phenomenc careful design of condensers must allow progressive cooling of the mixture with cryogenic fluids or simply, for high boiling compounds, cold water.

2 - COLUMN EFFICIENCY

Obtaining a good initial column efficiency (rated by highth equivalent to a theoretical plate, HETP) and maintaining this efficiency requires some skills and technology for solid support (packing) preparation, impregnation with stationary phases, packing into the column, optimizing separations without irreversibly flooding liquid phase at the inlet of the column and without exceeding temperature stability of stationary phase on the long run.

Column efficiency is here much more important than in other percolation processes as separations are usually difficult ($\alpha \simeq 1$).

2.1. Support selection and properties

Support of the stationary phase is an inert solid highly porous so that liquid can lay inside its pores without being washed out by carrier gas flow. Of course wettability of the support by stationary phase is essential in order to permit spontaneous penetration of stationary phase inside pores. Silico aluminas and particularly diatomaceous earth such as chromosorb (trademark) family are well suited to this purpose for the majority of stationary phases.

Beads of support should be resistant to attrition so that they withstand sieving and packing with minimal fines production. This quality appears rather opposed to first requirement of high internal porosity and some optimum has to be found.

Third necessity is absence of catalytic properties of the support. It should be reminded here that alumina content of supports such as chromosorb P is sufficient to develop strong surface acidity and subsequent catalytic activity. Resultant cracking and

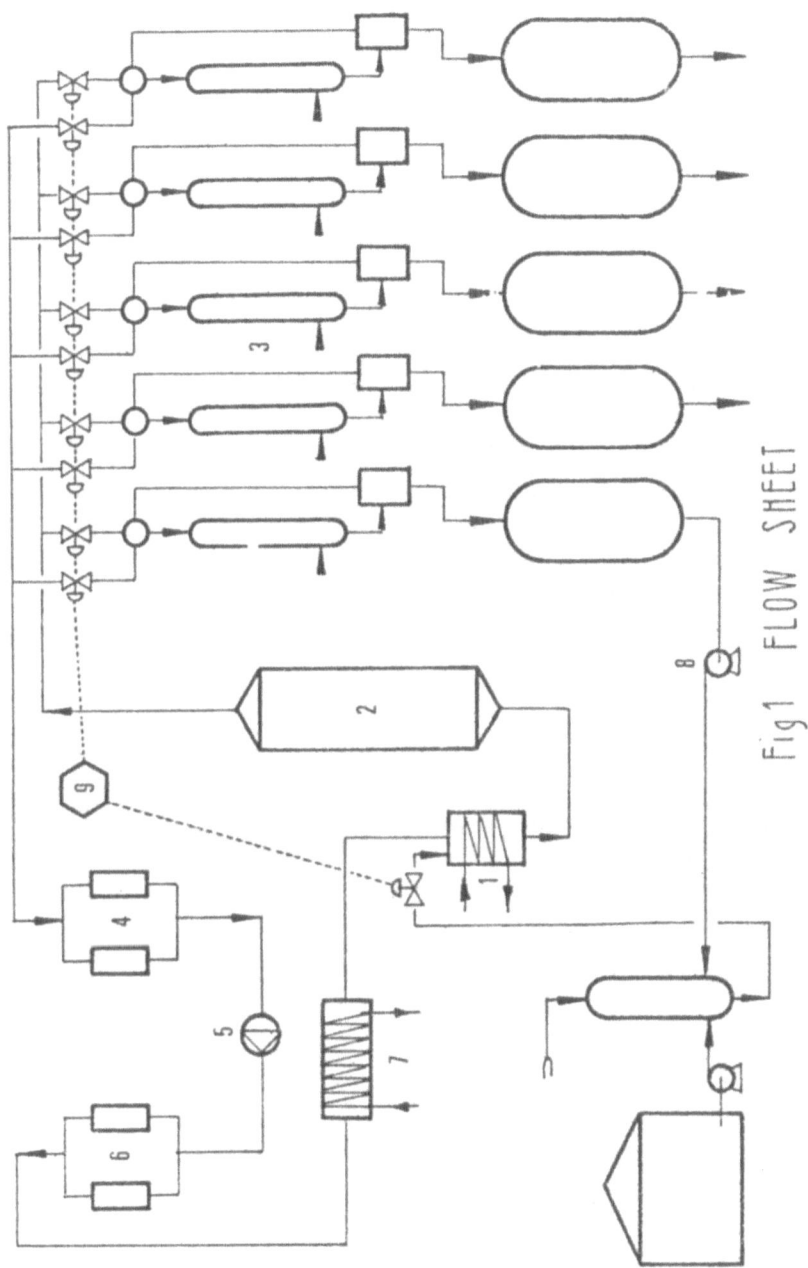

Fig1 FLOW SHEET

isomerizing properties are moderated only by low surface of support (about 4 m^2/g).

It is, in fact, absolutely necessary for most separations to inactivate totally support by appropriate long-lived treatment.

Best compromise between all these needed properties appear to be inactivated chromosorb P. Table I shows resultant distribution of solids, liquids and gases in a chromosorb P column (after impregnation with 20 % w/w stationary phase.

Table I

	VOLUME FRACTION	PORE DIAMETER
External voids	.33	5 - 100 μ
Internal voids	.35	0,4 - 2 μ
Liquid stationary phase (Density = 1)	.12	
Solid support	.20	
TOTAL	1	

Not obvious here is the large internal porosity of support (0.92 cm^3/g) filled partly with immobile gas phase ("stagnant phase") partly with liquid phase ("stationary phase").

Existence of a stagnant phase was often overlooked in GLC and in all percolation processes, but it has very important implication for cake washing, counter-flow operation and flooding, as stagnant phase communicates only by diffusion or slow convection with mobile phase flowing between beads.

2.2. Measuring efficiency of a column

Without going into controversy about formulation of efficiency of a packing by its Highth Equivalent to a Theoretical Plate (H.E.T.P.) we shall indicate major variables to be considered and their influence on HETP.

Considering a concentration wave propagating along column, HETP is defined (locally) as :

$$H(z) = \frac{d \; \sigma^2(z)}{d < z >}$$

where σ^2 is the variance of concentration distribution and $< z >$ is abscissa along the column of the mean of this distribution (3).

Such a definition stands only for a single component, at infinite dilution. So it is unsound to evaluate "variation of HETP" from peak widening when overloading : in fact, overloading itself is produced by non-linear phenomena which will be studied shortly.

In practice, HETP measured at infinite dilution appears a quite satisfactory evaluation of "efficiency" of the packing. In the usual case of Gaussian peaks following a Dirac injection pulse of the column, Equ.(1) can be transformed to give plate number of the column, N :

$$N = 16 \left(\frac{t_R}{\Delta t} \right)^2 \qquad\qquad (2)$$

where : t_R is retention time (peak maximum)

Δt is time with at intercept of inflexion tangents of the peak with base line.

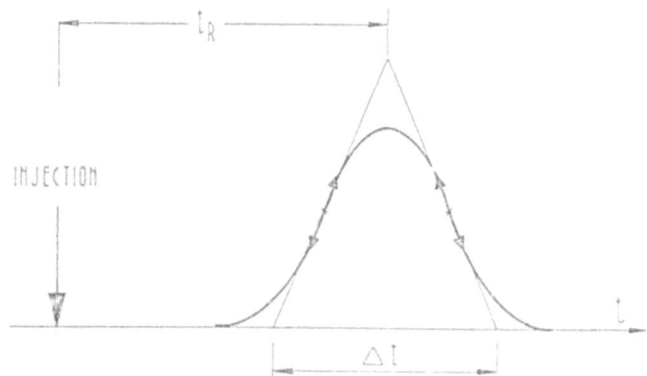

FIG.2 PLATE NUMBER

Then, (neglecting local variations due to pressure drop and carrier gas compressibility), H follows by

$$H = \frac{L}{N} \qquad\qquad (3)$$

with L, column length.

From Equ.(1) it is clear that every phenomenon contributing to increase $d\sigma^2(z)$, that is "rate of widening" of the peak, will have a detrimental effect on separation.

So we look for a minimum for H.

VAN DEEMTER equation describes effect of carrier gas velocity on H :

$$H = A + \frac{B}{u} + Cu \qquad (4)$$

u is defined as :

$$u = \frac{L}{t_o} \qquad (5)$$

where t_o is mean retention time of an inert tracer.

H has an hyperbolic shape with a minimum for

$$u = (\frac{B}{C})^{1/2}$$

and an infinite branch with asymptotic slope C. (fig. 3)

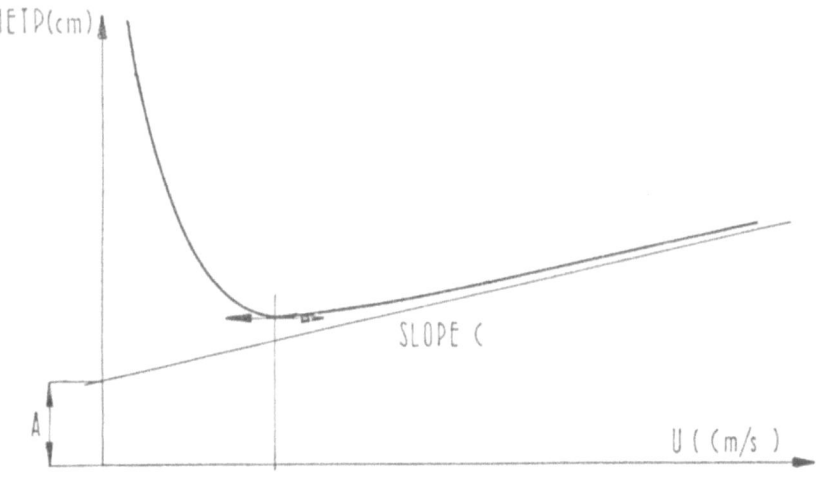

FIG 3 VAN DEEMTER CURVE

More refined theories can give a better insignt than Eq.(4) into important factors in H (see the paper from Pr. VILLERMAUX) but for preparative chromatography it is sufficient to note that (4) :

$$A = 2\lambda d_p \tag{6}$$

$$B = 2\gamma D_g \tag{7}$$

$$C = \delta\left(\frac{k}{1+k}\right)^2 \frac{d_p^2}{D_g} + \frac{k}{(1+k)^2} \frac{d_f^2}{D_1} \tag{8}$$

$$\text{with } k = \frac{t_R - t_o}{t_o} \tag{9}$$

d_p : particle diameter

d_f : liquid phase mean thickness

D_g , D_1 diffusion coefficients of solute in gas and liquid phase respectively.

λ, γ, δ, are constants, but in fact, due to inadequacy of the model, δ must be considered as an adjustable parameter, very sensitive to impregnation and packing technology.

δ is representative of local flow rate maldistribution and, a such, not predictable although the most influent parameter in Eq.(4).

Eq.(6) to (8) stress importance of using a :

1 - small diameter of beads, d_p (usually 60-80 mesh size, which is 200-250 μ to limit pressure drop, but 120-140 mesh size which is 105-125 μ, can be used).

2 - very diffusive carrier gas, such as He or H_2. It should be noted that u and D_g arise only as a ratio. All other things being equal and neglecting term in D_1 generally small, $H(u/D_g)$ will not depend on carrier gas nature.Thus throughput of the unit will be proportional to D_g. Another important conclusion which can be reached by inspection of this equation is that H is dependent only on mass or molar flow rate of carrier gas as u/D_g can be written $up/D_g.p$ and up and D_gp are independant of local pressure p in the column. So, at same molar flow rate of carrier gas H will be the same irrespective of oulet pressure : reduced outlet pressure will not impair efficiency of the column.

3 - very well packed column. δ factor is very sensitive to packing maldistribution, particularly at long range ; statistic fluctuation at short range are rapidly smoothed and do not contribute strongly on δ. Conversely even small regular variation in porosity on all the column section are very detrimental as they induce variations in local speed and radial unevenness of the concentration fronts.

Process is strongly affected by maldistribution of packing as high δ means high H, that is a long column and high investments, but also low u(u_{opt} is equal to $(B/C)^{1/2}$, that is $\delta^{-1/2}$) and consequent reduction in throughput.

Although wall effects have been advocated to explain abnormally high values of δ for a column packed without much effort, studies show that wall-effects do not extend farther than a few bead diameter from the wall. They are, in fact, negligible in a preparative column.

Devices habe been proposed (5) to promote mixing at regular intervals in the column, but their efficiency appears limited to not well packed columns and decreases strongly when diameter increases.

Packing maldistribution appears to be considered on three different grounds :

. diameter sorting out when beads are poured into the column

. variations in local space distribution of liquide phase due to unevenly coated beads

. variations in local porosity due to inequal distribution of mechanical constraints during packing

The two first influences can be reduced to an acceptable minimum by using narrow mesh range and analytical impregnation procedure followed by remixing (in situ coating appears to induce long range unevenness in stationary phase space distribution).

δ high values in preparative chromatography are generally due to variations in local porosity. Following figure shows strong difference between local porosity variations with two different packing procedures : clearly better efficiency stems from lower overall slope of porosity... Of interest is to note periodic variation of porosity superimposed on long range linear variation which could be produced by resonance vibrations in the column during packing operation.

These differences result in wide variations in efficiency as shown on Fig. 5 and, due to high carrier gas flow rate in highly efficient column, in a factor 10 in throughput.

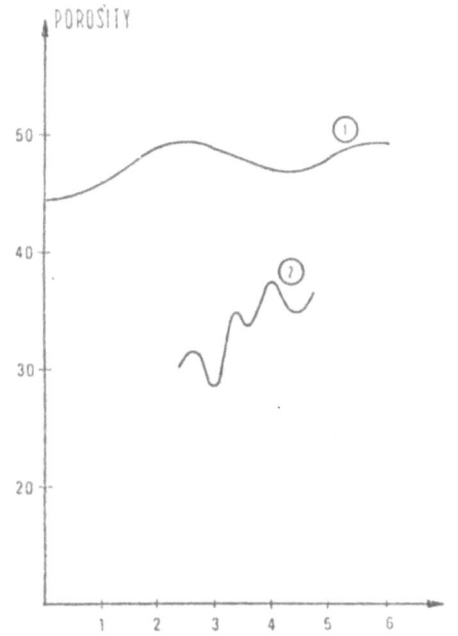

Fig.4 Porosity variations
Distance from column
axis in cm
1. good packing
2. bad packing

Fig.5 Packing methods
1. pouring
2. pouring + strong
   vibrations
3. pouring + vibrations
   + controlled shocks

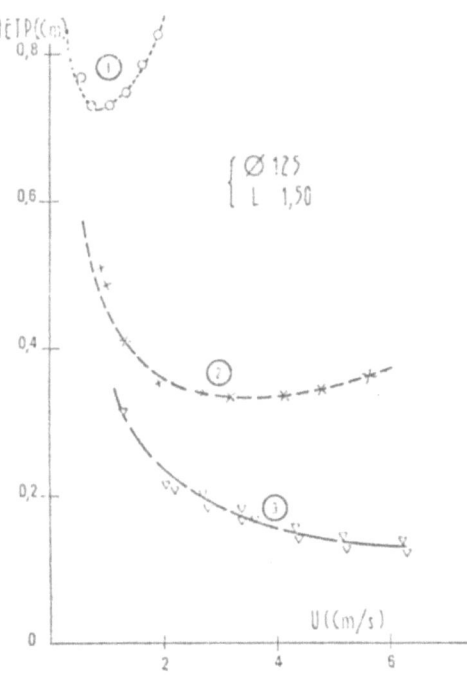

These differences would be even more drastic for larger columns, as almost same efficiency can be obtained with 12 and 40 cm
diameter columns with appropriate packing technology (fig. 6).

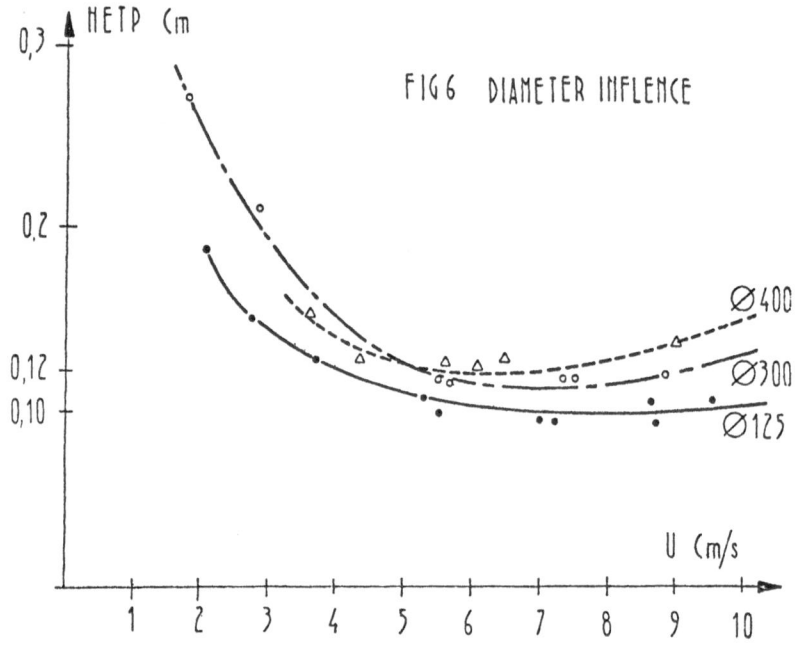

Fig.6   L = 1,5 m, chromosorb PNAW with squalan 20 %,
        solute : isopren, temperature 40°C

In contrast, with usual packing method efficiency lessens
rapidly when diameter increases.

2.4. Efficiency and throughput

We can appreciate influence of column efficiency on throughput with following example of αβ pinene separation. Fig. 7 gives
Van Deemter curves for pinenes when column is baffled and when
ELF-SRTI packing technology is used : in the former case, we find,
at best, 250 plates per meter, in the latter, 800. Consequently
throughput is much better in the latter case, as can be appreciated
from Table II.

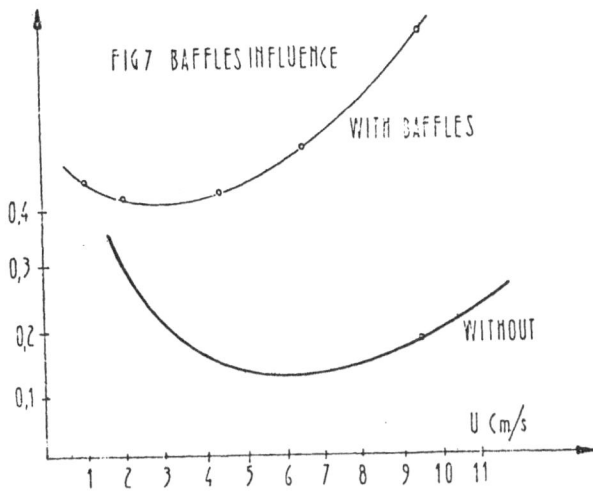

Fig. 7 Efficiency of columns for α-pinen
160°C, He carrier gas
1. with baffles, Ø 100 mm, L = 2.7 m
2. without baffles, Ø 125 mm, L = 15 m

Table II α/β pinen separation

OPERATING CONDITIONS	OTHER DATA (26)		ELF-SRTI Process	
Column diameter (mm) length (m) baffles	100 2.7 yes	305 unknown yes	125 1.5 no	400 1.5 no
Programmer	peak deflection		time-based	
Carrier gas velocity (cm/s)	He 9.2	He –	He 9.5	H_2 9
Temperature (°C)	160	160	160	160
Cycle (s)	80	–	80	65
Purity	98.5-98.6		99,1-97.8	
Productivity (kg/day)	19	160	40	540

2.5. Flooding effect

It has long been noted that large samples can be very detrimental to column efficiency. An irreversible exponential decrease in efficiency occurs (7). Such a decrease has been attributed to physical washing of stationary phase by carrier gas and sample and subsequent redeposition afterwards when liquid sample is diluted and evaporated. The resultant new distribution of stationary phase is unequal, not only in axial, but also in radial, direction and efficiency is affected. This effect is analogous to flooding which occurs in distillation, hence its name.

As a matter of fact, flooding limits not sample volume but sample injection rate by setting a maximum (that we shall call M.I.R., Maximum Injection Rate) not to be trespassed if column efficiency is to be maintained on the long run.

M.I.R. has to be calculated for every mixture to be separated and each operational conditions such as carrier gas flow rate, pressure and temperature at column inlet.

Such a calculation derives simply from the following physical model : when gaseous sample mixed with carrier gas reaches column inlet, it dissolves in stationary phase to such an extent that the resultant solution is in equilibrium with the incoming vapours. Volume of solution increases. If this volume is greater than the internal fixe volume in the beads, solution overflows and is drained away by carrier gas (fig. 8)

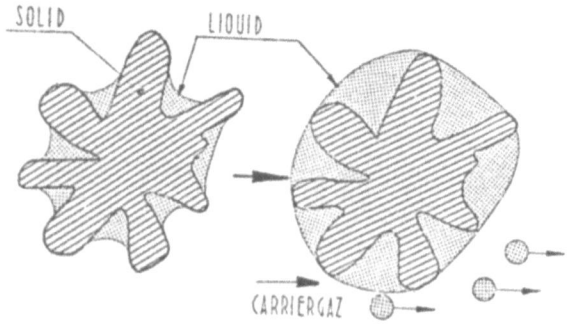

fig 8 FLOODING

If v_i^+ is internal free volume and v_s^+ volume of solution, per support mass unit :

$$\text{at M.I.R., } v_s^+ = v_i^+ \qquad (10)$$

Affecting with upper index $*$ molar quantities, we have, neglecting excess mixing volume

$$v_s^+ = n_o^+ v_o^* + n^+ v^* \qquad (11)$$

(lower index o for stationary phase, no index for solute).

This allows calculation of solute molar fraction, X_m at M.I.R :

$$\frac{X_m}{1 - X_m} = \frac{v_o^*}{v^*} \; \frac{v_i^+ - v_o^+}{v_o^+} \qquad (12)$$

In GLC, prevailing separation conditions are such that $X_m \simeq 1$:

v_o^*/v^* is at least 4 (so that phase stationary volatility be low compared to solute volatility).

$(v_i^+ - v_o^+)/v_o^+$ is about 3 for 20 % impregnation ratio.

Then, with a good approximation :

$$X_m = 1 - \frac{v_o^+ \, v^*}{(v_i^+ - v_o^+) \, v_o^*} \qquad (13)$$

In the upper part of isotherm curve, solution is nearly ideal (if no demixtion has occured), thus

$$X_m = \frac{p.Y_m}{P_o} \qquad (14)$$

Y_m : partial vapor pressure in the gas

P_o : vapor pressure of the solute

p : inlet pressure

with Y_m given by

$$Y_m = \frac{M.I.R.}{M.I.R. + F_v} \qquad (15)$$

F_v being molar flow rate of carrier gas during injection (lower than outside injection if carrier gas is pressure regulated).

Then

$$M.I.R. = F_v \frac{\dfrac{F_o}{p}}{1 - \dfrac{p_o}{p} X_m} \dfrac{X_m}{} \qquad (16)$$

A conservative value of X_m is 0,8 for chromosorb P and 20 % impregnation ratio. But for supports of low v_i^+, that is, not very porous, X_m can be drastically low and M.I.R. accordingly very small So Eq.(16) sets the utmost importance for productivity to use a very porous support to limit flooding at high injection rates.

We generally use Eq.(16) in the form :

$$M.I.R. = F_v \frac{0,8 \dfrac{p_o}{p}}{1 - 0,8 \dfrac{p_o}{p}} \qquad (17)$$

which stresses importance of getting a high p_o/p, that is, either a temperature of column about the sample boiling point at inlet pressure or, if column temperature is limited, an inlet pressure about the sample vapor pressure at column temperature.

Such a discussion points out absolute lack of interest of using a higher injector temperature than column temperature as is of common use in analytical GLC : MIR is not affected by precolumn conditions.

Eq.(17) allows also optimisation of injection to be simplified, injection rate being fixed by flooding limit. Only length of injection is left to be fixed up for maximum throughput.

2.6. Column useful lifetime

Column packing must be considered as is a catalyst, subject to remplacement after some time. When stationary phase properties are too much affected by vaporization and pyrolysis losses, column packing must be regenerated after unpacking.

We define column useful lifetime as the time necessary to halve stationary phase content.

Factor most prominent in determining useful life of a given stationary phase on a given support is temperature, with its exponential effect on vaporization and thermal degradation. In most cases, in preparative GLC, molecular weight of stationary phase is sufficiently high (≥ 1000) so that thermal degradation is the leading factor.

Each of these two factors produces different resultant distribution of stationary phase in column after some time as is
shown on Fig. 9 : vaporisation produces a sharp front in stationary phase loading ; pyrolysis induces a rather uniform decrease.
Slope of a straight line found is related to chromatography of
pyrolysis products.

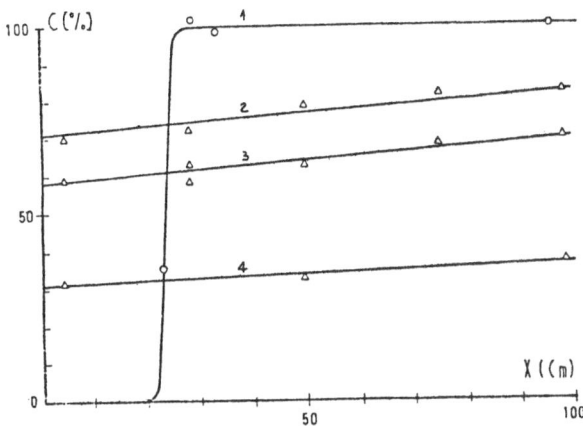

Fig. 9 - Stationary phase distribution in column after high temperature aging.
1. Squalane 10 %, Chromosorb P after 112 h. at 150° C
2.3.4. Ethylen-propylen copolymer, 10 %, chromosorb P
after 341 h. at 300° C, 34 h. at 350° C, 76 h. at 350° C

If we define now τ as ratio of stationary phase loss to stationary phase initially present, we can plot time versus column
temperature as shown on Fig. 10 and obtain useful column life at
any conditions (8).

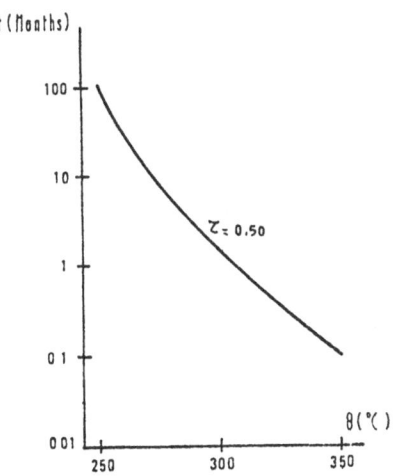

Fig 10 COLUMN LIFE

A conservative estimation of service maximal temperature is temperature which allows a two years useful life. Such a temperature can be directly read on Fig. 10. Investment costs associated with occasional overheating of the column can be also computed from there.

For instance 260°C is the service maximal temperature for a-polar phases such as ethylen-propylen copolymers.

Care should be taken that oxygen strongly affects useful life-time at any temperature, even at level as low as 10 ppm in carrier gas (8). Provision for continuous deoxygenation of carrier gas must be made to avoid rapid degradation.

3 - MODEL OF SEPARATION

We have discussed so far dispersion phenomena in the linear range, that is, in large column diameter but with very small in-jections. An elaborate discussion of this case could be possible (see for example (6)) but, has only, up to now, an academic inte-rest, due to lack of understanding of packing operations. So if we cannot derive theoretically a "good packing method", comprehensive theory describing influence of local space variables on efficiency has not its best utility.

It appears then that efficiency of column must then be considered as a technological landmark not easy to better, although work to understand packing operations and variables is clearly very desirable for the whole field of percolation processes, not only for separation but also for chemical reaction.

Our concern here will be to outline general concepts which can help operator to optimize a separation, overall technology being given. Most needed is some ideas about how to change condi-tions of a given separation to approach optimum throughput.

1. pressure drop in a GLC column is by no mean negligible : it is often around 1 atm. with a factor 2 consequent variation in carrier gas speed between inlet and outlet of the column. Thus, due to compressibility of carrier gas, it will be better to use molar fraction in gas as the primary variables and not concen-tration as is natural with incompressible carriers. Also equi-librium ratios will be taken as molar ratios rather than concen-tration ratios.

2. volume contraction due to sorption is very high as partial molar volume of a solute in gaseous and solution state are in ratio of about 200. Thus sorption phenomena will be accompanied by strong hydrodynamic effects : considering local velocity u as constant

along the column is not here a good approximation although usually made on other percolation systems where volume contraction is negligible.

3. existence of a stagnant gas phase must be taken in account for a correct definition of different velocities, although again, this fact if not often considered.

3.1. Model variables and definitions

Following figure outlines features of the model :

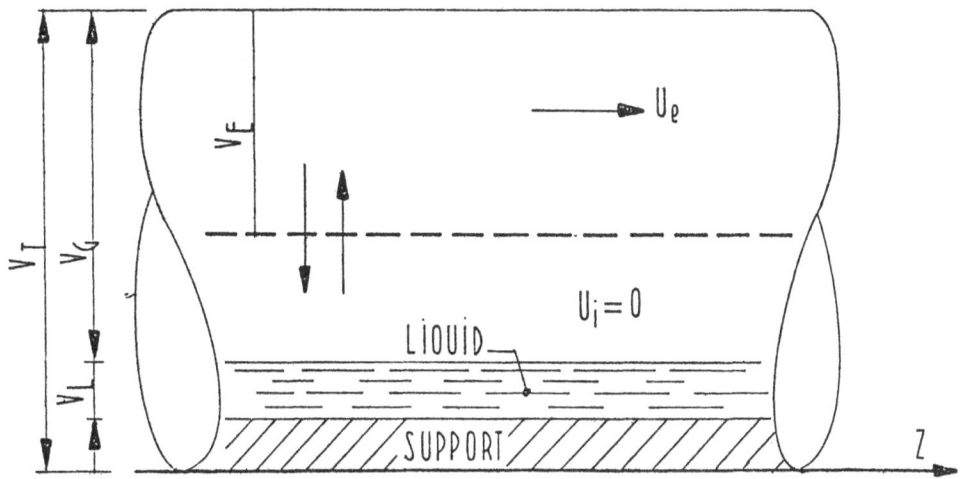

COLUMN MODEL

Diffusional exchanges between the two gas volumes and gas and liquid phases are considered instantaneous. Also axial dispersion is neglected. Column is isothermal.

V_G is total volume available to gas phase in the column. It will be assumed constant, that is liquid volume variations will be neglected.

We can write :

$$u_e = \frac{Q(z, t)}{\varepsilon_e\, S} \tag{18}$$

where Q is volumic flow rate, a function of column abscissa and time.

ε_e is external porosity ($\varepsilon_e = V_e / V_T$)

S overall column section.

We define now another velocity by :

$$u = \frac{u_e V_e}{V_G} \qquad (19)$$

which we shall identify afterwards as apparent velocity of an inert tracer.

Equilibrium between gas and liquid phase will be described by a molar partition ratio

$$K_*^i = \frac{n_L^i}{n_G^i} \qquad (20)$$

related to usual concentration partition ratio :

$$K^i = \frac{C_L^i}{C_G^i} \qquad (21)$$

by $\quad K_*^i = K^i \dfrac{V_L}{V_G} \qquad (22)$

Derivative of molar partition ratio will be extensively used

$$k_*^i = \frac{\partial n_L^i}{\partial n_G^i{}_{(T,p,n_G^{j\neq i})}} \qquad (23)$$

3.2. Molar balance equations

Local molar variations of each component, including carrier gas, are easily derived from balance in a small length dz of the column during time differential dt :

$$\frac{\partial n_m^i}{\partial t} + \frac{\partial n_s^i}{\partial t} + \frac{\partial n_L^i}{\partial t} + \frac{\partial (u_e n_m^i)}{\partial z} = 0 \qquad (24)$$

allowing for axial symmetry of column and absence of dispersion.

(m stands for mobile, s for stagnant)

Noting that $n_m^i + n_s^i = n_G^i$ and using (19), Eq. (24) can be recast as :

$$\frac{\partial n_G^i}{\partial t} + \frac{\partial n_L^i}{\partial t} + \frac{\partial (u\ n_G^i)}{\partial z} = 0. \qquad (25)$$

If local pressure p is dependent only on z, the two first terms can be rewritten as :

$$\left(\frac{\partial n_G^i}{\partial t}\right)_z \left(1 + \frac{\left(\dfrac{\partial n_L^i}{\partial t}\right)_z}{\left(\dfrac{\partial n_G^i}{\partial t}\right)_z}\right)$$

Use of (23) and of classical differential relations leads to

$$\frac{\partial n_G^i}{\partial t} (1 + \dot{k_*^i}) + \frac{\partial (u\ n_G^i)}{\partial z} = 0 \qquad (26)$$

Final form of molar balance equation stems then from equation of state of an ideal gas :

$$n_G^i = \frac{V_G\ p\ x_G^i}{RT} \qquad (27)$$

$$\frac{\partial x_G^i}{\partial t} (1 + \dot{k_*^i}) + \frac{\partial (u\ p\ x_G^i)}{\partial z} = 0 \qquad (28)$$

with p function of z only.

The system of n equations (28) has n unknown quantities x_G^1, x_G^{n-1} and u, labelling n the carrier gas.

Of course $x_G^n = 1 - \sum_1^{n-1} x_G^i$.

It should be emphasized that such a system is restricted to continuous and derivable functions. It must be completed by a system describing discontinuities breeding and propagation.

Let v_{12} be the migration velocity of a discontinuity of composition and   velocity. Dowstream values are denoted by index 1, upstream by index 2.

Again, balance on dz during dt, these two quantities being now such as :

$$v_{12} = \frac{dz}{dt} \qquad (29)$$

gives :

$$v_{12} \ (n^i_{G_1} - n^i_{G_2} + n^i_{L_1} - n^i_{L_2}) = u_1 \ n^i_{G_1} - u_2 \ n^i_{G_2} \qquad (30)$$

Taking in account Eq.(20) ; (27) and assuming that p is continuous across the discontinuity, Eq.(30) gives :

$$v_{12} \ \left[x^i_{G_1} \ (1 + K^i_{*_1}) - x^i_{G_2} \ (1 + K^i_{*_2}) \right] = u_1 \ x^i_{G_1} - u_2 \ x^i_{G_2} \qquad (31)$$

It must be noted that this equation stands for any discontinuity, its boundaries being not supposed to be constant.

3.3. Boundary conditions

Interaction of systems of equations (28) and (31) allow to solves any propagation of mass waves along column provided suitable boundary conditions are given.

Pressure is bound to be dependant on z only, with no variation during peak elution.

Mole fraction is bound to :

$$t = 0, \ x^i_G \ (z,0) = 0$$

$$0 < t \leqslant \Delta t, \ x^i_G \ (0,t) = x^i_{G0}(t) \qquad (32)$$

$$t > \Delta t, \ x^i_G \ (0,t) = 0$$

which corresponds to the injection of a band profile $x^i_{G0}(t)$ during time Δt.

3.4. General solution

System (28) is a first order partial differential quasi linear system and can be solved by method of characteristics (9). General studies on solution of such a system in the specific case of gas chromatography (although without considering pressure drop) have been completed (10,11).

We shall restrain here to discuss influence of thermodynamics quantities (T, p, mixture properties) on optimum separation (12).

To gain some insight in optimum conditions for separation without complete solution, we shall consider only a single solute with the idea that, if we can find conditions for minimum widening of the peak, they will be near separation optimum of a binary mixture.

3.5. 1-component propagation

We consider only one component (solute) injected into the column with carrier gas.

Then notation can be simplified, dropping i and G in all equations and recalling that molar fraction of carrier gas is $(1 - x)$.

Then system (28) reduces to :

$$p \frac{\partial x}{\partial t} (1 + k) + \frac{\partial (up\ x)}{\partial z} = 0$$

$$-p \frac{\partial x}{\partial t} + \frac{\partial \left[up(1-x) \right]}{\partial z} = 0$$

(33)

and after some elementary algebra, to :

$$p \frac{\partial x}{\partial t} \left[1 + k(1-x) \right] + up \frac{\partial x}{\partial z} = 0$$

$$kp \frac{\partial x}{\partial t} + \frac{\partial (up)}{\partial z} = 0$$

(34)

This system can be solved with the additional relation $p(z)$ (Darcy's Law) :

$$p^2 = P^2 - \frac{z}{L} (P^2 - 1)$$

(35)

with $P = p(0)/p(L)$, $p(L)$ being set to 1.

Features will be easier to outline if we suppose a very permeable column, that is $p = $ cste. For general case see ref. (14).

Then, we have :

$$\frac{\partial x}{\partial t} \left[1 + k(1 - x)\right] + u \frac{\partial x}{\partial z} = 0$$

$$\frac{\partial x}{\partial t} \left[k\right] + \frac{\partial u}{\partial z} = 0 \tag{36}$$

for u and x continuous functions of z and t.

For a discontinuity with downstream and upstream boundaries x_1 and x_2,

$$v_{12} \left[x_1 (1 + K_1) - x_2 (1 + K_2)\right] = u_1 x_1 - u_2 x_2$$

$$v_{12} \left[x_2 - x_1\right] = u_1 (1 - x_1) - u_2 (1 - x_2) \tag{37}$$

We discuss first solution of system (36), then stability and propagation of discontinuities (shock waves) with system (37).

Continuous solutions

System (36) has two homogeneous linear partial first order differential equations. Coefficients do not explicitely depend on variables t, z, but only on x and u. The system is then reducible and we can obtain solutions x, u fonctions of z, t or z, t functions of x, u.

x, u solution surfaces are represented by characteristic curves passing on boundary conditions.

Characteristic curves have local directions which satisfy quadratic equation

$$0 \; \zeta^2 + \left[1 + k(1-x)\right] \zeta - u = 0 \tag{39}$$

Characteristic directions are then :

$$\frac{dz}{dt} = \zeta^+ = + \infty, \quad \frac{dz}{dt} = \zeta^- = \frac{u}{1 + k(1-x)} \tag{40}$$

with variations of x and u given by :

$$\frac{dx}{du} = \xi^+ = 0 \quad \frac{dx}{du} = \xi^- = \frac{1 + k(1-x)}{ku} \tag{41}$$

From (41) we calculate, by integration of ξ^-

$$u_0 \exp \int_0^x \frac{kdx'}{1 + k(1-x')} = u \tag{42}$$

u_0 the local velocity of gaseous mixture prevailing when molar fraction of solute in the gas is nul.

From (40) we find apparent velocity of a constant molar fraction x.

$$v = \frac{u}{1 + k(1-x)} \qquad (43)$$

For small x, we find back the usual analytical result :

$$v = \frac{u_0}{1 + k} \qquad (44)$$

Comparison of these two results shows that local gaseous velocity variation has strong effect on migration velocity when x is not small. From (43), there is direct effect of u variation, but also "indirect" effect with factor (1-x) in denominator.

Of course such an analysis can be transposed to the case of two solutes, with necessity to solve a 3-equations system.

Analysis of shocks migration will give more profound insight into the phenomena, but before it will be interesting to discuss thermodynamic and sorption effects appearing in Eq. (43).

Isotherm effect

k dependance on x is called isotherm effect (on migration). Of course the first order effect of equilibrium isotherm between gas and liquid solute mixtures is to slow down migration of solute. Second, "curvature" of isotherm (in a vague meaning that we'll precise soon) produces a differential slow down on each concentration of solute with slackening of some parts of migration band and hightening of others.

We wish to relate k and thermodynamic expression of solute isotherm, namely :

$$a_G^i = \frac{p \, x_G^i}{p_o^i} = \gamma^i \, x_L^i \qquad (45)$$

Such an equation stems for equal activity of solute i in gas and liquid phase. γ is the activity coefficient, p_o^i vapor pressure of solute at column temperature and a_G^i activity of i in gas phase. From Eq. 23 :

$$k^i = \frac{RT}{p_o^i \, V_G^i} \frac{\partial n_L^i}{\partial a_G^i} \qquad (46)$$

The number of moles of solute i in solution is :

$$n_L^i = \frac{m^L}{M^L} \; \frac{x_L^i}{1 - x_L^i} \tag{47}$$

where m^L and M^L are the mass of stationary phase and its molecular weight respectively.

Differentiation of Eq.(47) and its combination with Eq.(46) gives :

$$k_*^i = \frac{RT}{p_o^i V_G} \; \frac{m^L}{M^L} \; \frac{1}{(1 - x_L^i)^2} \; \frac{d \, x_L^i}{d \, a_G^i} \tag{48}$$

a result obtained previously by Helfferich (15).

Eq.(48) can be further treated to introduce activity coefficient dependance on x_L^i :

$$k_*^i = k_o^i \; \frac{\gamma_\infty^i}{\gamma^i} \; \frac{1}{1 + x_L^i \frac{\partial \ln\gamma^i}{\partial x_L^i}} \; \frac{1}{(1 + x_L^i)^2} \tag{49}$$

where

$$k_o^i = \frac{RT}{p_o^i \gamma_\infty^i V_G} \; \frac{m^L}{M^L} \tag{50}$$

is the conventional value of k_*^i at zero concentration γ^i is a function of x_L^i (such as Margules or Wilson function) and γ_∞ is activity coefficient at infinite dilution.

A detailed discussion of k_*^i dependance on x_L^i (and, through Eq.(45), from x_G^i) has been given in (12). The most important result is that k_*^i is most often an increasing function of x_L^i and consequently of x_G^i in gas liquid chromatography. Only in a low activity coefficient situation($\gamma < 0,5$) could k_*^i be decreasing and this is not to be expected for usual systems (in gas solid chromatography such a simple result does not stay true).

From Eq.(44) we can then induce that v will be a decreasing function of x_G^i, or x.

Thus high concentration of solute will be differentially delayed leading to a blurred forward front of solute band in the hypothetical case of sorption effect being absent.

The exact expression (43) shows that such a conclusion is not correct in presence of sorption effect that we shall study now.

Sorption effect

To gain some insight in this rather subtle effect of sorption process on hydrodynamic we suppose first absence of non linear isotherm effects, that is $k_*^i = k_o^i = cste$

Then Eq.(43) can be integrated to give

$$v = u_o \frac{1+k}{[1 + k(1-x)]^2} \qquad (51)$$

an increasing function of x : thus sorption effect, in absence of isotherm effects leads to tailing band and blurred backward fronts.

We can thus conclude that sorption and isotherm effect have opposite consequences on band migration, and can then, to some extent, be "balanced" one against the other. Advantages of such a balance will be seen shortly for optimisation of separation.

Stability of concentration discontinuities

System of Eq.(37) is composed of two equations but has only one unknown, namely v_{12} . Thus v_{12} will exist only when x_1 and x_2 are linked, that is along a curve in the plane (x_1, x_2). Such an existence curve lies of course in the square limited by axises and straight lines $x_1 = 1$, $x_2 = 1$.

But in this square we can also draw stability curves, that is conditions in x_1 and x_2 for which :

$$v_{12} < v_2 \qquad (52)$$

$$v_{12} > v_1$$

separatly or together. This conditions are self-evident : v_{12} and v_1 and v_2 are velocities of discontinuity and of its boundaries. $v_{12} < v_2$ means that discontinuity will proceed slower than its upstream molar fraction. So in time course, mass coming from upstream will feed the discontinuity, stabilizing its upstream boundary.

Equations of system (52) define relations in boundaries x_1 and x_2 and consequently is represented in (x_1, x_2) plane by curves. These curves and regions delimited by them are called "stability diagram" any point of it being representative of the discontinuity with boundaries x_1 (downstream) and x_2 (upstream).

If $v_1 < v_{12} < v_2$, x_{12} discontinuity will be stable and called a shock.

When only one of these inequalities is true, discontinuity will partly stable and called a semi-shock.

If noone of these inequalities is true, discontinuity is unstable and will at once disappear or be reduced to a smaller strength shock or semi-shock, according to the case.

Following figure features stability diagram, when stability limiting curves C_1 and C_2 [equations $v_{12} = v_1$ and $v_{12} = v_2$] have arcs in the acceptable region ($x_1 \ \varepsilon \ [0, 1]$, $x_2 \ \varepsilon \ [0, 1]$).

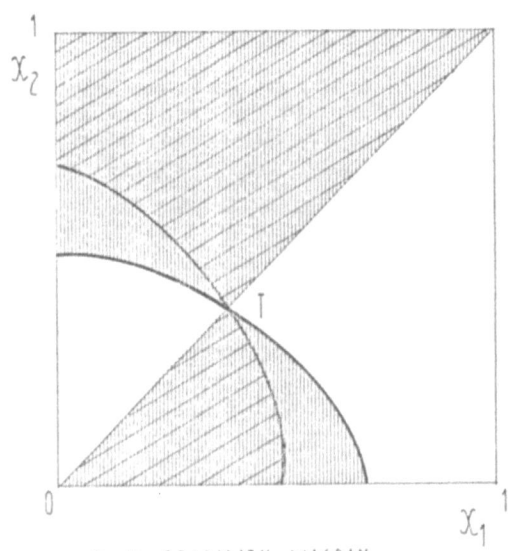

Fig 11 : STABILITY DIAGRAM

Equations of C_1 and C_2 follow steadily from Eq. (37) and (43) :

$$\frac{1 + k_1 \ (1 - x_1)}{1 + \dfrac{K_2 \ x_2 - K_1 \ x_1}{x_2 - x_1} \ (1 - x_2)} - 1 = 0 \qquad (53)$$

and a similar equation, indexes 1 and 2 being exchanged.

In fact Eq. (53) can be splitted in $x_1 - x_2 = 0$ and $f(x_1, x_2) = 0$, the latter equation being the correct equation for curves C_1 and C_2.

Exchange of 1 and 2 means that C_1 is the symmetric curve of C_2 with respect to first bisectrix. Thus if C_1 and C_2 have only one crossing point, it must be lie on first bisector also. We shall call such a point a transition point.

Transition point existence and physical meaning

Transition point location is made by setting $x_2 = x_1 + dx$ and expanding Eq. (53) in powers of dx. Vanishing term in dx will give transition point equation.

More elegantly it can be noted that appearance of a discontinuity must take place in some point of the first bisectrix, that is at a transition point. But this nascent discontinuity will be stable either at forward or backward part of concentration band

Then it must fulfill condition :

$$\frac{dv}{dx} = 0 \tag{54}$$

a very important equation for optimisation.

From Eq. (42) and (43) we derive usable expression for locating transition point :

$$2 k - (1 - x) \frac{dk}{dx} = o \tag{55}$$

which depends only on isotherm equation. Consequently, location of transition point will depend on activity coefficient expression pressure and temperature.

Recalling $a_G^i = \gamma^i x^i$ and dropping i and G letters, Eq. (55) can be rewritten with help of Eq. (48) as a non linear, second order différential equation.

$$2 p_o/p (1 - x_L) \left(\frac{da}{dx_L}\right)^2 - (1 - x) \left[2 \frac{da}{dx_L} - (1 - x_L)\frac{d^2 a}{dx_L^2}\right] = 0 \tag{56}$$

Roots of this equation are functions of (p_o/p) and T. In fact temperature does influence roots via isotherm derivatives da/dx_L and d^2a/dx_L^2 which are usually not very sensitive to T and via vapor pressure p_o which vary exponentially with T. Thus in first approximation we can consider temperature and pressure influence on transition point be condensed in the dimensionless ratio $\emptyset = p_o/p$.

An extensive study of isotherm and operating conditions influence on transition point location would be out of our present "qualitative optimization" goal. But very simple and helpful information can be gained considering limit transition point (origin in the stability diagram) and limit transition conditions $\emptyset$ associated.

For limit transition point, $x = x_1 = 0$, then (56) gives :

$$2 \, \emptyset_{::} \, \left(\frac{da}{dx_L}\right)_o^2 - \left[2 \left(\frac{da}{dx_L}\right)_o - \left(\frac{d^2 a}{dx_L^2}\right)_o \right] = 0 \qquad (57)$$

and

$$\frac{2 \left(\frac{da}{dx_L}\right)_o - \left(\frac{d^2 a}{dx_L^2}\right)_o}{2 \left(\frac{da}{dx_L}\right)_o^2} = \emptyset_x \qquad (58)$$

Experimental determination of $\emptyset_{::}$ can be made quite simply by plotting the variation of retention time of solute with sample size at various temperature and pressure as shown in (14).

Taking an account pressure drop in the column, we have derived (14) the more general formula :

$$j_3^4 \, \frac{p(L)}{P_o} = \frac{2 \left(\frac{da}{dx_L}\right)_o - \left(\frac{d^2 a}{dx_L^2}\right)_o}{2 \left(\frac{da}{dx_L}\right)_o^2} \qquad (59)$$

where

$$j_n^m = \frac{n}{m} \, \frac{P^m - 1}{P^n - 1}$$

is the EVERETT factor for pressure drop correction.

Eq. (59) has been tested in preparative ($\emptyset$ 40 mm) and analytical columns ($\emptyset$ 2 mm) (16).

In analytical columns righthand member of Eq. (59) calculated from isotherm measurement correlate well with experimental
limit transition point and also corresponding p (L)/p(o) values
are around 1 (on twenty systems studied no one gave p (L)/ɒ(o)
superior to 1.4, or inferior to 0.3. and mean was around 1).

On the other side, range of variation was more than 10 for
preparative columns with p (L)/p(o) at limit transition point
going to a maximum of 100 for methyllinolenate solute over poly-
ethylen glycol stationary phase at 190° C.

Departure between model and preparative result is probably
caused by thermal effects which can be important in large columns
when overloaded.

Present model could be extended to adiabatic operation which
is certainly more representative of sorption phenomena which occurs inside a large diameter column due to very low thermal conductivity of gases and solid support (diatomaceous earths are used
for insulation of industrial furnaces). But this extension would
necessit solution of a 3-equations system (x, u, T) which although
feasible, would be very heavy.

It can be foresaid that consideration of adiabatic thermal
effect will enhance sorption effect in the forward front, thus
leading to an optimal production temperature lower than in the
isothermal case. This has indeed been found as experimental p (L)
/p_0 for symmetry is greater than predicted or measured in analytical column and conversely, temperature at which symmetry occurs
in preparative GLC is lower (by 10 to 100° C term) than foresaid.
Of course this alleviates requirement on stationary phase thermal
stability.

Different migration types.

It is very important for the production engineer to understand from the chromatograms in which direction he should go sear-
ching for the optimum.

First it should be remembered that he must adjust two parameters : temperature and pressure. The latter is not often considered but it has been demonstrated that use of "reduced outlet
pressure"can increase throughput of a unit by a factor two (17).

Allowing the value of p_0/p, we can find different chromatograms (Fig 12).

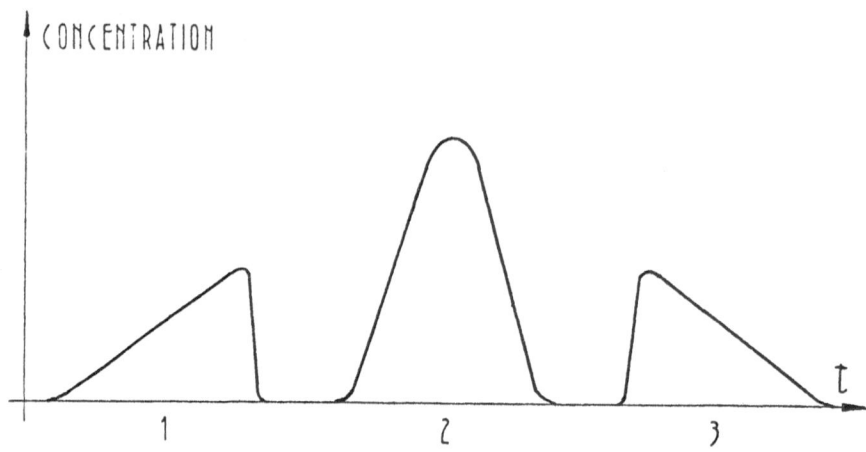

Fig 12 CHROMATOGRAMS IN OVERLOAD

These chromatograms, although blurred by dispersion in the column can be "recognised" on the figures 13 - 14 - 15 which explain different modes of migration in reference to chromatogram shape.

4 - CUT POINT OPTIMIZATION

We are now in a situation when we have selected temperature and pressure, but are left choice for cutting into elution bands with possibility of recycle of some "non separated core".

Up to now only single cut procedure has been analysed, but we have found that optimum throughput is obtained for recycling of some intermediate fraction with a two cuts procedure (Fig. 16)

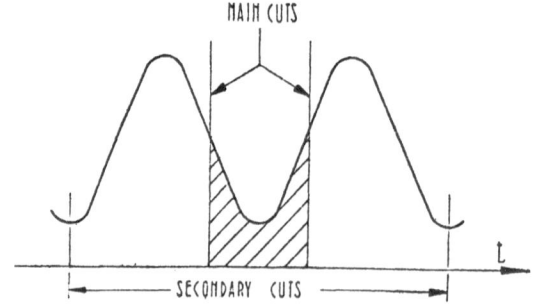

Fig16 CUTS POSITION

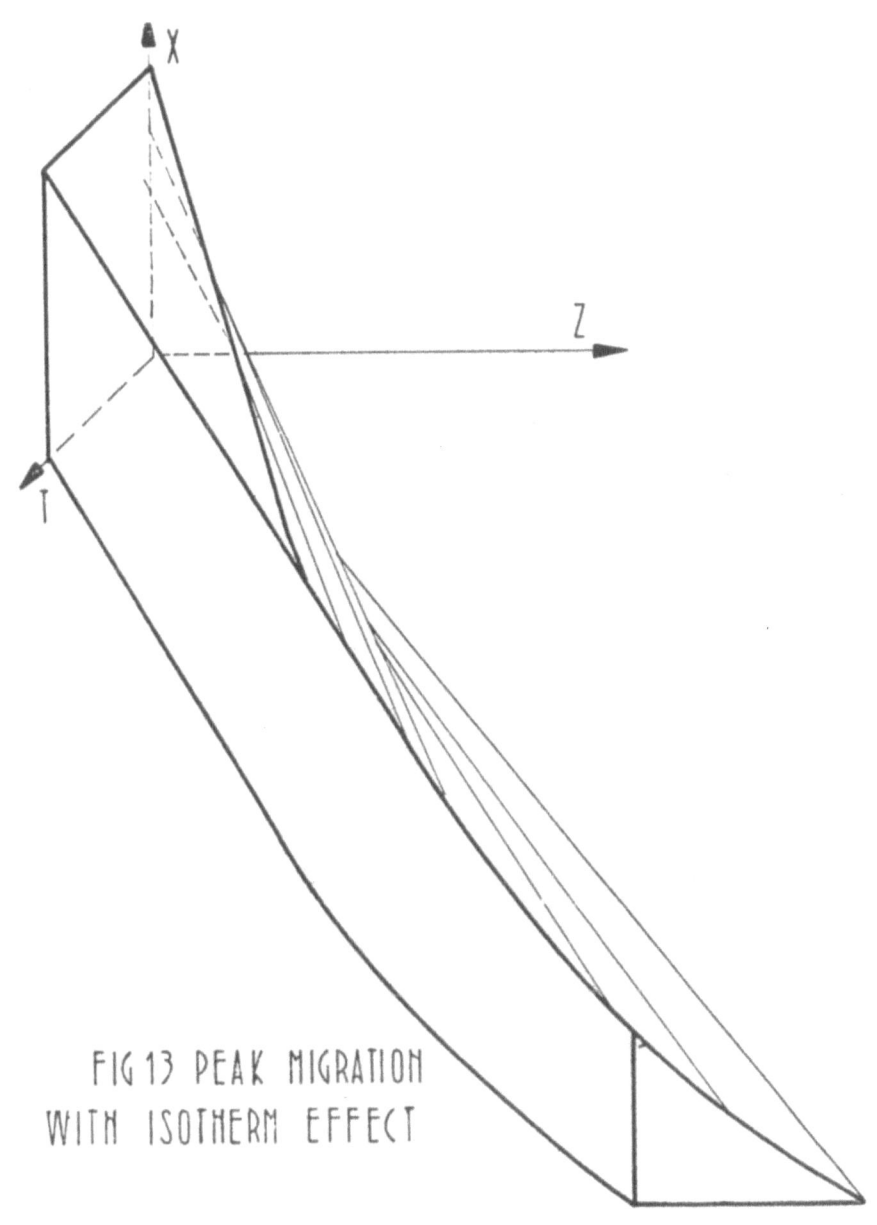

FIG 13 PEAK MIGRATION
WITH ISOTHERM EFFECT

174

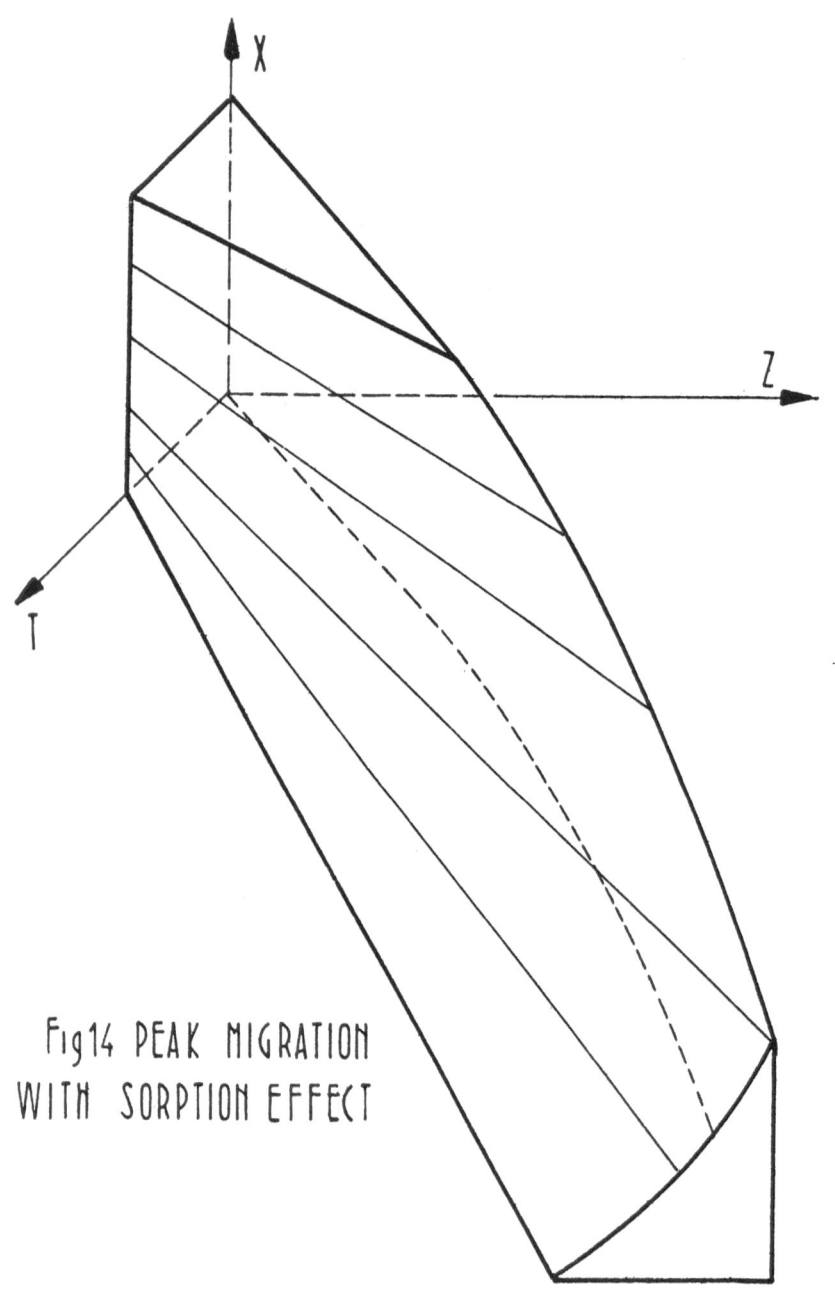

Fig14 PEAK MIGRATION
WITH SORPTION EFFECT

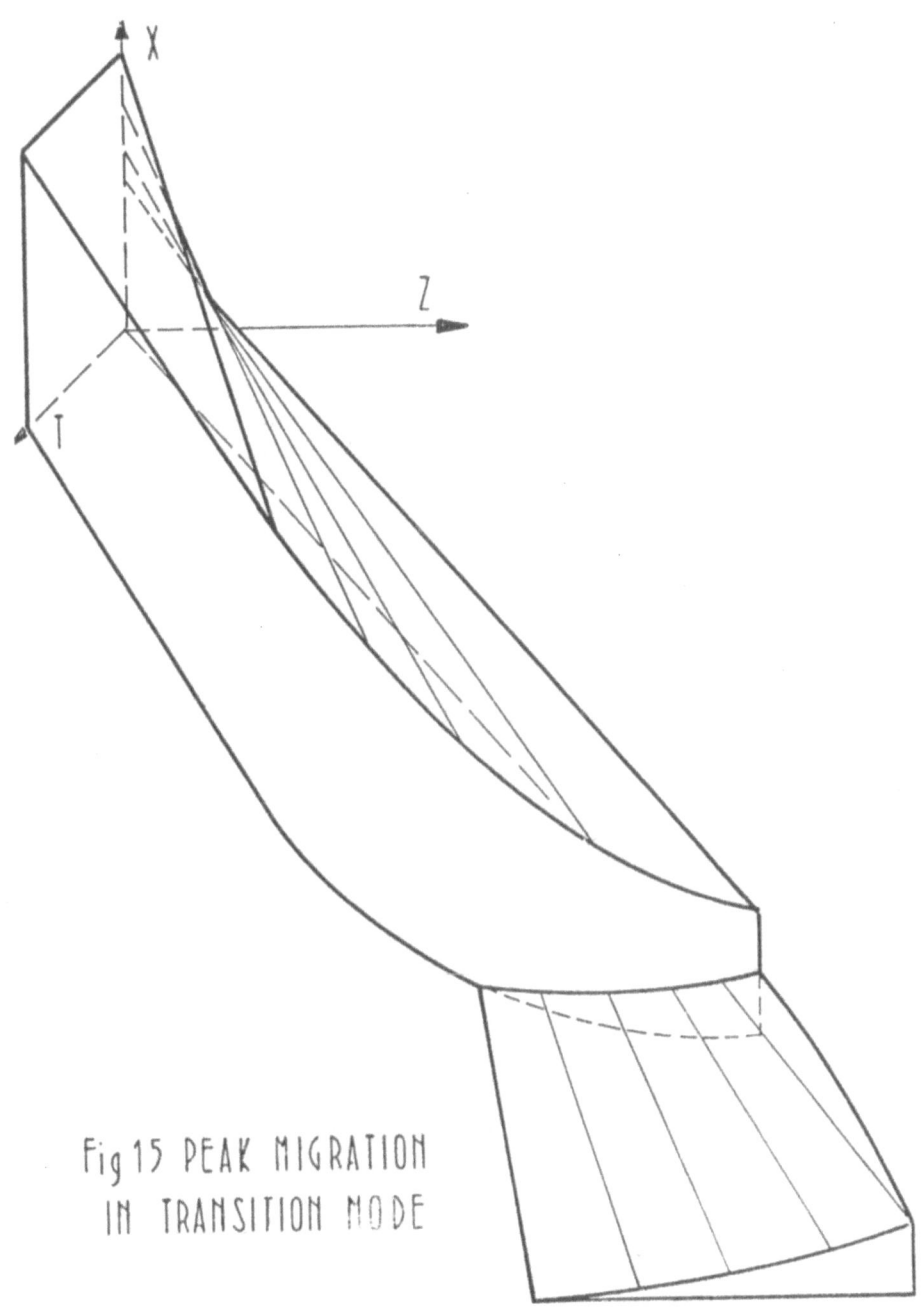

Fig 15 PEAK MIGRATION
IN TRANSITION MODE

In fact allowing for succession of cycles we have two more cuts to take in account between cycles, the secondary cuts, but those are not generally very influent on throughput.

Although all this situation could be calculated with help of our model, this would be very involved and necessitate isotherm measurement for each new separation. It appears cheaper to look directly for optimum cut-point location. But here also theoretical ideas can help to reduce experiments to a minimum. Then we shall derive first a theorem about single cut-point optimum, using the extent of separation, a new tensorial concept (18).

4.1. Tensorial extent of separation

Through a given surface S at the boundary of a separator, mass or molar flows can be known for any component of the mixture treated. If the mixture has n components and if no chemical reaction occurs in the system, at every moment the surface S is crossed by $(F^1, \ldots F^n)$ moles/second of components. It is equivalent to consider quantity (moles) having crossed the surface for a given time. Flow rate crossing the surface is then chracterised by n scalars which can be envisionned as the components of a vector in a n-dimensional vectorial space. The set of flow rates can be given such a vectorial space structure by giving him addition (physically that is mixing) and multiplication by a scalar (physically that is extrapolation).

Thus we can see now a separator as a "device" to convert a vector $\vec{F}$ to other vectors $\vec{F}_1 \ldots \vec{F}_p$ with mass balance which can be written :

$$\vec{F}_0 = \vec{F}_1 + \ldots + \vec{F}_p$$

in the steady state.

Of course we wish now to appreciate quality of the separation produced by this "device". First we note that if no separation has occured, all vectors $\vec{F}_1, \ldots, \vec{F}_p$ are colinear.

We shall restrict here to separators which give only two cuts $\vec{F}_1$ and $\vec{F}_2$. Such separators we call them 2-modules. More complete theory has been given elsewhere (18). When separation is performed, w'll define extent of the separation produced by the 2-vector (antisymmetric tensor).

$$\Xi = \vec{F}_1 \wedge \vec{F}_2$$

where $\wedge$ mean outer vectorial product

Such a definition of extent of separation is quite general, not restricted to binary separation as any other previous simila-ry parameters. Although difficult to understand by who is not fa-miliar with modern mathematics and tensor calculus, such a defi-nition has very useful properties.

One of these is that the modulus of extent of separation measures the "volume" generated by the vectors in the vectorial space : given $\vec{F}o$, it is evident that separation is associated with "opening of the umbrella" of the vectors and that the more opened, the greater the volume, better is the separation.

In case of a ternary mixture $\Xi_2 = \vec{F}_1 \wedge \vec{F}_2$ can be identified here with the vectorial product of the two vectors $\vec{F}_1$ and $\vec{F}_2$ in the usual Euclidian space : it is a vector orthogonal to these two and its length is two times area of the surface bounded by the two vectors.

Going to the simplest case of binary separation in two cuts, we have :

$$\Xi_2 = \begin{bmatrix} F_1^1 & F_2^1 \\ F_1^2 & F_2^2 \end{bmatrix} = F_1^1 \; F_2^2 - F_1^2 \; F_2^1$$

Dimension of Ξ_2 is $(\text{moles/sec})^2$.

Two related quantities can be defined :

ξ = molar extent of separation = $\dfrac{\Xi_2}{F_0^2}$ (dimension o).

ξ_0 = normed extent of separation

$\xi_0 = \dfrac{\Xi_2}{F_0^1 \; F_0^2}$ (dimension o)

F_j^i means flow of component i through the surface j.

It is easy to see that ξ_0 is nothing else that the "extent of separation" introduced by RONY in 1968, as :

$$\xi_0 = \begin{bmatrix} \dfrac{F_1^1}{F_0^1} & \dfrac{F_1^1}{F_0^2} \\ \dfrac{F_1^2}{F_0^1} & \dfrac{F_2^2}{F_0^2} \end{bmatrix} = \dfrac{1}{F_0^1 \; F_0^2} \begin{bmatrix} F_1^1 & F_2^1 \\ F_1^2 & F_2^2 \end{bmatrix}$$

Thus we have given a strong mathematical basis to the empirical mathematical index very cleverly introduced by RONY (21).

This definition generalises precedent RONY's treatment as multicomponent separation can be conveniently characterized and also, theorems of differential geometry and tensor calculus can be applied.

Although use of Ξ is not restricted to any peculiar separation process we shall use it to give demonstration of a very simple and important theorem for preparative chromatography (and all percolation processes).

A very interesting feature of Ξ is that it is zero for :

. same composition of cuts

. one cut has zero yield

Thus Ξ is zero for the two classical limits of functionment of a separator (minimal reflux and total reflux). So, there must be, at least, an extremum for Ξ in the rang of useful separations.

It can demonstrated than, for the binary separation Ξ measures the molar flow rate of exchange of 1 and 2 component between the flows.

Another important property of Ξ is that it is an intrinsec property : changing "name" of components affects only the reference frame in which Ξ is imbedded but not the geometric properties of Ξ.

Theorem : in a chromatographic 2-module, optimal cut-point is when outlet flow rate has the same composition as the feed.

Proof :

A chromatographic 2-module has only one inlet and one outlet further splitted by a three way valve timer-actuated.

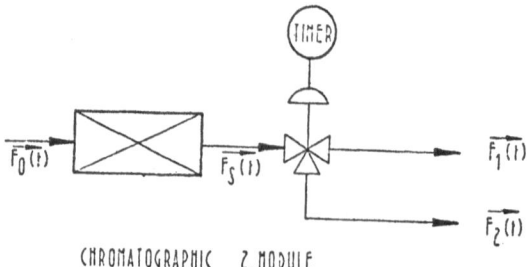

CHROMATOGRAPHIC 2-MODULE

A non zero separation can only be obtained in periodic, non-permanent operation. Principle of separation is selective accumulation of some components in the systeme, followed by depletion. Another name could be percolation 2-module.

Suppose the 2-module has the production period T we wish to know when valve must be actuated to get "the best performance". Here we choose to maximize extent of separation , that is to maximize the quantity of 1 and 2 transfered between the two cuts (totally mixed afterwards).

$\vec{F}_0$ (t) and $\vec{F}s$ (t) are fixed. $\int_0^T \vec{F}_0$ dt $= \vec{N}_o$ is the total amount introduced per cycle.

$\vec{N}_1$ and $\vec{N}_2$ are the amount of the two cuts produced (after mixing).

ζ is the reduced time, $= t/T \in [0,1]$, ζ' the cut-time,

$$\Xi_2 = \vec{N}_1 \wedge \vec{N}_2 \ [(moles)^2]$$

$$= \int_0^{\zeta'} \vec{F}_s \ d\zeta \wedge \int_{\zeta'}^1 \vec{F}_s \ d\zeta$$

$$= \int_0^{\zeta'} \vec{F}_s \ d\zeta \wedge \vec{N}_o$$

from $\vec{N}_0 = \vec{N}_1 + \vec{N}_2$ and properties of vectorial product.

$\vec{N}_0$ is a constant vector. Then by differentiation with respect to ζ', we can locate the optimum of Ξ .

$$\left(\frac{d\Xi}{d\zeta'}\right)_{opt} = 0 = \vec{F}s \wedge \vec{N}_o$$

This means that $\vec{F}s$ (ζ') and $\vec{N}_o$ are colinear, thus they have same composition : valve must be actuated when flow at the outlet of the 2- module has same composition as the feed.

This theorem is very general : it does depend of neither number of components nor time variation of outlet composition.

Intuitive demonstration is self-evident : as soon as composition "crosses" the feed one, we looses by mixing some of the separation accomplished if we don't actuate the valve.

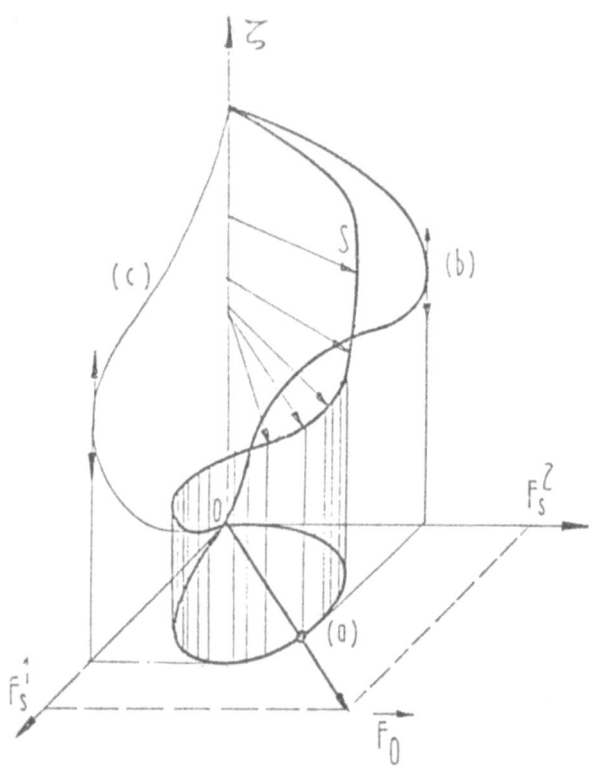

Fig 17 OUTLET FLOW VECTEUR TIME
VARIATION

FIG. 17 is a perspective drawing of vector $\vec{F}_s$ time evolution at column outlet. Extremity of this vector describes an helix traced on cylinder of vertical axis with curve (a) as projection on ground plane (F_s^1, F_s^2). Ξ is maximum for cut-point located at time when ground projection of $\vec{F}_s$ is crossing $\vec{F}_o$ projection.

Curves (b) and (c) are projection of helix on planes $(F_s^1 \ 0 \ t)$ and $F_s^2 \ 0 \ t)$. They would be the chromatograms obtained by detectors sensitive only to 1 or 2 components.

4.2. Double-cut optimization

The previous theorem cannot be used as such when we recycle some part of that sufficiently separated outlet mixture, But this suggest to locate the two cut-points such as the recycled part has same composition as the feed. As a rule this avoids undue remixing in the recycle and is thus probably not far from optimum double-cut point location,

Of course recycle will lower throughput by a factor $(1 - r)$ where r is recycled fraction but it will allow longer injection as separation does not used to be very good. Conversely, at given injected quantity, recycling will increase yield by a $1/(1-r)$ factor, which is very incentive to recycling.

5 - COMPARISON BETWEEN CHROMATOGRAPHY AND COUNTER CURRENT PROCESSES

A detailed comparison based on economical ground, is out of our present scope. But we wish here to outline caracteristic differences between these type of processes,

The first problem arising is the following : consider a mixture of equal parts of two components i and j to be separated as 99 % pure cuts. If relative volatility of these components is 1.10 we find that about 200 theoretical plates are necessary in counter-current processes and 2500 in chromatography.

Such a paradoxical inflation of plate number between these two types of processes, is often interpreted as arising from a different meaning of "theoretical plate" concept (19,20). W'll investigate this assumption later.

The second problem arising is fundamental : although two processes can be said equivalent if they give same separation they can behave very differently in certain cases : operating under transient conditions, with secondary chemical reaction etc, all phenomena which can be of outmost importance for choice of a process.

Thus we have first to consider stationary separation rate of a process, but also some other variable which could reflect operating conditions of process in practice.

5.1. The concept of HETP

Tiley (19) and his coworkers showed experimentally that operating in counter-current mode or in chromatographic mode does not have much influence on the HETP value. They measured the HETP of the same packing on a moving bed and on a fixed bed. In the first case it was 15 mm and in the second case 5 mm. The difference between these two values is probably due to a better uniformity of the bed that is obtained as soon as its motion is stopped. In any case, this ratio 3 far from explains the ratio 12 between the number of plates necessary in chromatography and in distillation in the case under review.

This difference is often explained by the definition of "effective theoretical plates" calculated under overload conditions (high injection rate) in chromatographic columns. As the peaks are much wider, the apparent number of theoretical plates of the column drops a lot.

But although this reasoning reaches the desired result, it does not have any sense, because, in actuel fact, one causes to intervene in the HETP thus calculated a number of phenomena which are not related to it, such as the non-linearity of the sorption isotherm, and these tend to widen the signals in the column.

Such a reasoning is just as incorrect as that which would consist in calculating the HETP of an extraction column, assuming constant separation coefficients, whereas they are not constant. (We know that MAC CABE AND THIELE's construction in distillation permits taking account of the shape of the balance curve for calculating the necessary number of theoretical plates).

We must therefore conclude that the HETP is a concept that is valid in both classes of process and that "chromatographic inflation" is due to something else than a different meaning of the HETP in these two classes of process.

We now have to find out what is the "forgotten variable" of the problem. To do this, we shall come back to the definition of a separation and deal with the "residence time" of the substances in the system.

5.2. The concept of residence time

Let us define more precisely what this "residence time" is. If we inject a pulse of a tracer of i, for example an isotope, we can define the average residence time of i as the time of transfer from the center of gravity of the isotope's fictive output curve, which one would obtain by mixing the flow at the top and at the foot of the equipment (after reflux and reboiling), so as to only have one input and one output in the system.

This average residence time is equal to the quotient, under

permanent conditions, of the quantity of i present in the system
to the flow of i entering (and leaving) the system :

$$\bar{t}^i = \frac{(\text{hold-up of } i)}{(\text{flow of } i)} \tag{72}$$

(The definition of "partial" residence times of i in various
parts of the equipment would be far more difficult, considering
the existence of counter-currents which lead to the existence of
two inlets and outlets in each part of the equipment) (10). In
chromatography, the residence time will be defined by sending
(without backmixing) the flows successively collected to the detector. One can also, obviously, place the detector directly at the
outlet of the column.

This is the way we define a residence time per component to
be separated.

5.3. Residence time in chromatography

A classical result in linear chromatography is :

$$\bar{t}^i = \frac{L\ (1 + k^i)}{u} \tag{73}$$

from what we can deduce :

$$\bar{t}^i = \frac{H\ (1 + k^i)}{u}\ N \tag{74}$$

with N, plate number of the column.

5.4. Residence time in counter current processes

We need here to calculate inventory of component i in the separator, a problem not solved on general grounds. But if we consider a gas-liquid or a two solvent liquid/liquid extraction column
we can derive simple results (22) :

$$\bar{t}^i = \frac{H\ (1 + k^i)}{u_G}\ \frac{(N_+\ \psi_+^i - N_-\ \psi_-^i)}{1 - r^i} \tag{75}$$

where $r^i = k^i \dfrac{u_L}{u_G}$ is the internal reflux, N is a number of theoretical plates and ψ^i is fraction of segregation of i.

184

Such an expression holds only when there is no external reflux at top+ or bottom- of column but could be modified to make
allowance of external reflux.

5.5. Efficiency factor and number of contacts

Eq. (74) and (75) are very similar. We can split all of them

a technological factor, the "efficiency factor" of the process

$$E^i = \frac{u}{H (1 + k^i)} \quad \text{(theoretical plates/hour)} \quad (76)$$

a theoretical factor, the "number of contacts"

$$f = f (N, r)$$

depending only on process type, reflux, plates number and separation to be performed.

5.6. Efficiency factor

For difficult separation $E^i \neq E^j$ as $k^i \neq k^j$ and we can choose
any component to evaluate efficiency factor of the process.

Multiplying top and bottom of Eq. (76) per section available
to gas phase in column, we find :

$$E^i = \frac{\text{rate of flow of i in gas phase}}{\text{quantity of i in gas and liquid phase on a height equivalent to a theoretical plate}} \quad (78)$$

Thus $1/E^i$ is the "residence time per plate".

Eq. (78) takes in account hold-up per plate and flow rates
it is thus representative of technological performance of contacting device, this is why we label it "efficiency factor" of the
process ; it will increase with troughput (flow rate) and decrease with hold-up in plate.

We can trace back such a concept as early as 1940 (23,24) to
compare different distillation technologies. Also $1/E^1$ is of current use in uranium isotopes separation field.

Following table shows efficiency factors of different processes :

T A B L E   III

EFFICIENCY FACTOR	DISTILLATION AND COUNTER CURRENT	CHROMATOGRAPHY
pL/hour	Industrial 30-200	20,000
	Small diameter knitted packings 500	100,000

Chromatography appears to be by far the process with the best "efficiency factor".

The reason for this ratio greater than 100 between these two types of processes is mainly due to the very small thickness of the layer of stationary liquid in chromatography : a few microns. On the other hand, in the counter-current processes, the liquid has to have a sufficient thickness to flow (a few tenths of a millimeter).

5.7. Number of contacts

In chromatography, $f(N,r) = N$ according to Eq.(74)

We can generalize this result in the case when we recycle a fraction r of mixture leaving the column :

$$f(N, r) = \frac{N}{1 - r} \qquad (79)$$

This result compares strikingly with counter current number of contacts :

$$f = \frac{N_+}{1 - r^i} \qquad (80)$$

when separation is $(\psi^i_+ = 1, \psi^i_- \simeq 0)$

Intuitive explanation for such a simple result is that
$f = \overline{t}^i E^i = \frac{N}{1 - r}$ represents the number of times a given molecule will go through a theoretical plate and suffer equilibrium. As

such, f is plates number N divided by (1 - r).

5.8. Existence of a minimum of residence time

Residence time is proportional to number of contact, for a given process, with efficiency factor as the scale factor.

We must now consider variation of f for a given separation in a given process with plate number (a variation of the usual procedure which considers N vs r).

A detailed comparison of N vs f relations in chromatography and counter-current separators is available elsewhere (25). We shall only give results for 50 % binary mixture to be splitted into two 99 % pure cuts, when α is 1,10 or 1,30 (fig. 18 and 19)

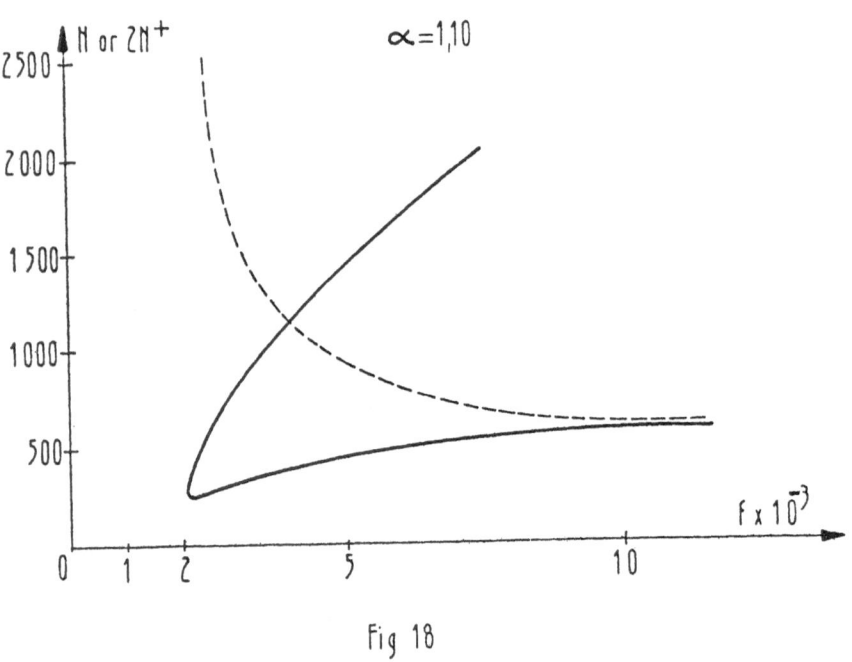

Fig 18

Number of contacts in chromatography (dotted curve) and counter-current (continuous curve).

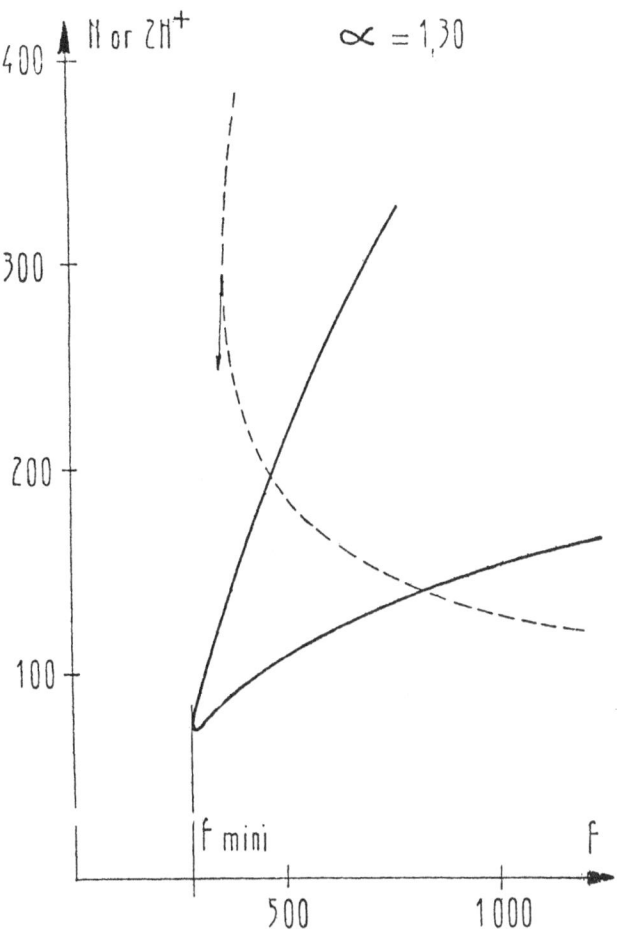

FIG 19 NUMBER OF CONTACTS IN
CHOMATOGRAPHY AND COUNTER-CURRENT

What is to note here is existence of a minimum of f, along with the known minimum in N.

It will be also noted that the value of these minima in f are about the same in chromatography and in counter-current techniques, although the values of the corresponding numbers of plates are very different (TABLE IV).

TABLE IV - Values of the minimums of number of contacts

	COUNTER CURRENT		CHROMATOGRAPHY	
	2 N + (1)	f	N	f
1.10	210	2 000	2 500	2 500
1.30	70	250	250	350

In the selected numerical example, the comparison of chromatography and the counter-current processes, each at its optimum value (minimum residence time) gives the following results :

Assuming efficiency factors of 500 and 50,000 plates per hour for the counter-current and the chromatography, the residence times are respectively four hours and three minutes if α equals 1.10. If α equals 1.30, the residence times are then 30 minutes and 25 seconds.

The minimum residence time, which depends mainly on the process efficiency factor, increases with the difficulty of the separation to be performed.

This time is very short in chromatography, but this gain is acquired in counter-part for a much larger number of plates.

This is not in fact an inconvenience since 2500 plates require a length of about 3 meters of column, whereas 210 plates (if we admit that the counter-current column is fed in its middle point) would require a column so high that is doubtful wether such a column can be constructed and operated.

To sum up, what matters for the quality of a separation is the number of times a molecule is subject to the equilibrium between 2 phases in the system, that is to say the number of theoretical plates of the system multiplied by the number of times a molecule is recycled or "refluxed", that is to say $N(1-r)$.

In chromatography without recycling, one needs a much higher number of theoretical plates than in the counter-current processes where the reflux is always relatively high (about 1).

This is owed merely to the fact that each molecule is only

subject to one balance per theoretical plate . But on the other hand, the very fact that the liquid phase is stationary permits the arranging of it with more regularity and specially, over a smaller thickness than when it is moving.

Whence a much higher efficiency factor in chromatography than in distillation and in the counter-current processes. Inversely, the residence time in a chromatographic column is 50 to 100 times shorter than in a counter-current column producing the same separation.

This has two consequences :

1 - The "time constant" of the chromatographic system when the installation is started or when its operation is changed, is about a few minutes. Starting is therefore much faster than with the counter-current processes.

2 - When there is a chemical reaction (polymerisation, decomposition), the progress of this reaction is, in chromatography, much slower since the residence time is much shorter. When the chemical reaction might become explosive, this factor is of particular interest.

In this case, dilution of the substances in the carrier gas and also the presence of a fine porous medium preventing the propagation of the shock waves, are also non negligible advantage.

From this we can derive following fields for application of Industrial Gas - Liquid Chromatography (TABLE V). See also (26,27)

TABLE V - Fields of Application of Gas Liquid Chromatography

N A T U R E	TO BE USED IF	FIELD OF APPLICATION
Quick start	Frequent change of separation	Pure substances Development of new substances
High residence time (high recycling)	Very difficult separations $(\alpha < 1.10)$	Isotopes Optical isomers
Low residence time (no recycling)	Fairly easy separations	Perfumes, tastes Heterocyclic intermediaries
	$(\alpha > 1.10)$ or unstable substances	Very pure monomers Explosive substances

SYNTAX OF LITTERAL SYMBOLS

Indexes and superscripts

Common notation rules are to place species or component identification letter in index position, as for exemple C_i for concentration of component i in a mixture. Location indicators (primes, asterisks, superscripts) are conversely generally set in exponent position.

Although such a syntax appears quite satisfying in absolute we have choosen the converse convention, unnatural at first glance : indexes are for location and exponents for species indication. In fact, use of tensorial notation leads unambiguously to the latter convention.

We can define a vectorial space of flows whose components of vectors are the n flow rates crossing a given surface, traditionnally written in mathematics with superscript as $(F^1, \ldots F^i, \ldots F^n)$. Such a flow vector can then be noted down :

$$\vec{F} = \sum_i F^i \, \vec{e}_i$$

with $\vec{e}_i$, unit vector $(0, \ldots 1, \ldots 0)$ associated to species i.

This notation allows use of EINSTEIN summation convention :

$$\vec{F} = F^i \, \vec{e}_i$$

(automatic summation on repeated index and superscripts, that is, in covariant and contravariant position, respectively).

For thermodynamic grounds also the proposed convention is desirable as formulas such as :

$$G = \sum_i n^i \, \mu_i$$

used in this fields are coherent with the latter convention and not with the former (according to tensorial calculus, μ_i being a partial derivate of G relatively to n^i must be written

$$\mu_i = \frac{\partial G}{\partial n^i} \quad \text{and not } \mu^i).$$

Thus proposed notation is F^i_j for a flow of F moles per second of species labelled i crossing the surface labelled j.

(use of such a notation is not limited to percolation processes but stands for all separation science and also chemical reaction science).

TABLE OF SYMBOLS

SYMBOL	QUANTITY	DIMENSION
a	Thermodynamic activity	
A	Constant in Van Deemter Eq .	cm
B	Constant in Van Deemter Eq.	$cm^2/sec.$
C	Constant in Van Deemter Eq . Concentration	$moles/cm^3$
d	Diameter	cm
D	Diffusion coefficient	$cm^2/sec.$
E	Efficiency factor	TP/hour
f	Number of contacts	
F	Molar flow rate	moles/sec.
G	Free enthalpy	cal
H	Height equivalent to a theoretical plate	cm
k	Molar sorption coefficient (differential)	
K	Sorption coefficient (integral)	
L	Column lenth	cm
m	Mass	g
M	Molar weight	g/mole
n, N	Number of moles	moles
N	Theorical plates number	
p	Local pressure	atm.

SYMBOL	QUANTITY	DIMENSION
P	Relative inlet pressure $p(o)/p(L)$	
Q	Volumic flow rate	cm^3/sec
r	Reflux or recycle ratio	
R	Ideal gas constant	$cal/°K$
S	Overall column section	cm^2
t	Time	sec
T	Temperature	$°K$
u	Velocity of an inlet tracer	cm/sec
v	Velocity of a constant molar fraction	cm/sec
	Or volume per ...	$cm^3/...$
V	Volume	
x	Molar fraction	
y	Molar fraction	
z	Abscissa along column	cm
γ	Activity coefficient	
ε	External porosity	
Λ	Exterior product of two vectors	
μ	Mean or free enthalpy per mole	cal/mole
ζ	Reduced time	
Ξ	Tensorial extent of separation	
ρ	Density	g/cm^3
σ	Variance	

SYMBOL	QUANTITY	DIMENSION
ψ	Molar yield of separation	
$\emptyset$	p_o/p at limit transition point	

SUPERSCRIPTS

i, j	Component i or j

INDEXES

0	Pure or at $x = 0$ or at $z = 0$	
∞	Infinite dilution	
m	Mobile phase	
s	Stagnant phase or at column outlet	
L	Liquid phase, stationary phase	
e	External	
G	Gas phase	
::	Per mole unit	$mole^{-1}$
+	Per mass unit	g^{-1}
T	Total	
1	Downstream (Shock boundary)	
2	Upstream (Shock boundary)	
$\pm$	The two extremities of separation column	

..

REFERENCES

(1) TARAMASSO, M. SAKODYNSKI, K. I.,
    Ouspekhi Kromatograffi, Naouka ed. 1972, P. 252

(2) BONMATI, R.,GUIOCHON,G., Parfums, cosmétiques, arômes
    11 37 (1976)

(3) LITTLEWOOD, A. G., Gas Chromatography, Academic Press,
    N. Y. 1962 p. 192

(4) ZLATKIS, A., PRETORIUS, V.,
    Preparative Gas Chromatography, ed. Wiley, 1971

(5) BADDOUR, R. F., U. S. Pat n° 3 250 058 (Mai 10, 1966)

(6) GIDDINGS, J. C., Dynamics of Chromatography
    Chromatographic Science Series, Dekker (1966)

(7) MIKKELSEN, L, DEBBRECHT. F. J., MARTIN A. J.,
    Gas Chromatography 4 263 (1966)

(8) THIZON, M., EON G., VALENTIN P., GUIOCHON G.,
    Anal. Chem. 48 1861 (1976)

(9) ARIS, R., AMUNDSON N. R., Mathematical Methods
    in Chemical Engineering, vol. 2, Prentice Hall (1973)

(10) JACOB, L., GUIOCHON, G.,
     J. Chim. Phys. 67 185 (1970)

(11) JACOB, L., GUIOCHON, G.,
     J. Chim. Phys. 67 291 (1970)

(12) VALENTIN P., GUIOCHON G., Sep. Sci. 10 245 (1975)

(13) VALENTIN, P., GUIOCHON G., Sep. Sci. 10 271 (1975)

(14) VALENTIN, P., GUIOCHON, G., Sep. Sci. 10 289 (1975)

(15) PETERSON, D. L., HELFFERICH, F., J. Phys. Chem.
     69 1283 (1965)

(16) LADURELLI, A., Thesis
     Paris, June 1976

(17) VALENTIN, P., International chromatography Meeting
     Barcelone, 1974,

(18) VALENTIN, P. Separation Engineering, Course at

Ecole Polytechnique, Palaiseau, France (1977)

(19) PRITCHARD, D. W., PROBERT, M. B., TILEY, P. F.,
Chem. Eng. Sci. 26 2063 (1971)

(20) RONY, P. R., Sep. Sci. 5 121 (1970)

(21) RONY, P. R., Sep. Sci. 3 239 (1968)

(22) ALDERS, L., Liquid-liquid extraction pp
149 - 155 Elsevier, 1959 (2 nd ed)

(23) BRAGG, L. B., A. I. Ch. E, New Orleans Meeting
2-4 dec. 1940

(24) PODBIELNIAK W. J., Ind. and Eng. Chem.
13 699 (1941)

(25) VALENTIN, P., ACS, 166th Meeting, Chicago
Sept. 1974

(26) GUIOCHON, G., BONMATI, R.,
Parf., Cosm., Arômes 11 37 (1976)

(27) GUIOCHON, G., BONMATI, R.,
VII Int. Congress of essential oils , KYOTO Japan 1977

(28) RYAN, J. M., DIENES , G. L., Drug. Cosm.
Ind. 99 60 (1966).

ENERGETICS AND COST OPTIMIZATION OF PREPARATIVE CHROMATOGRAPHY
COLUMNS

P. Le Goff and N. Midoux

Laboratoire des Sciences du Génie Chimique, CNRS-ENSIC,
Nancy, France

SUMMARY AND CONCLUSIONS

A binary mixture is to be separated into its components by passage through a chromatographic column.

For a given mass flowrate of the mixture, a given maximum impurity level in each effluent, and a given porous material as the
separating agent, the problem is to determine the optimum values
of the 3 parameters :

- the cross-sectional area of the column : Ω
- the length of the column : L
- the mass flowrate of the carrier fluid : $\dot{M}$

Optimization will be achieved by minimising one or the other of
the four following parameters :

-TECHNICAL parameters to be minimised :
  1. The volume of the column : $V = \Omega L$
  2. The mechanical power degraded by the pressure drop : $\dot{E}$

-ECONOMIC parameters to be minimised :
  3. The investment, that is the cost of buying and instal-
     ling the packed column and its ancillary equipment : C_{cp}
  4. The total cost of the separation, which is the sum of
     the investment and the operating costs over a given period

In each case, a double infinity of solutions will be obtained,
each solution being characterized by the values of two dimensionless variables : θ and B :
     θ is the relative injection time, that is the fraction of

time used for injection of the mixture. θ can vary between
0 for elution chromatography and 0.5 for eluto-frontal
chromatography.
B is the BODENSTEIN number : $B = ud_p/D$ (= particle PECLET
number) that is a dimensionless expression of the fluid
flowrate.

The results of the optimization procedures may be summarized as
follows:

1. The optimum value of θ, when minimizing technical parameters,
is 1/6 for imcompressible fluid (liquid chromatography), and
can vary between 1/8 and 1/4 for compressible fluids (gas chromatography).

2. The optimum value of θ, when minimizing economic parameters
should be always equal to or a little less than 1/6, for incompressible as well as for compressible fluids.

   The optimum value of θ is therefore practically independent
of all the parameters, whether scientific, technical, or even
economic.

3. The optimum value of B is zero, when minimizing the degradation of energy (or the operating costs). It follows that the
column should have as large as possible a cross-sectional area.

4. On the other hand, the optimum value of B is found to be
infinite, when the column volume for an imcompressible fluid is
minimised. It follows that the cross-sectional are should be as
small as possible. In cases involving compressible fluids, the
optimum value of B is a slightly greater (40%) than the value
corresponding to the minimum "Heigth of a Theoretical Stage".

5. Minimising the investment for the optimum value of B will not
give an overall conclusion usable for any case. But in the
special case of gas chromatography with very high pressure drop,
it is shown that the optimum value of B is of the order of 15% to
40% greater than the value corresponding to the minimum "Heigth
of a Theoretical Stage".

6. Minimising the total cost in the most general case, does not
give an analytical solution providing overall conclusions but
the equations to be solved numerically are not complicated.

Two further publications will give examples of application of the
present model that is the separation of phenotiazines by liquid
chromatography and the separation of α-β pinenes by gas
chromatography.

Contents
Introduction and presentation of the problem

INTRODUCTION

An engineer required to design and build a new plant for separating a molecular mixture into components of a given purity, generally aims to minimise the total cost of the project, that is the weighted sum of the operating costs and the construction costs taking into account the depreciation of the invested capital over the working life of the plant.

The operating costs mainly come from the amount of energy required by the plant to perform the separation.

In previous papers (ref. 1, 2, 3, 4, 5, 6, 7) we have made a quantitative study of the total cost of industrial processes as a function of the flux of energy (in kwh/year) which is degraded and we have proposed general method for the energetic optimisation of these processes.

The object of this paper is to apply this method to the optimization of a certain type of separation method namely preparative chromatography columns.

PRESENTATION OF THE PROBLEM

Let us consider a long packed column of length L and cross-sectional area Ω through which an inert carrier fluid flows at a

constant mass flowrate $\dot{M}$.
- A binary mixture is to be separated into its components A and B by being injected periodically as bands.
- The peaks coming out of the column, alternatively enriched in A and B are directed to two different outlets.

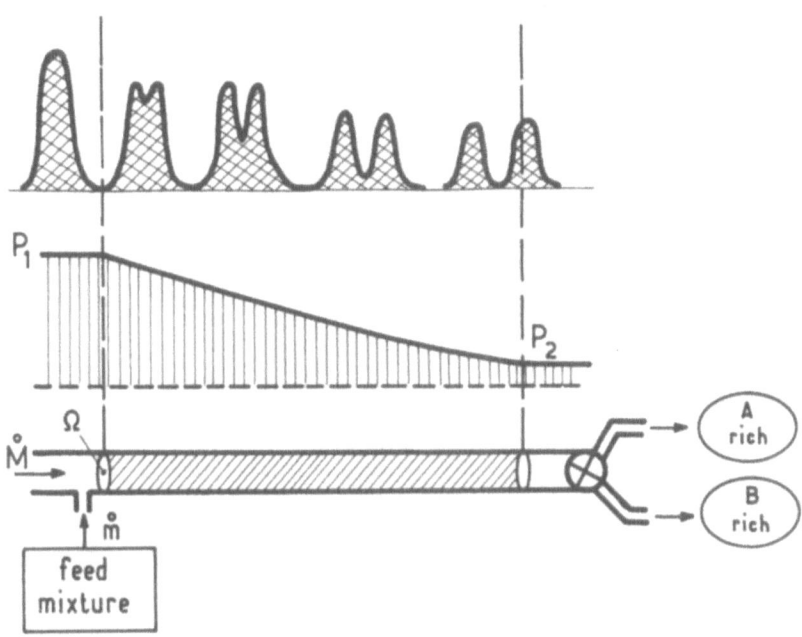

Figure 1

Let $\dot{m}$ be the average mass flowrate of the mixture to be separated. In addition we fix a maximum impurity level in each of the two separated effluents.

The problem is to determine the optimum values of the 3 parameters :
- The cross sectional area Ω and the length L of the column
- The flowrate $\dot{M}$ of the carrier fluid.

The criteria for the optimisation could be - either to minimise a technical parameter such as the volume of the column $V = \Omega L$ or the mechanical power degraded by the pressure drop over the packed bed, etc... - or to minimise an economic parameter such as the investment, that is the cost of buying and installing the column and its ancillary equipment or the total cost of the operation by adding the operating costs to the previous costs. In particular this includes the cost of the supplied energy.

We propose to compare the various optimum solutions for a same example corresponding to real industrial conditions.

1. THE QUALITY OF THE SEPARATION

11. THE SIMPLIFYING ASSUMPTIONS

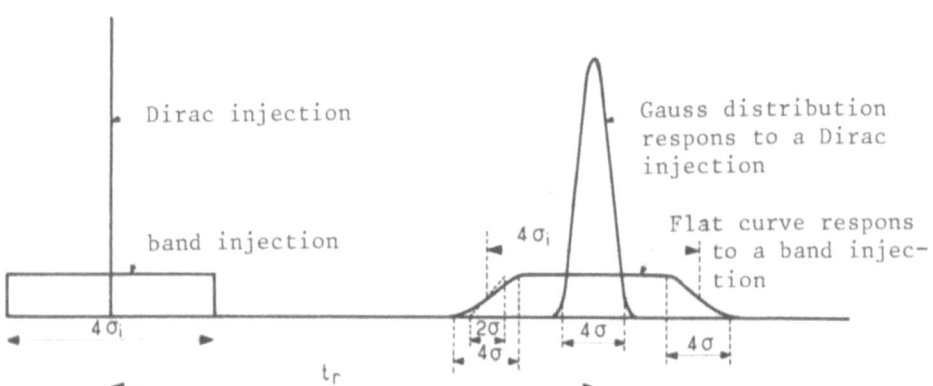

Figure 2

Let $4\sigma_i$ be the duration of the injection. We assume the injection is a rectangular band that is that the flowrate is constant over the whole period of injection. And we assume the system to be linear such that the output responses are symetrical.
Two extremes shown on figure 2, can be encountered.

- σ_i is infinitely short

The injection is a Dirac pulse and in the conditions of ideal chromatography the output response is a gaussian peak with a variance σ. This corresponds to elution chromatography.

- σ_i is greater than σ_A and σ_B

The response curve is then a plateau which ends in identical curves corresponding to the integrals of the gauss curve. The length of the response curve is, to a good approximation, given by :

$\qquad$ 4 $(\sigma_i - \sigma)$ at the upper part

$\qquad$ 4 $(\sigma_i + \sigma)$ at the lower part.

This corresponds to the case of an eluto-frontal chromatography.

12. MINIMUM LENGTH OF COLUMN FOR A COMPLETE SEPARATION OF TWO COMPONENTS

121. Dirac pulse injections - Elution chromatography

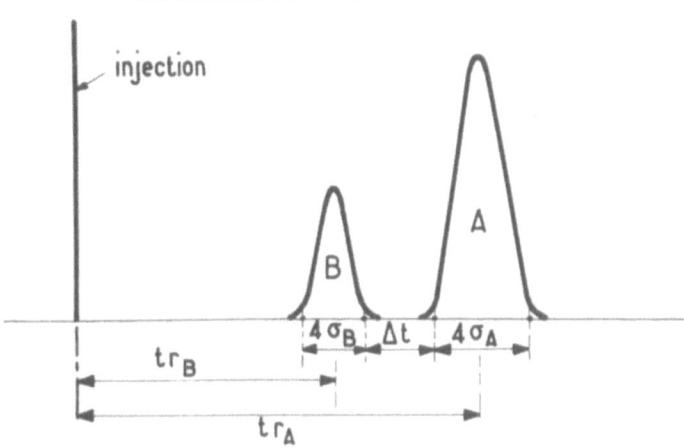

Figure 3

This case is shown schematically in figure 3. First we will summarise some definitions used in "ideal" chromatography.

As the variances of the injection peak and the response to a Dirac peak are additive it is easily shown (see annex 1) that the standard deviation of peak of A or B does not differ by more than 10 % from the ideal standard deviation σ_A or σ_B corresponding to an ideal Dirac pulse if the standard deviation σ_i of the band injection (that is width $4\sigma_i$) is such that :

$$\sigma_i \leq \begin{vmatrix} 0.4 \ \sigma_A \\ 0.4 \ \sigma_B \end{vmatrix} \tag{1}$$

Under these conditions :
- The column length L is taken to be composed of a stack of "theoretical stages" each having the same height HETP, thus :

$$L = NTP \ . \ HETP \tag{2}$$

the relation between the number of theoretical stages NTP is given by :

$$NTP = \left(\frac{t_{rA}}{\sigma_A} \right)^2 = \left(\frac{t_{rB}}{\sigma_B} \right)^2 \tag{3}$$

in which t_{rA} and t_{rB} are the retention times for the components A and B.
In this, we assume that the column is long enough for as to be able to neglect "dead volumes" (introduction, detection).

- The separation of the components A and B is generally characterized by a "resolution factor" R_{AB} such that :

$$R_{AB} = \frac{t_{rA} - t_{rB}}{4\bar{\sigma}} \qquad (4)$$

with
$$\bar{\sigma} = \frac{\sigma_A + \sigma_B}{2} \qquad (5)$$

A chromatographic separation is generally considered to be adequate if $R_{AB} \geq 1$ and complete if

$$\boxed{R_{AB} \geq 1.5} \qquad (6)$$

This corresponds to $\Delta t = 2\bar{\sigma}$ (see figure 3).
By combining expressions (3) (4) and (5) we can obtain an expression for the Number of Theoretical Stages required for a given value of the resolution factor :

$$\boxed{\frac{L}{H} \equiv N_o = 4\,R_{AB}^2 \left(\frac{\alpha + 1}{\alpha - 1}\right)^2} \qquad (7)$$

in which $\alpha = t_{rA}/t_{rB}$.

As from here, the symbols NTP and HETP will be simplified in the following forms : NTP $\equiv$ N and HETP $=$ H.

122. Rectangular injections : eluto-frontal chromatography

This case is shown schematically in figure 4. The resolution factor R_{AB} may be generalised by the expression :

$$R_{AB} = \frac{t_{rA} - t_{rB} - 4\sigma_i}{4\bar{\sigma}} \qquad (8)$$

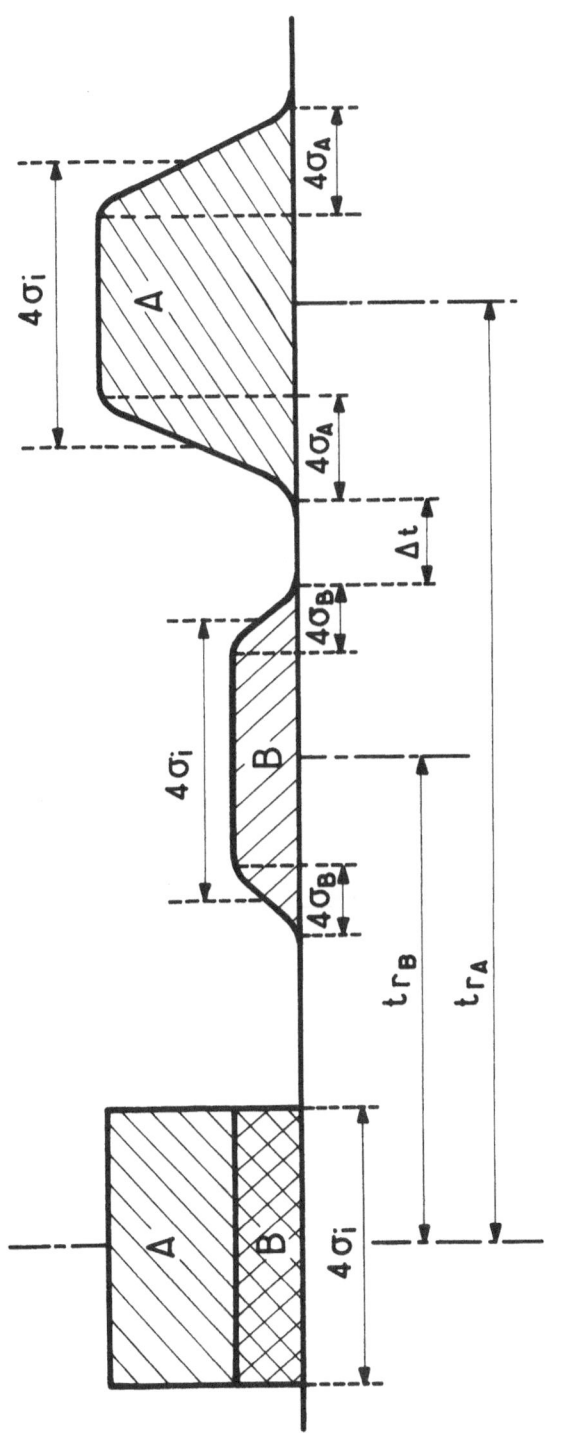

FIGURE 4

By analogy with expression (6) the separation may be considered to be complete if $\Delta t \geq 2\bar{\sigma}$ which corresponds to $R_{AB} \geq 1.5$ by reference to expression (8). Thus the separation can be considered to be complete if :

$$t_{rA} - t_{rB} \geq 4 \, (\sigma_i + \bar{\sigma}) + 2\bar{\sigma} \tag{9}$$

More generally refering to expressions (3) and (8) the Number of Theoretical Stages required to obtain a given value of R_{AB} is :

$$N = 4 \, R_{AB}^2 \left(\frac{\alpha + 1}{\alpha - 1 - \dfrac{4\sigma_i}{t_{rB}}} \right)^2 \tag{10}$$

This expression may be put in the following form :

$$\boxed{N = N_o \left(1 + \frac{\sigma_i}{\sigma \, R_{AB}} \right)^2} \quad \text{in which } N_o \equiv 4R_{AB}^2 \left(\frac{\alpha+1}{\alpha-1} \right)^2 \tag{11}$$

N_o is the minimum Number of Theoretical Stages for the case of elution chromatography (expression 7). It will be shown later that the ratio σ_i/σ is in practice much greater than unity which means that $N \gg N_o$.

13. PERIODICITY AND DURATION OF THE INJECTIONS - FLUID FLOW-RATE

131. Definitions

Let m be the mass of mixture (A + B) injected in each band in time $4\sigma_i$.
Let $\dot{M}$ be the mass flowrate of the carrier fluid which is assumed to be constant.

The concentration W of the mixture (A+B) during the injection in kg/kg of carrier fluid is :

$$W = \frac{m}{4\sigma_i \dot{M}} \tag{12}$$

(W is a mass ratio and is therefore dimensionless).

Let T be the period between successive injections. We put :

$$\theta \equiv \frac{4\sigma_i}{T} \qquad (13)$$

θ is the relative duration of the injections. Theoretically θ can vary between 0 in elution chromatography and 0.5 in eluto-frontal chromatography, as explained later.

The average mass flowrate $\dot{m}$ of mixture injected is :

$$\dot{m} = \frac{m}{T}$$

Therefore the relationship between the flowrates $\dot{M}$ and $\dot{m}$ is :

$$\boxed{\dot{M} = \frac{\dot{m}}{W\theta}} \qquad (14)$$

It should be noted that the flowrate $\dot{m}$ of the mixture to be separated is given.

In addition so as not to saturate the column the concentration W should be nowhere greater than a maximum value W_M (obviously the critical conditions are at the inlet). An expression for W_M as a function of the partition coefficient, the porosity of the solid and the compressibility of the carrier fluid is given in annex 3. In the case of an incompressible fluid, as in liquid chromatography, W_M is independent of the fluid flowrate. For a compressible fluid, as in gas chromatography, W_M is an increasing function of the flowrate.

For given values of $\dot{m}$ and W_M the flowrate $\dot{M}$ of the carrier fluid given by (14) is smaller, the longer is the relative duration of the injections θ.

132. Criteria for the separation of peaks

To fix the maximum value of θ we postulate that the separation criteria for two peaks, for two successive injections, is the same as the separation criteria for two peaks A and B from a same injection. In other words the difference Δt between all these peaks is the same. This is shown on figure 5. All the peaks are equidistant.

From expression (8) the period T between successive injections is

$$T = 2(t_{r_A} - t_{r_B}) = 8(\sigma_i + R_{AB}\bar{\sigma}) \qquad (15)$$

and the relative duration of the injections is :

$$\theta \equiv \frac{4\sigma_i}{T}$$

that is
$$\theta = \frac{1}{2} \cdot \frac{1}{1 + \dfrac{R_{AB}\overline{\sigma}}{\sigma_i}}$$
(16)

The maximum value of θ is therefore $\theta_m = 0.5$. This value is obtained when the standard deviation $\overline{\sigma}$ of the peaks is infinitely shorter than the duration $4\sigma_i$ of each injection (see figure 5a).

For a maximum productivity of a column the concentration W should be at its maximum value W_M. In the following, we will always made the assumption : $W = W_M$. The flowrate of carrier fluid $\dot{M}$ is then at its minimum $\dot{M}_o$:

$$\dot{M}_o \equiv \frac{2\dot{m}}{W_M}$$
(17)

By combining (14) (16) and (17), we obtain the following expression for $\dot{M}$ in the general case :

$$\dot{M} = \frac{\dot{M}_o}{2\theta} = \dot{M}_o \left(1 + \frac{R_{AB}\overline{\sigma}}{\sigma_i}\right)$$
(18)

Thus, $\dot{M}$ can vary from an infinite value in elution chromatography (for $\sigma_i \rightarrow 0$) to the minimum value $\dot{M}_o$ in eluto-frontal chromatography with $\overline{\sigma}/\sigma_i \rightarrow 0$ from which $\theta \rightarrow 0.5$.

Case of $\theta = 0.25$

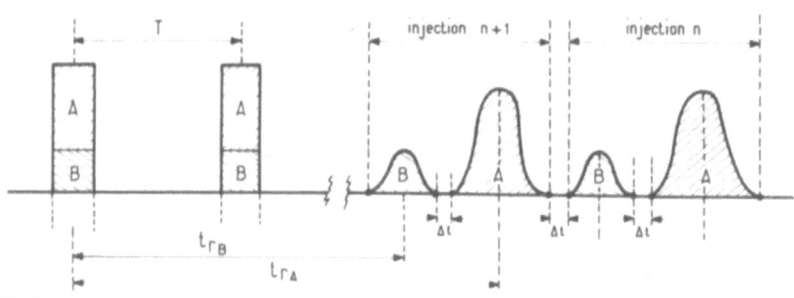

FIGURE 5a

Limiting case $\theta = 0.50$

FIGURE 5b

2. TECHNICAL OPTIMIZATION OF A COLUMN

21. DIMENSIONLESS EXPRESSIONS FOR THE HEIGHT OF A THEORETICAL STAGE

Going back to expression (2) for the length of a column

$$L = N \cdot H \tag{2}$$

and the expression (11) for the number of theoretical stages N

$$N = N_o \left(1 + \frac{\sigma_i}{\overline{\sigma}\, R_{AB}} \right)^2 \tag{11}$$

A dimensionless form of the Van Deemter equation, which is derived in annex 3, is used for the height of a theoretical stage, in gas chromatography

$$H = h_{-1} B^{-1} + h_o + h_1 B^1 \tag{19}$$

in which B is a dimensionless number usually called the Bodenstein number and defined as follows (see annex 3)

$$B \equiv \frac{u d_p}{D} = \frac{\dot{M}\, d_p}{\Omega \rho_o D_o} \tag{20}$$

This number gives a comparison between convective and diffusional transport. It is also possible to consider B to be a flowrate u, expressed in reduced coordinates which leads to the standard form of the Van Deemter equation.

The constant coefficients h_{-1}, h_o and h_1 are in length units.

For a small range of B it is always possible to replace the polynomial (19) by a monomial expression with HETP as a power function of B

$$H = h_c B^b \tag{21}$$

For example for a given column the exponent b is equal to -1 for very low flowrates and +1 for very high flowrates. b varies for intermediate flowrates but in a range of variation of a power of

10 of B, it is possible to use a constant value of b for example
b = 0.5 as shown in figure   in annex 3. The coefficient h_c
is in units of length. The same relationship (21) is often used
in liquid chromatography with $0.3 \leq b \leq 0.7$.

More generally the expressions (19) (21) and similar expressions
are represented by the symbolic expression for H as a function
of B

$$\boxed{H = H\ (B)} \tag{22}$$

22. LIQUID CHROMATOGRAPHY (case of incompressible fluid)

In this section it is assumed that the fluid is incompressible
($\rho = \rho_0$ = constant). It follows that the maximum concentration
W_M and the minimum flowrate M_0 from (17) are always constant.

221. The characteristic equation of the column

Using expressions (11) (16) and (18) it is a simple matter to derive expressions for N, B and L as a function of θ. This gives :

$$N = N_0\ (1 - 2\theta)^{-2} \tag{23}$$

and $\qquad B = \dfrac{\Omega_c}{2\ \theta\Omega} \qquad$ putting $\qquad \Omega_c \equiv \dfrac{\dot{M}_0\ d_p}{\rho_0\ D_0} \qquad$ (24)

whence $\qquad L = NH$

$$\boxed{L = N_0\ (1-2\theta)^{-2}.H(B)}$$

$$\left.\boxed{L = \dfrac{N_0}{(1-2\theta)^2}\ .\ H\left(\dfrac{\Omega_c}{2\ \theta\Omega}\right)}\right\} \tag{25}$$

We propose calling this expression the characteristic equation
of the column, as it is the fundamental relationship between the
parameters characterising the column and its performance that is:
- the construction variables { - the length L
                               - the cross-sectional area Ω

- the operating variable θ which is the flowrate of the carrier fluid in reduced form.

- the constants N_0 and Ω_c and the function H(B) which characterize - the properties of the mixture to be separated
            - the separating agent (the packing)
            - the technical data on the problem to be solved.

With 3 variables and a constraint between these variables there
is a double infinity of solutions to the problem : for example we

can make a "free" choice of the variables L and Ω for the column. Then by means of the characteristic equation we will determine the corresponding value of θ and therefore the flowrate $\dot{M}$ of the carrier fluid.

In actual fact due to technical constraints (loss of effectiveness) L and Ω cannot have just any values.

For example in the special case where the function H(B) reduces to a simple power function (21) the characteristic equation (23) can be put in the following dimensionless form

$$\boxed{\frac{L}{h_c} \cdot \left(\frac{\Omega}{\Omega_c}\right)^b \cdot (2\theta)^b \, (1-2\theta)^2 = N_o} \qquad (26)$$

This equation has the advantage of being especially simple and leading to a quasi-intuitive interpretation as the 3 variables L, Ω and θ are quite independent. Let us examine the origins of the different constraints which are involved.

h_c, which we call the "characteristic length" of a theoretical stage is defined by the expression (21). It is an experimental property of the packing,

Ω_c which we call the characteristic cross-sectional area of the column is defined by (24) and (17) which gives :

$$\Omega_c = \frac{2 \, \dot{m} \, d_p}{W_m \rho_o D_o} \qquad (27)$$

Ω_c is proportional to the flowrate of the mixture to be separated.

N_o defined by (7) is the minimum number of theoretical stages required for a given quality of separation.

In such a double infinity of solutions, let us now find those which require a minimum volume of packing and those which give a minimum degradation of mechanical energy by the flowing fluid.

222. The column with a minimum volume

We will use the Bodenstein number B as the 2nd variable next to θ. Expression (24) is then written :

$$\Omega = \frac{\Omega_c}{2 \, \theta B} \qquad (28)$$

The product of the two expressions (25) and (28) give an expression for the volume $V_i \equiv \Omega L$ of the column that is :

$$V_i = \frac{\Omega_c N_o}{2\theta(1-2\theta)^2} \cdot \frac{H(B)}{B} \qquad (29)$$

2221. Optimum value of θ for constant B

Expression (29) shows that for a given value of B, the volume V_i is infinitely great for $\theta = 0$ and $\theta = 1/2$ (Figure 6).

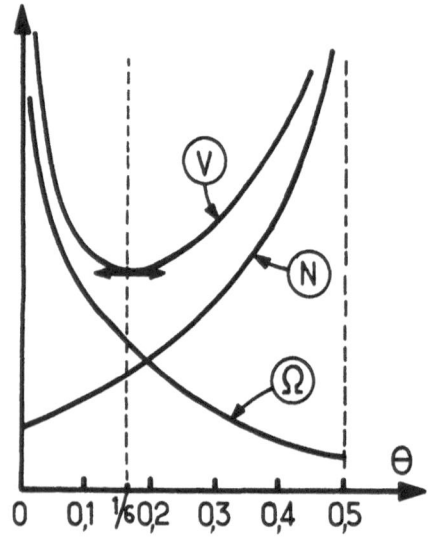

For $\theta=1/2$, the number of theoretical stages N, and therefore the length L, are infinitely great as given by (23).

For $\theta=0$ the cross-sectional area Ω is infinitely great as given by (28).

By putting the derivative $(\partial V/\partial \theta)_B$, equal to zero, we obtain the coordinates of the minimum of V_i that is :

$$\theta_v = 1/6$$

$$V_{i,min} = \frac{27}{4} \cdot \frac{\Omega_c N_o}{B} \cdot H(B)$$

(30)

Figure 6

from which the optimum number of theoretical stages is :

$$N_{opt} = \frac{9}{4} N_o$$

and the optimum cross-sectional area is $\Omega_{opt} = \frac{9\Omega_c}{B}$

2222. Optimum value of B at constant θ

By substituting H(B) from equation (21) into (29) we obtain :

$$V_i = \left[\frac{\Omega_c N_o}{2\theta(1-2\theta)^2}\right] \cdot h_c B^{-(1-b)} \qquad (31)$$

This gives the volume V as a continually <u>decreasing</u> function of B : for a minimum value of V, it is necessary to choose a value of the <u>Bodenstein number which is as large as possible</u>. This means a cross-sectional area Ω as small as possible as given by (28).

223. Column with a minimum degradation of energy

In the special case of laminar flow as considered here, the pressure drop is given by the Darcy equation

$$p_e - p_o = \frac{\mu}{k} u L \tag{32}$$

in which k is the permeability of the porous medium.

The mechanical power degraded by the fluid in passing through this medium is

$$\dot{E}_d = \frac{\dot{M}}{\rho_o} (p_e - p_o) \tag{33}$$

By combining this with (18) (24) and (25) we obtain

$$\dot{E}_d = \left(\frac{M_o \mu D_o}{k d_p^{\rho} o}\right) \cdot \frac{N_o}{2\theta(1-2\theta)^2} \cdot B \cdot H(B) \tag{34}$$

As with V_i the mechanical power degraded $\dot{E}_d$ is a function of the two free variables θ and B.

It can be seen that $\dot{E}_d$ is a function of θ in the same way as is V_i. This means that the minimum of $\dot{E}_d$ is obtained at the same value of θ that is :

$$\min \dot{E}_d \quad \text{for} \quad \theta_e = \theta_v = 1/6 \tag{35}$$

On the otherhand $\dot{E}_d$ is a function of B in a different way from V_i and the important point is that this function $\dot{E}d(B)$ is a continually increasing function of B.

For example let us consider the function H(B) to be the monomial power function (21). It follows

$$E_d = (\text{cste}) \cdot h_c B^{1+b} \tag{36}$$

To obtain a minimum value of $\dot{E}_d$ it is necessary to choose as small as possible a value of the Bodenstein Number and therefore as large as possible a cross-sectional area Ω by (28).

224. Conclusion : The incompatibility of the two optimizations

We therefore arrive at two conclusions which are contradictory depending on whether we seek to minimise the volume of the column or the mechanical power which is degraded. In one case B should be choosen to be as large as possible and in the other case it should be as small as possible.

In actual fact the construction costs are an increasing function of the mechanical power degraded. In section 3 we will obtain the economic optimum which will take here two phenomena into account at the same time.

23. GAS CHROMATOGRAPHY (case of a compressible fluid)

231. Expression for the pressure drop

In the case of a compressible fluid the pressure drop p_e-p_o between the inlet and the outlet of the column brings about a difference in the density of the carrier fluid. This means that the maximum admissible concentration W_e allowable at the entry to the column is not constant. It depends on the pressure p_e at the inlet and therefore the flowrate of the carrier fluid. We suppose that the outlet pressure p_o is constant and generally taken to be at atmospheric pressure.

If we assume that the carrier fluid is a perfect gas and the flow is isothermal, we obtain

$$W_e = W_M \cdot \frac{p_o}{p_e} = \frac{W_M}{P} \tag{37}$$

with $\qquad P = p_e/p_o \qquad$ P will be called the pressure factor

The expression for W_M as a function of the properties of the gas and the solid porous adsorbant is given in annex 3.

In addition in annex 4 is given the expression for P as a function of the permeability k of the porous medium using the Darcy equation (32). That is :

$$P = \sqrt{1 + \frac{2\,\mu}{k\rho_o p_o} \cdot \frac{\dot{M}L}{\Omega}} \tag{38}$$

232. Influence on the characteristic equation

The phenomenon of gas compressibility means that the minimum flowrate of the carrier fluid is no longer that given by expression (17) that is

$$\dot{M}_o = \frac{2\dot{m}}{W_M} \tag{17}$$

The minimum flowrate now becomes

$$\dot{M}'_o = \frac{2\dot{m}}{W_e} \qquad \text{that is} \qquad \dot{M}'_o = \dot{M}_o P \tag{39}$$

$\dot{M}_o$ must therefore be replaced by $\dot{M}'_o = \dot{M}_o P$ in expression (18) and those which are derived from it.

Thus expression (24) for the Bodenstein number is replaced by

$$B = \frac{\Omega_c P}{2\theta\Omega} \quad \text{with again} \quad \Omega_c \equiv \frac{\dot{M}_o d_p}{\rho_o D_o} \qquad (40)$$

With the proviso that the coefficient P is introduced in the definition of B, <u>the characteristic equation of the column is unchanged</u>. This is a large part of its interest :

$$\boxed{L = \frac{N_o}{(1-2\theta)^2} \cdot H\left(\frac{\Omega_c P}{2\theta\Omega}\right)} \qquad (41)$$

By substituting the values of Ω and L in (38) we obtain the following expression for the pressure factor :

$$\boxed{P = \left[1 + \frac{N_o}{(1-2\theta)^2} \cdot \frac{BH}{K}\right]^{1/2}} \qquad (42)$$

in which K is a characteristic constant of the hydrodynamics of the fluid in the porous medium and is defined as

$$K \equiv \frac{k \, p_o \, d_p}{2\mu \, D_o} \qquad (43)$$

Remarks

1. It can be seen that K also has the units of <u>length</u> which is useful for making a comparison with other coefficients h_{-1}, h_o, h, h_c as defined above.

2. In the special case where the function H(B) is reduced to the power function (21) the characteristic equation (26) is slightly modified to

$$\frac{L}{h_c} \cdot \left(\frac{\Omega}{\Omega_c}\right)^b (2\theta)^b (1-2\theta)^2 = N_o P^b = N_o\left(1 + \frac{BH}{K} \cdot \frac{N_o}{(1-2\theta)^2}\right)^{b/2} \qquad (44)$$

3. In the even more special case where the column causes a very high pressure drop, that is P >> 1, this expression may be reduced to

$$\frac{L}{h_c} \left(\frac{\Omega}{\Omega_c}\right)^b (2\theta)^b (1-2\theta)^{2-b} \cdot \left(\frac{BH}{K}\right)^{-b/2} = N_o^{1+(b/2)} \qquad (44 \text{ bis})$$

233. Column with the minimum volume

Here (as in 143) B and θ are considered to be free variables and we seek their optimum values for a minimum volume V.

From (29) (40) and (41) we obtain :

$$\boxed{V = V_i \cdot P}$$
(45)

in which V_i is the volume of the column given by (23) for an incompressible fluid. After development this becomes :

$$V = \frac{\Omega_c N_o}{2\theta(1-2\theta)^2} \cdot \frac{H}{B} \left[1 + \frac{N_o}{(1-2\theta)^2} \cdot \frac{BH}{K} \right]^{1/2}$$
(46)

2331. Optimum value of θ at constant B

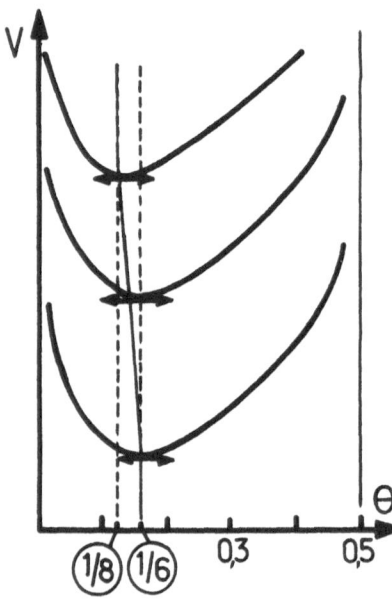

FIGURE 7

Here again the volume is infinitely great at the two limits $\theta = 0$ and $\theta = 1/2$ (figure 7).

But the calculation of the coordinates of the minimum is more complicated than in (1431).

Putting the derivative $(\partial V/\partial \theta)_B$ to zero leads to the following cubic equation :

$$24\theta_v^3 - 28\theta_v^2 + \left(10 + \frac{8N_o BH}{K} \right)\theta_v$$
$$- \left(1 + \frac{N_o BH}{K} \right) = 0$$
(47)

Figure 8 gives the exact solution of this equation as a function of the dimensionless parameter $N_o BH/K$.

It can be seen that θ_v does not vary very much. It has the value of 1/6 when $K \to \infty$, that is for an infinitely permeable column (ie $P \to 1$), and it has a value of 1/8 for $K \to 0$ for a column with a very small permeability.

For practical use in later calculations we will use these extreme values for a precision of more than 2 % on the determination of V:

$\theta_v = 1/6$	for	$\dfrac{N_o BH}{K} < 0.5$	incompressible fluids
$\theta_v = 1/8$	for	$\dfrac{N_o BH}{K} > 0.5$	compressible fluids

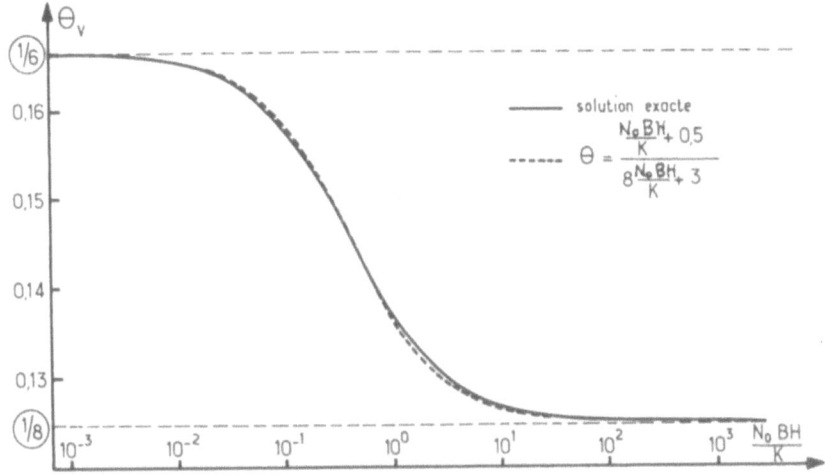

FIGURE 8

In addition to calculate θ_v it is possible to use the following relationship which is precise to better than 3 % on the value of θ (see the dotted line curve on figure 8).

$$\theta_v \simeq \frac{\dfrac{N_o BH}{K} + 0.5}{8 \dfrac{N_o BH}{K} + 3}$$

2332. Optimum value of B at constant θ

Let us go back to expression (45) that is $V = V_i P$.

In section (1432) we showed that the volume V_i is a continually decreasing function of B at constant θ.

For example in the case of the Van Deemter equation V_i is given by (31).

In addition expression (42) shows that the pressure factor P is a continually increasing function of B. For example using the Van Deemter equation we obtain :

$$P^2 = 1 + \frac{N_o}{(1-2\theta)^2} \cdot \frac{h_{-1} + h_o B + h_1 B^2}{K} \tag{49}$$

A priori the product $V = V_i P$ of these two functions can have a minimum.

Putting the derivative $(\partial V / \partial B)_\theta$ to zero gives the following equation

$$B_v^3 + \frac{1}{2}\frac{h_o}{h_1}B_v^2 - \left|\frac{1}{2}\frac{h_o^2}{h_1^2} + \frac{h_{-1}}{h_1}\right|B_v - \left|\frac{5}{2}\frac{h_o h_{-1}}{h_1^2} + \frac{h_o}{h_1^2}\frac{K(1-2\theta)^2}{N_o}\right| = 0 \quad (50)$$

2333. Special case of P >> 1

A special case, of great practical importance, is that where there is a large pressure drop (P >> 1).

In this case the expression (46) can be simplified to

$$V = \frac{\Omega_c N_o^2}{2\theta(1-2\theta)^3} H^{3/2} \cdot (BK)^{-1/2} \quad (51)$$

In this special case, equation (50) is reduced to a quadratic equation of which the solution is :

$$B_v = \frac{h_o}{4h_1}\left[1 + \sqrt{1 + 32\frac{h_1 h_{-1}}{h_o^2}}\right] \quad (52)$$

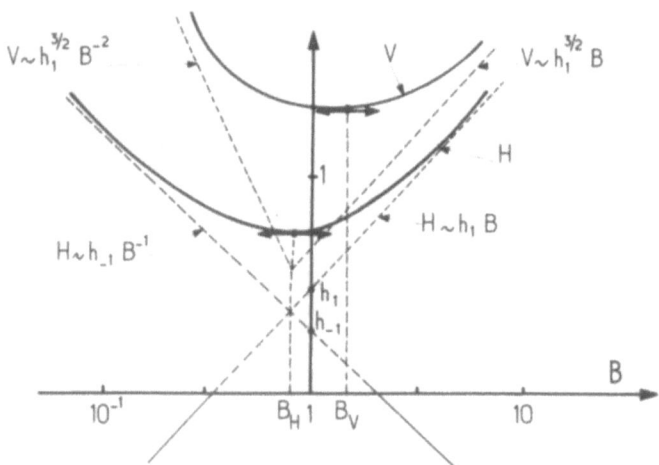

FIGURE 9

In the same log log coordinate figure 9 shows :

- The curve V(B) with its minimum B_v on the abcissa and the two asymptotes :

$$\left| \begin{array}{l} V = (cst) \, . \, h_{-1}^{3/2} \, K^{-1/2} \, B^{-2} \qquad \text{for B very small} \\[2ex] V = (cst) \, . \, h_{1}^{3/2} \, K^{-1/2} \, B \qquad \text{for B very large} \end{array} \right.$$

- The classical curve of the Van Deemter equation

$$H(B) = h_{-1} \, B^{-1} + h_o + h_1 \, B^{1}$$

with a minimum on the abcissa of $B_h = \sqrt{(h_{-1}/h_1)}$

It can be seen that in the general case the minima in the two curves have different abcissa. However it can be shown that, in the special case where h_o is 0 the abcissa of the two minima are linked by the very simple relation :

$$\left| B_v = \sqrt{2} \, . \, B_h \qquad \text{for} \quad h_o = 0 \right| \tag{53}$$

It is in fact often found that in the Van Deemter equation the constant term h_o is either zero or at least much smaller than the other terms.

The value B_v corresponding to the minimum column volume is therefore never very different from the value B_h corresponding to the minimum height of a theoretical stage.

234. Column with minimum degradation of energy

In the case of a compressible fluid, the degraded mechanical power is no longer given by the expression (33) but by the following expression in the case of isothermal expansion :

$$\dot{E}_d = \frac{\dot{M}}{\rho_o} \, P_o \, \text{Log} \, P \tag{54}$$

However in recirculating the carrier gas it is necessary to perform an adiabatic recompression such that the power supplied to the compressor is proportional to :

$$\dot{E} = \frac{\gamma}{\gamma-1} \frac{\dot{M}}{\rho_o} \, P_o \left[P^{\frac{\gamma-1}{\gamma}} - 1 \right] \tag{55}$$

This latter value will be used to determine the optimum column.

It is known that if the pressure factor P is too high ($\simeq$ greater than 5) then the gas is heated too much in adiabatic compression.

A multistage compressor with interstage cooling, is then required. Let s be the number of stages, each stage having a recompression factor of 5. s is determined by the inequality :

$$5^s \leq P \leq 5^{s+1} \tag{56}$$

The total amount of energy required by the s stages of compression is then proportional to :

$$E = \frac{\gamma}{\gamma-1} \frac{\dot{M}_o P_o}{\rho_o} \cdot \frac{sP}{2\theta} \left[P^{s \frac{\gamma-1}{\gamma}} - 1 \right] \tag{57}$$

This expression placed in (42) for the pressure factor P leads to :

$$\dot{E} = \frac{\gamma}{\gamma-1} \cdot \frac{\dot{M}_o P_o}{\rho_o} \cdot \frac{s}{2\theta} \left[1 + \frac{N_o}{(1-2\theta)^2} \frac{HB}{K} \right]^{\frac{1}{2}} \left\{ \left[1 + \frac{N_o}{(1-2\theta)^2} \frac{HB}{K} \right]^{\frac{\gamma-1}{2s\gamma}} - 1 \right\} \tag{58}$$

2341. Optimum value of θ at constant B

As for the case of an incompressible fluid it is found that $\dot{E}$ is infinitely great at the two limits $\theta = 0$ and $\theta = 0.5$ (fig. 10).

Putting the derivative $(\partial\dot{E}/\partial\theta)_B$ to zero does not lead to a simple equation and a solution can only be obtained by numerical analysis.

As an example figure (10) shows the energy (in reduced coordinates) as a function of θ for $\gamma = 1.4$ and for three values of the dimensionless group $N_o HB/K$.

The accidents in the curve correspond to increases in the number of compression stages s. Thus on the curve of $N_o HB/K = 5$ we pass from 1 to 2 then to 3 stages as θ increases.

It can be seen that the abcissa of the minimum vary only slightly with the parameter θ. The range is from 1/6 to 1/4.

We may therefore conclude that :

$$\frac{1}{6} \leq \theta_e \leq \frac{1}{4} \tag{59}$$

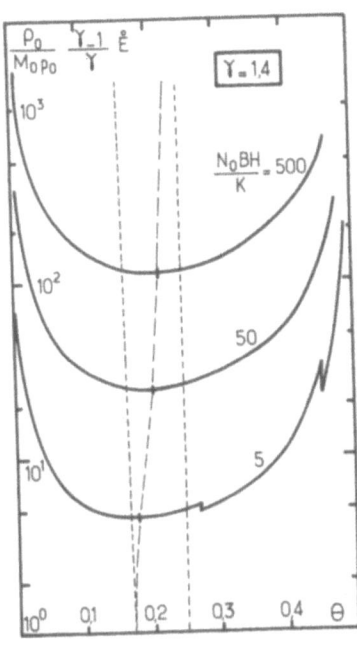

FIGURE 10

Figure (11) shows the value of θ_e which gives a minimum $\dot{E}$ as a function of the dimensionless group $N_0 BH/K$ for several values of γ. These values include those for helium (1.66), nitrogen and air (1.4). We have taken a limit of 3 stages corresponding to a $P_{max} = 125$ which is greater than the values usually encountered.

In practice θ_e will not vary in the whole domain (0.166 - 0.25) but stay less than 0.20.

We will take :

$$\theta_e \approx 1/6 \quad \text{and} \quad s=1 \quad \text{for} \quad \frac{N_o BH}{K} \leq 10$$

$$\theta_e \quad 1/5 \quad \text{and} \quad s=2 \text{ or } 3 \quad \text{for}$$

$$10 < \frac{N_o BH}{K} < 5000$$

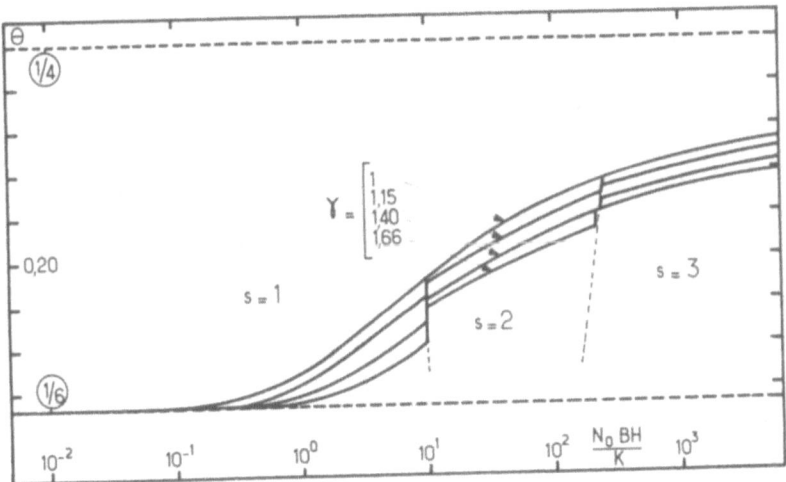

FIGURE 11

2342. Optimum value of B at constant θ

Expression (57) shows that $\dot{E}$ is a continually increasing function of P at constant θ. In addition we saw previously that P is a continually increasing function of B.

This means that the mechanical power $\dot{E}$ is a continually increasing function of B.

The conclusion is therefore qualitatively the same as for an incompressible fluid : as small a value as possible of the Bodenstein number should be chosen to obtain a minimum value of the power $\dot{E}$.

Nevertheless the function $\dot{E}(B)$ is very different from the expression (36) for an incompressible fluid.

2343. Special case P >> 1

In the special case of a very high pressure drop across the column, that is P >> 1, the expression (58) can be reduced to

$$\dot{E} \simeq \frac{\gamma}{\gamma-1} \cdot \frac{\dot{M}_o P_o}{\rho_o} \cdot \frac{s P}{\theta}^{\frac{\gamma(s+1)-1}{s\,\gamma}} \quad \text{with} \quad P = \left[\frac{N_o BH}{(1-2\theta)^2 K}\right]^{1/2} \quad (61)$$

For example, for $\gamma = 1.4$ and $s = 3$ we obtain :

$$\dot{E} = (cst) \cdot (BH)^{0.55} \quad (61 \text{ bis})$$

In the case of an incompressible fluid $\dot{E}$ was found to be proportional to BH (see expression 34). Here we find an exponent of 0.55. The energy which is degraded increases much less rapidly as a function of the variable B.

24. CONCLUSION : COMPARISON OF THE TECHNICAL OPTIMA

The table given below presents a recapitulation of the various optimum solutions obtained.

It may be concluded that the optimum value of θ never varies much from 1/6. On the other hand, there is a large variation in the values of B. However, in all the cases considered minimising the energy degraded implies $B_e \to 0$, that is a column with as large as possible a cross sectional area which was probably evident to start with.

Table. Optimum values of the relative duration of the injections θ (at constant B) and of the Bodenstein number B (at constant θ).

OPTIMUM VALUES.	OPERATING CONDITIONS		
	Incompressible fluid or infinitely permeable column	Compressible fluid column with a finite permeability	Compressible fluid column with zero permeability $N_o/BH \to \infty$
Minimum column volume $\theta_v =$	1/6	$\theta_v \approx \dfrac{\dfrac{N_o BH}{K} + 0.5}{8\,\dfrac{N_o BH}{K} + 3}$ Fig. 8	1/8
$B_v =$	∞	relation (50)	$\dfrac{h_o}{4h_1}\left[1 + \sqrt{1 + 32\,\dfrac{h_1 h_{-1}}{h_o^2}}\right]$
Minimum Degradation of energy $\theta_e =$	1/6	Fig. 11	1/4
$B_e =$	0	0	0

3. ECONOMIC OPTIMIZATION OF THE SEPARATION

31. THE COMPONENTS OF THE TOTAL COST

In the preceeding publications (ref. 1, 2, 3, 4) we have proposed a mathematical model of the economic optimization of industrial processes as a function of the flux of energy which is degraded in them.

We decomposed the total cost of an operation into the four following terms :

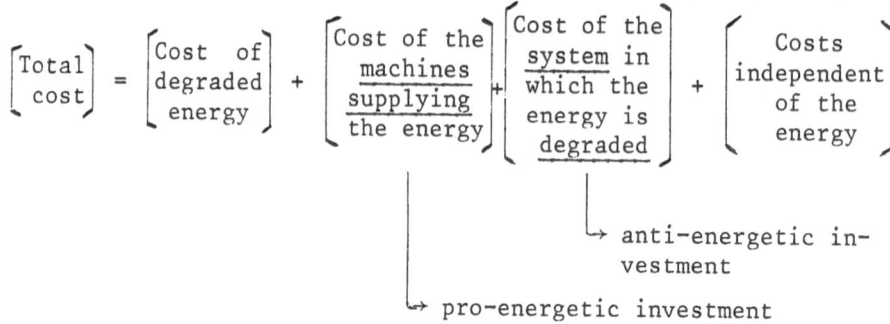

$$\begin{bmatrix} \text{Total} \\ \text{cost} \end{bmatrix} = \begin{bmatrix} \text{Cost of} \\ \text{degraded} \\ \text{energy} \end{bmatrix} + \begin{bmatrix} \text{Cost of the} \\ \underline{\text{machines}} \\ \underline{\text{supplying}} \\ \text{the energy} \end{bmatrix} + \begin{bmatrix} \text{Cost of the} \\ \underline{\text{system}} \text{ in} \\ \text{which the} \\ \text{energy is} \\ \underline{\text{degraded}} \end{bmatrix} + \begin{bmatrix} \text{Costs} \\ \text{independent} \\ \text{of the} \\ \text{energy} \end{bmatrix}$$

↳ anti-energetic in-
vestment

↳ pro-energetic investment

Let us decompose the total cost of a separation system for a given mixture in the same way.

. The operation is to have a working life of n years and separate
the mass nṁ of a mixture (in which ṁ is the constant flowrate of
the mixture in kg/year). It will consume nĖ of energy (in which
Ė is in kwh/year).

. We have proposed calling "pro-energetic" the investment into
machines which supply the energy to the system (pumps, motors,
fans, boilers, etc...). Their cost is an increasing function of
their nominal power and therefore an increasing function of the
flux of energy degraded Ė.
    In the present case this is the motor-compressor unit which sup-
plies the column with fluid under pressure.

. In the same way, the "anti-energetic" investment is that
equipment where the energy is degraded (tanks, pipes in which the
mechanical energy is degraded into heat, walls which are more or
less well insulated through which heat energy is dissipated to
the surroundings, etc...).
    As a general rule the cost of such a system is a decreasing
function of the flux of energy which is degraded in it. (There
are however some exceptions to this rule, see further on).
    In the present case, this item is the chromatographic column
itself, in which mechanical energy is degraded by friction. We
will make a distinction between the cost of the column and its
ancillary equipment and the cost of the packing in the column.

Remark. In the problem under consideration here, we neglect the
cost of the heat exchangers which are often necessary for interstage cooling in recompression of the carrier gas.

$$\begin{bmatrix} \text{Total} \\ \text{cost} \end{bmatrix} = \begin{bmatrix} \text{Cost of} \\ \text{degraded} \\ \text{energy} \end{bmatrix} + \begin{bmatrix} \text{Cost of} \\ \text{motor-compressor} \\ \text{unit} \end{bmatrix}$$

$$+ \underbrace{\begin{bmatrix} \text{Cost of} \\ \text{the} \\ \text{column} \\ \text{body} \end{bmatrix} + \begin{bmatrix} \text{Cost of} \\ \text{the} \\ \text{packing} \end{bmatrix}}_{\begin{bmatrix} \text{Cost of} \\ \text{the packed} \\ \text{column} \end{bmatrix}} + \begin{bmatrix} \text{Costs which} \\ \text{are independent} \\ \text{of energy} \end{bmatrix}$$

Let us examine each of these cost components in turn.

32. <u>MINIMUM TOTAL COST OF THE PACKED COLUMN</u> (anti-energetic investment)

321. <u>Expression of the cost</u>

Let us assume that the cost of the packed column is the sum of two terms :

- <u>The cost of the packing</u> $C_p = A'_p \gamma_p (1-\varepsilon) \, \Omega L$ (63)

- C_p is proportional to the volume of the solid therefore to the mass of column packing,
- γ_p is the price of 1 kilogramme of the packing material
- A'_p is a coefficient such that the difference $(A'_p - 1)$ is the cost of transport and filling the column.

- <u>The cost of the column body</u> $C_c = A'_c \gamma_c \, \Omega^i L^j + I_o$ (64)

The term C_c is the cost of the empty cylindrical column and its accessories. This cost is obviously an increasing function of the dimensions Ω and L, but it increases less rapidly than proportional to these two parameters :

The exponents i and j are less than or at most, equal to 1.

As there is no industrial data available on large chromatographic columns, we must use values for similar cylindrical equipment (packed columns for distillation, gas-liquid absorption, etc...).

Experience shows that the most usual values of the exponents i and j are :

$$\boxed{i \simeq 0.3 \ \text{ and } \ j \simeq 0.6 \quad \text{that is} \quad \frac{i}{j} \simeq 0.5} \quad (65)$$

However this does not exclude the possibility of a change in the ratio i/j over quite a large range of the order of 0.1 to 1.

γ_c is the price of a unit quantity of the construction materials (for example a m^2 of sheet steel) and A'_c is a coefficient which includes the construction costs. I_0 is a constant term which is independent of the size of the column (miscellaneous equipment).

By expressing (45) the two costs as a function of the volume V and the pressure factor are :

cost of the packing $\}\rightarrow$ $C_p = A''_p V = A''_p V_i P$

cost of the body of the column $\}\rightarrow$ $C_c = A''_c V^i L^{j-i} = A''_c . V^i P^i L^{j-i} + I_0$

with $\qquad A''_p = A'_p \gamma_p (1-\varepsilon) \qquad A''_c = A'_c \gamma_e$

The cost of the packed column is the sum of these two costs :

$$C_{cp} = A''_p . V_i P + A''_c . V_i^i P^i L^{j-i} + I_0 \qquad (66)$$

By developing V, L and P as a function of the free variables B and θ, using expressions (29) (41) and (42) we finally obtain C_{cp}.

322. Case of liquid chromatography (P = 1)

The cost of the packed column is

$$C_{cp} = \underbrace{\frac{A_p}{2\theta(1-2\theta)^2} \cdot \frac{H}{B}}_{\text{packing}} + \underbrace{\frac{A_c}{(2\theta)^i(1-2\theta)^{2j}} \cdot \frac{H^j}{B^i}}_{\text{column}} + I_0 \qquad (67)$$

with $A_p = A''_p \Omega_c N_0$ and $A_c = A''_c \Omega^i N_0^j$

Let us find the values of B and θ which give a minimum cost C_p.

In using equation (21) for H(B) expression (67) becomes :

$$C_{cp} = \frac{A_p}{2\theta(1-2\theta)^2} \cdot h_c B^{b-1} + \frac{A_c}{(2\theta)^i(1-2\theta)^{2j}} h_c^j B^{jb-i} + I_0 \qquad (68)$$

Two cases may then be considered depending on the sign of $(jb-i)$.

3221. $\underline{b < i/j}$

In this case where C_c and C_p are both decreasing functions of B
we must choose a value of B as large as possible, that is a
cross-sectional area Ω as small as is practically possible.

The value of θ which at constant B gives a minimum V_i namely
$\theta_v = 1/6$, also gives a minimum C_p. It is easily shown that the
minimum of C_c is obtained for a θ of

$$\theta_c = \frac{i/j}{2i/j + 4} \qquad (69)$$

In addition the value θ_{cp} which gives a minimum value of C_p at
constant B is such that :

$$\frac{i/j}{2i/j + 4} < \theta_{cp} < 1/6 \qquad (70)$$

for example for $i/j = 0.5$ we have $0.1 < \theta_{cp} < 1/6$.

3222. $\underline{b > i/j}$

In this case, C_c is a decreasing function and C_p an increasing
function of B giving an absolute minimum at the coordinates
$\left[\theta_{cp}, B_{cp}\right]$ which may be determined analytically.
Putting the partial derivative of expression (68) to zero
$(\partial C_p/\partial B)_\theta = 0$ gives :

$$B = \left[\left|\frac{1-b}{jb-i}\right| \cdot \frac{A_p}{A_c} \cdot \frac{h_c^{1-j}}{(2\theta)^{1-i}(1-2\theta)^{2(1-j)}} \right]^{\frac{1}{1-b+jb-i}} \qquad (71)$$

and the optimum of optimums is at :

$$\left| \begin{aligned} \theta_{cp} &= \frac{b}{2b+4} \\ B_{cp} &= \left[\left|\frac{1-b}{jb-i}\right| \frac{A_p}{A_c} h_c^{1-j} (1 + \frac{2}{b})^{1-i} (1 + \frac{b}{2})^{2(1-j)} \right]^{\frac{1}{1-b+jb-i}} \end{aligned} \right| \qquad (72)$$

323. Case of gas chromatography (compressible fluid)

By developing expression (66) for V_i, P and L, using expressions
(25), (29) and (42), we obtain :

$$\text{cost of the packing } C_p = \frac{A_p''}{2\theta(1-2\theta)^2} \cdot \frac{H}{B} \left[1 + \frac{N_o}{(1-2\theta)^2} \frac{BH}{K} \right]^{1/2} \qquad (73)$$

$$\text{cost of the column } C_c = \frac{A_c''}{(2\theta)^i(1-2\theta)^{2j}} \frac{H^j}{B^i} \left[1 + \frac{N_o}{(1-2\theta)^2} \frac{BH}{K}\right]^{i/2} \quad (74)$$

3231. Cost of the packing

We have already found the minimum for this term in paragraph 232 in obtaining the minimum volume by using the Van Deemter equation $\frac{\theta}{v} \, \epsilon \, (1/8, \, 1/6)$ and B_v is the solution of equation (50).

3232. C_c cost of the column itself

The value of θ which minimises the cost of the column C_c at constant B is the solution of the following expression

$$(8i+16j)\theta^3 - (12i+16j)\theta^2 + \left[6i+4j+2(i+j)N_o BH/K\right]\theta - i(1+N_o BH/K)=0 \quad (75)$$

Figure 12 gives the solution of this equation as a function of the ratio i/j and for the two limiting cases : $N_o BH/K \to 0$ and $N_o BH/K \to \infty$.

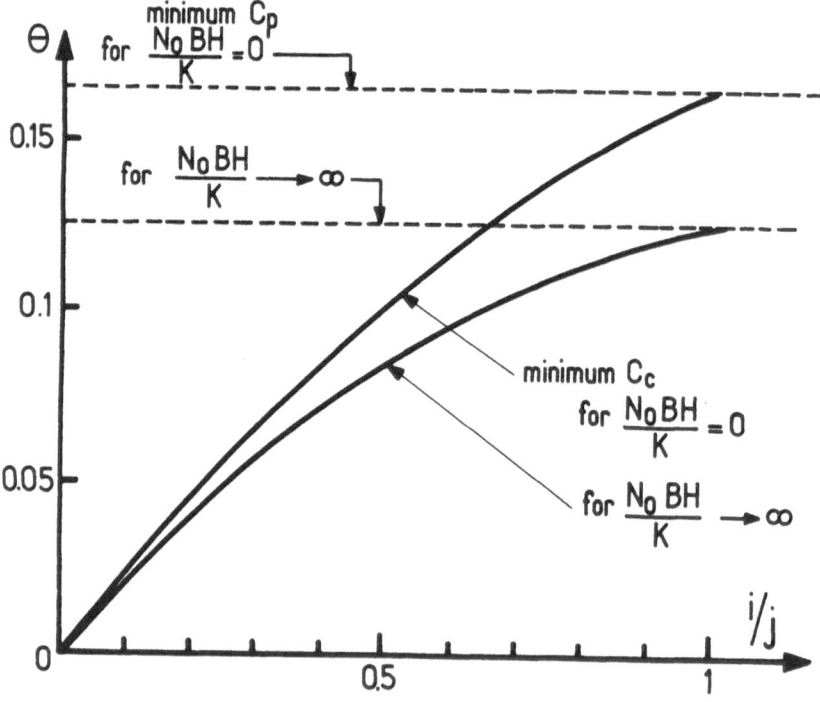

FIGURE 12

It can be seen that the value of θ which gives a minimum C_c is always less than that which gives a minimum C_p.

For example for the most frequently encountered $i/j = 0.5$ and for a column with a high pressure drop (that is $N_0BH/K \rightarrow \infty$), we have a minimum C_c for $\theta_c \simeq 0.083$ whilst the minimum C_p is for $\theta = 0.125$.

Remark. In addition it would be possible to calculate the value of B which gives a minimum of the cost C_c at constant θ. We would obtain an equation similar to (50) which would require a numerical solution.

3233. Special case : P >> 1

We have seen (expression 52) that the value of B which gives a minimum value for V, that is C_p, for a given θ is

$$B_p = B_v = \frac{h_o}{4h_1}\left[1 + \sqrt{1 + \frac{32h_1 h_{-1}}{h_o^2}}\right] \qquad (52)$$

It can also be shown that the value B_c which gives a minimum cost of the column, is

$$B_c = \frac{i}{4j}\frac{h_o}{5h_1}\left[1 + \sqrt{1 + \frac{16j(i+j)}{i^2}\frac{h_{-1}h_1}{h_o^2}}\right] \qquad (76)$$

In particular for the frequently encountered case where $h_o \simeq 0$, we have :

$$B_p \simeq \sqrt{2} \cdot B_h$$

and $\qquad B_c \simeq \sqrt{1 + (i/j)} \cdot B_h$

It may be seen that for $i/j = 0.5$, B_c only differs from B_p by about 15 % thus giving a good approximation of the optimum B_{cp}, for the cost of the packed column.

33. MINIMUM COST OF THE MOTOR-COMPRESSOR UNIT (pro-energetic investment)

In annex 5 we give some cost values for motor-compressors as a function of their nominal power.

This cost function can always be put in the following form :

$$C_m = C_{mo} + A'_m \gamma_m \left(\frac{P_n}{P_{no}}\right)^m \qquad (77)$$

The exponent m generally lies between 0.6 and 1, γ_m is a unit cost (the price of a motor-compressor of power P_{no}). The coefficient A'_m is such that the difference (A'_m-1) represents the cost of installing the machine.

If η is the mechanical efficiency of the motor-compressor and Y is the annual operating time (in hours/year) the power P_n (in kw) can be expressed as a function of the amount of energy degraded per year $\dot{E}$ (in kwh/year) that is :

$$P_n = \frac{\dot{E}}{\eta Y} \tag{78}$$

Finally the expression (77) for the cost of the motor pump becomes :

$$\boxed{C_m = C_{mo} + A_m \dot{E}^m} \tag{79}$$

with $\qquad A_m \equiv \dfrac{A'_m \gamma_m}{\eta Y}$

The minimum value of the cost of the motor-pump as a function of the variables B and θ is reduced to finding the minimum value of $\dot{E}$ which was given in paragraph 234.

In the rest of this discussion we will use a value of 1 for the exponent m. This only slightly reduces the generality of the conclusions.

34. MINIMUM VALUE OF THE TOTAL COST : Investment + operating costs

341. Expression for the total cost

As it was explained in paragraph (31) the total cost over a working life of n years comprises 5 terms :

$$C_t = A_e . \dot{E} + C_{mo} + A_m \dot{E} + A_p V_i . P + A_c (V_i . P)^i L^{j-i} + I_o \tag{80}$$

energy   the motor   the packed column
degraded compressor (anti-energetic
(pro-ener- investment)
getic
investment)

By grouping the two terms which are proportional to $\dot{E}$ and those which are independent, the total cost becomes :

$$\boxed{C_t \equiv (A_e + A_m) \dot{E} + A_p V_i . P + A_c (V_i . P)^i L^{j-i} + C_o} \tag{81}$$

Therefore the total cost is the sum of two components, the minima of which have already been studied separately :

- the <u>pro-energetic component</u> : $(A_e + A_m)\dot{E}$ discussed in paragraph 234.
- the <u>anti-energetic component</u> : $A_p V_i P + A_c (V_i P)^i L^{j-i}$, discussed in paragraph 32.

We must now make a synthesis.

342. <u>Case of an incompressible fluid (P = 1)</u>

The minimum value of the pro-energetic component (cf. § 144) is obtained for

$$\theta = 1/6 \qquad \text{and} \qquad B_e \rightarrow 0$$

The minimum value of the anti-energetic component is obtained for

$$\begin{cases} \theta \, \mathcal{E} \left[1/6, \dfrac{i/j}{2i/j + 4} \right] \\ B \, \mathcal{E} \left[B_c, \infty \right] \end{cases}$$

3421. <u>Optimization of θ</u>

In practice the value used for θ should be always equal to or just a little less than $1/6$: $1/10 \leq \theta \leq 1/6$.

The optimum value of θ is practically constant and independent of all the parameters whether scientific technical or even economic. This result may be consider as very interesting and useful for practical applications.

3422. <u>Optimization of B</u>

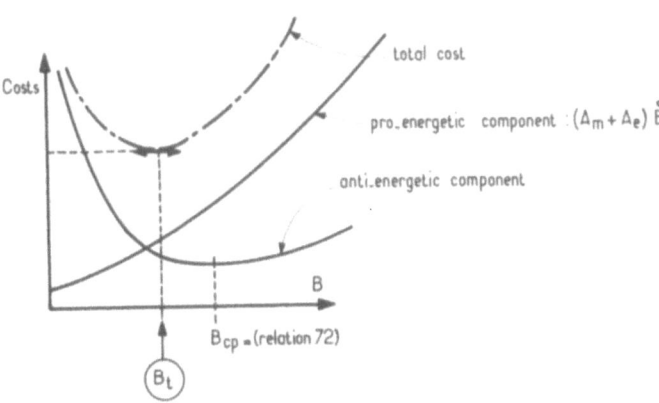

FIGURE 13

Figure 13 shows the sum of the 2 components of the total cost C_t,

. The anti-energetic component is a minimum for $B = B_{cp}$ and we have shown that the value obtained by numerical minimisation of (67) is certainly greater than the value B_{co} given by (71). This latter value is itself slightly greater than the value B_h corresponding to the minimum value of the height of a theoretical stage.

. The optimum value B_t corresponding to the minimum of the total cost has to be calculated numerically for each case. We may only say that it lies between 0 and B_{cp}.

$$B_t \quad \left[0, B_{cp}\right]$$

343. Case of a compressible fluid

Optimization of θ

The minimum value of the pro-energetic component (cf. § 234) is obtained for $\theta_e \simeq 1/6$ to $1/4$.

The anti-energetic component is a minimum (cf. § 323) for a value of θ between 0 and 1/6 and probably of the order of 1/8 to 1/12 as is shown by figure 12.

We may therefore conclude that for the most general case of a compressible fluid, the optimum value of θ varies only slightly with the various parameters and probably remains always of the order of 0.08 to 0.16.

Optimization of B

It is not possible to draw any general conclusions about the optimum value of B. We may only state that there is always an optimum value of B which is neither zero nor infinity. This value must be obtained numerically for each case.

FINAL REMARK :

The present paper has only presented a mathematical model for the optimization of preparative chromatography columns. Some general conclusions were derived from this model but no numerical example was given.

A further publication will be devoted to the separation of two phenotiazines by liquid chromatography. Another publication will treat the separation of α- and β-pinene by gas chromatography.

Nomenclature

A_c, A_c', A_c''	Cost coefficients for the column defined in § 32
A_e	Energy cost coefficient defined in § 33
A_m, A_m''	Cost coefficient for the motor-compressor unit defined in § 33
A_p, A_p', A_p''	Cost coefficients for the column packing defined in § 32
B	$= u d_p/D$ Bodenstein number dimensionless
B_c, B_e, B_h, B_{cp}, B_v }	Values of B which give a minimum for C_c, C_e, H, C_{cp}, V
b	Exponent for the Bodenstein number in the definition of H
C_c, C_{cp}, C_e, C_m, C_p, C_t	Respective costs of the column, the anti-energetic investment, the energy, the motor-compressor unit, the packing and the total cost
D	Diffusion coefficient of the solute
d_p	Diameter of a particle of packing
$\dot{E}$	Mechanical power degraded by or supplied to the carrier fluid (in kwh per year)
$H \equiv HETP$	Height of a theoretical stage
h_{-1}, h_o, h_1, h_c }	Coefficients in the definition of H
I_o, I_o', I_o''	Fixed costs, total, anti-energertic, pro-energetic
i	Exponent of Ω in the definition of C_c
j	Exponent of L in the definition of C_c
K	$= (k p_o d_p/2\mu D_o)$, permeability coefficient
k	Permeability of the column
L	Length of the column
$\dot{M}$	Mass flowrate of the carrier fluid
$\dot{M}_o$, $\dot{M}_o'$	$= 2\dot{m}/W_M$, $= \dot{M}_o P$
$\dot{m}$	Mass flowrate of the mixture to be separated
m	Exponent of $\dot{E}$ in the definition of C_m
$N \equiv NTP$ $N_o \equiv NTP_o$	Number of theoretical stages in eluto-frontal chromatography and elution chromatography respectively

P	pressure factor $= p_e/p_o$
p_e, p_o	fluid pressure at the inlet and exit from the column
R_{AB}	resolution factor of the exit peaks
T	periodicity of the injections
t_R	retention time
u	superficial velocity of the carrier fluid defined on the basis of the cross sectional area of the column
V, V_i	volume of the column for a compressible and an incompressible fluids respectively
W, W_M	mass concentration of the solute in the mixture to be separated in the carrier fluid W_M is its maximum value
Y	annual operating time (in hours per year)

Greek letters

α	$= t_{r_A}/t_{r_B}$ ratio of the retention times for the products A and B
γ	ratio of the specific heats at constant pressure and volume of the carrier gas
γ_c, γ_e, γ_m, γ_p	respective unit costs of the column, the energy, the motor-pump unit and the packing
η	mechanical efficiency of the motor-compressor unit
μ	dynamic viscosity
ρ	density
σ_A, σ_B	variances of the elution peaks of the products A and B
$\bar{\sigma}$	$(\sigma_A + \sigma_B)/2$
σ_i	one fourth of the injection time
θ, θ_v, θ_e	$= 4\sigma_i/T$, relative injection time, minimising V, minimising E
Ω	cross-sectional area of the column
Ω_c	$\dot{M}_o d_p / \rho_o D_o$

References

1. LE GOFF P. and N. MIDOUX "Investissements et economies d'énergie" Techniques de l'Ingénieur, n° J. 7100 (1977).

2. LE GOFF P. and N. MIDOUX "Optimization of energy and material costs in heat and mass exchangers", Chimica et Industria, 59 (1977) p. 523.

3. LE GOFF P. and N. MIDOUX "Interpretation of heat mass and momentum transfer analogy, in terms of energy - application to the optimization of heterogeneous reactors", Proceedings of the "Particle Technology Congress", Nuremberg, mars 1977.

4. LE GOFF P. "Optimisation énergétique des procédés industriels", Revue Générale de Thermique (France) 1978.
   1ère partie - Investissements et coûts d'exploitation, janv. 78, p. 9
   2ème partie - Actualisation - inflation et enchérissement de l'énergie, fév. 78, p. 89.

5. LE GOFF P. and N. MIDOUX "Energétique et optimisation économique", Livre à paraître en 1978.

6. LE GOFF P. "Analyse des systèmes et valeurs de l'énergie", Réunion annuelle de la Société Française des Thermiciens, Bordeaux, mai 1978.

7. LE GOFF P. "Les rendements d'utilisation de l'énergie par et pour les êtres humains", Revue Générale de Thermique (France) (1977), 181, p. 15.

8. JONES W.L., Anal. Chem., 33, 829, 1961.

9. GIDDINGS J.C., SEAGER S.L., STUCKI L.R. and G.H. STEWART, Anal., 32, 867, 1960.

10. SHERWOOD T.K., PIGFORD R.L. and WILKE C.R. "Mass transfer", Mc Graw Hill (1975).

11. VAN DEEMTER J.J., ZUIDERWEG F.J. and A. KLINKENBERG, Chem. Eng. Sci., 5, 271, 1956.

12. TRANCHANT J. and coll. "Manuel pratique de chromatographie en phase gazeuse" Masson, 1968.

13. KIRKLAND J.J. "Chromatographie en phase liquide", Gauthier Villars (1973).

14. CAUDE M. and P. LE XUAN, J. Chromato. Sci., 13, 390, 1975.

15. GRUSHKA E., SNYDER L.R. and J.H. KNOX, J. Chromato. Sci.,

16. CHAUVEL A., LEPRINCE P., BARTHEL Y., RAIMBAULT C. and J.P. ARLIE "Manuel d'évaluation économique des procédés" TECHNIP, 1976.

ANNEX 1

Injection conditions for elution chromatography

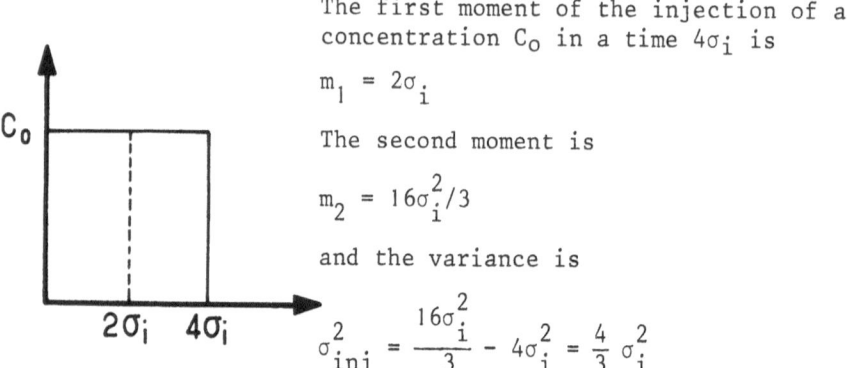

The first moment of the injection of a concentration C_0 in a time $4\sigma_i$ is

$$m_1 = 2\sigma_i$$

The second moment is

$$m_2 = 16\sigma_i^2/3$$

and the variance is

$$\sigma_{inj}^2 = \frac{16\sigma_i^2}{3} - 4\sigma_i^2 = \frac{4}{3}\sigma_i^2$$

The variance of the response to this injection is such that :

$$\sigma^2 = \sigma_{inj}^2 + \sigma_A^2 = \frac{4}{3}\sigma_i^2 + \sigma_A^2$$

and $\quad \sigma = \sigma_A \left[1 + \frac{4}{3}\left(\frac{\sigma_i}{\sigma_A}\right)^2\right]^{1/2}$

Clearly $\sigma_A \leq \sigma \leq 1.1\,\sigma_A$ if $0 < \sigma_i \leq 0.4\,\sigma_A$.

Conclusion : the model of ideal elution chromatography may be considered to be valid to within 10 % if the injection time $4\sigma_i$ is such that its variance σ_i is less than 40 % of the smallest of the variances of the products to be separated :

$$\sigma_i \leq \left| \begin{array}{l} 0.4\,\sigma_A \\ \\ 0.4\,\sigma_B \end{array} \right.$$

ANNEX 2

Expression for the height of a theoretical stage

In gas phase chromatography in packed columns, we may use the general expression for the height of a theoretical stage proposed by Jones and Kieselback |8| and Giddings |9|

$$H = 2\lambda d_p + 2\frac{D}{u} + \frac{K_A}{(1+K_A)^2}\frac{d_p^2}{D_s}u + \frac{d_p^2}{D}u \qquad (1)$$

In this expression λ is a coefficient which is close to 0.5 and which takes into account the irregularities in the packing.

- K_A is the capacity factor of the column with respect to product A
- d_p is the diameter of a particle of packing
- u is the velocity of the mobile phase
- D and D_s are the molecular diffusion coefficients in the mobile phase and the fixed phase.

By introducing the Bodenstein number $B = (ud_p/D)$, which is practically independent of the pressure, since both u and D are inversely proportional to p, the expression (1) becomes

$$H/d_p = 2\lambda + 2B^{-1} + \left[1 + \frac{K_A}{(1+K_A)^2} \cdot \frac{D}{D_s}\right] B \qquad (2)$$

For the calculations, we simply write this expression in the form

$$\boxed{H = h_{-1} B^{-1} + h_o + h_1 B^1} \qquad (3)$$

h_{-1}, h_o and h_1 are coefficients which are proportional to d_p.

As shown by figure (15) from the book by SHERWOOD et Al |10| expression (1) is particularly suitable for representing H as a function of u.

In the case of capillary columns of diameter d, H can be written |12|

$$H = a \frac{24(1+k')^2}{1+6k'+11k'^2} \frac{D}{u} + \frac{1+6k'+11k'^2}{96(1+k')^2} \frac{d^2}{D} u + \frac{2k'}{3(1+k')^2} \frac{e^2}{D_s} u \qquad (4)$$

in which

a is a constant

$k' = \frac{4e}{d} B$ is the capacity of the column

β is the partition coefficient

e is the thickness of the film of stationary phase.

This time putting $B = ud/D$, (4) becomes

$$H/d = a \frac{24(1+k')^2}{1+6k'+11k'^2} B^{-1} + \left[\frac{1+6k'+11k'^2}{96(1+k')^2} + \frac{2k'}{3(1+k')^2} \left(\frac{e}{d}\right)^2 \frac{D}{D_s}\right] B \qquad (5)$$

which is formally analogous to expression (3) with $h_o = 0$.

238

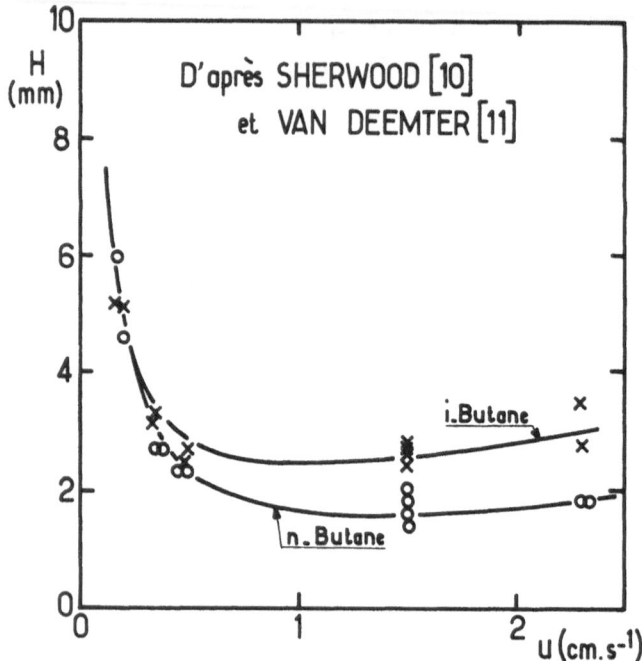

FIGURE 1

Height of a theoretical plate in a gas chromatographic column.
Data of Van Deemter, Zuiderweg and Klinkenberg |11| for a tube
18 cm long and 6 mm inside diameter filled with Celite particles
(around 80 microns average diameter) wetted with a hydrocarbon
oil of molecular weight 312. For i-butane, HETP = $0.8/(\varepsilon v)$ +
$0.8 + 0.08(\varepsilon v)$; for n-butane, HETP = $0.3 + 0.9(\varepsilon v) + 0.5(\varepsilon v)$.

In the case of liquid chromatography the form of the curves are
quite different as is shown in figure 2 taken from the book by
KIRKLAND |13|.

Caude et al. |14| have shown that this sort of curve can be repre-
sented (over its whole domain) by expressions of the following
type

$$H \sim u^b \, d_p^c \qquad (6)$$

By analogy with the preceeding we put this expression in the form

$$H = K_b \, B^b \qquad (7)$$

In general $0.3 \leq b \leq 0.7$, the most frequently encountered values
being 0.4 or 0.5. Expressions of this type have also been proposed
by GRUSHKA, SYNDER or KNOX |15|.

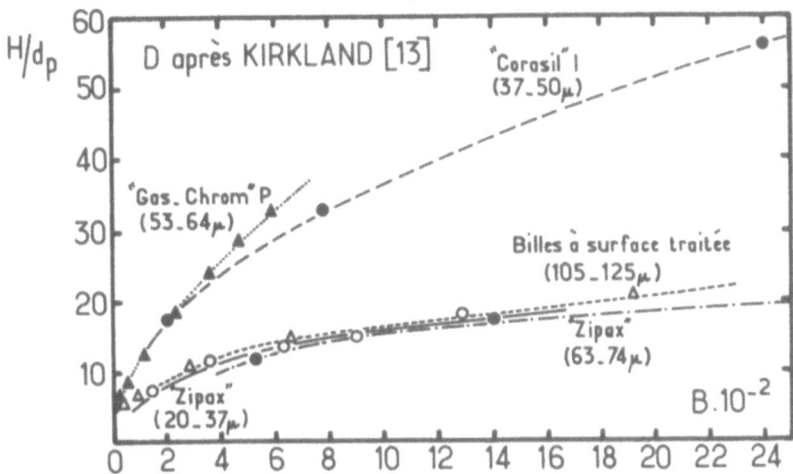

FIGURE 2

ANNEX 3

Injection rate and minimum flowrate of carrier fluid

We will treat the general case of a compressible gas.

Let m be the mass of mixture (A+B) injected in time $4\sigma_i$ at each pulse. If M is the molecular weight of the solute the molar flowrate of an injection is $(m/M)/4\sigma_i$. In the injection conditions with a mass flowrate $\dot{M}$ of carrier gas the volumetric flowrate is $\dot{V} = (\dot{M}/\rho_E) = (\dot{M}/\rho_o P)$ and the concentration of the solute in the carrier gas is

$$C_{mi} = \frac{\rho_o \, m \, P}{4\sigma_i \, M \, \dot{M}} \qquad \text{(in kmoles/vol)} \qquad (1)$$

At equilibrium this concentration is distributed between the fixed and mobile phases such that

$$\varepsilon C_{mi} = \varepsilon C_m + (1-\varepsilon) \, C_f \qquad (2)$$

in which ε is the void fraction of the packing
C_m and C_f as the respective concentrations in moles per unit volume of the mobile and the fixed phases.

If $\beta = C_f/C_m$ is the partition coefficient expression (2) becomes

$$C_{mi} = \left[(1/\beta) + (1-\varepsilon)/\varepsilon\right] C_f \qquad (3)$$

So as not to saturate the column, we assume that C_f must remain less than a certain limiting value C_f^* leading to the following conditions for the injection

$$C_{mi} = \frac{\rho_o \, m \, P}{4\sigma_i \, M \, \dot{M}} < \left[\frac{1}{\beta} + \frac{1-\varepsilon}{\varepsilon}\right] C_f^* \qquad (4)$$

It is easily shown that if two solutes A and B are introduced in the column with a fraction x of A the expression (4) becomes

$$\frac{m}{x\,M_A + (1-x)\,M_B} \cdot \frac{\rho_o P_E}{4\sigma_i \dot{M}} < \frac{C_f^*}{\dfrac{x}{\dfrac{1}{\beta_A} + \dfrac{\varepsilon}{1-\varepsilon}} + \dfrac{1-x}{\dfrac{1}{\beta_B} + \dfrac{\varepsilon}{1-\varepsilon}}} \qquad (5)$$

that is
$$\frac{\rho_o \, m \, P_E}{4\sigma_i \, \dot{M}} < \left[\frac{1}{\dfrac{1}{\beta} + \dfrac{\varepsilon}{1-\varepsilon}}\right]^{-1} \overline{M} \, C_f^* \qquad (6)$$

Introducing the concentration W which is a mass ratio, that is expressed in kg of solute (A+B) per kg of carrier fluid

$$W = \frac{m}{4\sigma_i \dot{M}} \qquad (7)$$

We may deduce the value of W_e that should not be exceeded

$$\left|\begin{array}{l} W_e = \dfrac{W_M}{P} \\[4mm] W_M = \dfrac{1}{\rho_o}\left[\dfrac{1}{\dfrac{1}{\beta} + \dfrac{\varepsilon}{1-\varepsilon}}\right]^{-1} \overline{M}\, C_f^* \end{array}\right. \qquad (8)$$

ANNEX 4

Expression for the pressure drop of the fluid

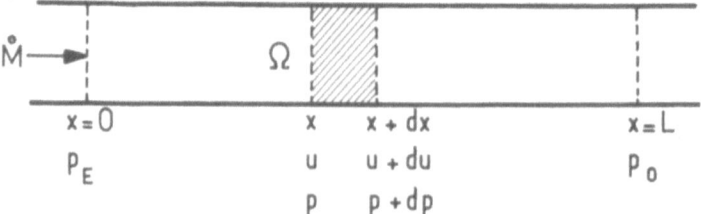

Let us consider the case of a perfect gas in isothermal flow in a packed tube. Euler's theorem allows us to write that in an elemental slice, the inertial and external forces are equal.

If the pressure drop by friction is developed by the Kozeny equation for laminar flow ($\rho u d_p / \mu < 1$) :

$$\frac{dp}{dx} = - \frac{\mu u}{k} \qquad (1)$$

In this expression k is the permeability of the column in units of L^2. This leads to :

$$\dot{M} \frac{du}{dx} = - \Omega \frac{dp}{dx} - \Omega \frac{\mu u}{k} \qquad (2)$$

For an ideal gas :

$$p/p_o = \rho/\rho_o \qquad (3)$$

The conservation equation allows us to write :

$$u = \frac{\dot{M}}{\rho \Omega} = \frac{\dot{M} p_o}{\Omega \rho_o} \frac{1}{p} \qquad (4)$$

Combining expressions (4) and (2) we obtain :

$$\frac{\rho_o}{P_o} p \, dp = \frac{\dot{M}^2}{\Omega^2} \frac{dp}{p} - \frac{\mu}{k} \frac{\dot{M}}{\Omega} dx \qquad (5)$$

Integrating this differential equation between the inlet the exit gives :

$$p_E^2 - p_o^2 = 2 \left(\frac{\dot{M}}{\Omega}\right)^2 \frac{P_o}{\rho_o} \text{Log} \frac{P_E}{P_o} + \frac{2\mu}{k} \frac{\dot{M}}{\Omega} \frac{P_o}{\rho_o} L \qquad (6)$$

In general :

$$\frac{\dot{M}}{\Omega} \text{Log} \frac{P_E}{P_o} << \frac{\mu}{k} L$$

We retain

$$\boxed{P^2 \equiv (p_E/p_o)^2 = 1 + \frac{2\mu}{k} \frac{\dot{M}}{\Omega \rho_o P_o} L} \qquad (7)$$

In the case of an incompressible fluid $du/dx = 0$ in expression (2) leading simply to

$$\boxed{P = 1 + \frac{\mu}{k} \frac{\dot{M}}{\Omega \rho_o P_o} L} \qquad (8)$$

ANNEX 5

Cost of the pumps and compressors

The next figure gives the costs of compressors (curve 2) and their driving motor (curve 1). Curve (3) represents the sum of these two costs taken from the book by Chauvel et al. |16|. In fact the costs of the compressors are for P > 10 HP but we have extrapolated these results to smaller values.

In the same way, we have plotted the cost of positive displacement pumps for lower (curve 4) or higher (curve 3) pressures than 100 bars.

For the calculations, we will use the numerical functions given below which are in good agreement with the practical results and represent the before-tax costs at mid-1975 in Fr, if P is in HP.

Compressor	$0.5 \leq P \leq 100$ HP	$C = 7270 + 6870\ P$
Pump	$0.05 \leq P \leq 100$ HP,	$\Delta p < 100$ bars
	$C = 4000 + 12600\ P^{0.754}$	
	$2 \leq P \leq 20$ HP,	$\Delta p > 100$ bars,
	$C = 23150 + 20000\ P$	

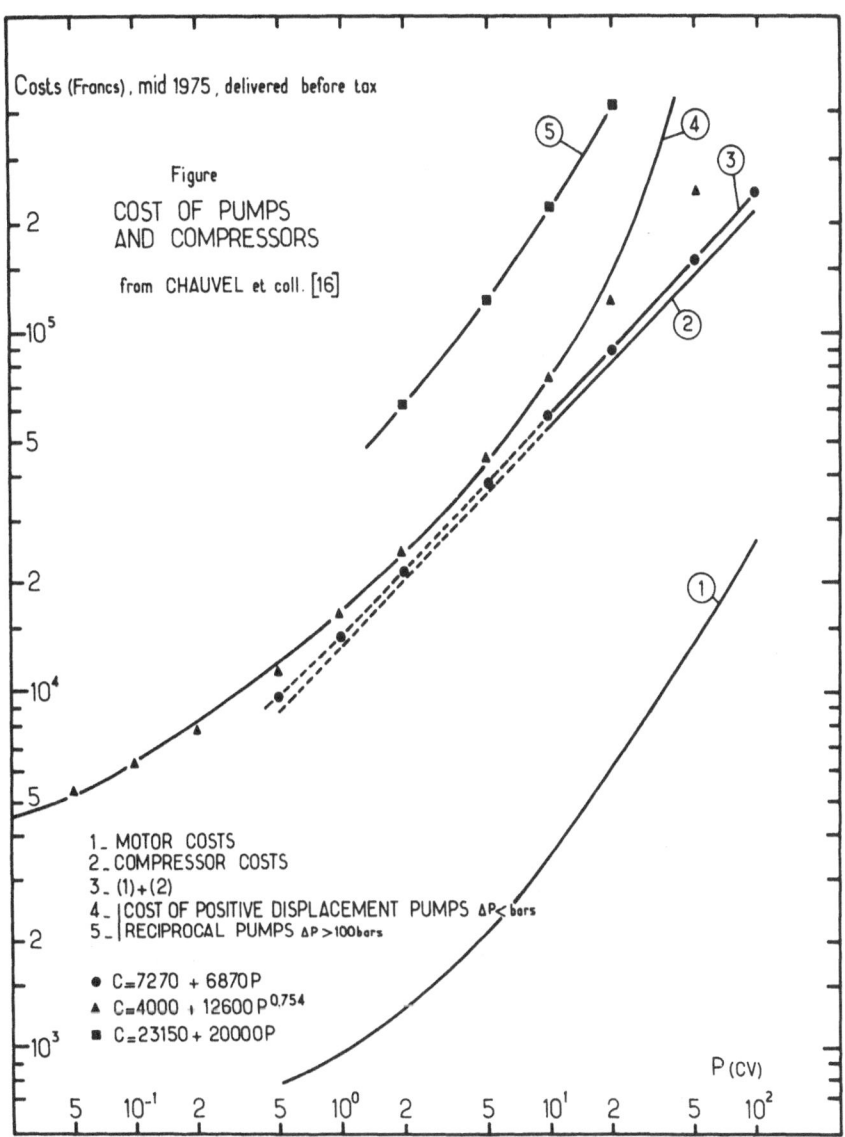

Costs (Francs), mid 1975, delivered before tax

Figure

COST OF PUMPS
AND COMPRESSORS

from CHAUVEL et coll. [16]

1 _ MOTOR COSTS
2 _ COMPRESSOR COSTS
3 _ (1)+(2)
4 _ | COST OF POSITIVE DISPLACEMENT PUMPS ΔP< bars
5 _ | RECIPROCAL PUMPS ΔP >100bars

● $C = 7270 + 6870P$
▲ $C = 4000 + 12600 P^{0.754}$
■ $C = 23150 + 20000P$

P (CV)

AN EXAMPLE OF ECONOMIC OPTIMISATION OF COLUMNS FOR PREPARATIVE
CHROMATOGRAPHY: SEPARATION OF TWO PHENOTIAZINES BY LIQUID
CHROMATOGRAPHY

N. Midoux and P. Le Goff

STATEMENT OF THE PROBLEM

MAJORS [1] and CAUDE et al. [2] have made a laboratory scale inves
tigation of the separation of phenotiazines on various adsorbants.
We propose using their results to make an economic optimisation of a
preparative scale separation of two phenotiazines, A an B. The
objective is to find the optimum operating conditions for treating
50 kg/24 hr of a half and half mixture of the two phenotiazines.
The laboratory investigation showed that the separation could be
considered to be satisfactory for a resolution factor of $R_{AB} = 3$
which we will adapt for this example.

Physico-chemical data (in SI units)

- mean density of the liquid

$\rho = 900$ kg/m^3

- composition of the mixture

$x_A = x_B = 0.50$

- maximum concentrations on a weight
  basis which should not be exceeded
  to remain in the ideal linear range
  (estimation)

100 kg/m^3

- porosity of the bed

$\epsilon = 0.35$

- partition coefficients

$\begin{vmatrix} \beta_A = 3.5 \\ \beta_B = 1.6 \end{vmatrix}$

- flow resistance of the bed (appro
  ximate)

$$\frac{P_e - P_s}{L} = \frac{\mu}{k' d_p^a} \cdot u$$

with d_p = partical diameter

   $a = 1.8$

   $\dfrac{\mu}{k'} = 2.0$ Pa m$^{-0.2}$. s

Height of a theoretical stage (approximate)

- according to Caude : $H = 1.8 \times 10^6 \ d_p^{1.75} \times u^{0.40}$

- according to Majors : $H = 5.2 \times 10^6 \ d_p^{1.75} \times u^{0.60}$

On the contrary to these workers we shall assume that the efficiency of the column is independent of its length. This is justifiable where the cross-sectional are is large enough to give a homogenous filling.

Economic data

As there is practically no cost data available for columns for pre perative chromatography, for the purposes of this example, we shall use cost data for similar columns used in other chemical engineering operations.

- cost of an empty column and its ancillary equipment with Ω in m^2 and L in m $\qquad C_c = C_{co} + 1.5 \times 10^5 \ \Omega^{0.3} \ L^{0.6}$

- cost of the packing $\qquad C_p = 2 \times 10^5$ frcs/m^3

- cost of the motor-pump unit (P is given in watts) $\qquad C_m = 13\ 000 + 17$ P

- cost of energy $\qquad Y_e = 0.2$ frcs/kwh

- overall efficiency of the motor-pump unit $\qquad \eta = 0.8$

- working life of the plant   25 000 hours

- working life of the packing   700 hours

The costs of assembly maintenance, etc... are assumed to be included in the above costs.

(Ref. 1) CAUDE, M., Le Xuan PHAN, TERLAIN, B et J.J.THOMAS, J. Chromatog.Sci.,13,390,1975.
(Ref. 2) MAJORS, R.E., J.Chromatog.Sci.,11, 88, 1973.

1st QUESTION - PRELIMINARY CALCULATIONS

Calculate:
- The maximum allowable concentration in terms of a mass ratio W_M, in kg of solute per kg ov carrier fluid.
- The minimum flow rate $\dot{M}_0$ of the carrier fluid.
- The minimum number of theoretical stages N_0 for elution chromatography.

2nd QUESTION - EXPRESSIONS FOR THE VOLUME OF THE COLUMN, THE ENERGY WHICH IS DEGRADED AND THEIR COSTS

Derive expressions for the following and substitute the appropriate numerical values:
- The volume of the column,
- The energy degraded and the pressure drop,
- The anti-energetic investment (column and packing),
- The pro-energetic costs (the motor-pump unit and the energy required).

3rd QUESTION - OPTIMISATION OF THE PACKED COLUMN (anti-energetic investment)

31. Optimum value of θ (relative duration of the injections)

Determine the value of θ which gives a minimum anti-energetic investment (that is θ_{cp}) and the minimum degradation of mechanical power (that is θ_e).

32. Effects of the size of the particles in the packing : d_p, for given values of θ and Ω

For the optimum value $\theta=\theta_{cp}$, and an arbitrary value of Ω, (for example $\Omega=2\times10^{-3}$ m^2 corresponding to a column diameter of 7.5 cm), determine the effects of the particle size on the various parameters (the mechanical power degraded and the anti-energetic investment). Use the following values of d_p : 5, 10, 20, 30 and 40 μm.

33. Effect of θ for given values of d_p and Ω

For a given set of values of d_p and Ω (10 μm and 2×10^{-3} m^2) determine the effect of θ on the column length, the mechanical power degraded and the investment.

34. Optimisation of the optimums

For a given optimum value θ_{cp} which corresponds to a minimum investment find the value of d_p and therefore Ω which also gives a minimum degradation of mechanical power.

Determine those cases where b is greater or less than i/j.
Bearing in mind the technological constraints, compare these theoretical minima with those which could be of <u>practical</u> use.

4th QUESTION - <u>MINIMUM TOTAL COST (INVESTMENT AND OPERATING COSTS)</u>

- Using the appropriate numerical values optimise C_t as given by
  the general expressions derived in part 2.

    41. For the 5 values of d_p (5, 10, 20, 30 and 40 µm) and the
two expressions for the HTS (that of Caude and that of Majors) de-
termine the optimum values of θ, Ω, L, Ê and C_t, that is θ^*, Ω^*,
L^*, $\dot{E}^*$ and C_t^*

    Discuss the practicality of the theoretical optimum condi-
tions determined in this fashion.

    42. Give a detailed description of the column and its opera-
tion for a given value of particle size d_p:10µm.

5th QUESTION - <u>THE ISO-COST DIAGRAM AS A FUNCTION OF Ω AND θ</u>

The intention is to examine how the total cost of the operation
varies when conditions are used different to the optimum determined in the previous question.
The domain of variation is however limited by the following techno
logical constraints:

- the pressure drop should not exceed 500 bars

- the column should not be longer than 1.5 to ensure a correct
  packing

- the cross-sectional area of the column should not be greater
  than 0.02 m^2 to avoid the possibility of a drop in efficiency
  due to bad liquid distribution.

Use Ω and θ independent variables. Delimit the possible zone of ope
ration on a diagram of $\Omega = f(\theta)$ and draw a set of iso-total cost
curves.
Use a dimensionless diagram with the following reduced coordinates
defined with respect to the optimum values

$$\Omega^+ = \frac{\Omega}{\Omega^*} \qquad \theta^+ = \frac{\theta}{\theta^*} \qquad C^+ = \frac{C}{C_t^* - I_o}$$

In particular examine the 2 "<u>principal sections</u>" through this
diagram that corresponding to the effect of Ω for $\theta^+ = 1$ and the
influence of θ for $\Omega^+ = 1$.
Discuss the effect of an extra investment and the economy of energy
obtained by using values different from the optimum.
Use only one value of particle size for example $d_p = 10$ µm.

INDUSTRIAL APPLICATIONS OF PREPARATIVE CHROMATOGRAPHY

A. J. de Rosset, R. W. Neuzil, D. B. Broughton

UOP Inc.
Des Plaines, Illinois

ABSTRACT

Adsorptive separation processes have assumed a major industrial
importance in the past ten years. The principal expansion has
been in simulated countercurrent operation, using highly specific
adsorbents. The engineering techniques involved, and methodology
for developing adsorbent/desorbent systems are discussed. Examples are submitted of a number of commercial applications.

1.   INTRODUCTION

Classical equilibrium separation processes employed in industrial
chemistry include continuous fractionation, azeotropic and extractive distillation, solvent extraction and crystallization.
These separations are based on selective transfer of components
of a mixture between two discriminate phases: gas liquid, liquid-
liquid or crystal-melt.

Equilibrium separations based on selective transfer between a
fluid and a solid separating agent also have been used for a long
time, generally for removal of minor components. Examples include drying with desiccants, decolorizing with charcoals and
demineralization with water softeners. The mechanisms may include adsorption, absorption and ion exchange. Insofar as these
separations have been carried out by percolation of fluids
through fixed beds of the solid separating agent, they bear some
analogy to chromatography.

Application of such percolation processes to bulk separation of

major components of a mixture is a more recent development.
Adsorptive separations in particular have exhibited a spectacular
industrial growth in the past ten years. Over forty adsorptive
separation units have been licensed in four different specific
applications. These units have a total capacity of 3.5 million
tons per year of extracted product. They utilize an aggregate
adsorbent loading of above 12,000 tons. In many cases these
separations would be impossible to accomplish in any other way.

2. BACKGROUND

2.1 Arosorb Process

An early extension of chromatography to a large scale industrial
separation is exemplified by the Sun Oil Arosorb process.[1]
This was operated in the early 1950's to extract aromatic hydrocarbons from refinery streams. Arosorb was an outgrowth of the
successful use of silica-gel adsorption as an analytical method
for separating organic compounds.

The analytical method is a column development type of analysis,
where the separated components remain in the column. The gel is
discarded after each test, as the relatively low cost of the gel
does not warrant regeneration. In any commercial application of
silica gel for selective adsorption, however, discarding the gel
or regenerating with heat to drive out the contained hydrocarbon
or organic liquid would be prohibitively expensive.

The problem was solved by converting the chromatographic operation from column development to column elution. A desorbent was
employed to completely remove the adsorbed species in chromatographic sequence and to prepare the gel for reuse. The desorbent was chosen from the same chemical type as the material
being selectively adsorbed from the charge stock, but of a
sufficiently different boiling point to permit subsequent separation by distillation.

The Arosorb unit was designed to process 2500 barrels per day of
catalytic reformate for the production of nitration-grade benzene
and toluene. The charge stock is pretreated by passing over
activated alumina or other suitable material for the removal of
water and other gel poisons, and then passes into one of several
silica-gel cases for a period of 30 minutes. The charge is then
diverted to another case, and the first case is fed with desorbent for a period of 70 minutes, at the end of which time the
gel is ready to receive charge again.

As the charge stock passes down through the bed, the charge
aromatics are adsorbed, leaving the bulk of the charge saturates

in the voids around the gel particles. The desorbent, which
consists of a crude xylene stream containing about 65% xylenes,
pushes the saturates out of the silica gel bed and displaces the
charge aromatics adsorbed on the gel.

The effluent from the bed is divided three ways. The first
effluent, containing the charge saturates, goes to a tower for
recovery of the saturated product overhead and desorbent as
bottoms, the latter being returned to the system for re-use. The
next portion of the effluent may be recycled to the charge stream
if particularly high purity products are desired. The final
portion of the effluent containing the charge aromatics flows to
a fractionator for recovery of the aromatic product as overhead
and the desorbent again as bottoms.

The silica gel cases had diameters of 3 to 10 feet and heights of
15 to 25 feet. In retrospect, Arosorb appears to be a simple
case of preparative liquid column chromatography. However, the
scale of operation represented a real breakthrough.

2.2 Hypersorption

Continuous adsorptive separation on an industrial scale actually
predates Arosorb. The Hypersorption process for separation of
gases with activated carbon was commercialized in 1947. [2]
Continuous operation was achieved by countercurrent contacting of
upflowing gas versus a downflowing bed of adsorbent. The process
concentrated ethylene from 5% in a hydrogen/methane feed mixture,
to a 95% purity ethylene make gas containing less than 0.1%
methane. The hydrogen-methane discharge stream contained less
than 0.1% ethylene.

Feed gas is charged near the center of the Hypersorber column,
into a slowly downward moving bed of activated carbon, which
tends to adsorb heavy components of the mixture more strongly
than the light components. The carbon containing the adsorbed
components is then contacted in the rectifying section below the
feed point with a countercurrent reflux of the bottoms product,
which serves to displace any of the lighter overhead components
adsorbed from the feed.

Now saturated with the bottoms product, the carbon passes through
the stripping section, a vertical tube bundle heated externally
by condensing Dowtherm vapors. Steam is introduced into the bed
below the heating section, passes up through the tube bundle
countercurrent to the carbon and strips the adsorbed components.
The steam and desorbed bottoms product or "make gas" are disengaged from the bed at a point above the stripping section and
just below the rectifying section.

252

The hot stripped carbon leaving the Dowtherm heater drops at a controlled rate into a sealing leg which restricts flow of steam into a lift system. A gas lift transports the carbon to the top of the Hypersorber tower.

The operating carbon bed level is maintained in a hopper at the top of the unit by addition from an external storage bin as necessary. Below the hopper the carbon is cooled in a vertical tube-in-shell water-exchanger above the feed point to complete the cycle.

The lighter components of the feed, which are not adsorbed or displaced from the carbon by the bottoms product reflux, pass up through the adsorption section and are split into two streams. The major stream is disengaged below the cooler, and constitutes the overhead product or "discharge gas". A small part of the overhead gas serves to dehydrate the stripping carbon by passing up through the tubes of the cooler countercurrent to the carbon flow. This "purge gas" then joins the lift-gas circulation stream.

The Hypersorber tower is 4.5 feet in diameter and 85 feet high. It was designed to process 1,800,000 standard cubic feet of gas per day at 75 psig pressure. The maximum carbon circulation rate is about 32,000 lb/hr, and the carbon stripping temperature is 265°C.

The Hypersorber bears some analogies to a temperature programmed gas/solid chromatograph with steam being used as carrier gas. However, the countercurrent operation has converted the transient temperature and concentration pulses of the chromatograph to steady state temperature and composition profiles, which allow for continuous feed introduction and continuous product withdrawal.

One clear shortcoming of a moving bed operation, such as Hypersorption, is attrition of the adsorbent. This would be even more critical were the process to use a modern "tailored" adsorbent such as those required for difficult separations, rather than a relatively inexpensive carbon.

2.3 Ion Exchange

The moving bed, continuous, countercurrent mode of operation has been applied to another important percolation process, ion exchange. This application is known as countercurrent ion exchange (CCIX).[3] The process appeared in the patent literature as far back as 1922, but did not realize industrial importance until the 1960's, with the development of a rugged polystyrene resin bead type ion exchanger. The first large scale demonstration was in

Japan for recovery of copper and ammonia from wastes. Instal-
lations handling 8,000 lb/day of copper have been reported.
Water treatment units range up to 8,400 gpm in capacity.

Several process variations are commercial. They are all based on
passing the aqueous liquid to be treated up through a vertical
packed column of ion-exchange resins in an adsorption tank.
Periodically (every 10 to 60 minutes) the liquid flow to the unit
is stopped for less than one minute, and a small portion (5 to
10%) of exhausted ion-exchange resin is removed from the bottom
of the adsorption tank and transferred to a regeneration tank,
while an equal amount of regenerated and rinsed resin is transferred and returned to the top of the adsorption tank. Because
the exhausted resin is withdrawn from the liquid inlet end as the
freshly regenerated resin is introduced at the liquid outlet end,
countercurrent flow of liquid and resin is achieved. Similarly,
the flow of regenerating liquid is countercurrent to the resin
flow in the regeneration tank.

Advantages claimed for the process are lower resin inventory,
lower consumption of regenerating chemicals and better product
quality. Except for brief interruptions for resin transfer,
operation is continuous. Resin attrition in CCIX is greater than
in fixed bed plants, and rugged resins are essential.

2.4 Summary

A recapitulation of the features of these three processes is
given in Table 1. All three use conventional bed packings --
silica gel, carbon or ion exchange resin. They are operated
either in conventional chromatographic mode (Arosorb), or employ
countercurrent contacting of fluid and solid. In the latter
case, the bed is physically moved to effect the countercurrent
contact. Finally, they may operate isothermally, as in the case
of Arosorb and countercurrent ion exchange; or they may involve
a temperature change, as in the case of Hypersorption.

The last entry in Table 1, the Sorbex [R] process, has contributed
most heavily in recent years to industrial applications of adsorptive separation. It will be the major topic of the remainder
of this discussion. Sorbex is a general name for a number of
processes effecting a variety of commercial separations, all
using the same engineering technology. The chemical technology,
that is, the nature of the associated adsorbent varies. Adsorbent composition and structure are tailored to the demands of
the particular separation to be made. The contacting mode is
countercurrent and continuous, but this is achieved without
physical movement of the bed. Finally, the process is isothermal
and liquid phase, analogous to liquid column chromatography, but

PROCESS	FUNCTION OR PRODUCT	BED PACKING	CONTACT MODE	TEMPERATURE VARIATION
AROSORB	AROMATICS	SiO$_2$ GEL	CONVENTIONAL CHROMATOGRAPHIC FIXED BED	ISOTHERMAL
HYPERSORPTION	ETHYLENE	CARBON	COUNTERCURRENT MOVING BED	2-LEVEL
CCIX	DEMINERAL- IZATION	IX RESIN	COUNTERCURRENT MOVING BED	ISOTHERMAL
SORBEX	VARIED	SPECIALIZED	COUNTERCURRENT FIXED BED	ISOTHERMAL

UOP 168A-1

Table 1. Characteristics of Commercial Separation Processes.

HYDROCARBON	CAPACITY cc/g ADSORBENT
n-HEXANE	0.309
n-HEPTANE	0.311
n-OCTANE	0.300
ISOOCTANE	0.282
BENZENE	0.295
TOLUENE	0.301
CYCLOPENTANE	0.334
CYCLOHEXANE	0.288

UOP 168A-2

Table 2. Capacity of Zeolite X for Hydrocarbons.

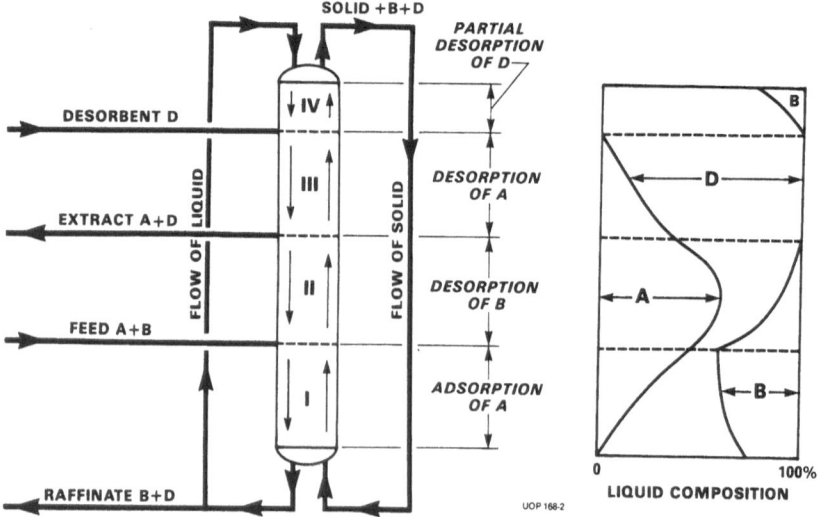

Fig. 1. Adsorptive Separation with Moving Bed.

in a number of respects more efficient for large scale operations.

3. SORBEX PRINCIPLE

The Sorbex principle[4] has been preeminently useful in commercializing difficult bulk separations. It takes advantage of adsorbents containing pores which are specifically selective for a particular component of a mixture, and of the large number of separation stages available in a chromatographic column. The Sorbex operation is designed to produce the same results and process relations that would be obtained by continuous countercurrent flow of solid and liquid. Feed and product streams enter and leave the adsorbent bed continuously and at substantially constant composition. This result is achieved, however, without actual movement of the solid.

3.1 Description of Sorbex in Terms of Moving-Bed Operation

Sorbex does not actually employ a moving bed. However, it is clearer to first describe the principles of operation in terms of a moving-bed system. The means by which the same result is obtained without bed movement will then be outlined.

A sketch of a hypothetical moving-bed system is shown in Figure 1. The adsorbent is circulating continuously, in a closed cycle, and moves up the adsorbent chamber from bottom to top. Liquid streams flow down through the bed, countercurrently to solid.

For simplicity, the feed is assumed to be a binary mixture of A and B, with component A being selectively adsorbed relative to B. It is introduced to the bed as shown.

Desorbent D is introduced to the bed at a higher level as shown. It is a liquid, of different boiling point from the feed components, that is capable of displacing feed components from the pores. Conversely, it is also possible for feed components to displace desorbent from the pores when the relative flow rates of solid and liquid are properly adjusted.

Raffinate product, consisting of the less strongly adsorbed Component B, mixed with desorbent, is withdrawn from the bottom of the bed. A portion of this stream is recirculated to the top of the bed.

Extract product, consisting of the more strongly adsorbed Component A, mixed with desorbent, is withdrawn from an intermediate point in the bed as shown.

The positions of introduction and withdrawal of net streams divide the bed into four zones, each of which performs a different function.

Zone I -- The primary function of this zone is to adsorb A from the liquid. The solid entering the bottom of this zone carries only B and D in its pores. As the liquid stream flows downward, countercurrent to this solid, Component A is transferred from the liquid stream into the pores of the solid. At the same time, Component D is desorbed -- i.e., transferred from the pores to the liquid stream -- to make room for A in the pores.

Zone II -- The primary function of this zone is to remove B from the pores of the solid. When the solid arrives at the fresh feed point, the pores will contain the quantity of A that was adsorbed in Zone I. However, the pores will also contain a large quantity of B, because the solid has just been in contact with fresh feed.

The liquid entering the top of Zone II contains no B -- only Components A and D. As the solid moves upward, countercurrent to this stream, Component B is gradually displaced from the pores and is replaced by Components A and D. Thus, when the solid arrives at the top of Zone II, the pores will contain only A and D.

By proper regulation of the liquid rate in Zone II, B can be completely desorbed from the pores. This can be done without simultaneously desorbing all of Component A, because A is more strongly adsorbed than B.

Zone III -- The function of this zone is to desorb A from the pores. The solid entering the zone carries A and D in the pores. The liquid entering the top of the zone consists of pure D. As the solid rises, Component A in the pores is displaced by D.

A portion of the liquid leaving the bottom of Zone III is withdrawn as extract; the remainder flows downward into Zone II as reflux.

Zone IV -- The practical purpose of this zone is to reduce the required circulation rate of fresh desorbent. This is desirable in order to reduce the load on the fractionators which separate desorbent from the net products. This is accomplished in the following manner.

When the adsorbent leaves Zone III, the pores are completely filled with desorbent. The liquid entering the top of Zone IV consists of B and D. If the liquid flow rate in Zone IV is properly regulated, Component B will be completely readsorbed from the liquid and an equal quantity of D will be displaced from the pores. This quantity of D then flows as liquid into Zone III, where it functions effectively as desorbent.

The liquid phase composition profile shown in Figure 1 serves to clarify the action taking place.

3.2 Difficulties of Moving-Bed Operation

In the early stages of the Sorbex development, the possibility of an actual moving-bed operation, as pictured in Figure 1, was considered. However, a number of serious difficulties soon became apparent. In the first place, the adsorbents available were not strong enough to resist the abrasion that would be encountered in a moving bed.

Another serious problem was that of obtaining uniform flow of both liquid and solid phases in a moving-bed operation. The performance of this type of operation can be greatly degraded by non-uniform flow of either phase. It was anticipated that great difficulty would be eccountered in obtaining uniform flow of solid over a bed of large diameter, particularly since the solid must be removed from one end of the vessel and reintroduced at the other end. In addition, it was anticipated that an attempt to move the bed would lead to non-uniformity in packing, which would lead to channeling of the liquid.

Another arrangement considered was the use of a series of fluidized beds, in which solid would overflow from each bed to the next. This was abandoned because of the sacrifice in mass-transfer efficiency that would have resulted. In a series of fluidized beds, the number of theoretical equilibrium stages cannot exceed the number of physical beds. The analogy in distillation is the fact that the number of theoretical trays in a trayed fractionating column is never significantly greater than the number of actual trays.

In contrast, it is known that very high mass-transfer efficiencies can be achieved in the flow of fluids through fixed beds of adsorbent. For example, laboratory chromatography is known to provide thousands of theoretical stages of separation in beds of modest length.

At this point, attention was redirected to the problem of retaining the advantages of continuous countercurrent operation, without introducing the disadvantages of a moving bed. A description of the scheme developed follows.

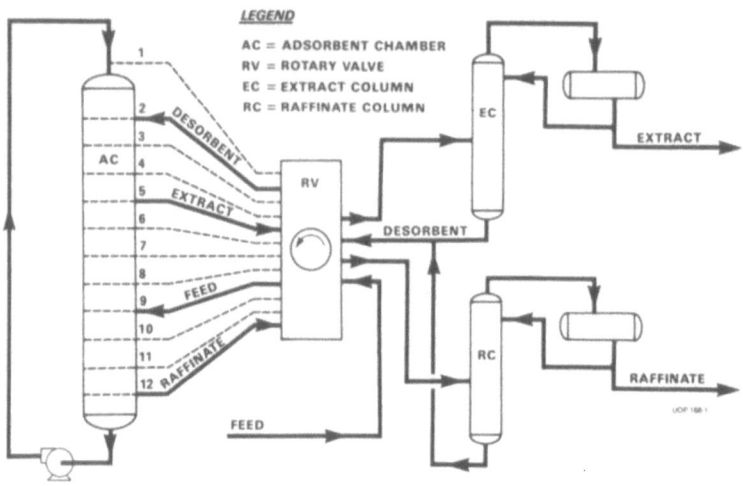

Fig. 2. Sorbex - Simulated Moving Bed for Adsorptive Separation.

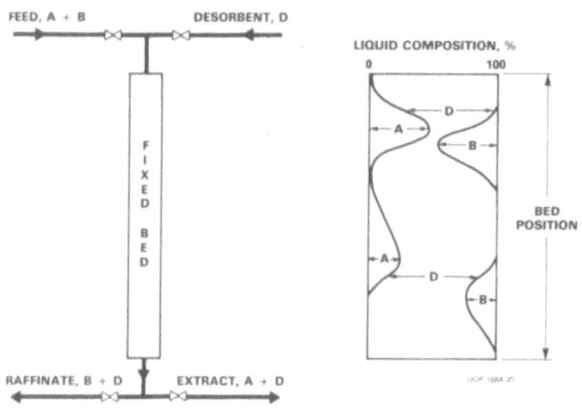

Fig. 3. Batch Adsorption.

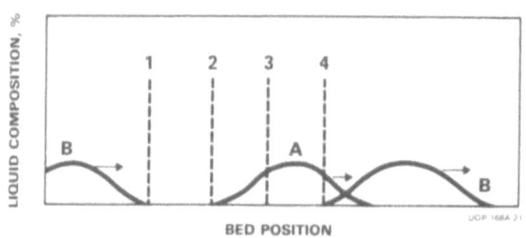

Fig. 4. Band Positions in Batch Column.

3.3 Simulated Moving Bed - The Sorbex Technique

In the moving bed system of Figure 1, solid is moving continuously in a closed circuit, past fixed positions of introduction and withdrawal of liquid. The same process results can be obtained by holding the bed stationary and periodically moving the positions at which the various streams enter and leave. A shift in the positions of liquid feed and withdrawal, in the direction of fluid flow through the bed, simulates the movement of solid in the opposite direction.

It is, of course, impossible to move the liquid feed and withdrawal points continuously. However, approximately the same effect can be produced by providing multiple liquid access lines to the bed, and periodically switching each net stream to the next adjacent line.

The commercial embodiment of this concept is shown in Figure 2. Here, the adsorbent is maintained as a stationary bed. A liquid circulating pump is provided to pump liquid from the bottom outlet to the top inlet of the adsorbent chamber.

A fluid-directing device known as a "rotary valve" is provided. This functions on the same principle as a multiport stopcock. At the right-hand face of the valve, the four net streams to and from the process are continuously fed and withdrawn. From the left-hand of the valve, a number of lines are connected, which terminate in distributors within the adsorbent bed.

At any particular moment, only four of the lines from the rotary valve to the adsorbent chamber are active. Figure 2 shows the flows at a time when lines 2, 5, 9 and 12 are active. Then the rotating element of the rotary valve is moved to its next position, each net flow is switched to the next adjacent line. Thus, desorbent will enter line 3 instead of line 2, extract will be drawn from line 6 instead of 5, feed will enter line 10 instead of 9 and raffinate will be drawn from line 1 instead of 12.

Functionally, the adsorbent bed has no top or bottom; it is equivalent to an annular bed. Therefore, the four liquid access positions can be moved around the bed continually, always with the same distance between the various net streams.

From the moving-bed operation of Figure 1, it can be seen that the liquid flow rate in each of the four zones is different, because of the addition or withdrawal of the various net streams. In the simulated moving bed of Figure 2, liquid rate is controlled by the circulating pump. At the position shown in Figure 2, the pump is between the raffinate and desorbent ports and, therefore, should be pumping at a rate appropriate for Zone IV.

However, after the next switch in position of the rotary valve, the pump will be between the feed and raffinate ports, and should, therefore, be pumping at a rate appropriate for Zone I.

Stated briefly, the circulating pump must be programmed to pump at four different rates. The control point will be altered each time that a net stream is transferred from line 12 to line 1.

To complete the simulation, it is obviously necessary that the liquid flow rate relative to the solid must be the same in both the moving-bed and simulated moving-bed operations. Since the solid is stationary in the simulated moving-bed operation, the liquid velocity relative to the vessel wall must be higher than in an actual moving-bed operation.

3.4 Mathematical Modelling

It is known that the theoretical performance of this operation, as commercially designed is practically identical to that of a system in which solids flow completely continuously, as a dense bed, countercurrent to the liquid, as depicted in Figure 1. The operation is modelled[5] in terms of theoretical equilibrium trays having the same significance as in fractionating columns. Solid and liquid are assumed to flow continuously through hypothetical well-mixed theoretical trays in which equilibrium is attained. The number of trays is determined by bed height, mass transfer coefficient, and flow rates. Axial mixing is generally of much greater significance in liquid systems than in vapor systems because of the greater mass of process fluid in the voids relative to that in the selective pores. In order to allow for axial mixing in the liquid phase, it is necessary to assume that the solid entrains with itself from tray to tray, a certain amount of interstitial fluid. This model of equilibrium theoretical trays with entrainment is readily implemented by computer by methods analogous to those used in the design of fractionating columns.

Two parameters are adjusted in the model to reproduce experimental concentration profiles:

Number of theoretical trays: $n = KkH/Lz$ (1)

Axial mixing ratio: $e = E/L = Kk/DL^2$ (2)

where: K = linear equilibrium constant
k = mass transfer coefficient
H = bed height
L = liquid rate
$z = m \ln (m/m-1)$

m = adsorption factor = pore circulation rate K/L
E = entrainment rate
D = axial diffusion coefficient

Each zone is treated separately according to the component whose concentration is changing most rapidly. Values of n and e corresponding to the equilibrium properties of the critical component for each zone are used.

For a simple binary feed, using a desorbent intermediate between the feed components in selectivity, it is possible to calculate the minimum circulation rates of adsorbent and desorbent required to achieve perfect separation when the mass transfer coefficient is infinite, the axial diffusion coefficient is zero, and the equilibrium enrichment factors are constant. For this case:

$$\frac{\text{Pore circulation rate}}{\text{Feed Rate}} = \frac{1}{\beta_{AB}-1} + X_{AF} \qquad (3)$$

$$\frac{\text{Desorbent circulation rate}}{\text{Feed Rate}} = \frac{1 + (\beta_{AB}-1)X_{AF}}{\beta_{DB}} \qquad (4)$$

where: β_{AB} = selectivity for strongly adsorbed feed Component A, relative to weakly adsorbed feed Component B.

β_{DB} = selectivity for Desorbent D, relative to weakly adsorbed feed Component B.

X_{AF} = volume fraction strongly adsorbed component in feed.

3.5   Comparison with Fixed-Bed Operation

It is of interest to compare the characteristics of the continuous Sorbex operation with those of the conventional liquid chromatography, or batch operation. The batch operation is illustrated in Figure 3. It consists simply of charging increments of feed and desorbent alternately to a fixed bed. As the feed components are eluted through the bed, they gradually separate into bands, which travel at different rates, and are withdrawn alternately as raffinate and extract. Band separation as the pulse of feed travels through the bed is illustrated by the composition profiles. A second increment of feed must be delayed long enough to ensure that the least strongly adsorbed component does not overtake the most strongly adsorbed component in the first increment.

A comparative mathematical modelling of the two operations has shown that the batch operation requires more adsorbent inventory by a factor of 3-4, and more desorbent circulation by a factor of 2. Without going into the details of these mathematical analyses, it is possible to explain the large difference in adsorbent requirement in physical terms.

In the continuous system, every part of the bed at all times can be identified as performing useful work with respect to the primary function of each zone. In the batch system, however, various parts of the bed at various times can be identified as doing either nothing or something useless.

This is most clearly seen near the entrance of the bed in the batch system. As feed enters, the adsorbent near the inlet rapidly comes to complete equilibrium with the feed; as feed continues to enter, this section performs no further function except that of a pipe, carrying the feed down into the part of the bed where action is occurring. A similar situation exists when desorbent is introduced. For instance, the first theoretical tray in the example used here is practically in complete equilibrium with feed or desorbent for over 95% of the time.

Non-useful zones can also be identified further down the bed. Figure 4 shows the position of the bands in an intermediate section of the column. The B-bands are moving to the right faster than the A-bands, and both are broadening as they travel through the column. An obvious "dead zone" exists between positions 1 and 2, where nothing but desorbent is present. This dead zone must be present through the entire column, ideally diminishing to zero at the bed exist.

The A-component in the interval 2-4 of Figure 4 is already pure, and the passage of this material through the rest of the column serves no useful function. It is true that pure Component A must be desorbed from the solid in the continuous operation also. However, in the batch system, this pure material will be readsorbed and desorbed many times before reaching the bed exit.

Consider the history of the narrow segment of bed as the band of Component A passes through it from position 3 to 4. The solid first accumulates, then loses A and undergoes no net change in condition. During this time, no net action of any kind has been accomplished by this bed segment.

4.   ADSORBENTS

4.1 Generally Desired Properties

The success of commercial chromatographic separation processes depends heavily on the development of suitable column packings. These packings may function as absorbents, adsorbents or ion exchangers. The present discussion will focus mainly on adsorbents. However, any column packing, be it absorbent, adsorbent or ion exchanger, used in a Sorbex operation should have the following characteristics: (1) high capacity, (2) high selectivity, (3) good kinetics, (4) good packing qualities, (5) chemical stability, (6) physical stability and (7) commercial availability in large quantities.

The first three requirements are basic, and determine the adsorbent inventory needed to meet the design capacity of a separation process. In the absence of kinetic control, the pore circulation rate, as stated in Equation 3, is related to plant capacity.

Capacity may be defined as selective pore volume per volume of adsorbent. Clearly, all other factors -- selectivity, valve rotation time, feed rate and feed composition -- being the same, the volume of adsorbent required will decrease as its pore volume, or capacity, is increased.

The selectivity of an adsorbent for Component A of a mixture relative to Component B is defined as:

$$\beta_{AB} = \frac{X_A}{Y_A} \Big/ \frac{X_B}{Y_B} \qquad (5)$$

where X_A and Y_A are concentrations of Component A in the adsorbed and liquid phases, respectively, and X_B and Y_B are the corresponding values for Component B. Under conditions of equilibrium control, β is the major determinant of pore circulation rate. As indicated by Equation 3, its effect is overwhelming at values close to unity. The higher the selectivity, the lower the adsorbent inventory required.

The kinetics of an adsorptive separation process, as in chromatography, involve mass transfer of the adsorbed species of feed and desorbent into and out of the pores of the adsorbent. Good kinetics, as in chromatography, are characterized by sharp concentration profiles and the absence of "tailing".

Good packing is as essential in commercial chromatographic separation processes as in laboratory chromatography. It minimizes axial mixing due to channeling.

Chemical stability implies that the adsorbent is not dissolved, eluted or corroded by the process fluid. It also implies that the adsorbent does not promote any catalytic transformations of the process fluids. Since adsorbents and catalysts share the

common characteristics of high surface area and active sites,
one must ensure that the adsorptive sites are only strong enough
to perform their proper function, and do not activate the adsorbed molecules into a chemical change.

Commercial availability becomes a problem in the light of the
current trend in column packing technology to employ more selective and specific solid separating agents. Standard resin ion
exchangers are readily available. However, many new resin type
ion exchangers have been developed, and continue to be developed,
in analytical laboratories for selective removal of specific
ions -- copper, cobalt, nickel, platinum and mercury. These are
only beginning to appear on the market for commercial applications. Similarly, the standard, readily available adsorbent
packings such as activated carbon, silica and alumina, do not
meet the selectivity requirements for difficult separations.
New zeolitic adsorbents have had to be brought from laboratory
formulation to commercial production before these separations
could be realized on an industrial scale.

4.2 Zeolites

Zeolites[6,7] are crystalline alumino-silicates. They can be
considered as a cross linked inorganic polymer whose units are
$Si (0/2)_4$ and $Al (0/2)_4^-$ tetrahedra, where $0/2$ represents an
oxygen atom shared between two connected tetrahedra. The tetra-
valent silica is electrically neutral in such a tetrahedral
structure. However, the trivalent aluminum bears a formal
charge of -1. This requires the presence of a counter cation,
such as Na^+, one for each aluminum atom. These cations are, to
varying degrees, exchangeable and, as a matter of fact, the
early industrial use of zeolites was for ion exchange and water
softening.

The silica and alumina tetrahedra can be combined in many ways --
as rings, prisms and higher polyhedra. A ring of six tetrahedra
is represented by a 6-ring where the vertices represent silica
and alumina atoms, and the midpoints of the connecting lines
represent bridging oxygens.

Similarly, a hexagonal prism of tetrahedras is represented by
two 6-rings joined by bridging oxygens. Various combinations of
such structures generate the pores or cavities that confer on
zeolites their remarkable adsorptive properties.

Sodalite -- An important polyhedral structure is the truncated
octahedron, bounded by 6-rings and 4-rings. This polyhedron is
generated from an octahedron by slicing off the six vertices.
The six slices yield six square faces, and eight hexagonal

faces, corresponding to the six vertices and eight sides of the original tetrahedron. This is known as the sodalite structure, and the cavity within the structure is known as the sodalite cavity. A close packing of truncated octahedra in space represents the silica-alumina structure of minerals such as sodalite and ultramarine, which differ according to the nature of the ions within the cavities. Zeolites of this type show limited adsorptive properties because of their close packed structure and relatively small cavities.

Zeolite A -- Zeolites with more desirable adsorptive properties have a more open structure. These can be built up by stacking the sodalite structures in different fashions. For example, if they are connected at their square faces by square prisms, in a cubic array, one obtains the Zeolite A lattice. This is a simple cubic lattice with a sodalite cage at each corner, and an 8-ring on each face. The enclosed volume is called the α-cage and has a diameter of 11.4A. The free diameter of the six 8-ring apertures is 4.2A. This is large enough to admit normal paraffinic hydrocarbons, but not isoparaffins or cyclics.

Zeolite A is generally synthesized in the sodium form, 4A. The sodium ions partially block the apertures, and exclude normal hydrocarbons. Exchange with calcium yields the 5A form. Since each calcium ion provides the equivalent positive charge of two sodium ions, the apertures are unblocked. This is the form of Zeolite A used in the UOP Molex [R] process for adsorptive separation of normal paraffin hydrocarbons.

Faujasite -- The faujasite structure can be built up from sodalite units by stacking them in a tetrahedral array with the 6-ring faces connected by hexagonal prisms. The enclosed volume, or supercage, has 12-ring apertures with a maximum free diameter of 7.4A. At saturation the cavity will hold 4-5 molecules of light hydrocarbons like benzene. A bulky hydrocarbon like 1,3,5-triethylbenzene may or may not be able to penetrate into the cavity, depending on temperature and the nature of the cation.

Mordenite -- Another commercially important zeolite is mordenite. This is built from 5-rings, while the large apertures are 12-rings. Because of the high degree of distortion of the 12-rings, their free diameter is only about 6.8A instead of 7.4A as in a faujasite. Unlike faujasite and Zeolite A, which have a three dimensional network of channels, the large channels in mordenite are parallel to one axis, and are usually interconnected by smaller channels.

4.3 Zeolitic Adsorbents

Crystalline zeolite adsorbents have shown marked advantages over the conventional amorphous column packings, such as silica, alumina or activated carbon. These advantages pertain to all three of the basic requirements of a good adsorbent -- capacity, selectivity and kinetics.

Capacity -- The adsorptive capacity of dry faujasitic zeolites for C_6-C_8 hydrocarbons has been measured by vapor adsorption. The saturation capacity is reached at relative pressure, and the isotherm is normal to the vertical line at $p/po = 1$. The volume of various hydrocarbons adsorbed is shown in Table 2. The capacity is around 0.3 cc/g with some variations attributed to the ease of packing of the different types of molecules in the supercage.

For amorphous adsorbents, where the adsorptive capacity is modelled as the volume of the adsorbed phase, Va, on a plane surface, the customary chromatographic correlation is:[8]

$$Va = 0.00035 \times (\text{surface area, } M^2/g) \tag{6}$$

Therefore, the zeolite for which the data in Table 2 were secured corresponds in capacity to an amorphous adsorbent of 850 M^2/g surface area. This is considerably higher than the surface areas of silica gels or activated aluminas used industrially. Activated carbons have higher surface area. However, some of this area, as measured by nitrogen adsorption, is in pores too small to permit access to larger organic molecules.

Commercial zeolitic adsorbents may not exhibit the high capacities quoted above. They may contain amorphous material or non-adsorbing crystalline components, produced as impurities in their synthesis. They may contain a relatively non-absorptive binder required for pelletizing, or they may contain varying amounts of water. Nevertheless, satisfactorily high capacities are commercially attainable.

Selectivity -- Zeolitic adsorbents can be tailored to provide selectivities that are not attainable with conventional adsorbents, such as silica and alumina, or with other separating agents, such as selective solvents.

At least three compositional variations are possible -- the zeolite structure, the silica to alumina ratio, and the nature of the replaceable cation. Previously mentioned was the conversion of the sodium form of Zeolite A by calcium ion exchange, to give the Molex adsorbent, selective for normal paraffins.

Table 3 illustrates the variety of adsorption sequences among the four C_8- aromatics which can be obtained by modification of

1 = MOST SELECTIVELY ADSORBED
4 = LEAST SELECTIVELY ADSORBED

	ADSORBENT NO. 1	ADSORBENT NO. 2	ADSORBENT NO. 3
p-XYLENE	1	4	3
ETHYLBENZENE	2	1	4
o-XYLENE	3	2	1
m-XYLENE	4	3	2

Table 3.   Selectivity of Adsorbents in C_8-Aromatic System.

ADSORBENT	COMPONENT A/COMPONENT B	β
NO. 1	FRUCTOSE/GLUCOSE	7.0
	FRUCTOSE/SUCROSE	> 50
NO. 2	GLUCOSE/FRUCTOSE	2.1
	GLUCOSE/MALTOSE	> 50

Table 4.   Enrichment Factor (β) of Adsorbents for Separation of Saccharides.

	FEED, WT-%	EXTRACT, WT-%	RAFFINATE, WT-%	EXTRACTION, %
p-XYLENE	14.3	98.6	0.9	93.8
ETHYLBENZENE	32.6	0.5	37.5	0.2
m-XYLENE	35.5	0.5	38.8	0.2
o-XYLENE	17.6	0.4	22.8	0.2

Table 5.   Bench Scale Parex Separation of *p*-Xylene.

structure and composition. Adsorbent No. 1 is selective for p-xylene, and is used in the UOP Parex ^{R} process for p-xylene separation. Adsorbent No. 2 shows an entirely different order of adsorptive selectivity, and is a candidate for ethylbenzene extraction. Adsorbent No. 3 will deliver ethylbenzene as a raffinate, and has been utilized on a pilot plant scale in the UOP Ebex ^{R} process for separation of ethylbenzene.

Table 4 illustrates adsorbents that have been studied for commercial separation of fructose from glucose. Adsorbent 1 shows a selectivity of 7 for fructose over glucose. This adsorbent is used in the UOP Sarex ^{SM} process. A change of composition and structure gives Adsorbent 2 with reversed selectivity. Both adsorbents reject the oligosaccharides, maltose and sucrose, probably on account of steric hindrance to adsorption of the larger molecules.

Kinetics -- In the crystalline zeolites, the adsorption isotherms do not exhibit hysteresis as do isotherms on many non-crystalline, microporous solids. Adsorption and desorption are completely reversible. In terms of a chromatographic operation, this implies absence of tailing. If mass transfer is good through any portion of the pore structure, it is good throughout, because of the uniformity of the structure. This is in contrast to, for example, activated carbons. These may contain a number of pores only marginally large enough to admit adsorbate molecules. Slow diffusion into and out of these portions of the adsorbent causes excessive band broadening.

5. EVALUATION OF ADSORBENT/DESORBENT SYSTEMS

The development of an adsorbent for a given separation problem, and the selection of a suitable desorbent involves a large number of candidate systems. These must be evaluated for capacity, selectivity and kinetics. To do this in a pilot plant, or even in bench scale continuous equipment, is expensive and time consuming. Convenient and dependable screening techniques are needed.

5.1 Static Tests

A simple test for the selectivity of an adsorbent between Components A and B of a binary liquid mixture is to equilibrate a sample of the mixture, of known composition, with a sample of adsorbent. The change in composition of the mixture after equilibration is an indication of selectivity of the adsorbent. Concentration of the more strongly adsorbed component will decrease, and that of the less strongly adsorbed component will increase. If the capacity of the adsorbent is known and the sample weights are measured, it is possible to calculate the selectivity.

Desorbents can be screened in the same way. A mixture of candidate desorbent with either Component A or B is equilibrated with the solid. If the liquid composition after equilibration does not change much, the test indicates a balance between desorbent and feed, which is desirable in the Sorbex operation.

This technique is a simplification of that used by many workers in the study of thermodynamics of adsorption from solution. In general, complete isotherms are developed for the entire range of composition of the binary mixture. These have been termed "composite isotherms".[9]

One useful isotherm was given by Everett:[10]

$$\frac{X_L(1-X_L)}{W_c \Delta X/A} = \frac{A}{W_A} \left(X_L + \frac{1}{\beta-1} \right) \tag{7}$$

In terms of mixture of p-xylene with a second hydrocarbon component, the quantities measured in the experiment are A, weight of adsorbent charged; W_c, weight of the hydrocarbon charged; X_c, weight fraction of p-xylene in the original mixture; X_L, weight fraction of p-xylene in the raffinate liquid; and $\Delta X = X_c - X_L$. Quantities that are to be derived are W_A/A, the specific capacity of the adsorbent in grams of hydrocarbon adsorbed per gram of adsorbent; and β, the selectivity, defined as before in Equation 5. In the case of the binary mixture, weight fractions of the second hydrocarbon are simply one minus the weight fraction of p-xylene.

The Everett isotherm can be derived by combining the definition of β with equations for mass balance and p-xylene balance between charge and the raffinate and adsorbed phases. A plot of the expression on the left, which involves ΔX, the change in p-xylene concentration, in the denominator, versus raffinate composition, should yield a linear isotherm. The slope A/W_A is the reciprocal of the desired specific capacity. The value of β can be calculated from the intercept.

Considerable data have been accumulated by equilibrating mixtures of p-xylene and other C$_8$- aromatics with zeolitic adsorbents over the complete concentration range. In no case have the Everett plots been linear. This was disappointing, since it was hoped that the uniformity of the zeolites would simplify the mechanism of adsorption and lead to selectivities independent of composition.

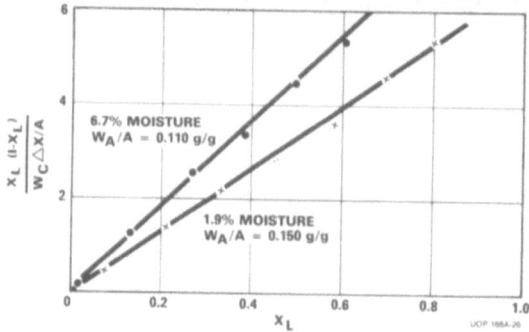

Fig. 5.   Everett Plot for *p*-Xylene vs. *n*-Octane.

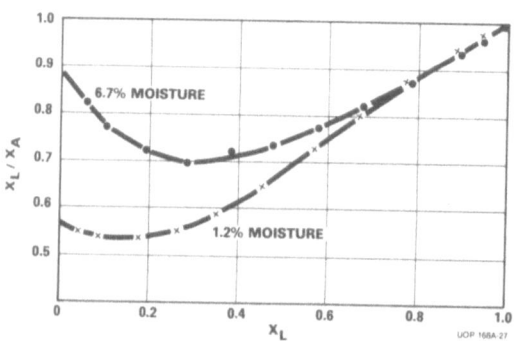

Fig. 6.   Isotherms for *p*-Xylene vs. Ethylbenzene.

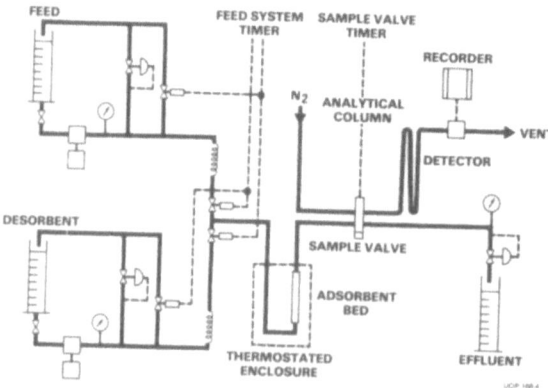

Fig. 7.   Dynamic Adsorption Test Unit.

Where the second component is a paraffin, rather than another aromatic, the results are more useful. In this case, β is very large. The term $1/(β-1)$ vanishes. Figure 5 shows the Everett plot of the composition isotherm of a p-xylene vs. n-octane mixture over a Parex type zeolitic adsorbent. Two levels of hydration of the adsorbent were used. The isotherms are linear and pass through the origin, confirming that the value of β is large. The reciprocal of the slopes gives the specific capacities of 0.110 g/g for the wet adsorbent, and 0.150 g/g for the dry adsorbent. It is clear that in the former case water is occupying a portion of the volume of the zeolite that normally is available to p-xylene.

When the capacity of the adsorbent has been determined, a composite isotherm of aromatic pairs can be developed even if the selectivity is not independent of concentration. The composition of the adsorbed phase is calculated from that of the liquid raffinate phase by mass and component balance, from the change in composition of the liquid, $ΔX$, the weight of liquid mixture charged, W_c, and the weight of the adsorbed phase, known from a previously measured capacity.

$$X_A = X_L + \frac{W_c \; ΔX}{W_A} \tag{8}$$

The selectivity, β, can be calculated directly from X_L and X_A. Alternatively, the definition of selectivity (Equation 4) can be transformed into an isotherm, and β is the reciprocal of the intercept.

$$\frac{X_L}{X_A} = \frac{1}{β} + (1-\frac{1}{β})X_L \tag{9}$$

Figure 6 shows composite isotherms of p-xylene ethylbenzene mixtures on the zeolitic adsorbent cited previously at the two levels of hydration. They are far from ideal. At the p-xylene rich end, selectivity is constant, down to about 60% ethylbenzene. The extrapolated intercept on the ordinate is 0.4 corresponding to a β of 2.5 favoring p-xylene. At the p-xylene lean end, the selectivity deteriorates, more so for the wet adsorbent than for the dry.

Measurements of binary isotherms are pertinent to studies of the thermodynamics of competitive adsorption. For evaluating a practical adsorbent system these isotherms have found only limited use because they give no information on kinetics, and neglect the possible modifying effect of the desorbent on selectivities. Ternary isotherms, in which the desorbent is included as a third component, can, of course, be developed, but this complicates and lengthens the experiment considerably.

5.2 Dynamic Testing[11]

The principal tool used in evaluating adsorbent/desorbent systems for Sorbex is a form of liquid column chromatography. The candidate adsorbent is the column packing, and the candidate desorbent is the eluent. The feed, or sample, is not necessarily the commercial feed, but is generally a mixture prepared by blending individual compounds, representative of the major critical components in the commercial feed. It is usually diluted with desorbent or with an unadsorbed diluent.

The procedure differs from conventional analytical chromatography in that the adsorbent is of commercial mesh size, offering negligible pressure drop; the injection of feed sample is relatively large; and on-stream GC analysis of the column effluent provides specific detection of each feed component as it emerges from the column.

Figure 7 illustrates the equipment. Feed or desorbent is pumped under capillary flow control from graduated cylinders through the thermostated adsorbent bed. The automatic GC sample valve injects samples of effluent into the high speed analytical GC column sufficiently often to define the effluent concentration profiles generated as the column input is switched from desorbent to feed and back to desorbent. If no on-stream analysis is available, one must resort to sample collection and off-site analysis.

The recorder output from the analytical GC is a series of clusters of GC peaks, each representing analysis of the effluent at a given time.

Breakthrough Test -- The dynamic test may be operated in two modes, as a breakthrough test or as a pulse test. In the breakthrough test the adsorbent is filled to equilibrium with the feed. The feed pump is operated until the composition of effluent is constant and the same as the feed. At some convenient time the feed pump is stopped and the desorbent pump is started. This is continued until all feed components are eluted from the column.

The breakthrough mode is generally used for measurement of adsorbent capacity. In this case, this test may be simplified by including only one adsorbed component in the feed.

Pulse Test -- The pulse test resembles conventional liquid chromatography. A pulse of feed is injected into the column that is full of desorbent. The pulse is eluted, and the concentration profiles developed.

Selectivity of the adsorbent for the most strongly held component (i.e., the last to emerge), with respect to each of the other components is given by the ratio of its retention volume to the respective retention volume of each other component. The value of the lowest selectivity provides a basis for appraising the practicability of the adsorbent-desorbent system for Sorbex application. In the case of elution of a mixture of C_8- aromatics from a Parex adsorbent, the critical selectivity is that of p-xylene vs. ethylbenzene. The value of β by pulse test measurement was 2.2. This compares with 2.5 previously obtained by the isotherm technique.

Continuous Bench Scale Operation -- The next step in evaluating a candidate adsorbent-desorbent system is actual Sorbex operation, using bench scale continuous equipment. This test unit operates on the same principle as pilot and commercial scale Sorbex units. The adsorbent is loaded in a number of modules. The modules are circularly disposed around the rotary valve. They are connected in series by transfer lines to form an annular adsorbent column. The annulus is intercepted at the top of each module by a line to the valve. By means of the valve, two liquid streams are introduced to the bed, feed and desorbent; and two are withdrawn, raffinate and extract. Points of introduction and withdrawal alternate around the bed, and are all simultaneously shifted. Liquid flow through the bed is constrained to a clock-wise direction by check valves at the top of each module. These substitute for the function of the liquid circulating pump used in the commercial units.

Steady-state composition and concentration profiles around the annulus can be obtained by sampling the liquid phase at a fixed point in the bed each time the valve steps. These profiles are constantly moving around the bed, but are stationary with respect to the points of introduction and withdrawal of the liquid streams.

The bench scale unit has no distillation column for recovery of solvent. Results are evaluated on the basis of chromatographic analyses of extract and raffinate. Table 5 gives the product distribution obtained in the bench scale adsorptive separation of p-xylene, using toluene desorbent, normalized to a desorbent free basis.

5.3   Correlation of Pulse Test with Continuous Operation

There is a qualitative similarity between concentration profiles in pulse and continuous operation. To derive a quantitative relation between the two, it is necessary to proceed indirectly. A mathematical model of the pulse test has been developed.[5]

The value of k, the mass transfer coefficient, and D, the axial diffusion coefficient, may be derived from parameters of this model. These coefficients can then be used in Equations 1 and 2, together with β values to predict continuous performance.

In the pulse test model, it is assumed that the selective pore volume and non-selective pore volume have been measured by independent methods. The bed is divided into a number of equal segments, each denoted as a theoretical tray. It is initially filled with a single liquid, and then subjected to a step change in liquid feed.

This feed is divided into equivolume increments, each having a volume equal to fraction, f, of the void volume of one tray. It is assumed that, when one increment of liquid enters the bed, liquid in the voids is displaced in plug flow, while liquid in the pores remains stationary. Complete mixing and equilibration are then assumed to occur within each tray, and the operation is repeated as successive increments of liquid enter the bed.

This model is readily implemented by computer, to calculate the cyclic history of liquid composition from each tray in succession, from the bed entrance to the exist. It contains two kinetic parameters -- the number of theoretical trays, n, and the fractional filling of each tray, f. The use of two kinetic parameters allows the dispersions of any two components of the feed to be matched. This is done, preferably, for the components of lowest and highest β values. It has been found that if this is done, the model closely reproduces the dispersion of the intermediate components.

In the case of linear equilibria, with a reasonably large number of theoretical trays, the following approximations hold:

$$\text{Mass transfer Coefficient} = k = \frac{2\,L\,pn}{H\,(v + Kp)} \tag{10}$$

$$\text{Axial diffusion Coefficient} = D = \frac{LH\,(1-f)}{2n\,(v + Kp)} \tag{11}$$

where: L = liquid rate
         p = volume fraction selective pores
         n = number of theoretical trays
         H = bed height
         v = volume fraction of non-selective pores
         K = linear equilibrium constant
         f = fraction filling per tray per increment of feed

It is useful to note that the position of mass centers of the pulse test bands are controlled by the selectivity value, β, assigned to each component. The dispersion of a non-adsorbed species is determined in Equation 11 solely by the ratio, $(1-f)/n$, independently of the individual values of n and f. Also, the value of K is zero. The dispersion of an adsorbed species is determined both by the ratio $(1-f)/n$ and by the actual value of n.

6. COMMERCIAL SORBEX PROCESSES

As the preceding discussion suggests, commercialization[4,12] of an adsorption process proceeds through a number of phases:

1. Exploratory research on testing of adsorbent-desorbent systems for equilibrium and kinetics by chromatographic techniques and model studies.

2. System feasibility testing in a small continuous countercurrent bench scale unit, together with preliminary engineering and economic studies.

3. Longer term testing for design optimization in large continuous countercurrent pilot plant.

This discussion will conclude with a survey of a number of adsorptive separation processes that are actually in commercial use, or that have reached at least the third phase of development.

6.1 Molex

The UOP Molex process[12] uses a 5A zeolite as a molecular sieve to extract normal paraffins from kerosine. The linear chain paraffins are used for production of degradable detergents.

Since the 5A sieves adsorb only normal paraffins and exclude cyclic or branched hydrocarbons, the selectivity in the Molex process is infinite. Therefore, consideration of mass transfer and axial mixing assumes more importance in the design of this process than adsorption equilibria.

Table 6 shows results obtained in pilot plant extraction of a wide boiling range feedstock, C_{10} to C_{23}. A total of 93.5% of the normal paraffins were extracted at a purity of 99.5%. Typical commercial performance in a kerosine range stock is 97% extraction at 98.7% purity.

	FEED, WT-%	EXTRACT, WT-%	RAFFINATE, WT-%	EXTRACTION, (%)
n-PARAFFINS				
$C_{10} - C_{13}$	2.0	3.9	0.0	100.0
$C_{14} - C_{17}$	16.1	29.2	2.4	93.1
$C_{18} - C_{21}$	33.5	61.8	4.6	93.1
$C_{22} - C_{23}$	2.4	4.6	0.1	(98)
SUBTOTAL	54.0	99.5	7.1	93.5
OTHER COMPONENTS	46.0	0.5	92.9	
TOTAL	100.0	100.0	100.0	

UOP 168A-10

Table 6.   Pilot-Plant Molex Operation C_{10}-C_{23} n-Paraffin Extraction.

	FEED, WT-%	EXTRACT, WT-%	RAFFINATE, WT-%	EXTRACTION, (%)
n-OLEFINS	9.0	96.2	0.6	93.9
n-PARAFFINS	90.1	1.1	98.5	
OTHER COMPONENTS	0.9	2.7	0.9	
TOTAL	100.0	100.0	100.0	

Table 7.   Commercial Olex Operation Linear C_{11}-C_{14} Olefin Extraction.

WT-%	FRESH FEED	AFTER BUTADIENE SATURATION	RAFFINATE	EXTRACTION
1-BUTENE	31.2	30.4	3.3	99.2
ISOBUTENE	50.5	50.5	70.1	0.7
trans-2-BUTENE	8.1	8.6	12.0	0.1
cis-2-BUTENE	1.1	1.4	1.9	TR
ISOBUTANE	3.7	3.6	5.0	TR
n-BUTANE	5.1	5.5	7.7	TR
1, 3-BUTADIENE	0.3	(35 ppm)		(50 ppm est.)
ETHYL- AND VINYLACETYLENES	(30 ppm)	(4 ppm)		(6 ppm est.)
SULFUR	(1 ppm)	—	—	—
TOTAL	100.0	100.0	100.0	100.0
1-BUTENE EXTRACTION EFFICIENCY				92.0

UOP 1(

Table 8.   Pilot Plant Extraction of 1-Butene.

6.2 Olex

A mixture of normal paraffins, such as Molex extract, is commercially dehydrogenated to give corresponding linear olefins. They are useful as chemical intermediates for plasticizers and detergents. Separation of the olefins from unreacted paraffins is done by the Olex ^{R} process.[12]

Since the mixture spans a range of about four carbon numbers, for instance, C_{11} to C_{14}, it is impossible to accomplish this separation by either fractional distillation or crystallization.

The use of polar solvents in a liquid-liquid extraction process might be considered. However, it is found that the relative solubilities of paraffins and olefins in these solvents are as shown in Figure 8. It would be possible, although difficult, to extract an olefin from a paraffin of the same chain length. However, in a broad-boiling mixture, there is an overlap in the relative solubilities of the various olefins and paraffins present, and separation becomes impossible.

In contrast, UOP's Olex adsorbent has selectivities as shown in Figure 9, when adsorption is carried out from a liquid-phase mixture of hydrocarbons. Here, all of the olefins are more selectively adsorbed than any of the paraffins, and the separation is readily accomplished.

Table 7 shows commercial results for Olex separation of olefins, a C_{11}-C_{14} dehydrogenate containing 9% olefins. The olefins were extracted at 95.2% purity, 93.9% recovery.

Olefin separations involving a split between two double bond isomers have also been handled by Sorbex. One example is the separation of α- and β-pinene,[13] involving a discrimination between exo- and endo-olefins. A second is the separation of 1-butene from all other C_4 paraffins and monoolefins.[14] In the latter case, the Olex adsorbent was modified to improve the separation between 1-butene and the other C_4 olefins. The difficult separation is that between 1-butene and isobutylene which cannot be separated by either normal or extractive distillation. The standard Olex adsorbent gave a selectivity of only 1.37 for 1-butene vs. isobutylene. The improved Sorbutene ^{SM} adsorbent increased this value to 2.26. Continuous pilot plant operation gave the results shown in Table 8, 92% 1-butene extraction at 99.2% purity. A preliminary selective hydrogenation of the commercial feedstock was employed to remove dienes and acetylenes.

6.3 Parex

The Parex process[12] for separation of p-xylene from a C_8

(WT-%)	FEED	EXTRACT	EXTRACT AFTER TOLUENE REMOVAL	RAFFINATE	RECOVERY AS PRODUCT (%)
NON-AROMATICS	0.29	0.00	0.00	0.29	
TOLUENE	0.45	1.54	0.00	0.28	
ETHYLBENZENE	12.38	0.34	0.35	14.10	
p-XYLENE	11.76	97.94	99.48	0.53	96.7
m-XYLENE	62.96	0.09	0.09	70.93	
o-XYLENE	12.16	0.09	0.08	13.87	
TOTAL	100.00	100.00	100.00	100.00	

UOP 168A-13

Table 9. Commercial Parex Operation with p-DEB Heavy Desorbent Separation of p-Xylene from Crystallizer Mother Liquor.

FEED: 62% m-CRESOL, 38% p-CRESOL

p-CRESOL EXTRACTION EFFICIENCY, %	60	75	85	95
EXTRACT, % p-CRESOL	99.0	98.5	98.0	90.0
RAFFINATE, % m-CRESOL	80.6	85.5	90.5	97.0

Table 10. Bench-Scale Continuous Extraction of p-Cresol.

(WT-%, DRY BASIS)	FEED	EXTRACT	RAFFINATE	EXTRACTION
FRUCTOSE	41.1	94.3	5.8	91.5
GLUCOSE	51.8	5.1	81.3	
PSICOSE	TRACE	0.2	TRACE	
SORBOSE + MANNOSE	0.3	0.2	0.9	
DP-2	3.4	0.2	5.8	
DP-3[+]	3.4	TRACE	6.2	
TOTAL	100.0	100.0	100.0	

Table 11. Separation of Fructose from High Fructose Corn Syrup.

aromatic extract, or from C_8 catalytic reformate, is in worldwide use.

Most Parex units are operated with an aromatics isomerizer to convert the C_8-aromatics in the raffinate to additional p-xylene for recycle. This conversion produces naphthenes which boil in the toluene range. To circumvent contamination of the desorbent, the toluene is replaced by a heavy desorbent, p-diethylbenzene. This desorbent itself is produced commercially in a Sorbex unit, using commercially available mixed diethylbenzenes as feed -- an operation designated as the Debex [SM] process.

Commercial results in a Parex plant using p-xylene crystallizer mother liquor as feed and p-diethylbenzene as desorbent are shown in Table 9. A recovery of 96.7% p-xylene at 99.5% purity was achieved in spite of the lean feed.

Extraction of *para*-substituted compounds from isomeric mixtures is not limited to hydrocarbons. *para*-Cresol has been successfully purified in both bench scale and pilot plant operations. Typical bench scale results are shown in Table 10. A 99% purity p-cresol was obtained at 60% extraction efficiency and 98% purity at 85% extraction efficiency. At the highest efficiency, the *m*-cresol in the raffinate was 97% pure. This suggests a blocked out operation for production of either isomer from the mixture.

6.4 Sarex

The most recent Sorbex process to be commercialized is the Sarex process[15] for the extraction of fructose. This is the first Sorbex process to operate in the aqueous phase. Feedstock to the separation can be high fructose corn syrup. This is produced from starch, first by conversion of the starch to glucose, or corn syrup; and then by enzymatic isomerization of glucose to a mixture containing about 42% fructose, 50-55% glucose and 3.8% or higher saccharides. An alternative feed might be invert sugar derived from sucrose inversion. Fructose is sweeter than either glucose or sucrose, and high level fructose concentrate has potential as a new industrial sweetener.

In Sarex, the fructose is extracted at over 90% purity and over 90% yield. Actual performance curves in bench scale continuous operation are shown in Figure 10. This shows the purity/efficiency curve with various adsorbents at various operating conditions. The inorganic adsorbents generally gave higher yield and purity than ion exchange resin, at similar conditions. Best results with Adsorbent A gave 95% fructose at 95% extraction. The effect of doubling the feed rate is shown by comparison of the second and third curves. Table 11 gives complete analysis of feed, extract and raffinate in a typical operation.

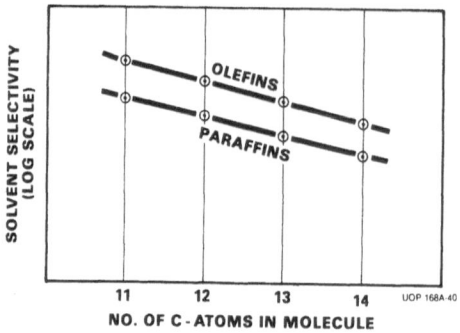

Fig. 8. Liquid-Liquid Extraction Selectivity of Polar Solvents in Paraffin-Olefin System.

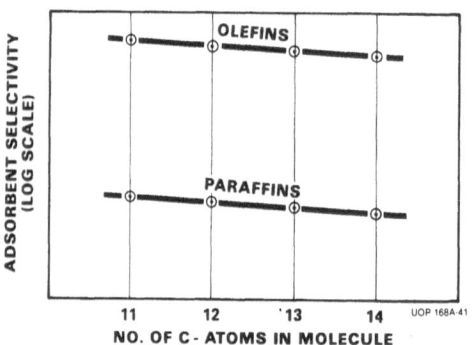

Fig. 9. Adsorption from Liquid Phase Selectivity of Olex Adsorbent.

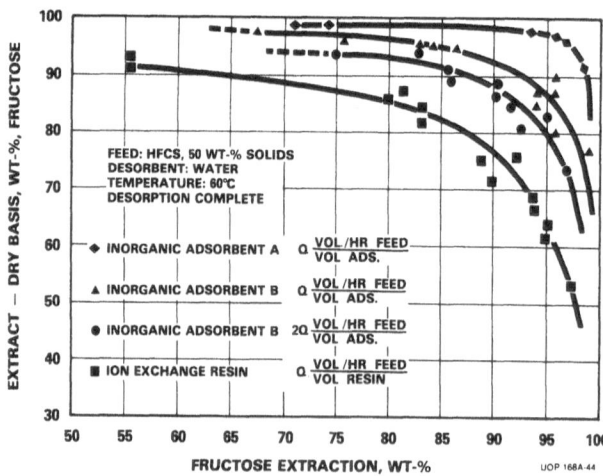

Fig. 10. Performance Functions Sarex Process.

REFERENCES

1. W. H. Davis, J. I. Harper, E. R. Weatherly, Petr. Ref., 31 109 (1952).

2. H. Kehde, R. G. Fairfield, J. C. Frank, L. W. Zahnstecher, Chem. Eng. Prog., 44 575 (1948).

3. M. L. Gilwood, Chem. Eng. Dec. 18, 1967.

4. D. B. Broughton, H. J. Bieser, M. C. Anderson, Petrolieri International (Milan), 23 (3) 91 (1976).

5. D. B. Broughton, R. W. Neuzil, J. M. Pharis, C. S. Brearley, Chem. Eng. Prog., 66 70 (1969).

6. H. S. Sherry, in "Ion Exchange", Vol. 2, J. A. Marinsky ed., Marcel Dekker, New York, 1969.

7. D. B. Breck, "Zeolite Molecular Sieves", John A. Wiley & Sons, New York, 1974.

8. L. R. Snyder, "Principles of Adsorption Chromatography", Marcel Dekker, New York, p. 131, 1968.

9. J. J. Kipling, "Adsorption from Solutions of Non-Electrolytes", Academic Press, London, 1965.

10. D. H. Everett, Trans. Farad. Soc., 60 1803 (1964).

11. A. J. deRosset, R. W. Neuzil, D. J. Korous, Ind. Eng. Chem., Process Des. Dev., 15 250 (1976).

12. D. B. Broughton, H. J. Bieser, R. A. Persak, Petrolieri International (Milan), 23 (5) 36 (1976).

13. J. G. Kaufman, A. J. deRosset, Assoc. Official Anal. Chem., 88th AM Meeting, Washington, D. C., October 14, 1974.

14. A. J. deRosset, J. W. Priegnitz, D. J. Korous, D. B. Broughton, Amer. Chem. Soc., 175th Nat'l. Meeting, Anaheim, Cal., March 12-17, 1978.

15. H. J. Bieser, A. J. deRosset, Die Starke, 29 (11) 392 (1977).

Part 2

MULTICOMPONENT PERCOLATION PROCESSES

EQUILIBRIUM THEORY OF MULTICOMPONENT CHROMATOGRAPHY

Hyun-Ku Rhee

Department of Chemical Engineering
Seoul National University
Seoul, Korea

1. INTRODUCTION

The chromatography in its application may be regarded as a process in which a bed packed with adsorbent or ion exchange particles is brought into contact with a fluid stream for the separation of solute species present in the stream. The principle has found its applications in a variety of areas of science and the process has been an engineering practice of long standing.

Here it is intended to present a mathematical theory of mul - ticomponent chromatography with an ideal column. The term 'ideal', as used here, implies the following conditions: (1) The system is one-dimensional in the direction of flow with uniform cross-sectional area; (2) The volumetric flow-rate and the bed voidage are constant; (3) Effects of diffusion are negligible compared with the convective transport and there is no channeling; (4) Local equilibrium is established between two phases everywhere at any time; (5) The process is isothermal and isochoric.

The theoretical study of such a system has been the goal of numerous publications. The first mathematical description was given by Wilson(1940) who, however, overlooked the interactions among the concentrations of various species. Several conclusions, qualitatively correct, were derived by DeVault(1943) while Walter (1945) obtained explicit solutions for the formation of chromatograms, assuming a chemical equilibrium between adsorbed materials and dissolved ones. With Langmuir isotherms Glueckauf(1946, 1949) accomplished a rigorous analysis of two solute problem though his treatment did not bring out the connection with the classical theory of quasilinear equations (Courant & Friedrichs 1948).

An independent study was carried out by Sillen(1950) who presented analogous results by using the 'ψ- condition'. Bayle and Klinkenberg (1954) made a critical review of the previous results and claimed that the experimental results were not in conflict with the relation between the coexisting concentrations.

A theoretical analysis of multicomponent ion exchange in fixed beds was presented by Klein, Tondeur, and Vermeulen(1967). Tondeur and Klein(1967) also extended the approach originated by Walter(1945) to multicomponent systems. However, their discussions were confined to some special cases. Helfferich(1967, 1968) considered similar problems to achieve a further advance by introducing the so-called 'h-transformation' and presented a full discussion in a separate monograph (Helfferich & Klein 1970).

Since only the convective transport is important in the system, the mathematical description gives rise to a quasilinear system of partial differential equations of first order and thus the mathematical model has also attracted the interest of many mathematicians. A uniqueness proof was established by Oleinik(1957, 1959) and construction of solution was discussed through finite difference schemes (Lax 1957; Oleinik 1957; Glimm 1965) whereas a numerical scheme was first developed by Courant, Isaacson and Rees(1952) and has been refined by various authors (cf. Jeffrey & Taniuti 1964). By assuming the existence of generalized Riemann invariants, Lax(1957) was able to generalize the simple wave theory as well as the theory of discontinuities originated by Courant and Friedrichs(1948). This, however, is subject to a severe restriction because, for a system of more than two equations, a Riemann invariant strictly analogous to the one for a system of two equations cannot exist in general (Jeffrey & Taniuti 1964;Glimm 1965).

Based on the Lax approach, Rhee, Aris and Amundson examine the class of Riemann's problem in which the boundary data are constant, and developed an equilibrium theory of multicomponent chromatography with mathematical rigor (Rhee 1968; Rhee, Aris & Amundson 1970, 1971; Rhee & Amundson 1970). The present discussion is aimed to closely follow the theoretical development established in these studies so that readers may fully appreciate the connection with the mathematical theory of quasilinear equations

2. BASIC FORMULATIONS

Consider an ideal chromatographic column of voidage ϵ through which a fluid mixture containing m different solutes $\{A_i\}$ flows with linear velocity v. Let c_i and n_i denote the concentrations of A_i in the fluid and solid phases, respectively, both being expressed in moles per unit volume of their own phase. Then the material balance for each solute component A_i in a section between planes distant x and $x + \Delta x$ from the entrance to the bed over a time interval θ to $\theta + \Delta\theta$ yields, in the limit, the following quasi-linear system of m partial differential equations of first order:

$$\frac{\partial c_i}{\partial z} \quad + \quad \frac{\partial c_i}{\partial t} + \nu \frac{\partial n_i}{\partial t} \; = \; 0 \qquad\qquad (2.1)$$

$$n_i \; = \; f_i (c_1, \, c_2, \, \dots, \, c_m) \qquad\qquad (2.2)$$

for $i = 1, 2, \dots, m$. Here ν represents the volume ratio of the two phases : i.e.,

$$\nu \; = \; (1 - \epsilon)/\epsilon \qquad\qquad (2.3)$$

while z and t are the dimensionless independent variables defined as

$$z \; = \; x/L, \qquad t \; = \; v\theta /L \qquad\qquad (2.4)$$

in which L is the characteristic length of the column.
In Eq. (2.2) the functions f_i, representing the equilibrium relationship between the adsorbed phase and the fluid phase, may be regarded as continuous functions of $\{ c_i \}$ with as many derivatives as may be required.

The system (2.1) with Eq. (2.2) may be put into a vector form

$$\frac{\partial \mathbf{c}}{\partial z} \quad + \quad (\mathbf{I} + \nu \, \boldsymbol{\nabla} \mathbf{f}) \frac{\partial \mathbf{c}}{\partial t} \; = \; 0 \qquad\qquad (2.5)$$

where $\mathbf{c}$ and $\mathbf{f}$ represent the vector-valued functions of m elements, $\{ c_i \}$ and $\{ f_i \}$, respectively, and $\boldsymbol{\nabla}$ denotes the gradient in the m-dimensional concentration space.

According to the mathematical theory, the characteristic directions in the (z, t)-plane are given by the eigenvalues of the matrix of coefficients in Eq. (2.5); i.e., the characteristic direction λ must satisfy the equation

$$\left| \, (\mathbf{I} + \nu \, \boldsymbol{\nabla} \mathbf{f}) - \lambda \, \mathbf{I} \, \right| \quad = \; 0 \qquad\qquad (2.6)$$

If Eq. (2.6) has m real, distinct roots, the system is called totally hyperbolic. It is then required to determine the right eigenvectors $\{ r^k \}$ of the matrix. These are all functions of $\{ c_i \}$ and lead to the generalized Riemann invariants $\{ J^k \}$ by the solution of the equation

$$\boldsymbol{\nabla} J^k \cdot r^k \; = \; 0 \qquad k = 1, 2, \dots, m \qquad\qquad (2.7)$$

The characteristic directions $\{\lambda_k\}$ and the Riemann invariants $\{J^k\}$ so determined play the key role in the construction of solutions. In practice, however, the application of the method is by no means promising except for some special cases for which the eigenvalues can be obtained explicitly.

We shall therefore confine our discussion to a class of special, yet standard, initial and entry value problems and follow an independent approach. Suppose the initial and boundary conditions are specified by two different constant states of concentrations with a jump discontinuity at the origin of the (z, t)-plane:

$$
\begin{cases}
\text{at } t = 0, & c_i = c_i^I \text{ (a constant)} \\[2mm]
\text{at } z = 0, & c_i = c_i^F \text{ (a constant)}
\end{cases}
\qquad (2.8)
$$

for $i = 1, 2, \ldots, m$ and $c_i^I \neq c_i^F$ for some i. A problem of this class is called a Riemann's problem.

It is then noticed that the solution $c_i(z, t)$ is a function of z/t only if it is unique (Sillen 1950; Lax 1957). This fact can be shown as follows : If $c_i(z, t)$ is a solution of the system (2.1) satisfying the conditions of Eq. (2.8), the function $(c_i)_p = c_i(pz, pt)$ where p is any positive constant is also a solution and takes on the same initial and boundary values. Hence, in order that the solution be unique, we must have $(c_i)_p = c_i(z, t)$ for all i. This is true if and only if every $c_i(z, t)$ is a function fo z/t only.

Let us now suppose that the solution has been obtained in the form

$$
c_i = c_i(z/t) \qquad i = 1, 2, \ldots, m \qquad (2.9)
$$

and consider the inverse function I_i of Eq. (2.9); i.e.,

$$
z/t = I_i(c_i) = I_j(c_j) \qquad i, j = 1, 2, \ldots, m
$$

This equation implies that c_j for $j \neq i$ can be expressed as a function of c_i and consequently, there exists a unique one-parameter representation of the solution:

$$
c_i = c_i(h) , \qquad i = 1, 2, \ldots, m \qquad (2.10)
$$

where the parameter h is yet to be determined in terms of c_i. In the m-dimensional concentration space Eq. (2.10) is represented by a single curve with the parameter h running along it. Such a curve is the image of the solution and will be called a Γ.

Denoting the differentiation with h as the independent variable along the Γ by $\mathcal{D}/\mathcal{D}h$, we may introduce the directional derivative

$$\frac{\partial f_i}{\partial c_i} = \frac{\Delta f_i / \Delta h}{\Delta c_i / \Delta h} = \sum_{j=1}^{m} f_{i,j} \frac{\Delta c_j / \Delta h}{\Delta c_i / \Delta h} \qquad (2.11)$$

where $f_{i,j} = \partial f_i / \partial c_j$ $i, j = 1, 2, \ldots, m,$ (2.12)

and rearrange the system (2.1) in the form

$$\left\{ \frac{\partial h}{\partial z} + (1 + \nu \frac{\partial f_i}{\partial c_i}) \frac{\partial h}{\partial t} \right\} \frac{\Delta c_i}{\Delta h} = 0 \qquad (2.13)$$

It then follows that

$$s = \left(\frac{dt}{dz} \right)_h = 1 + \nu \frac{\partial f_i}{\partial c_i} \qquad (2.14)$$

since $\Delta c_i / \Delta h = 0$ due to the existence of Γ.

We notice that s represents a specific direction in the (z, t)-plane along which the parameter h is held constant. Eq. (2.14) must be independent of the choice of i and hence we have

$$\frac{\partial f_1}{\partial c_1} = \frac{\partial f_2}{\partial c_2} = \ldots = \frac{\partial f_m}{\partial c_m} \qquad (2.15)$$

Eq. (2.15) is called the fundamental differential equation of the Riemann's problem, the solution of which generates the one-parameter family (2.10); i.e., the curve Γ in the m-dimensional concentration space. It was given by Glueckauf (1946, 1949) for m=2 and by Bayle and Klinkenberg for m>2.

On the other hand, Eq. (2.14) may be expanded by using Eq. (2.11) in the form

$$\nu f_{i,1} \frac{\Delta c_1}{\Delta h} + \ldots + (1 + \nu f_{i,i} - s) \frac{\Delta c_i}{\Delta h} + \ldots + \nu f_{i,m} \frac{\Delta c_m}{\Delta h} = 0$$

$$i = 1, 2, \ldots, m$$

or briefly in matrix form

$$(\mathbf{I} + \nu \nabla \mathbf{f} - s\mathbf{I}) \Delta \mathbf{c} / \Delta h = 0$$

Eq. (2.10) guarantees that there exists a unique non-trivial solution for $\partial \mathbf{c} / \partial h$ and this entails the condition that

$$\left| \mathbf{I} + \nu \nabla \mathbf{f} - s\mathbf{I} \right| = 0 \qquad (2.16)$$

Consequently, s defined by Eq. (2.14) is identical to the characteristic direction λ in the (z, t)-plane.

In the following sections we shall be concerned with a particular form of Eq. (2.2), the Langmuir isotherm, for which a completely analytical treatment can be accomplished. For an arbitrary equilibrium relationship, the readers may be referred to the general development due to Rhee and Amundson (1970). Applications have been made to the analysis of adiabatic adsorption column (Rhee, Heerdt & Amundson 1970, 1972) and also to the analysis of heterovalent ion exchange systems (Tondeur 1970).

3. EQUILIBRIUM RELATIONS

The equilibrium relation f_i which is usually called the adsorption isotherm is, in general, a complicated nonlinear function of $\{c_i\}$ in which mutual influences among different solutes are taken into account. Since the ideal localized monolayer model was introduced by Langmuir (1916), the Langmuir relation has been extensively employed not only for single solute systems but also for multiple solute systems (see, for example, Wilson 1940; and Gluekauf 1946, 1949). A rigorous discussion was given by DeBoer (1953).

We shall introduce the Langmuir isotherms for Eq. (2.5); i.e., for each solute component A_i we have

$$n_i = N_i K_i c_i / (1 + \sum_{j=1}^{m} K_j c_j) \quad i = 1, 2, \ldots, m \quad (3.1)$$

Here K_i is the reciprocal value of c_i when half the sites are occupied by molecules of A_i and the other half are vacant. The parameter N_i denotes the limiting value of n_i which is defined as the maximum number of moles of A_i that can be adsorbed per unit volume of adsorbent. According to Kembell, Rideal and Guggenheim (1944), the value of N_i must be the same for all solutes: otherwise, there results in a contradiction to the Gibb's adsorption isotherm. In practice, however, the experimental value of N_i varies from one solute species to the other (James & Phillips 1954; Weber & Keinath 1966; Shen & Smith 1968; Hsieh 1974).

For convenience, we shall assume that the solute components are arranged in such a way that

$$N_1 K_1 < N_2 K_2 < \cdots < N_m K_m \quad (3.2)$$

This is equivalent to numbering the components in the order of the adsorptivity from the smallest to the largest since the relative adsorptivity (or the separation factor) may be defined as

$$\gamma_{ir} = \frac{n_i/c_i}{n_r/c_r} = \frac{N_i K_i}{N_r K_r} \qquad i = 1, 2, \ldots, m \quad (3.3)$$

where the subscript r denotes the reference species. It may be noticed that γ_{ir} is independent of concentrations.

The Langmuir isotherm carries an underlying assumption that the solvent or the carrier behaves as an inert component. If the solvent has a non-zero adsorptivity but one which is smaller than any of the solutes, Eq. (3.1) becomes

$$n_i = N_i K_i c_i / (1 + \sum_{j=0}^{m} K_j c_j) \qquad i = 0, 1, 2, \ldots, m \quad (3.4)$$

in which the subscript 0 denotes the solvent. Under the conditions of constant temperature and pressure, we claim that

$$c_t = \sum_{j=0}^{m} c_j = \text{constant}$$

and thus Eq. (3.4) may be rearranged in the form

$$n_i = N_i' K_i' c_i / (1 + \sum_{i=1}^{m} K_i' c_i) \qquad (3.5)$$

where

$$N_i' = N_i K_i / (K_i - K_0) \qquad (3.6)$$

$$K_i' = (K_i - K_0) / (1 + K_0 c_t) \qquad (3.7)$$

for $i = 1, 2, \ldots, m$. Therefore, the Langmuir relation can be reduced to a form which excludes the solvent.

The constant-separation-factor equilibrium relations introduced for the ion exchange systems (Tondeur & Klein 1967; Helfferich & Klein 1970) can also be rewritten in the form of Eq. (3.5) excluding the species with the smallest separation factor.

4. RIEMANN INVARIANTS AND CHARACTERISTIC PARAMETERS

In view of the form of the Langmuir isotherms, Eq. (3.1), we shall introduce the following dimensionless variables:

$$X_i = K_i c_i \qquad i = 1, 2, \ldots, m \quad (4.1)$$

and

$$D = 1 + \sum_{i=1}^{m} X_i \qquad (4.2)$$

It then follows from Eq. (2.10) that D may be expressed as a function of h or h as a function of D and thus X_i as well as n_i may be regarded as a function of D alone; i.e.,

$$X_i = X_i(D), \qquad n_i = N_i X_i(D)/D \qquad (4.3)$$

for i = 1, 2, ..., m.

By using D in place of h in Eq. (2.11) and introducing Eq. (4.3), the directional derivative may be expanded in the form

$$\frac{\partial f_i}{\partial c_i} = K_i \frac{\partial n_i / \partial D}{\partial X_i / \partial D} = N_i K_i \frac{\partial}{\partial D}\left(\frac{X_i}{D}\right) / \frac{\partial X_i}{\partial D}$$

$$= N_i K_i (D - X_i / \frac{\partial X_i}{\partial D}) / D^2 \qquad i = 1, 2, \ldots, m \qquad (4.4)$$

Substituting into the fundamental differential equation (2.15), multiplying the equation through by D^2, and differentiating once more with respect to D, we obtain

$$\frac{\partial^2 X_1 / \partial D^2}{\frac{1}{N_1 K_1 X_1}\left(\frac{\partial X_1}{\partial D}\right)^2} = \frac{\partial^2 X_2 / \partial D^2}{\frac{1}{N_2 K_2 X_2}\left(\frac{\partial X_2}{\partial D}\right)^2} = \ldots = \frac{\partial^2 X_m / \partial D^2}{\frac{1}{N_m K_m X_m}\left(\frac{\partial X_m}{\partial D}\right)^2}$$

$$= \frac{\sum\limits_{i=1}^{m} \partial^2 X_i / \partial D^2}{\sum\limits_{i=1}^{m} \frac{1}{N_i K_i X_i}\left(\frac{\partial X_i}{\partial D}\right)^2} \qquad (4.5)$$

Since it follows from Eq. (4.2) that $\sum\limits_{i=1}^{m} \partial^2 X_i / \partial D^2 = 0$, any solution to Eq. (4.5) must satisfy either

$$\sum\limits_{i=1}^{m} (N_i K_i X_i)^{-1} (\partial X_i / \partial D)^2 = 0 \qquad (4.6)$$

or

$$\partial^2 X_i / \partial D^2 = 0 \qquad i = 1, 2, \ldots, m \qquad (4.7)$$

Eq. (4.6) cannot have any physical meaning since it requires at least one of the $\{X_i\}$ to be negative. Consequently, it is from Eq. (4.7) that the physically relevant solution may be determined.

Direct integration of Eq. (4.6) yields

$$X_i - X_i^{\circ} = J_i(D - D^{\circ}) \qquad i = 1, 2, \ldots, m \qquad (4.8)$$

in which the superscript o denotes the fixed state of concentrations and J_i is the integration constant. Adding the m equations in the system (4.8) together, we find that

$$\sum\limits_{i=1}^{m} J_i = 1 \qquad (4.9)$$

Eq. (4.9) implies that there are $(m-1)$ linearly independent J_i's. The system (4.8) represents a straight line passing through the fixed point $\{X_i^0\}$ in the m-dimensional concentration space, $\mathcal{X}(m)$. This line is the Γ and has a direction given by the invariant set $\{J_i\}$.

The constant J_i can be determined by substituting Eq. (4.8) back into the original differential equation (2.15). Since Eq. (4.4) yields

$$\frac{\partial f_i}{\partial c_i} = K_i(N_i - n_i/J_i)/D \qquad i = 1,2,\ldots,m \qquad (4.10)$$

we obtain

$$K_1(N_1 - n_1/J_1) = K_2(N_2 - n_2/J_2) = \cdots = K_m(N_m - n_m/J_m)$$

$$= h \qquad (4.11)$$

or

$$J_i = K_i n_i/(N_i K_i - h) \qquad i = 1,2,\ldots,m.$$

which, upon substitution into Eq. (4.9), gives the following equation

$$\sum_{i=1}^{m} \frac{K_i n_i}{N_i K_i - h} = 1 \qquad (4.12)$$

Here the new parameter h, representing the common value of the directional derivatives, can be determined for a given state of concentrations, $\{c_i\}$ or $\{n_i\}$, by solving the m-th order algebraic equation (4.12). As shown in Fig. 1, there exist precisely m real, distinct, positive roots which may be arranged as

$$0 \leqslant h_1 \leqslant N_1 K_1 \leqslant h_2 \leqslant N_2 K_2 \leqslant h_3 \cdots h_m \leqslant N_m K_m. \qquad (4.13)$$

Consequently, there exist m different sets of $\{J_i\}$, to each of which there corresponds a different Γ. We shall denote the k-th one as Γ^k and the corresponding invariant set as $\{J_i^k\}$, where

$$J_i^k = K_i n_i/(N_i K_i - h_k) \qquad i,k = 1,2,\ldots,m \qquad (4.14)$$

It should be pointed out that $\{J_i^k\}$ so determined corresponds to the k-Riemann invariants (Lax 1957). From Eqs. (4.13) and (4.14), one can readily observe that

$$0 < J_m^1 < J_m^2 < J_m^3 \cdots < J_m^{m-1} < J_m^m \qquad (4.15)$$

and that

$$J_i^k \begin{cases} < 0, & i < k \\ > 0, & i \geqslant k \end{cases} \qquad (4.16)$$

294

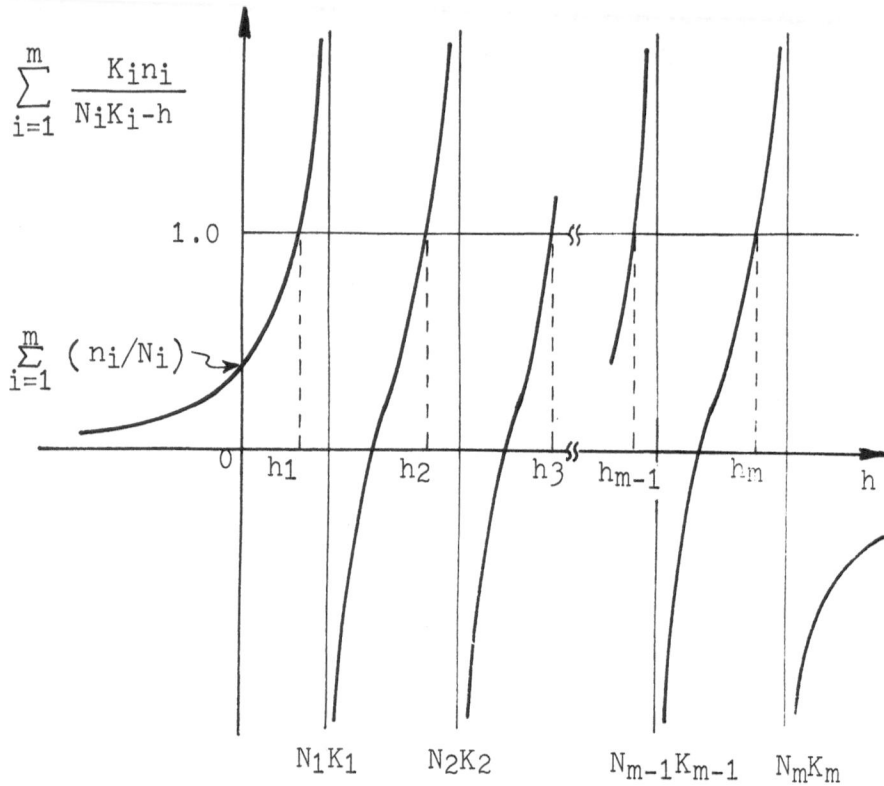

$$\sum_{i=1}^{m} \frac{K_i n_i}{N_i K_i - h}$$

$$\sum_{i=1}^{m} (n_i/N_i)$$

1.0

h_1 h_2 h_3 h_{m-1} h_m h

$N_1 K_1$ $N_2 K_2$ $N_{m-1} K_{m-1}$ $N_m K_m$

Fig. 1. Graphical solution of Eq. (4. 12).

for $i, k = 1, 2,,..., m$.

Also, it can be shown that

$$W_k = h_k D = N_i K_i (D - X_i/J_i^{\,k}) \quad k=1,2,\ldots,m \quad (4.17)$$

which, being independent of the choice of i, remains invariant along a Γ^k. (cf. Eq. (4.8)) Furthermore, it follows from Eq. (4. 15) that

$$0 < W_1 < W_2 < \cdots < W_{m-1} < W_m \qquad (4.18)$$

On the other hand, we notice that Eq. (4. 12) is a one-to-one, continuous mapping of the m-dimensional concentration space $\mathcal{X}(m)$ onto the h-space $\mathcal{H}(m)$. It can be shown that the inverse mapping is also continuous and given by

$$X_i = \left(\frac{N_i K_i}{h_i} - 1\right) \prod_{j=1,i}^{m} \frac{(N_i K_i / h_j - 1)}{(N_i K_i / N_j K_j - 1)}$$

$$i = 1, 2, \ldots, m \qquad (4.19)$$

Hence, the two spaces $\mathcal{X}(m)$ and $\mathcal{H}(m)$ are homeomorphic with the homeomorphism defined by Eqs. (4.12) and (4.19). Here we remark that this homeomorphism is really identical to the h-transformation of Helfferich and Klein (1970).

In order to find the image of a Γ^k in $\mathcal{H}(m)$, let us consider Eq. (4.12) with $h = h_j$ and substitute $K_i n_i = J_i^k (N_i K_i - h_k)$ from Eq. (4.14) to obtain

$$\sum_{i=1}^{m} J_i^k \frac{N_i K_i - h_k}{N_i K_i - h_j} = 1 = \sum_{i=1}^{m} J_i^k$$

in which Eq. (4.9) has been applied. This equation can be reduced to

$$(h_j - h_k) \sum_{i=1}^{m} \frac{J_i^k}{N_i K_i - h_j} = 0 \qquad (4.20)$$

It then follows that h_j, if $j \neq k$, remains unchanged along a Γ^k. In other words, only h_k varies along a Γ^k and thus its image in $\mathcal{H}(m)$ falls on a straight line parallel to the h_k-axis.

We summarize the pertinent parameters in Table 1. Incidentally, the h-coordinate system is equivalent to the characteristic coordinate system associated with a system of two equations (Courant & Friedrichs 1948) and, in this sense, we shall call h_k the characteristic parameter.

In the space $\mathcal{H}(m)$, the physically relevant portion is finite and bounded by the inequalities in Eq. (4.13). If one of the concentrations is zero; e.g., $X_j = 0$, Eq. (4.12) requires that

Table 1. Pertinent parameters along a Γ^k.

parameter	$j = k$	$j \neq k$
Riemann invariant, J_i^j	invariant	variable
W_j	invariant	variable
Characteristic parameter, h_j	variable	invariant

one of h_k must be equal to $N_j K_j$. It follows from Eq. (4.13) that $h_j = N_j K_j$ or $h_{j+1} = N_j K_j$. For $X_m = 0$, however, we always have $h_m = N_m K_m$. For the pure state; i.e., $X_i = 0$ for all i, Eq. (4.12) generates m roots $\{N_k K_k\}$. Consequently, the image of the pure state in $\mathcal{H}(m)$ is given by the point $h_k = N_k K_k$, $k = 1, 2, \ldots, m$.

5. SIMPLE WAVE THEORY

From Eq. (4.10) and (4.11), we notice that $\partial f_i / \partial c_i = h_k / D$ for $k = 1, 2, \ldots, m$ and hence, there exist m different values for the characteristic direction defined by Eq. (2.14). These are determined by the equation

$$s_k = \left(\frac{dt}{dz}\right)_{h_k} = 1 + \nu\, h_k / D \qquad (5.1)$$

$$= 1 + \nu\, W_k / D^2 \qquad (5.2)$$

for $k = 1, 2, \ldots, m$. A curve whose direction is given by s_k will be called a k-characteristic and denoted as C^k.

Applying Eq. (4.13) or (4.18), we observe that

$$1 < s_1 < s_2 < \cdots < s_{m-1} < s_m \qquad (5.3)$$

Note that the unity corresponds to the reciprocal of the fluid velocity whereas s_k represents the reciprocal of the propagation speed of a disturbance.

It is obvious that a region of constant state in the (z, t)-plane is bounded by straight characteristics because s_k remains constant for all k. Since the image of a constant state in $\mathcal{X}(m)$ falls on a single point, the solution adjacent to a constant state must have its image along a Γ emanating from the point image and thus one set of Riemann invariants is constant. It can be further shown that, if a portion of the boundary of a constant state is a C^k, then all k-Riemann invariants on the other side of the C^k are constant (Lax 1957; Rhee, Aris & Amundson 1970).

According to Lax(1957), a continuous solution in a region of the (z, t)-plane for which all k-Riemann invariants are constant is called a k-simple wave. An immediate consequence is that the solution adjacent to a constant state bounded by a C^k is a k-simple wave and has its image along the Γ^k issuing from the point image of the constant state.

In a k-simple wave region, $\{J_i^k\}$ remain constant and thus there exists a one-parameter representation of solutions which is given by Eq. (4.8) :

$$X_i - X_i^o = J_i^k (D - D^o) \quad i = 1, 2, \ldots, m \qquad (5.4)$$

Also s_k is a monotone function of D because W_k is constant. In addition, h_k is held constant along a C^k and hence D as well as all X_i's remains constant along it. (cf. Eq. (4.17)) This implies that every C^k is straight. Consequently, a k-simple wave region is covered by a family of straight C^k.

On the other hand, it is required in a k-simple wave region that

$$\frac{\partial s_k}{\partial z} = \frac{ds_k}{dD} \cdot \frac{\partial D}{\partial z} < 0 \qquad (5.5)$$

for otherwise the characteristic C^k would overlap among themselves. We notice from Eq. (5.2) that $ds_k/dD < 0$ and also that $h_k D = W_k = $ constant. It then follows that, in a k-simple wave, D increases in the z-direction while h_k decreases in the z-direction.

6. DISCONTINUITIES (SHOCKS)

If it happens that

$$\partial s_k / \partial z > 0 \quad \text{or} \quad \partial D / \partial z < 0 \qquad (6.1)$$

then the characteristic C^k would overlap in a k-simple wave region and thus the solution cannot be determined uniquely. This can be resolved by allowing discontinuities in the solution itself.

At a discontinuity Eq. (2.1) is no longer valid and must be replaced by conservation equations expressing the fact that the discontinuity propagates with such a speed that there is no accumulation of material at the discontinuity. Formulation gives the equation

$$s^d = (dt/dz)^d = 1 + \nu \Delta f_i / \Delta c_i \qquad (6.2)$$

for any i, where s^d is the reciprocal of the propagation speed of the discontinuity and the symbol Δ denotes the jump of the quantity across the discontinuity.

Since the adsorption equilibrium is established, s^d is the same for all i and therefore the following equality, which is called the compatibility condition, must be satisfied :

$$\frac{\Delta f_1}{\Delta c_1} = \frac{\Delta f_2}{\Delta c_2} = \ldots = \frac{\Delta f_m}{\Delta c_m} \qquad (6.3)$$

Eqs. (6.2) and (6.3) together form a system of m algebraic equations. Therefore, given the state on one side of a discontinuity the state on the other side can be determined if one of the concentrations or the value of s^d is known.

These conditions, however, are not sufficient to determine the physically relevant solution since an ambiguity still exists concerning the direction of jumps. More generally, with the allowance of discontinuities, the solution of the system (2.1) is defined in the sense of weak solutions which are not uniquely determined by the data, Eq. (2.8) (Lax 1954). Therefore, we need a criterion for selecting the physically relevant solution from weak solutions. Such a criterion may be given by an additional condition which, being usually called the 'entropy' condition regulates the direction of jumps at discontinuities. This condition is not obtained from the conservation law for a discontinuity but from arguments based on the conservation law for a continuous solution.

For the present problem the 'entropy' condition can be derived from Eq. (6.1) which originated from Eq. (2.1) : At a discontinuity the parameter D decreases from the left-hand side to the right.

In the limit as a discontinuity becomes very weak, Eq. (6.3) is reduced to the fundamental differential equation (2.15). This implies that the image of a discontinuity in $\mathcal{X}(m)$ is tangent to a Γ at each end and thus there may exist m different kinds. Recalling the fact that Γ's are straight, we assert that the image of a discontinuity falls on a single Γ and so Eq. (6.3) is an integral of Eq. (2.15). Along a Γ^k, for instance, we have from Eq. (5.4) that

$$K_i \Delta c_i / J_i^{\ k} = K_j \Delta c_j / J_j^{\ k}$$

and

$$\partial f_i = \frac{\partial c_i}{\partial c_j} \partial f_j = \frac{K_j J_i^{\ k}}{K_i J_j^{\ k}} \partial f_j$$

Integration of the latter relation along the Γ^k yields

$$K_i \Delta f_i / J_i^{\ k} = K_j \Delta f_j / J_j^{\ k}$$

and comparing with the former, we find that Eq. (6.3) is satisfied. This argument holds valid for any k and hence there exist m different solutions of Eq. (6.3). Consequently, a discontinuity necessarily has its image along a Γ in $\mathcal{X}(m)$ and there are m different kinds.

Consider a discontinuity with its image on a Γ^k. Since then the set $\{J_i^k\}$ remains invariant across it, the states on both sides are connected by Eq. (5.4) which may be written in the form

$$X_i^l - X_i^r = J_i^k (D^l - D^r) \qquad (6.4)$$

where the superscripts l and r denote the left- and right-hand sides of the discontinuity, respectively. The propagation direction can be determined as follows :

$$\frac{\Delta f_i}{\Delta c_i} = (n_i^l - n_i^r)/(c_i^l - c_i^r) = (N_i K_i / D^l D^r)(D^l - X_i^l \Delta D / \Delta X_i)$$

$$= (N_i K_i / D^l D^r)(D^l - X_i^l / J_i^k) = W_k / D^l D^r \qquad (6.5)$$

in which W_k is invariant across the discontinuity ; i.e.,

$$W_k = h_k^l D^l = N_i K_i (D^l - X_i^l / J_i^k)$$

$$= h_k^r D^r = N_i K_i (D^r - X_i^r / J_i^k) \qquad (6.6)$$

Eq. (6.2) with Eq. (6.5) gives

$$s_k^d = (dt/dz)_k^d = 1 + \nu W_k / D^l D^r, \ k=1,2,\ldots,m \qquad (6.7)$$

Since $D^l > D^r$ due to the 'entropy' condition, it follows from Eqs. (5.2) and (6.7) that

$$s_k^l < s_k^d < s_k^r \qquad (6.8)$$

where s_k^l or s_k^r represents the value of s_k for D^l or D^r, respectively. In addition, by applying Eqs. (5.2), (6.6) and (6.7) we notice that

$$s_{k+1}^l - s_k^d = (N_k K_k / D^l)(X_k^r / D^r J_k^r - X_k^l / D^l J_k^{k+1}) > 0$$

because $J_k^k > 0$ and $J_k^{k+1} < 0$ from Eq. (4.16). Likewise, $s_k^d - s_{k-1}^r \geqslant 0$ and thus we obtain

$$s_{k+1}^l > s_k^d > s_{k-1}^r \qquad (6.9)$$

According to Lax (1957), Eqs. (6.8) and (6.9) are precisely the inequalities characterizing shocks. Therefore, every discontinuity appearing in the present study is a shock.

A shock, if its image falls on a Γ^k, will be called a k-shock

and its propagation path a k-shock line S^k. The concentration
field along an S^k is called a k-shock wave. It is obvious from
Table 1 that, in a k-shock wave, W_k and the set $\{J_i^k\}$ remain
constant and h_k is the only variable among the characteristic
parameters $\{h_i\}$. Based on Eq. (6.6), we may now rephrase the
'entropy' condition as follows : i.e., in a k-shock wave,

$$D^l > D^r \quad \text{and} \quad h_k^l < h_k^r \tag{6.10}$$

for $k = 1, 2, \ldots, m$.

Suppose that a k-shock and a (k+1)-shock are adjacent to
each other with the states D^l, D^m and D^r from the left-hand
side to the right. If the k-shock wave is followed by the (k+1)-
shock, it can be shown that

$$s_{k+1}^d - s_k^d = (N_k K_k / D^m)\{X_k^r / D^r J_k^k - X_k^l / D^l J_k^{k+1}\} > 0$$

because $J_k^k > 0$ and $J_k^{k+1} < 0$ from Eq. (4.16). This is true
for any k and thus

$$1 < s_1^d < s_2^d < \cdots \qquad < s_{m-1}^d < s_m^d \tag{6.11}$$

7. SOLUTION OF RIEMANN'S PROBLEM

We are now in a position to combine the results developed in the
previous sections to construct the solution of a Riemann's problem.
Given the constant initial data $\{X_i^I\}$ and the constant feed data
$\{X_i^F\}$ with a discontinuity at the origin in the (z, t)-plane,
we begin with the homeomorphism (4.12) to determine the sets
of characteristic parameters $\{h_k^I\}$ and $\{h_k^F\}$, which are the
images of data in the m-dimensional space $\mathcal{H}(m)$.

7.1 Existence and uniqueness

In the space $\mathcal{H}(m)$ the images of the initial data and the feed data
would fall on two different points, I and F, respectively, and
hence the image of a solution is given by a continuous path that
connects the points I and F by a sets of Γ's. A Γ^k, being
parallel to the h_k-axis in the space $\mathcal{H}(m)$, such a path can always
be formed and so there exists a solution to the Riemann's problem.

In general, the path from F to I may consist of many seg-
ments of Γ's of various kinds. To a Γ^k there corresponds either
a k-simple wave centered at the origin in the (z, t)-plane, if the
parameter h_k decreases in the direction from F to I, or a
k-shock wave with the shock line emanating from the origin, if
h_k increases in the direction from F to I. Between any pair of

adjacent waves, there appears a region of constant state which is often called a plateau region in some references.

According to the inequalities (5.3), (6.8), (6.9) and (6.11), however, the path from F to I must consist of at most m segments each of which is part of a Γ of distinct kind and which are arranged in the order of descending k as one passes from F to I. Since a Γ^k is straight and parallel to the h_k-axis of the h-coordinate system, such a path always exists and unique. This establishes the existence of a unique solution.

7.2 Construction of solution

It is clear that the initial discontinuity appearing at the origin of the (z, t)-plane has its range of influence centered at the point and this region is of our main interest because outside we simply have two constant states corresponding to the initial and feed data, respectively. It is, therefore, convenient to construct the wave solutions in the (z, t)-plane, from which the concentration profiles for various solute species can be established without difficulty. For the sake of convenience, the (z, t)-plane will be called the physical plane.

The general situation in the physical plane is depicted in Fig. 2, where it is seen that there is a succession of m waves which are arranged in the order of descending k as we proceed from the feed state to the initial state. These waves are either centered simple waves or shock waves and each wave separates regions of constant states. A k-simple wave is represented by a family of straight characteristics C^k while a k-shock wave is given by a straight shock line S^k.

The constant state in the region between the (k+1)-wave and the k-wave will be called the k-constant state and denoted by the superscript k ; i.e., $\{X_i^k\}$. Thus $\{X_i^I\}$, the initial data, corresponds to $\{X_i^0\}$ whereas $\{X_i^F\}$, the feed data, corresponds to $\{X_i^m\}$. The angular region between the boundaries of 1-constant state (the initial state) and the m-constant state (the feed state) is the range of influence of the discontinuity at the origin.

Since h_k varies only across a k-wave, be it a simple wave or a shock wave, the k-constant state must have its image on the vortex

$$(h_1^F , h_2^F , \ldots , h_k^F, h_{k+1}^I , \ldots , h_m^I) \qquad (7.1)$$

of the path between F and I in $\mathcal{H}(m)$. This set of the characteristic parameters may then be used in the inverse mapping given by Eq. (4.19) to determine the concentrations $\{X_i^k\}$ in the

302

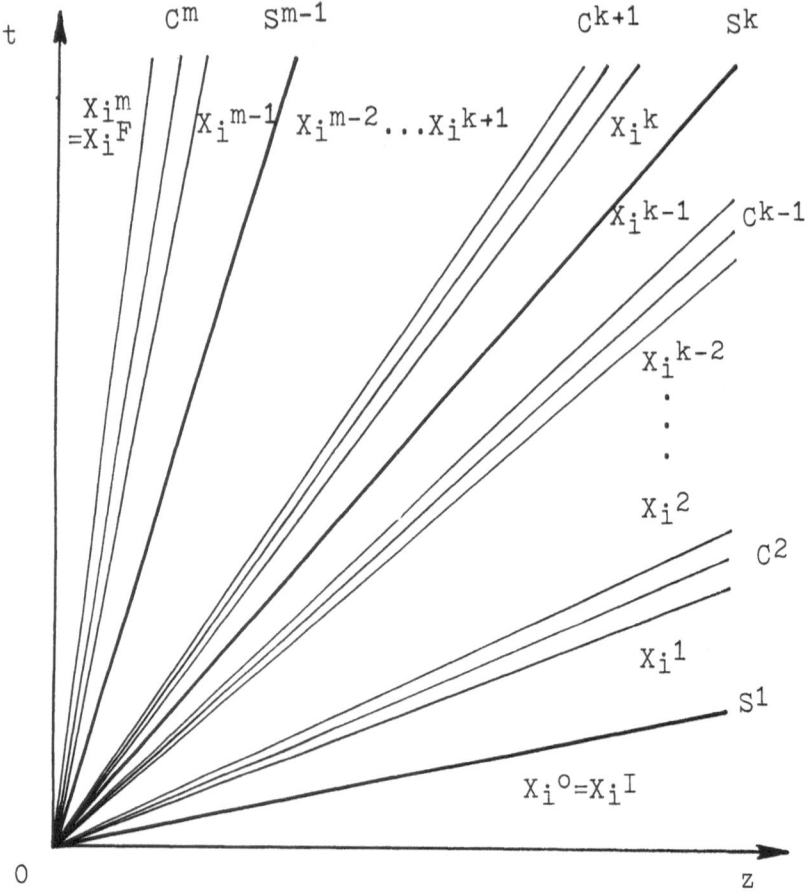

Fig. 2. Schematic portrait of a solution in the physical plane.
———— Characteristic ; ▬▬▬ Shock line.

k-constant state. Another expression for X_i^k can be formulated as follows (cf. Rhee, Aris & Amundson 1970) :

$$X_i^k = X_i^I \prod_{j=1}^{k} (1 - N_i K_i/h_j^F)/(1 - N_i K_i/h_j^I)$$

$$= X_i^F \prod_{j=k+1}^{m} (1 - N_i K_i/h_j^I)/(1 - N_i K_i/h_j^F) \qquad (7.2)$$

for $k = 1, 2, \ldots, m$.

Once every constant state is determined, it is straghtforward to construct the wave solutions. For the k-wave, where $1 \leqslant k \leqslant m$, we proceed as follows:

(1) If $h_k^F > h_k^I$, the k-wave is a simple wave. We then apply Eq. (5.1)

$$s_k = 1 + \vartheta h_k/D \quad \text{for} \quad h_k^I \leqslant h_k \leqslant h_k^F \qquad (7.3)$$

and draw as many straight characteristics C^k's as desired from the origin to generate the centered k-simple wave. The constant values of $\{X_i\}$ along each C^k are determined from Eq. (5.4):
i. e.,

$$X_i = X_i^k - J_i^k (D - D^k) \quad i = 1, 2, \dots, m \qquad (7.4)$$

with $\quad J_i^k = K_i n_i^k / (N_i K_i - h_k^F)$.

(2) If $h_k^F = h_k^I$, the k-wave does not appear in the solution. For such a degenerate case, the image in $\mathcal{H}(m)$ will not contain a segment of Γ^k.

(3) If $h_k^F < h_k^I$, the k-wave is a shock wave: we then apply Eq. (6.7) in the form of

$$s_k^d = 1 + \vartheta h_k^I/D^k = 1 + \vartheta h_k^F/D^{k-1} \qquad (7.5)$$

and draw a straight shock line S^k to give the k-shock wave.

7.3 Special cases

If one or more of the solute species are absent from the feed mixture or from the initial bed of adsorbent, it is expected that some interesting features may appear. These are certainly of potential interest in the practice of multicomponent chromatography and hence worth examining in detail.

In Section 4 it was noticed that, if $X_j = 0$, we have either $h_j = N_j K_j$ or $h_{j+1} = N_j K_j$. The case when $X_m = 0$ is exceptional to give $h_m = N_m K_m$. We further observed that, among the characteristic parameters $\{h_k\}$, h_j is the only variable across a j-wave. When associated with the discussions in §7.2, these arguments lead us to the following conclusions in regard to the practice of chromatography :

(1) The species A_m can be exhausted through an m-shock wave and emerge through an m-simple wave.
(2) A particular species A_j can be exhausted through a j-shock wave or a (j+1)-simple wave.

(3) A particular species A_j can emerge through a j-simple wave or a (j+1)-shock wave.

Consider now the conventional adsorption process (saturation of a column) for which the initial bed is clean. The feed mixture contains m solute species and each species must vanish somewher in the column. It is now obvious that A_j must be exhausted through the j-shock wave for $1 \leqslant j \leqslant m$. In other words, each species disappears successively in the order of decreasing adsorptivity along the column. This fact was proved for an arbitrary equilibrium relation by DeVault (1943). The physical plane portrait is shown in Fig. 3(a).

For the conventional desorption process (elution of a column) we have the pure solvent flowing into the column, whereas the initial bed is uniformly saturated with m solute species. There-fore, each species must emerge somewhere in the column. It is not difficult to realize that A_j must emerge through the j-simple wave for $1 \leqslant j \leqslant m$ and thus the physical plane portrait would be as shown in Fig. 3(b).

Another interesting example is produced by the alternating data; i.e., $X_j^I = 0$ if j is even and $X_j^F = 0$ if j is odd or vice versa. It is clear that along the column, exhaustion of species

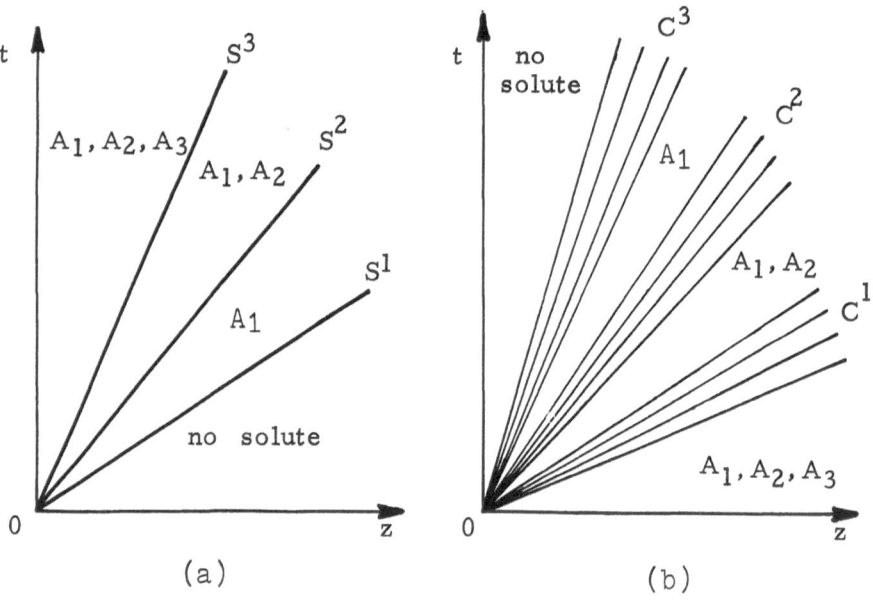

(a)   (b)

Fig. 3. Typical solutions in the physical plane (a) for the saturation of a column and (b) for the elution of a column (m = 3).

A_j alternates with the emergence of species of A_{j-1} . The species A_j may emerge through the j-simple wave or be exhausted through the j-shock wave while in these two kinds of waves A_{j-1} is respectively exhausted or emerges. A number of different situations can occur depending upon the data specified. These are suggested in the basic profile patterns for m = 3 by Klein, Tondeur and Vermeulen (1967). In any case, once the sets $\{h_k{}^F\}$ and $\{h_k{}^I\}$ are determined, construction of the solution would be straight-forward. Moreover, half of $\{h_k\}$ are given by N_iK_i where $X_i = 0$.

8. NUMERICAL EXAMPLES

In order to illustrate the applications, several numerical examples will be treated in this section. Since each case was considered in a general fashion in Section 7, we shall add a brief discussion and present the solution in the form of graphs.

For simplicity, it will be assumed in all examples that the limiting values N_i are all identical and given by

$$N_i = N = 1.0 \text{ mole/liter of adsorbent}$$

for i = 1, 2, ..., m. The bed voidage is specified as

$$\epsilon = 0.4 \qquad \text{or} \qquad \nu = 1.5$$

The concentrations $\{c_i\}$ are given in moles per liter of fluid phase and the concentrations n_i in moles per liter of adsorbent phase. The unit of K_i is, of course, the reciprocal of concentration.

Example 1 : General case (m = 3).

i	1	2	3
K_i	5.0	7.5	10.0
$c_i{}^F$	0.150	0.060	0.020
$c_i{}^I$	0.032	0.114	0.075
$h_i{}^F$	2.445	6.667	9.586
$h_i{}^I$	2.797	5.409	8.974

Since all h_i's vary from the feed state to the initial state we expect to have three waves and two additional constant states. The 1-wave is a shock wave because $h_1{}^F < h_1{}^I$ whereas the 2- and 3-waves are centered simple waves because $h_2{}^F > h_2{}^I$ and $h_3{}^F > h_3{}^I$, respectively.

The two constant states are characterized by the set of characteristic parameters, respectively; i.e., $(h_1{}^F, h_2{}^I, h_3{}^I)$ for the 1-constant state and $(h_1{}^F, h_2{}^F, h_3{}^I)$ for the 2-constant state. Therefore, the concentrations at each constant state can be readily determined by using Eq. (4.19) or Eq. (7.2): for example, from Eq. (7.2)

$$X_3{}^1 = X_3{}^I (1 - N_3K_3/h_1{}^F)/(1 - N_3K_3/h_1{}^I) = 0.90$$

$$X_3{}^2 = X_3{}^F (1 - N_3K_3/h_3{}^I)/(1 - N_3K_3/h_3{}^F) = 0.53$$

Once all the concentrations are determined, it is fairly straightforward to construct the solution in the physical plane, whence the concentration profiles can be obtained.

The image of the solution in the concentration space consists of three segments of Γ's arranged in the order of Γ^3, Γ^2 and Γ^1 from the feed state to the initial state. The projections of the image on the (X_1, X_3)-plane and on the (X_2, X_3)-plane are shown in the upper part of Fig. 4. Presented also in Fig. 4 are the physical plane portrait of the solution and the concentration profiles when $t = 4.0$.

Example 2: Saturation of a clean bed ($m = 3$).

i	1	2	3
K_i	5.0	10.0	15.0
$c_i{}^F$	0.05	0.05	0.05
$c_i{}^I$	0	0	0
$h_i{}^F$	3.387	7.050	12.563
$h_i{}^I$	5.0	10.0	15.0

This is the conventional process for the saturation of a clean bed of adsorbent. All h_i's increase from the feed state to the

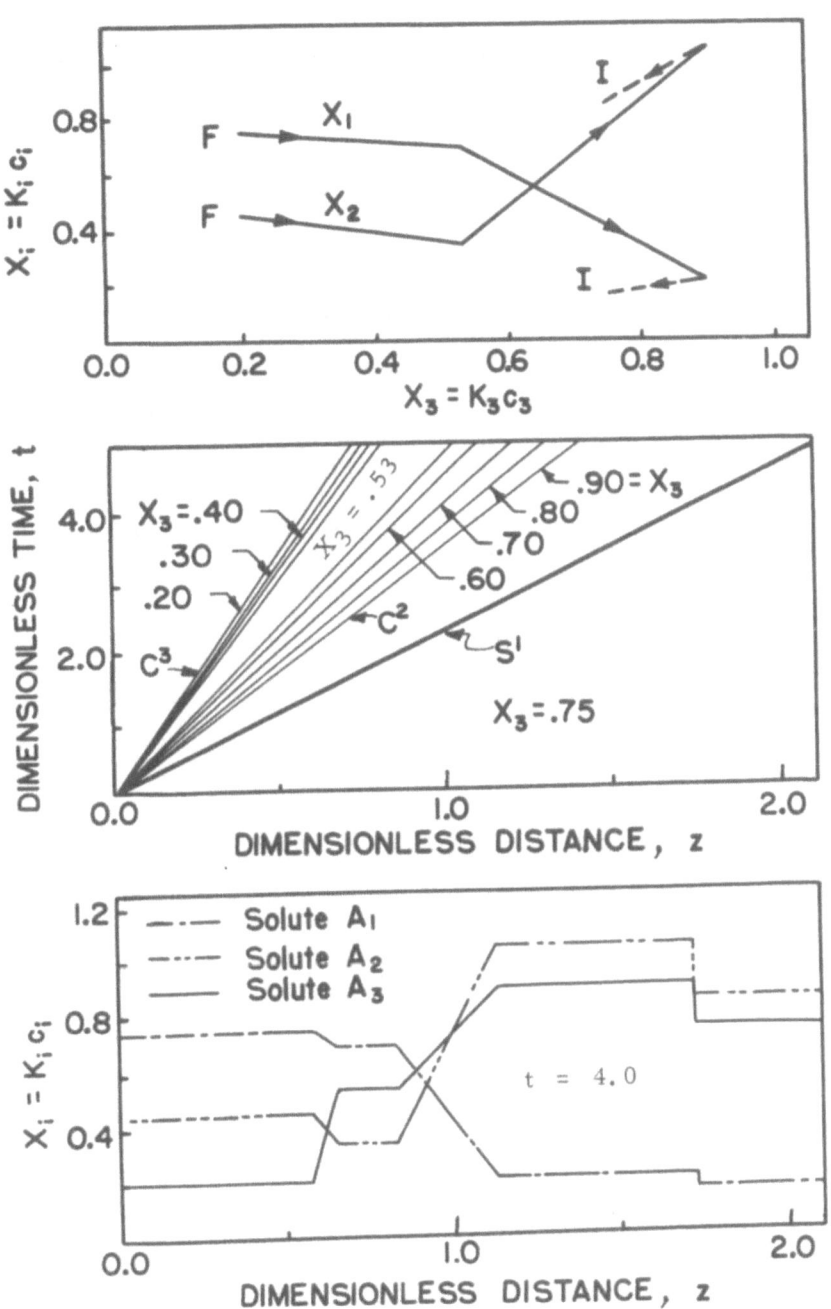

Fig. 4. Solution of Example 1.

initial state and thus we have three shock waves. Three shock lines, S^3, S^2 and S^1, divide the physical plane in the manner as shown in Fig. 5.

For the 2-constant state, we have the set (h_1^F, h_2^F, h_3^I). Here it is noticed that $h_3^I = N_3K_3 = 15.0$ and hence $X_3^2 = 0$. This implies that A_3 would be exhausted across the 3-shock wave. From Eq. (7.2),

$$X_i^2 = X_i^F(1 - N_iK_i/h_3^I)/(1 - N_iK_i/h_3^F)$$

for $i = 1, 2$. It is obvious from the relation $h_3^F < h_3^I$ that both X_1 and X_2 will be increased across the 3-shock wave.

Since we have the set (h_1^F, h_2^I, h_3^I) for the 1-constant state and $h_2^I = N_2K_2 = 10.0$, it follows that $X_2^1 = 0$ and so A_2 would be exhausted while X_1 increases across the 2-shock wave. Likewise, A_1 would be exhausted across the 1-shock wave.

Fig. 5 shows the profiles of $\{X_i\}$ as well as $\{n_i\}$ when $t = 5.0$ and 12.0, respectively.

Example 3: Elution of a saturated bed ($m = 3$).

i	1	2	3
K_i	5.0	10.0	15.0
c_i^F	0	0	0
c_i^I	0.05	0.05	0.05
h_i^F	5.0	10.0	15.0
h_i^I	3.387	7.050	12.563

This is just the reverse of Example 2, representing the conventional process for the elution of a previously saturated bed of adsorbent. Here the feed state and the initial state are interchanged so that the arguments on the parameter h_i are simply reversed from the privious case. Therefore, there appear three simple waves, one solute emerging through each of the waves. Fig. 6 shows the physical plane portrait of the solution as well as the profiles of $\{X_i\}$ and $\{n_i\}$ at $t = 6.0$.

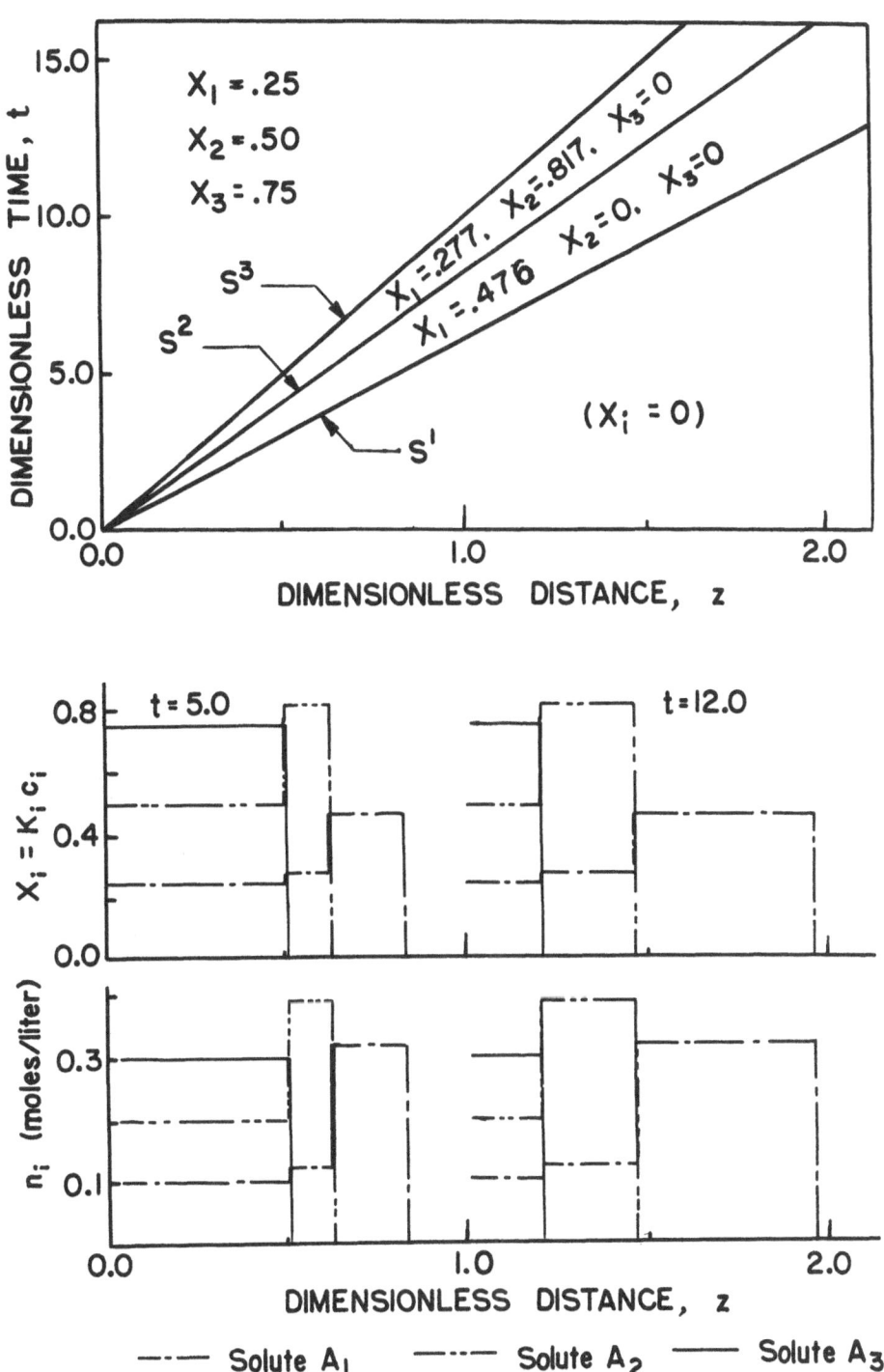

Fig. 5. Solution of Example 2.

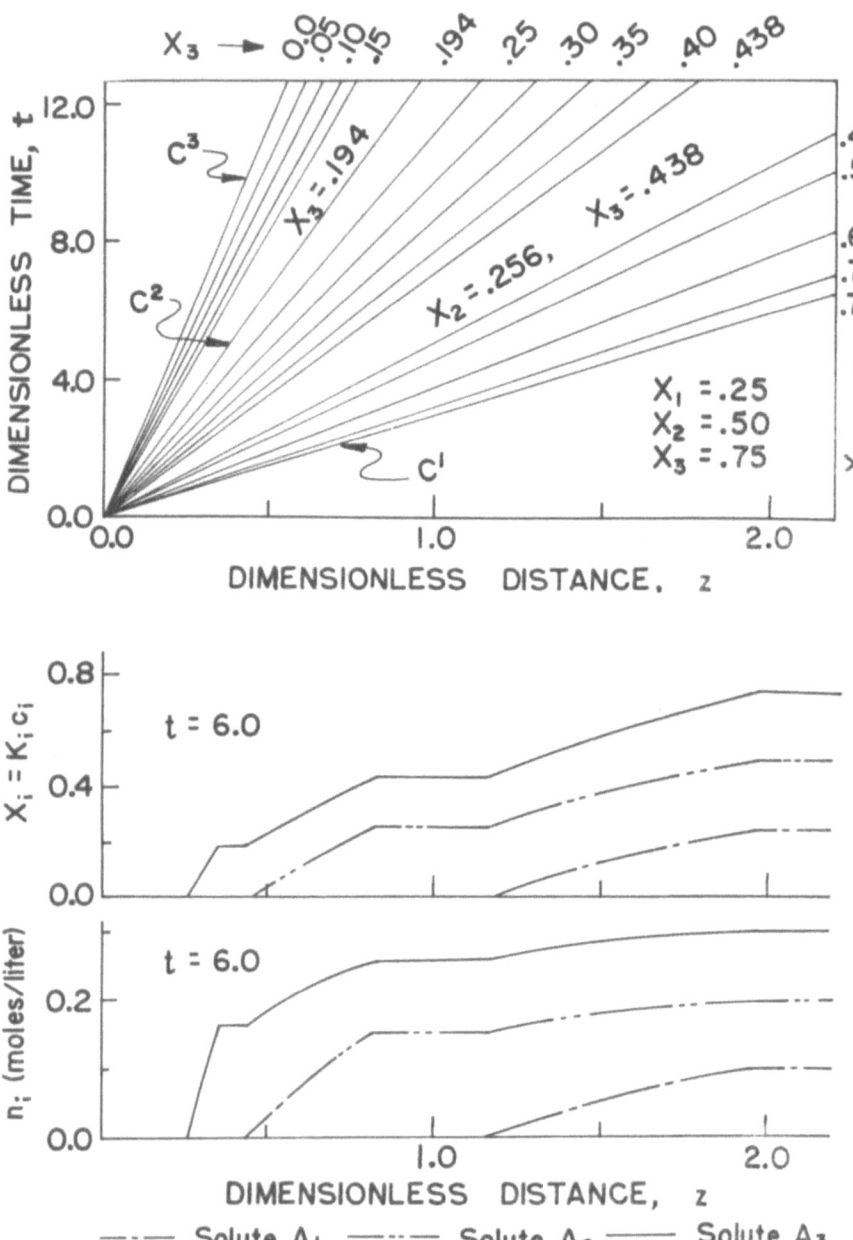

Fig.6. Solution of Example 3.

Example 4 : Alternating data (m = 3).

i	1	2	3
K_i	5.0	7.5	10.0
c_i^F	0	0.04	0
c_i^I	0.04	0	0.02
h_i^F	5.0	5.769	10.0
h_i^I	4.060	7.5	8.978

In this case a bed of adsorbent saturated by two solutes is being eluted by a stream carrying one solute of intermediate adsorptivity. For the feed state, Eq. (4.12) is reduced to a linear equation for h whereas, for the initial state, it becomes a quadratic equation.

Comparing the values of h_i^F and h_i^I, we find that the solution consists of the 1-simple wave, the 2-shock wave, and the 3-simple wave as depicted in Fig. 7. The 2-constant state, being characterized by the set (h_1^F, h_2^F, h_3^I) with $h_1^F = N_1 K_1 = 5.0$, would not contain the species A_1. For the 1-constant state we have (h_1^F, h_2^I, h_3^I), where $h_1^F = N_1 K_1 = 5.0$ and $h_2^I = N_2 K_2 = 7.5$, and hence both the species A_1 and A_2 are absent from the 1-constant state. Consequently, the emergence and exhaustion of solute species are alternating along the column.

Fig. 7 shows the physical plane portrait and the concentration profiles when t = 10.0.

Example 5 : Alternating data (m = 4).

i	1	2	3	4
K_i	5.0	10.0	12.5	20.0
c_i^F	0	0.06	0	0.04
c_i^I	0.04	0	0.02	0
h_i^F	5.0	5.40	12.5	15.434
h_i^I	4.075	10.0	10.575	20.0

312

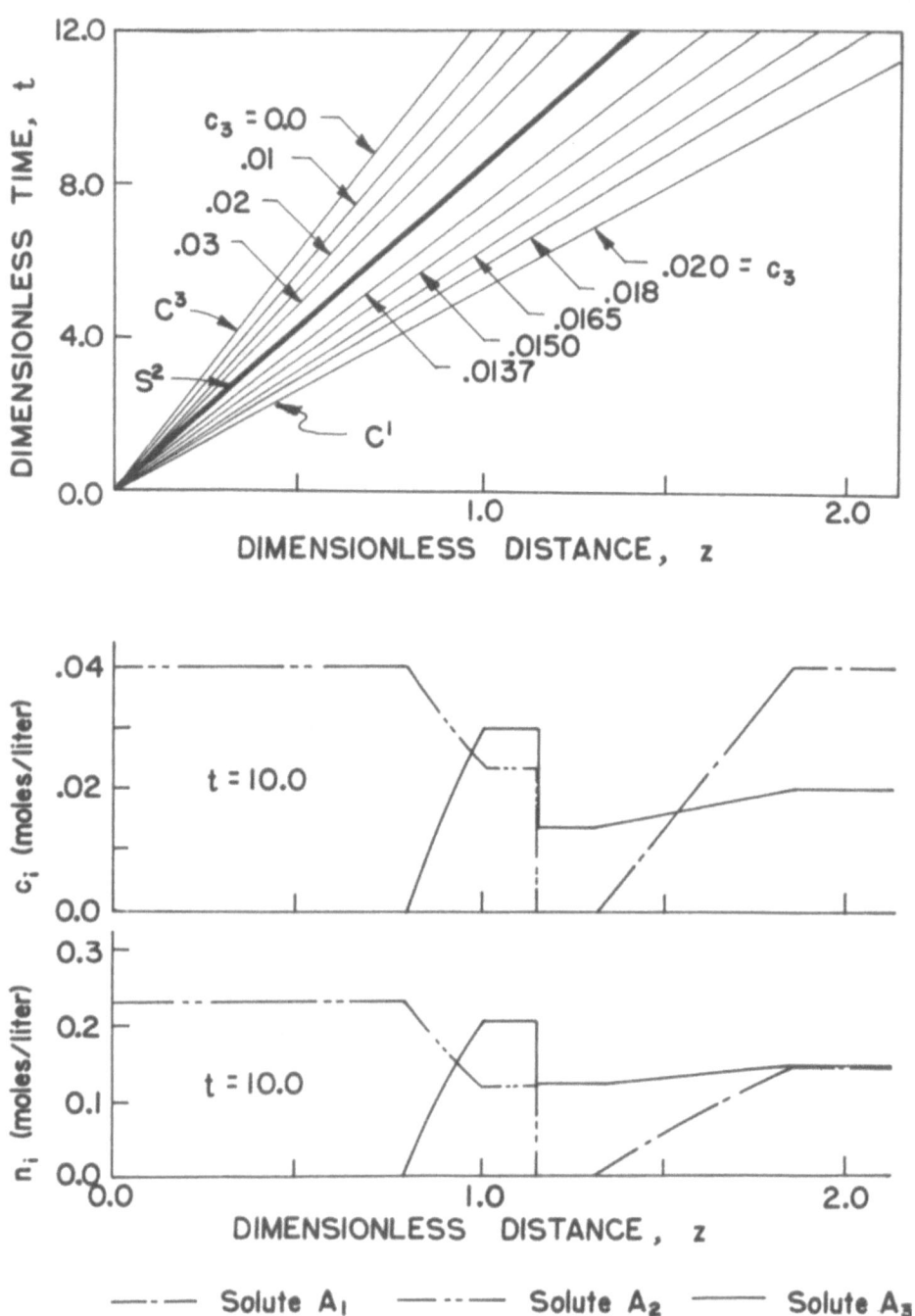

Fig. 7. Solution of Example 4.

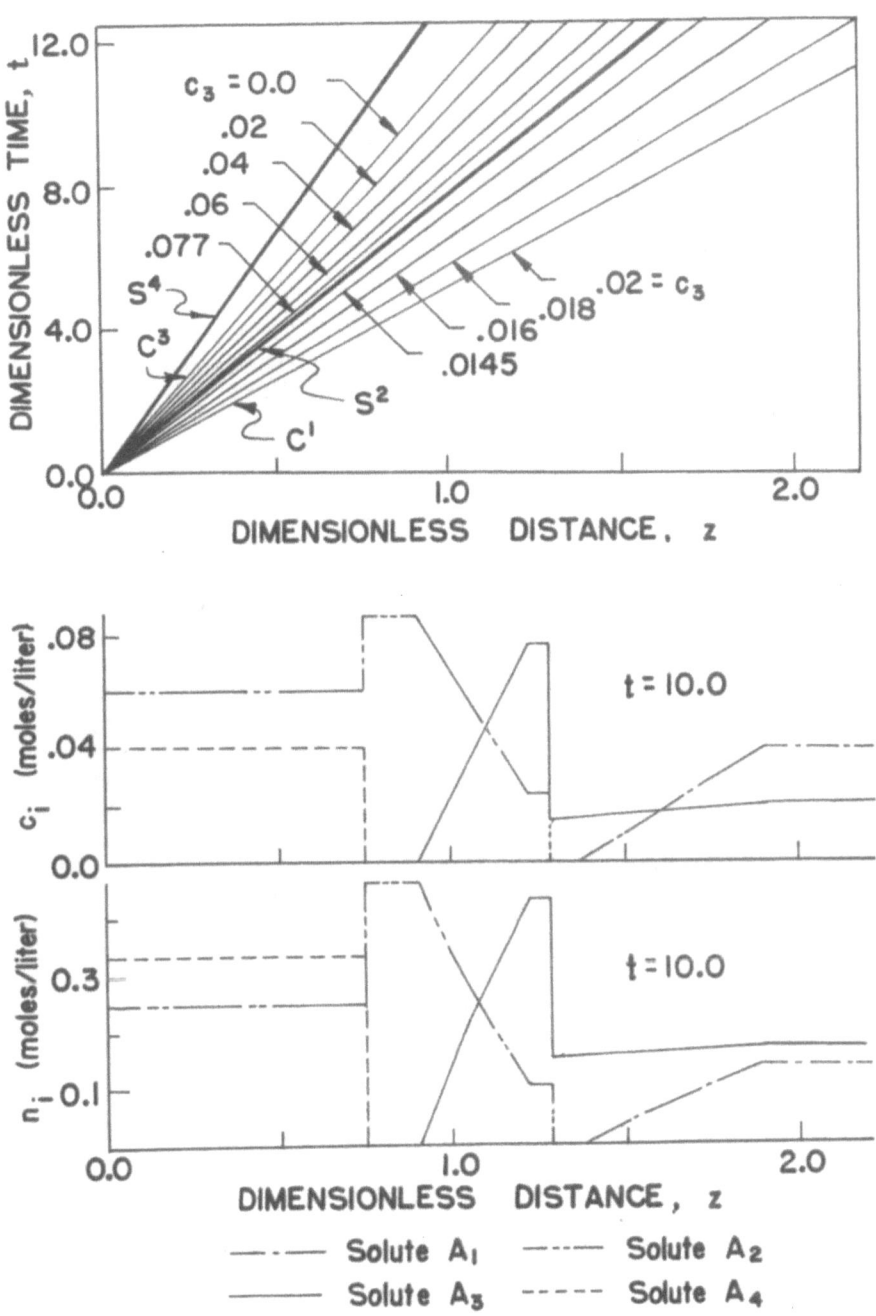

Fig. 8. Solution of Example 5.

This is the same as the previous example except for the fact that the feed stream contains an additional solute species having the highest adsorptivity. Here again Eq. (4.12) is reduced to a quadratic equation of h.

Based on the values of h_i^F and h_i^I, we expect to have the 4-shock wave, the 3-simple wave, the 2-shock wave and the 1-simple wave. In addition, it is not difficult to realize that only A_2 would appear in the 3-constant state whereas A_2 and A_3 would be present in the 2-constant state, and finally the 1-constant state would contain A_3 only. Therefore, alternating through the column are not only the wave patterns but also the exhaustion and the emergence of solutes.

Fig. 8 presents the physical plane portrait of the solution along with the distribution of solutes both in the fluid and adsorbent phases at $t = 10.0$.

9. INTERACTION BETWEEN WAVES

Both the fundamental differential equation (2.15) and the compatibility condition (6.3) remain unchanged if the origin of the (z, t)-coordinate system is translated arbitrarily. Furthermore, the Riemann invariants as well as the characteristic parameters are determined in terms of concentrations only. It then follows that, given a jump discontinuity in the physical plane, the image of solution in $\mathcal{X}(m)$ or $\mathcal{H}(m)$ is independent of the location of the discontinuity in the physical plane. Consequently, the simple wave theory as well as the shock wave theory can be equally well applied with the same homeomorphism (4.12) to a jump discontinuity located at any point in the physical plane.

The present theory, therefore, finds itself applicable to a class of problems associated with stepwise constant initial and/or feed data. The solution as a whole can be established by constructing separately the wave solutions centered at each point of discontinuity. This is the characteristic of hyperbolic systems (Courant & Friedrichs 1948).

After a finite period of time, however, any two wave solutions centered at two different, but adjacent, points of discontinuity will meet each other so that an overlapped region appears in which the solution is influenced by two different sets of data at the same time. Such a phenomenon is called an interaction between waves. Among various cases we will be concerned here with the interaction between a simple wave and a shock wave because those are the situations we would encounter in the next section in discussing the separation of solutes. Full discussions may be found from references (Rhee, Aris & Amundson 1970, 1971).

9.1 Interaction between a k-simple wave and a k-shock wave

Let us consider a k-simple wave overtaking a k-shock wave. For a more general discussion we assume that the simple wave is not centered but based on distributed data given along the t-axis as shown in Fig. 9. Since both waves have their images on the same Γ^k in the space $\mathcal{X}(m)$, the simple wave does not transmit across the shock wave : instead, the former is blocked by the latter and the net effect is the continuous decrease in the strength of the shock.

The k-Riemann invariants remaining constant everywhere, there exists a one-parameter representation of the whole solution. The state on the right-hand side of the shock is constant and so D^r remains constant along the shock line. We may now identify D^l with D along the shock line which has a one-to-one correspondence to the feed data along the t-axis. We also assume that the feed condition is prescribed by an invertible function so that the

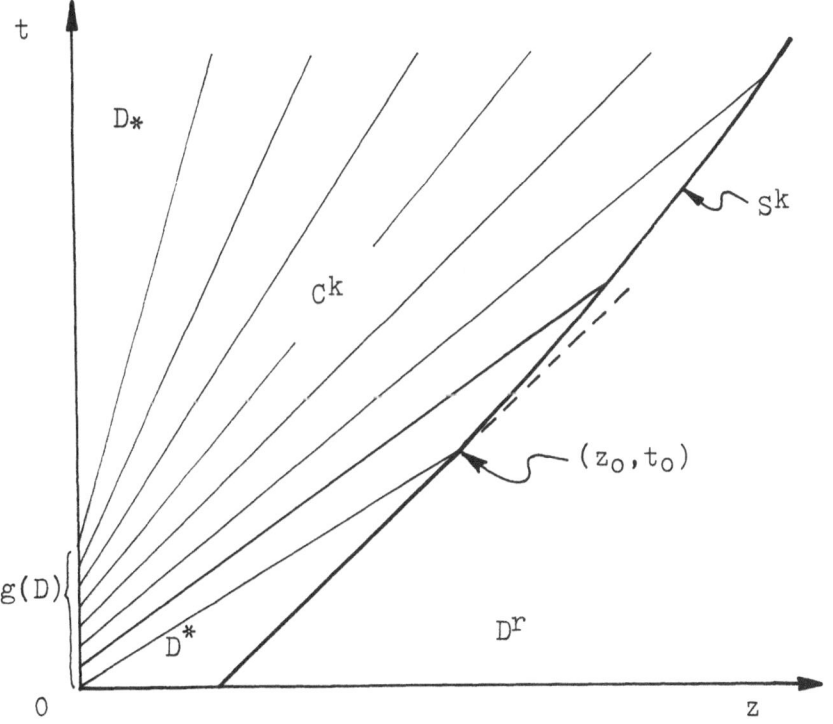

Fig. 9. Interaction between a k-simple wave and a k-shock wave.

intercept of the characteristics C^k may be obtained in terms of D. Along the shock line (z, t), therefore, we have two equations

$$\begin{cases} t - g(D) = s_k z & (9.1) \\[2mm] \dfrac{dt}{dz} = \dfrac{dt}{dD} \Big/ \dfrac{dz}{dD} = s_k^d & (9.2) \end{cases}$$

which may be combined to yield

$$(s_k^d - s_k) \frac{dz}{dD} - \frac{ds_k}{dD} z = \frac{dg}{dD} \qquad (9.3)$$

This is an ordinary differential equation which, if solved with the initial condition

$$z = z_0 \quad \text{at} \quad D = D^* \qquad (9.4)$$

will generate z as a function of D. Eq. (9.1) is then used to determine t as a function of D. The set (z, t) so determined is the parametric representation of the curved shock line.

Substituting Eqs. (5.2) and (6.7) with $D^l = D$ into Eq. (9.3) and rearranging, we obtain the equation

$$(D - D^r) \frac{dz}{dD} + \frac{2D^r}{D} z = (D^r / \nu W_k) D^2 \frac{dg}{dD} \qquad (9.5)$$

which can be integrated with the initial condition (9.4) to give the solution in the form

$$z(D) = z_0 \left\{ \frac{1 - D^r/D^*}{1 - D^r/D} \right\}^2 + \frac{D^r}{\nu W_k} \left\{ \frac{D}{D - D^r} \right\}^2 \left\{ (D-D^r)g(D) \right.$$

$$\left. - \int_{D^*}^{D} g(D)dD \right\} \qquad (9.6)$$

If the simple wave is centered at the origin, we may put $g = 0$ to obtain

$$z(D) = z_0 \left\{ \frac{1 - D^r/D^*}{1 - D^r/D} \right\}^2 \qquad (9.7)$$

Otherwise, we shall take the point (z_0, t_0) as the new origin of the coordinate system so that z is given by the second term in the right-hand side of Eq. (9.6).

In addition, we notice from Eq. (9.6) that z diverges in the limit as D approaches to D^r and hence the interaction will go on indefinitely unless $D_* > D^r$. This implies that a shock can never disappear completely. Since D ($= D^1$) decreases as the interaction goes on, it follows from Eq. (6.7) that the slope of the shock line increases as the shock propagates. Consequently, the shock is continuously decelerated due to the interaction.

9.2. Interaction between a k-simple wave and a j-shock wave

Here the two waves involved in the interaction are of differnt kinds and thus their images in the space $\mathcal{X}(m)$ fall on two different Γ's joined together. The interaction, therefore, causes the simple wave to transmit across the shock wave. Since neither Riemann invariants remain constant throughout the whole region and the state of concentrations changes simultaneously on both sides of the shock wave, the situation appears to be complicated. However, a rather complete analysis can be performed by using the Riemann invariants together with the characteristic parameters.

Consider the interaction between a k-simple wave and a j-shock wave where $j > k$. Again we assume that the simple wave is based on data distributed along the t-axis so that the intercepts of the characteristics C^k are defined by a function $g(D)$ as shown in Fig. 10.

Here we notice that both W_k and h_j remain constant on each side of the j-shock wave. We shall identify D^1 with D along the shock line which, in turn, has a one-to-one correspondence to the feed data $g(D)$ and write the equations

$$
\begin{cases}
t - g(D) = s_k^1 z & \text{(9.8)} \\[2mm]
\dfrac{dt}{dz} = \dfrac{dt}{dD} \Big/ \dfrac{dz}{dD} = s_j^d & \text{(9.9)}
\end{cases}
$$

along the shock line, where

$$ s_k^1 = 1 + \partial\, W_k^1/D^2 \tag{9.10} $$

$$ s_j^d = 1 + \partial\, h_j^r/D \tag{9.11} $$

The four equations together form a one-parameter representation of the interaction. Combination of Eq. (9.8) and (9.9) gives

$$ (s_j^d - s_k^1)\, \frac{dz}{dD} - \frac{ds_k^1}{dD}\, z = \frac{dg}{dD} \tag{9.12} $$

318

which is subject to the initial condition

$$z = z_0 \qquad \text{at} \qquad D = D^* \qquad (9.13)$$

Substituting Eqs. (9.10) and (9.11) into Eq. (9.12), we obtain

$$(D - R_{kj}) \frac{dz}{dD} + 2(R_{kj}/D)z = (D^2/\nu\, h_j{}^r) \frac{dg}{dD} \qquad (9.14)$$

in which

$$R_{kj} = W_k{}^l/h_j{}^r \qquad (9.15)$$

Eq. (9.14) is an ordinary differential equation which, if solved for a given g(D), will generate z as a function of D, because both

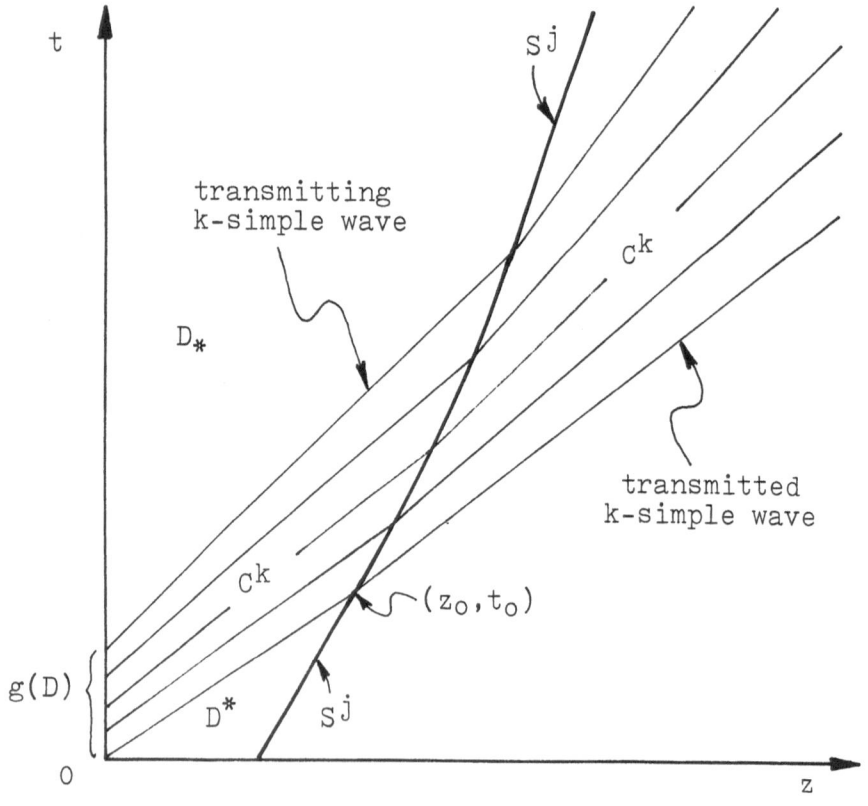

Fig. 10. Interaction between a k-simple wave and a j-shock
wave where j > k.

$W_k{}^1$ and $h_j{}^r$ remain constant along the shock line. Integration with the initial condition (9.13) yields the solution in the form

$$z(D) = z_0 \left\{ \frac{D}{D^*} \cdot \frac{D^* - R_{kj}}{D - R_{kj}} \right\}^2 + \frac{1}{\nu\, h_j{}^r} \left\{ \frac{D}{D - R_{kj}} \right\}^2 \left\{ (D - R_{kj}) g(D) \right. $$
$$\left. - \int_{D^*}^{D} g(D)\, dD \right\} \qquad (9.16)$$

If the transmitting k-simple wave is centered at the origin, we can take $g = 0$ to obtain

$$z(D) = z_0 \left\{ \frac{D}{D^*} \cdot \frac{D^* - R_{kj}}{D - R_{kj}} \right\}^2 \qquad (9.17)$$

Otherwise, we shall take the point (z_0, t_0) as the new origin of the coordinate system so that z is given by the second term in the right-hand side of Eq. (9.16).

As we cross the j-shock line along a C^k, both h_k and W_j remain constant. Therefore, the transmitted k-simple wave can be constructed by drawing straight characteristics C^k of slope

$$s_k{}^r = 1 + \nu\, (W_k{}^1 h_j{}^r / h_j{}^1) / D^2 \qquad (9.18)$$

and the state of concentrations is determined by the relation

$$X_i{}^r = X_i{}^1 (1 - N_i K_i / h_i{}^r) / (1 - N_i K_i / h_i{}^1) \qquad (9.19)$$

for $i = 1, 2, \ldots, m$.

The interaction will terminate when $D = D_*$. Since both D^1 and D^r decrease as the interaction proceeds, it follows from Eq. (6.7) that the slope of the shock line increases as the shock propagates. This implies that the shock is continuously decelerated while interacting

10. SEPARATION OF SOLUTES

It is well known that separation of different solute species can be accomplished by applying the conventional processes of saturation and elution successively to the same bed of adsorbent. Suppose an initially clean bed of adsorbent is first saturated with a fluid mixture containing m different solutes $\{A_i\}$, laying down the sample in the form of a chromatogram. After a finite period of time,

t_0 , the inlet stream will be changed from the mixture to the pure solvent so that the chromatogram starts to be eluted. Since the data here contain two discontinuities, one at the origin and the other at the point ($0, t_0$), two sets of waves will develop in the physical plane and hence there will naturally occur interactions between waves.

We shall discuss the problem for $m = 3$ but the same procedur is certainly applicable for any m greater than three. Fig. 11(a) shows the image of the solution in the concentration space $\mathcal{X}(3)$. Here it is interesting to note that the image in $\mathcal{X}(3)$ or $\mathcal{H}(3)$ of the state along the bed at any instant is given by a closed loop. Such a loop will be called a chromatographic cycle. While inter-actions take place between pairs of waves, the cycle will deform continuously. By pursuing the course of deformation of the chro-matographic cycle, one can clearly visualize how different solute species are separated.

By applying the results of the previous sections, a complete analysis can be accomplished, giving the solution as shown in Fig. 11. The image in $\mathcal{X}(3)$ is presented to show how the chro-matographic cycle deforms but, though this is certainly a useful visualization, it is not necessary for the construction of a solution. In the following, we shall discuss in detail how the solution can be determined.

Until the moment $t = t_1$ when the earliest interaction starts, the solution may be given by the combination of Examples 2 and 3 of Section 8. The distribution of solutes is represented by the chro-matographic cycle ($O \rightarrow E \rightarrow F \rightarrow P \rightarrow Q \rightarrow R \rightarrow O$), in which the arrow denotes the direction from the feed state to the initial state.

At $t = t_1$ the 1-simple wave starts to overtake the 3-shock wave so that the state P disappears and the 3-shock finds its image ($\overline{PQ}$) receding toward $\overline{FG}$. During the interaction which is characterized by $k = 1$ and $j = 3$, the 3-shock line is located by Eq. (9.17) whereas the transmitted 1-simple wave is determined by Eqs. (9.18) and (9.19). Meanwhile, the chromatographic cycle becomes ($O \rightarrow E \rightarrow F \rightarrow a \rightarrow b \rightarrow Q \rightarrow R \rightarrow O$) for example. When the interaction is over at $t = t_a$, the 3-shock line separates the mixture containing A_2 and A_3 from that containing A_1 and A_2 from the left to the right.

At $t = t_2$ the once-transmitted 1-simple wave starts again an interaction with the 2-shock wave. The 2-shock line and the twice-transmitted 1-simple wave are given by Eqs. (9.16) and (9.18), respectively, with k=1 and j=2. The chromatographic cycle is now reduced to ($O \rightarrow E \rightarrow F \rightarrow G \rightarrow b \rightarrow e \rightarrow R \rightarrow O$). When the interactior terminates at $t = t_b$, the 2-shock line separates the mixture con-taining A_2 only from that containing A_1 alone from the left to the

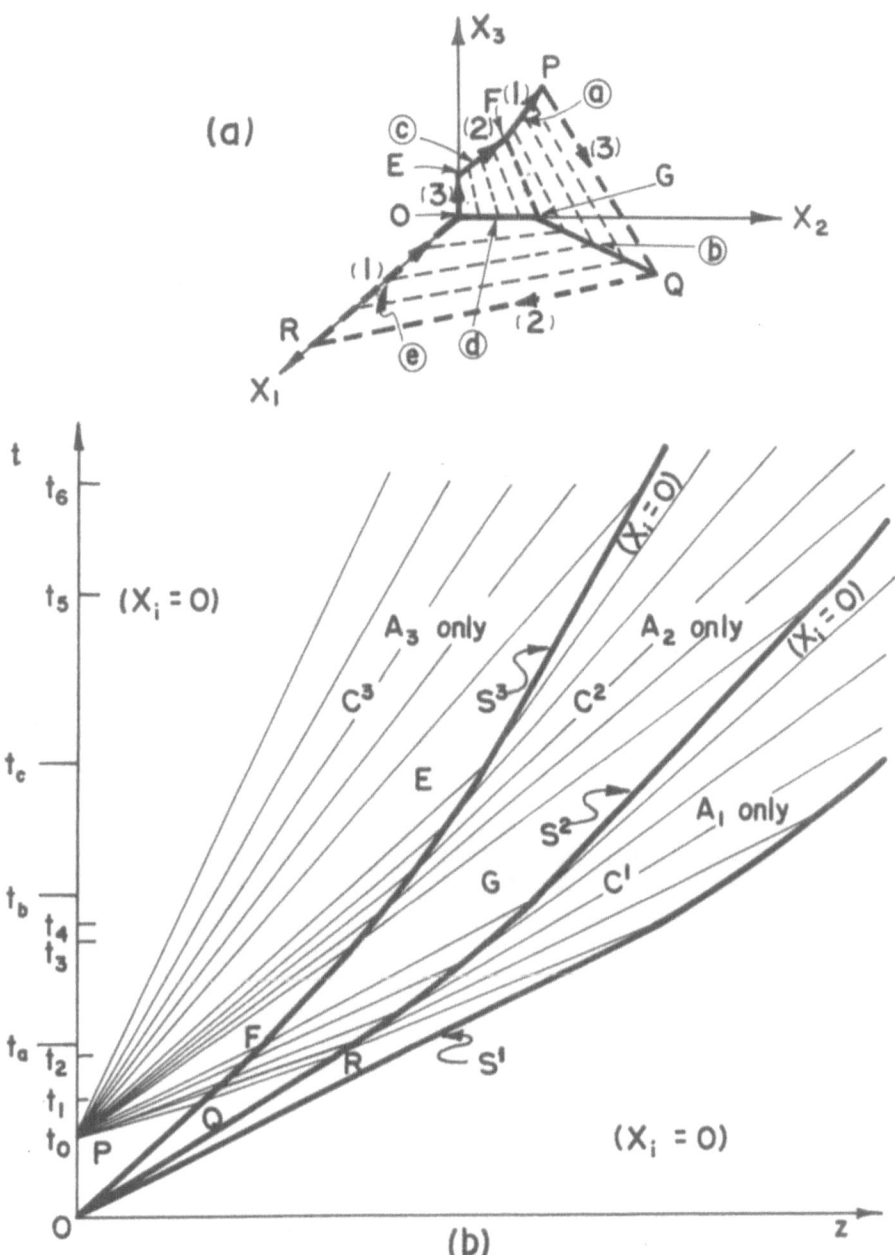

Fig. 11. Separation of three solutes by succesive operation of
saturation and elution processes. (a) Concentration space
$\chi(3)$; (k) image of k-simple wave; (k) image of k-shock
wave. (b) physical plane portrait of the solution.

right; that is, chromatograms of pure A_1 and pure A_2 are obtained

On the other hand, the 2-simple wave also starts an interaction with the 3-shock wave at $t = t_3$. The 3-shock line is located by Eq. (9.17) with $k = 2$ and $j = 3$, and the transmitted 2-simple wave is determined by Eqs. (9.18) and (9.19). The chromatographic cycle now becomes ($O \rightarrow E \rightarrow c \rightarrow d \rightarrow G \rightarrow O \rightarrow R \rightarrow O$). When this interaction is over at $t = t_c$, the 3-shock line separates the mixture containing A_3 only from that containing A_2 alone from the left to the right. Hence, the complete separation of three solutes, A_1, A_2 , and A_3 is accomplished at $t = t_c$.

Meanwhile, the twice-transmitted 1-simple wave may begin to overtake the 1-shock wave at $t = t_4$. As t increases, the once-transmitted 2-simple wave also starts an interaction with the 2-shock wave at $t = t_5$. The corresponding shock line can be located by applying Eq. (9.6) with $k = 1$ or $k = 2$, respectively. The 3-simple wave will also start to overtake the 3-shock wave at $t = t_6$ and this interaction is described by Eq. (9.7).

Each interaction between waves of the same kind in the above is accompanied with the disappearance of the constant state R, G, or E, respectively. Therefore, the chromatographic cycle will be reduced to ($O \rightarrow E' \rightarrow O \rightarrow G' \rightarrow O \rightarrow R' \rightarrow O$), where R', for example, denotes an intermediate point on $\overline{OR}$. Further operation of the chromatography will lead only to the lengthening of each chromatogram of a single solute which is not desirable although it results in a more distinct separation.

Finally, it is possible to construct the physical plane portrait as shown in Fig. 11(b), from which the concentration profiles can be determined at any moment. It is interesting to notice that the 3-shock line may be regarded as a screen which holds the solute A_3 only and passes other solutes. Similarly, the 2-shock line acts as a screen that holds the solute A_2 only and passes A_1. The 1-shock line holds the solute A_1 which has passed through the two screens. Exactly the same procedure can be applied to a system involved with more than three solutes.

Example 6 : Separation of three solutes ($m = 3$).

i	1	2	3	Remark
K_i	5.0	10.0	15.0	
$c_i{}^F$	0.05	0.05	0.05	for $0 < t < 3.0$
	0	0	0	for $t > 3.0$
$c_i{}^I$	0	0	0	
$h_i{}^F$	3.387	7.050	12.563	for $0 < t < 3.0$
	5.0	10.0	15.0	for $t > 3.0$
$h_i{}^I$	5.0	10.0	15.0	

Other parameters are specified as

$$\epsilon = 0.4 \qquad \text{or} \qquad \nu = 1.5$$

and $\qquad N_i = N = 1.0$ mole/liter of adsorbent
for $i = 1, 2, 3$.

This example may be regarded as the superposition and interaction of Examples 2 and 3. Presented in Fig. 12 is the physical plane portrait of the solution showing the interactions between the simple waves and the shock waves. On the left-hand side of the 3-shock line S^3, the set of three simple waves represents the same situation as Example 3 whereas the three shock waves below the lowest characteristic C^1 are identical to those in Example 2.

All the interactions involved here can be analyzed without difficulty by following the procedure explained in the above. In particular, the interaction between the 1-simple wave and the 3-shock wave terminates at $z_a = 0.58$ when $t = t_a = 6.0$. The interaction between the 1-simple wave and the 2-shock wave is over at $z_b = 0.9$ when $t = t_b = 8.0$. Here, the solute A_1 is completely separated to form a chromatogram of pure A_1 on the right-hand side of the 2-shock line.

Finally, the interaction between the 2-simple wave and the 3-shock wave is completed at $z_c = 1.88$ when $t = t_c = 28.4$. The three solutes are then completely separated. Since the dimensionless time t was defined by Eq. (2.4), the volume of sample introduced during a period of $t_0 = 3$ amounts to 1.2 ($= \epsilon t_0$) volume units in terms of the bed volume. Consequently, the ratio of the sample volume to the bed volume required for a complete separation is about 0.64 which is relatively large. Such a remarkable efficiency is attributed to the particular values of K_i that were chosen, but this example demonstrates that with strong adsorption and very different values of K_i the solutes may be separated very easily.

In Fig. 13 the distribution of solutes along the bed is shown at various times. In the first part ($t = 3.0$) of the figure the sample has just been put on the bed and the next two parts ($t = 5.5$ and $t = 7.0$) show the development of the chromatogram proceeding. By $t = 10$, the solute A_1 has been separated from A_2 and A_3. The chromatographic mixture of these two continues to develop ($t = 15$) and finally, the two have also separated as shown in the last part ($t = 30$) of Fig. 13.

324

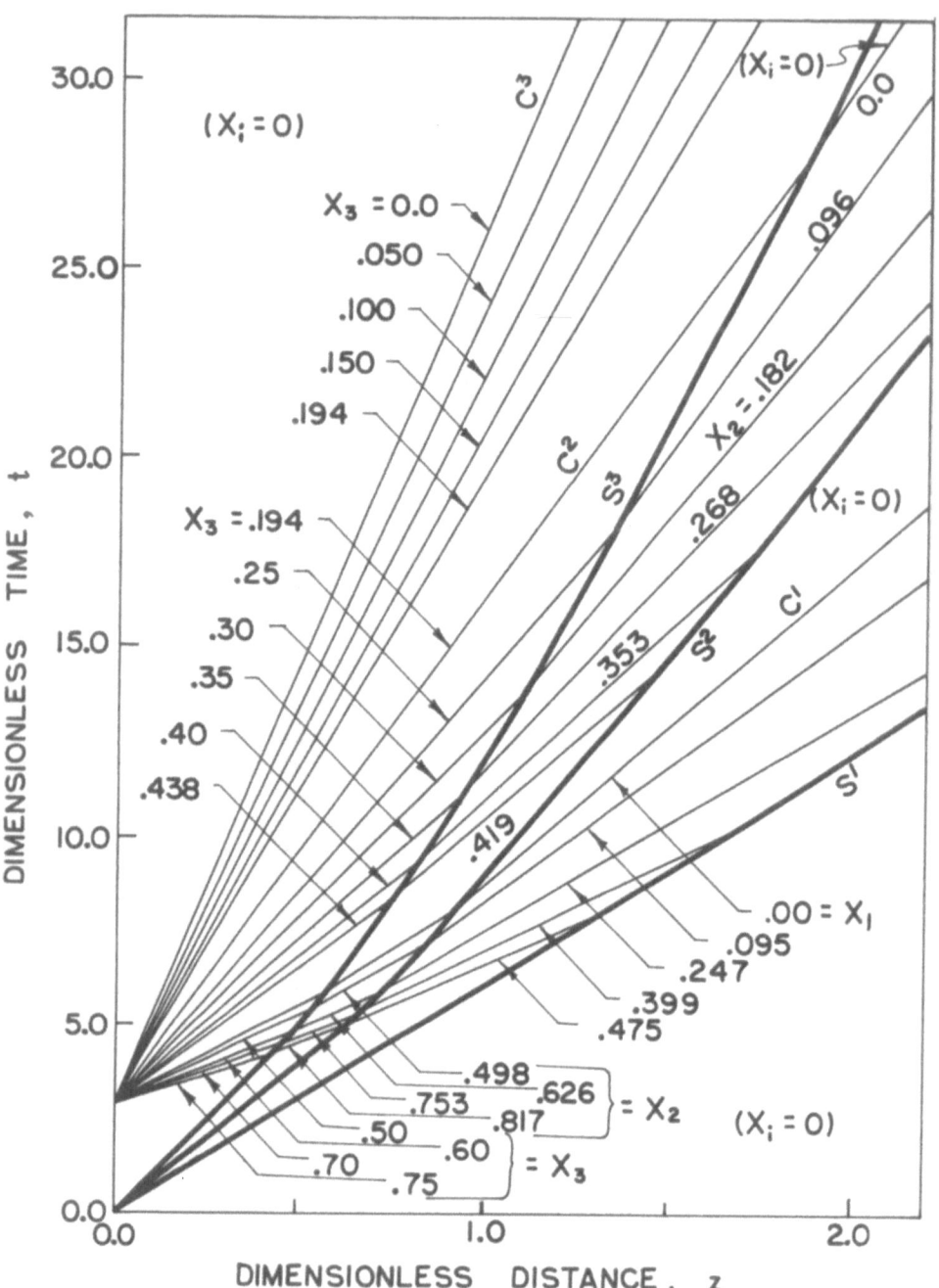

Fig. 12. Solution of Example 6 in the physical plane.

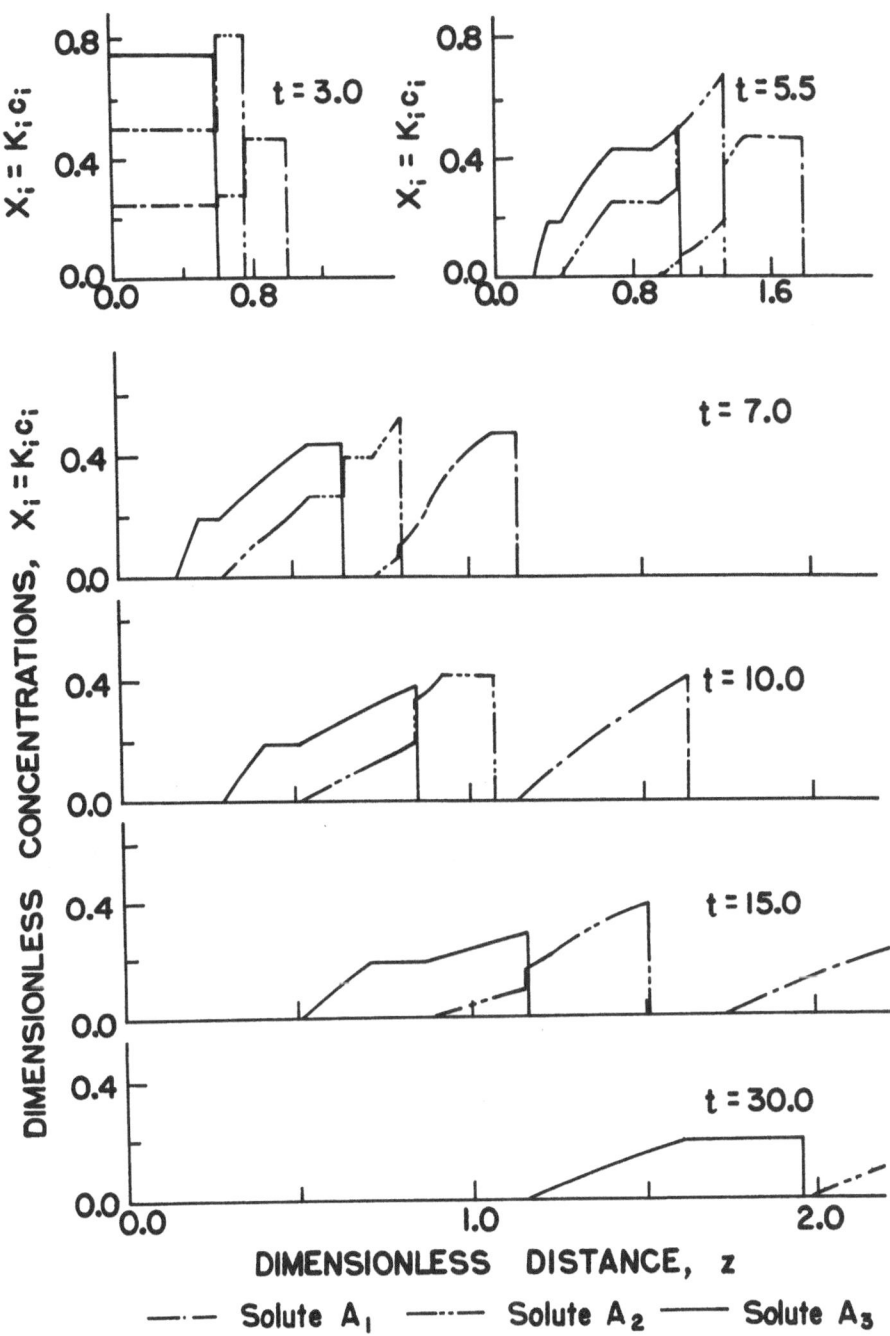

Fig. 13. Concentration profiles at successive times for Example 6.

NOTATION

A_i	i-th solute species
C^k	characteristic of the k-th kind
c_i	molar concentration of species A_i in the fluid phase (mol L^{-3})
D	$= 1 + \sum_{i=1}^{m} K_i c_i$, dimensionless
f_i	equilibrium relation
g	dimensionless time variable along the t-axis
$\mathcal{H}(m)$	m-dimensional h-space
h_k	characteristic parameter of the k-th kind, dimensionless
J_i^k	k-Riemann invariant, dimensionless
K_i	Langmuir isotherm parameter (L^3 mol^{-1})
L	characteristic length of the system (L)
m	number of solute species appearing in the system
N_i	limiting concentration of adsorbed solute (mol L^{-3})
n_i	molar concentration of adsorbed solute (mol L^{-3})
R_{kj}	$= W_k^I / h_j^r$, dimensionless
S^k	k-shock line
s	characteristic direction in the physical plane, dimensionless
s^d	direction of shock propagation, dimensionless
t	$= v\theta/L$, dimensionless time variable
v	interstitial velocity of fluid phase (Lt^{-1})
W_k	$= h_k D = N_i K_i (D - X_i / J_i^k)$, dimensionless
X_i	$= K_i c_i$, dimensionless concentration
x	distance in flow direction (L)
z	$= x/L$, dimensionless position variable
Γ^k	path of the k-th kind in $\mathcal{X}(m)$ or $\mathcal{H}(m)$
γ_{ij}	relative adsorptivity, dimensionless
ϵ	fractional void space of fixed bed, dimensionless
θ	time (t)
ν	$= (1 - \epsilon)/\epsilon$, volume ratio between phases, dimensionless
$\mathcal{X}(m)$	m-dimensional concentration space

Brackets

Δc_i	jump of c_i across a discontinuity
$\{a_i\}$	collection of m elements a_i associated with the subscrip

Superscripts

d	discontinuity
F	feed condition
I	initial condition

k	k-th kind
l	left-hand side of a discontinuity
o	fixed
r	right-hand side of a discontinuity
*	higher value

Subscripts

i, j	solute species in multicomponent system
j, k	j-th kind, k-th kind
m	the most adsorbable species
o	fixed
1, 2, 3	solute species , A_1, A_2, A_3
*	lower value

REFERENCES

1. Bayle, G. G and A. Klinkenberg, Recl. Trav. Chim. Pays-Bas Belg. 73, 1037(1954).
2. Courant, R. and K. O. Friedrichs , Supersonic Flow and Shock Waves, Interscience, New York, 1948.
3. Courant, R. , W. Isaacson, and M. Rees, Communs Pure Appl. Math. , 5 , 243(1952).
4. DeBoer, J. H. , The Dynamical Character of Adsorption , 2nd ed. , Clarendon Press, Oxford, 1968.
5. DeVault, D. , J. Amer. Chem. Soc. , 65, 532(1943).
6. Glimm, J. , Communs Pure Appl. Math. , 18, 697(1965).
7. Glueckauf, E. , Proc. Roy. Soc. Lond. A, 186, 35(1946).
8. Glueckauf, E. , Discuss. Faraday Soc. , 7, 12(1949).
9. Helfferich, F. , Ind. Eng. Chem. Fundam. , 6, 362(1967).
10. Helfferich, F. , Adv. Chem. Ser. no. 79, 30(1968).
11. Helfferich, F. and G. Klein, Multicomponent Chromatography, Marcel Dekker, New York, 1970.
12. Hsieh, J. S. , Ph. D. Thesis, Syracuse University, 1974.
13. James, D. H. and C. S. G. Phillips, J. Chem. Soc. , 1066(1954).
14. Jeffrey, A. and T. Taniuti, Nonlinear Wave Propagation, Academic Press, New York, 1964.
15. Kemball, C. , E. K. Rideal, and E. A. Guggenheim, Trans. Faraday Soc. , 44, 948(1944).
16. Klein, G. , D. Tondeur, and T. Vermeulen, Ind Eng. Chem. Fundam. , 6, 339(1967).
17. Langmuir, I. , J. Amer. , Chem. Soc. , 38, 2221(1916).
18. Lax, P. D. , Communs Pure Appl. Math. , 7, 159(1954).
19. Lax, P. D. , Communs Puee Appl. Math. , 10, 537(1957).
20. Oleinik, O. A. , Uspehi Math. Nauk, 12, 3(1957);translated in Amer. Math. Soc. Transl. , Ser. 2, no. 26, 95 (1963).
21. Oleinik, O. A. , Uspehi Math. Nauk, 14, 159 & 165 (1959);

translated in Amer. Math. Soc. Transl., Ser. 2, no. 33, 277 & 285(1963).

22. Rhee, H., Ph. D. Thesis, University of Minnesota, 1968.

23. Rhee, H. and N. R. Amundson, Chem. Eng. J., 1, 241(1970).

24. Rhee, H., R. Aris, and N. R. Amundson, Phil. Trans. Roy. Soc. Lond. A., 267, 419(1970).

25. Rhee, H., R. Aris, and N. R. Amundson, Phil. Trans. Roy. Soc. Lond. A., 269, 187(1971).

26. Rhee, H., E. D. Heerdt, and N. R. Amundson, Chem. Eng. J., 1, 279(1970).

27. Rhee, H., E. D. Heerdt, and N. R. Amundson, Chem. Eng. J., 3, 22(1972).

28. Shen, J. and J. M. Smith, Ind. Eng. Chem. Fundam., 7, 100(1968).

29. Sillen, L. G., Arkiv Kemi Miner. Geol., 2, 477(1950).

30. Tondeur, D., Chem. Eng. J., 1, 337(1970).

31. Tondeur, D. and G. Klein, Ind. Eng. Chem. Fundam., 6, 351(1967).

32. Walter, J. E., J. Chem. Phys., 13, 229(1945).

33. Weber, W. J. and T. M. Keinath, Chem. Eng. Prog. Symp. Serie no. 74, 63, 79(1966).

34. Wilson, J. N., J. Amer. Chem. Soc., 62, 1583(1940).

NONISOTHERMAL AND NONEQUILIBRIUM FIXED BED SORPTION

Norman H. Sweed

Oxirane International Technical Center
201 College Road East
Princeton, New Jersey   08540

The theory of multicomponent chromatography that
Professor Rhee presented in the previous chapter contains two
important assumptions which limit its applicability. One is that
local equilibrium exists between the solid and fluid phases. The
other is that the bed temperature is constant. In this chapter
these restrictions are removed, and the theory is extended to
nonequilibrium and nonisothermal chromatography.

Nonisothermal behavior in chromatography columns can have
several causes. The temperature of a bed can rise, for example,
if heat is liberated during the adsorption process. This
commonly occurs during the adiabatic adsorption of gases.
Temperature changes also occur whenever the temperature of the
feed is different from that initially in the bed. Thermal
regeneration is an example of this nonisothermal behavior.
Temperatures also change in parametric pumping and cycling zone
adsorption where heat is intentionally added or removed through
the walls of a packed bed to effect a separation.

Nonequilibrium effects occur in all real chromatography
systems. The causes are the same as for single solute
chromatography: axial diffusion and dispersion, finite mass
transfer rates between fluid and solid and/or within the solid
(adsorbent) phase, and nonuniform flow rates in the bed due to
channelling, poor packing technique, or inadequate flow
distribution.

In the first part of this chapter the equilibrium theory is
expanded to include adiabatic adsorption and thermal regeneration.

The development here follows closely that of Rhee and Amundson and their coworkers. Following this is a brief discussion of thermal parametric pumping. The last part of the chapter deals with nonequilibrium effects in multisolute and nonisothermal columns, and the advantages and disadvantages of using the equilibrium theory for real systems.

ADIABATIC CHROMATOGRAPHY

Nonisothermal effects in packed beds often can be modelled using the multicomponent equilibrium theory provided that <u>enthalpy</u> (i.e., energy content) is treated as one of the components. The use of enthalpy as a component is valid only if enthalpy is a conserved quantity. The First Law of Thermodynamics teaches that in a flow system enthalpy will be conserved if the system is adiabatic and if the flowing fluid does no work on the surroundings. Since flow in a packed bed does not produce work, the multicomponent theory will apply to adiabatic chromatography.

The term <u>adiabatic chromatography</u> refers to flow in a bed of adsorbent in which no heat enters or leaves through the walls of the bed. Temperature effects can occur here because of heat released on adsorption or because of differences between the feed and initial bed temperatures. Adiabatic chromatography can produce very unusual and interesting breakthrough curves.

An Example of Adiabatic Chromatography

Consider a simple illustration which shows how temperature effects can appear in what, at first, might seem like an isothermal problem. A column packed initially with clean adsorbent is fed with a mixture of carrier gas and adsorbate. Both feed and initial bed temperatures are at T_O. The feed concentration is C_F; the initial concentration is 0. The bed is well insulated and so exchanges no heat with its surroundings.

The effluent history for this example is sketched in Figure 1. The initial part of the breakthrough curve mirrors the initial state and has concentration of zero and temperature of T_O. The feed will eventually break through and the effluent after a long time will be the same as the feed, $C=C_F$, $T=T_O$. One might be tempted to sketch the temperature between the initial and feed states as being T_O throughout. This is not always correct however.

During adsorption adsorbate molecules lose energy to their surroundings in the form of heat. (This assumes that the heat of adsorption is exothermic which is the most common situation.)

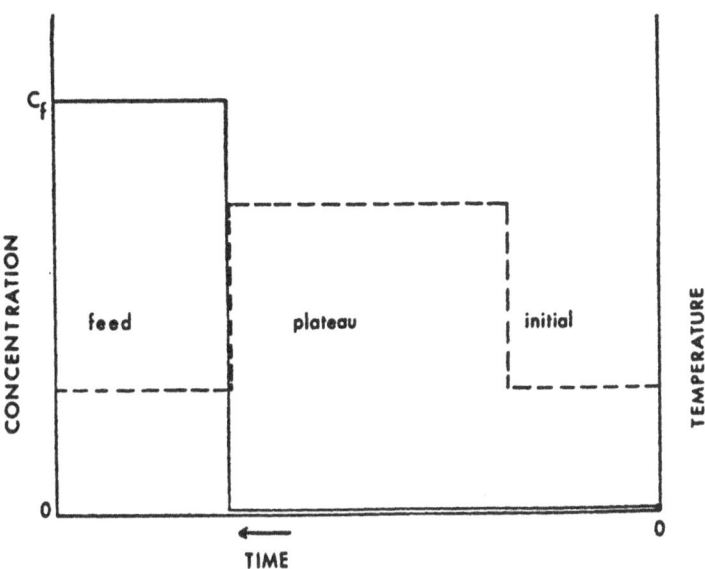

Figure 1.   Effluent history: saturation of a clean bed

The effluent temperature therefore must be higher than T_0 somewhere between the initial and final states to account for the heat release. The theory below shows that there is a plateau of higher temperature extending for some distance between the initial and feed states. On either side of this plateau there is an abrupt transition to T_0.

In Dr. Rhee's discussion of multisolute chromatography we saw that the concentration profiles of all species usually undergo transition at the same locations. Thus we might expect that the concentration profile undergoes abrupt transition at the two places where the temperature (i.e., enthalpy) does. However, in some cases--this being one--one of the concentration transitions is degenerate, i.e., no transition occurs. Only the temperature wave marks the transition. This pure thermal wave moves faster than the combined concentration and temperature wave that follows it. In other cases, the pure thermal wave trails the combined wave. It can even move slower than one combined wave and faster than another in the same column.

Mathematical Formulation of Adiabatic Adsorption in Columns

Let us now develop the mathematical basis for treating adiabatic adsorption using the multicomponent theory. Since enthalpy is conserved we write an enthalpy balance for an adiabatic adsorption column:

$$v\varepsilon \, \frac{\partial H}{\partial x} + \frac{\partial H}{\partial t} + (1-\varepsilon) \, \frac{\partial H_s}{\partial t} = 0 \qquad (1)$$

where
H = enthalpy of fluid, J/m^3 of fluid
H_s = enthalpy of solid phase including adsorbed molecules, J/m^3 of solid
t = time, s
v = interstitial velocity of fluid, m/s
x = distance from column inlet, m
ε = void fraction in the bed, m^3 of fluid/m^3 of bed

Equation 1 assumes that no enthalpy is carried along the column by thermal conduction.

The fluid enthalpy, H, can be written as a sum of the contributing species enthalpies:

$$H = \sum_{j=1}^{M-1} c_j h_j \qquad (2)$$

where
c_j = concentration of species j in fluid, mol/m^3 of fluid
h_j = partial molar enthalpy of species j in fluid, J/mol
M-1 = number of species in fluid phase

The solid phase enthalpy is made up of two parts--the enthalpy of the solid adsorbent and the enthalpy of the adsorbed molecules. Thus

$$H_s = C_{p_s} (T_s - T_R) + \sum_{j=1}^{M-1} n_j h_j^* \qquad (3)$$

where
C_{p_s} = heat capacity of solid support, J/m^3 K
h_j^* = partial molar enthalpy of species j in solid, J/mol
n_j = concentration of solute j in the sorbent phase, mol/m^3 of solid
T_R = arbitrary reference temperature, K
T_s = temperature of solid phase, K

<u>Assumptions for Equilibrium Model.</u> Let us now make
these assumptions:

- •The fluid is an ideal mixture. The fluid phase partial
  molar enthalpies therefore depend on temperature
  alone. Thus

$$h_j(T) = h_j(T_R) = C_{p_j}(T-T_R).$$

- •The adsorbed phase is an ideal mixture so that

$$h_j^*(T) = h_j(T) + \Delta H_j(T) = h_j(T_R) + C_{p_j}(T-T_R) + \Delta H_j(T)$$

  where $\Delta H_j(T)$ = heat of adsorption of species j, J/mol of j

- •The heat of adsorption is independent of temperature.
- •By definition, $h_j(T_R) = 0$ when $T_R = 0$
- •The fluid velocity is constant and uniform.

Combining Equations 2 and 3 with the assumptions, and
inserting into Equation 1 gives

$$\frac{\partial T}{\partial t}\left[\varepsilon \sum_{j=1}^{M-1} c_j C_{p_j} + (1-\varepsilon) C_{p_S} + (1-\varepsilon) \sum_{j=1}^{M-1} n_j C_{p_j}\right]$$

$$+ \frac{\partial T}{\partial x}\left[v\varepsilon \sum_{j=1}^{M-1} c_j C_{p_j}\right] + T\varepsilon\left[\sum_{j=1}^{M-1} C_{p_j}\left[v\frac{\partial c_j}{\partial x} + \frac{\partial c_j}{\partial t} + \left(\frac{1-\varepsilon}{\varepsilon}\right)\frac{\partial n_j}{\partial t}\right]\right]$$

$$+ (1-\varepsilon) \sum_{j=1}^{M-1} \Delta H_j\frac{\partial n_j}{\partial t} = 0 \tag{4}$$

The conservation of mass for species j was given by
Professor Rhee as

$$v\frac{\partial c_j}{\partial x} + \frac{\partial c_j}{\partial t} + \left(\frac{1-\varepsilon}{\varepsilon}\right)\frac{\partial n_j}{\partial t} = 0 \tag{5}$$

Inserting Equation 5 into the third term in Equation 4 eliminates
that term.

Let us now define a new species, M, such that

$$c_M = T \tag{6a}$$

$$n_M = \frac{C_{p_s}}{C_{p_f}} T - \sum_{j=1}^{M-1} \frac{(-\Delta H_j)n_j}{C_{p_f}} \tag{6b}$$

where $C_{p_f} = \sum_{j=1}^{M-1} c_j C_{p_j}$ is assumed to be a constant.

Also define $\tau = \dfrac{Lt}{v}$

$$z = \frac{x}{L}$$

$$\nu = \frac{1-\varepsilon}{\varepsilon}$$

With these definitions Equation 4, the energy balance, now takes on precisely the same form as Equation 5. Thus the conservation equations that describe adiabatic adsorption are

$$\frac{\partial c_j}{\partial z} + \frac{\partial c_j}{\partial \tau} + \nu \frac{\partial n_j}{\partial \tau} = 0 \quad j = 1, 2, \ldots, M \tag{7}$$

and the adsorption equilibrium equations are

$$n_j = f_j (c_1, c_2, \ldots, c_j, \ldots c_M)$$
$$j = 1, 2, \ldots, M \tag{8}$$

Solving the Model Equations

Equations 7 and 8 are 2M simultaneous equations in as many unknowns, and are solved in the same way described by Rhee. That is, first Equation 8 is differentiated with respect· to time,

$$\frac{\partial n_j}{\partial \tau} = \frac{\partial f_j}{\partial c_1} \frac{\partial c_1}{\partial \tau} + \frac{\partial f_j}{\partial c_2} \frac{\partial c_2}{\partial \tau} + \ldots + \frac{\partial f_j}{\partial c_i} \frac{\partial c_i}{\partial \tau} + \ldots + \frac{\partial f_j}{\partial c_M} \frac{\partial c_M}{\partial \tau}$$

$$\tag{9}$$

and then Equation 9 is substituted into Equation 7:

$$\frac{\partial c_j}{\partial z} + \frac{\partial c_j}{\partial \tau} + \nu \sum_{i=1}^{M} \frac{\partial f_j}{\partial c_i} \frac{\partial c_i}{\partial \tau} = 0$$

$$j = 1, 2, \ldots, M \qquad (10)$$

If the initial and feed conditions are constant states, then Equation 10 forms a Riemann's problem and the concentrations are functions of (z/τ). As Rhee has shown this implies that

$$\frac{\partial c_i}{\partial \tau} = \frac{dc_i}{dc_j} \frac{\partial c_j}{\partial \tau} \qquad (11)$$

Combining Equations 10 and 11 gives

$$\frac{\partial c_j}{\partial z} + \frac{\partial c_j}{\partial \tau} \left\{ 1 + \nu \sum_{i=1}^{M} \frac{\partial f_j}{\partial c_i} \frac{dc_i}{dc_j} \right\} = 0 \quad j = 1, 2, \ldots, M \qquad (12)$$

Defining

$$\frac{\mathcal{D}f_j}{\mathcal{D}c_j} = \sum_{i=1}^{M} \frac{\partial f_j}{\partial c_i} \frac{dc_i}{dc_j} \qquad (13)$$

gives

$$\frac{\partial c_j}{\partial z} + \frac{\partial c_j}{\partial \tau} \left\{ 1 + \nu \frac{\mathcal{D}f_j}{\mathcal{D}c_j} \right\} = 0 \qquad (14)$$

Equation 14 is same derived by Rhee for multisolute adsorption. The only difference between that and adiabatic adsorption are in the definitions of c_M and n_M and in the fact that the equilibrium function, $n_M = f_M (c_1, c_2, \ldots, c_M)$ usually has a form different from the other functions.

The solution of Equation 14 shows that a concentration wave propagates through the bed at a constant speed $1/\sigma$ where

$$\frac{dz}{d\tau} = \frac{1}{1 + \nu \dfrac{\mathcal{D}f_j}{\mathcal{D}c_j}} = \frac{1}{\sigma} \qquad (15)$$

where σ is defined by the equation.

Only certain sets of concentrations are allowable in each Reimann's problem. These are determined by solving Equation 16

$$\frac{\mathcal{D}f_1}{\mathcal{D}c_1} = \frac{\mathcal{D}f_2}{\mathcal{D}c_2} = \cdots = \frac{\mathcal{D}f_M}{\mathcal{D}c_M} \tag{16}$$

which is really (M-1) (M/2) separate equations. If there is one adsorbing species and one carrier gas, then for adiabatic adsorption M = 3 and three equations are needed, i.e.,

$$\frac{\mathcal{D}f_1}{\mathcal{D}c_1} = \frac{\mathcal{D}f_2}{\mathcal{D}c_2}, \quad \frac{\mathcal{D}f_1}{\mathcal{D}c_1} = \frac{\mathcal{D}f_3}{\mathcal{D}c_3}, \quad \frac{\mathcal{D}f_2}{\mathcal{D}c_2} = \frac{\mathcal{D}f_3}{\mathcal{D}c_3} \tag{17}$$

Here let species 1 be the adsorbing one, species 2 the enthalpy, and species 3 the nonadsorbing carrier.

Note that $\dfrac{\mathcal{D}f_3}{\mathcal{D}c_3} = 1$ since $f_3 = 0$, i.e., the carrier does not adsorb.

Therefore, combining (13) and (17) gives a single equation

$$\frac{\partial f_1}{\partial c_1} + \frac{\partial f_1}{\partial c_2}\frac{dc_2}{dc_1} = \frac{\partial f_2}{\partial c_1}\frac{dc_1}{dc_2} + \frac{\partial f_2}{\partial c_2} \tag{18}$$

Since $\dfrac{dc_2}{dc_1} = \left(\dfrac{dc_1}{dc_2}\right)^{-1}$, Equation 18 is simply a

quadratic equation with $\dfrac{dc_2}{dc_1}$ as its unknown variable:

$$a\left(\frac{dc_1}{dc_2}\right)^2 + b\left(\frac{dc_1}{dc_2}\right) + c = 0 \tag{19}$$

where $a = \dfrac{\partial f_2}{\partial c_1}$

$b = \dfrac{\partial f_2}{\partial c_2} - \dfrac{\partial f_1}{\partial c_1}$

$c = -\dfrac{\partial f_1}{\partial c_2}$

The coefficients of this equation, a, b, and c, are <u>known</u>
<u>functions</u>. Solving the quadratic gives two solutions

$$\left(\frac{dC_1}{dC_2}\right)_1 = \frac{-b + \sqrt{b^2 - 4ac}}{2a} \tag{20a}$$

$$\left(\frac{dC_1}{dC_2}\right)_2 = \frac{-b - \sqrt{b^2 - 4ac}}{2a} \tag{20b}$$

The two roots of Equation 19 are two ordinary differential
equations, 20a and 20b, whose solutions map out curves in the c_1
vs c_2 plane (i.e., hodograph plane). These curves are called
$\bar{\Gamma}(1)$ and $\Gamma(2)$ depending on which root is used.

From Equations 13, 15, and 20 is it clear that there will be
two expressions for the speed of propagation of a concentration
wave through the adsorption bed

$1/\sigma^{(1)}$ and $1/\sigma^{(2)}$ where

$$\frac{1}{\sigma^{(1)}} = \frac{1}{1 + \nu \left[\frac{\partial f_2}{\partial c_2} + \left[\frac{-b + \sqrt{b^2 - 4ac}}{2} \right] \right]} \tag{21a}$$

$$\frac{1}{\sigma^{(2)}} = \frac{1}{1 + \nu \left[\frac{\partial f_2}{\partial c_2} - \left[\frac{-b - \sqrt{b^2 - 4ac}}{2} \right] \right]} \tag{21b}$$

Shock Transitions Appear in Adiabatic Adsorption

Professor Rhee has shown for multisolute adsorption that the
differential equation material balance, Equation 5, is not valid if
discontinuities or shocks exist in the bed. With adiabatic
adsorption the same is true for the energy balance, Equation 1. If
shocks exist, the correct balance is a difference equation. The
difference equation formulation which gives the speed of a shock,
$1/\sigma_S$ is

$$\frac{1}{\sigma_S} = \frac{1}{1 + \nu \frac{\Delta f_j}{\Delta c_j}} \tag{22}$$

where Δ denotes the difference between the concentrations upstream
and downstream of the shock.

Since all species--including enthalpy--move at the same speed,

$$\frac{\Delta f_1}{\Delta C_1} = \frac{\Delta f_2}{\Delta C_2} = \cdots = \frac{\Delta f_m}{\Delta C_m} \qquad (23)$$

Equation 23 is analogous to Equation 16 except that it applies to shock, not simple waves.

For adiabatic adsorption of a single solute,

$$\frac{f_u - f_d}{c_u - c_d} = \frac{C_{ps}}{C_{pf}} - \frac{(-\Delta H)}{C_{pf}} \frac{n_u - n_d}{T_u - T_d} \qquad (24)$$

where u and d refer to the upstream and downstream states respectively. Equation 24 is an algebraic analog to Equation 20. Given any upstream conditions, Equation 24 allows calculation-- with the equilibrium expression--of all possible downstream conditions. Such solutions of Equation 24 map out the hodograph plane with curves called Σ.

Adsorption of Benzene on Charcoal

To make this mathematics concrete, let us consider an example from Rhee, et al (1970)--the adsorption of benzene vapor onto charcoal where nitrogen is the inert carrier gas and the total pressure is 10 atm. Adsorption is according to the Langmuir isotherm. Table I lists data for this adsorption system.

For this adsorption system it can be shown from Equation 20a and b that, in the concentration-temperature plane (i.e., hodograph plane), the $\Gamma(1)$ curves always have a slope that is positive or zero and the $\Gamma(2)$ curves always have a slope that is negative or zero.

Figure 2 shows the hodograph plot for the benzene-nitrogen-charcoal system over a wide range of temperatures. The plot has several very important features, the most notable one being the point T_R, located on the T-axis at 505.5K. To the left of T_R, i.e., in the low temperature part of the plot along the T-axis, the family of $\Gamma(1)$ curves appears to be asymptotic to the T-axis at low concentrations. To the right of T_R, the $\Gamma(1)$ curves abruptly intersect the axis. The $\Gamma(1)$ that goes through T_R, $\Gamma_c(1)$ separates the $\Gamma(1)$ family into two parts--the asymptotic and intersecting parts. To the left of T_R, the T-axis is a $\Gamma(1)$.

TABLE I

Data for Benzene Adsorption on Charcoal

ε = 0.5

C_{ps} = 405 cal/liter K

C_{pf} = 2.7 cal/liter K

The adsorption isotherm is

$$n = \frac{NKc}{1+Kc}$$

where c = concentration of benzene in nitrogen,
moles benzene/liter gas

K = $K_0 T^{\frac{1}{2}} e^{-\Delta H/RT}$, liters/mole

K_0 = 3.88×10^{-5} liter/mole $K^{\frac{1}{2}}$

ΔH = -10,400 cal/mole

N = 5.5 mol/liter

n = concentration of benzene adsorbed,
moles benzene/liter of charcoal

T = temperatures, K

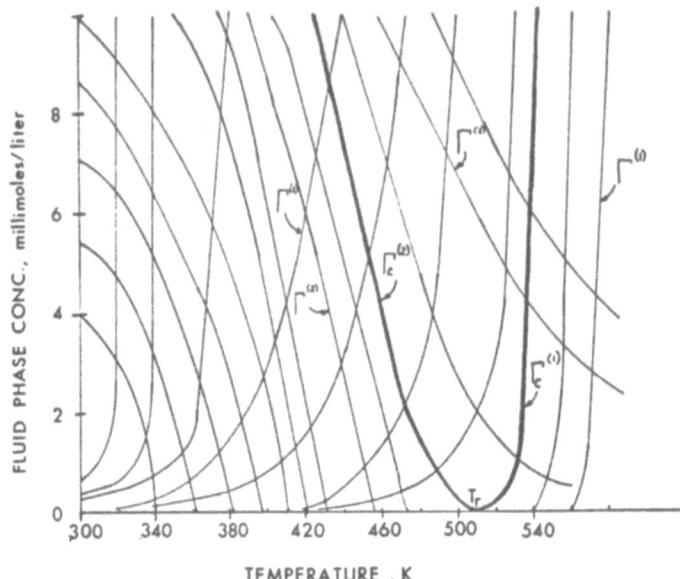

Figure 2. Hodograph plot for benzene-nitrogen-charcoal system.
Watershed point is T_R; $\Gamma_c(1)$ and $\Gamma_c(2)$ divide their
respective Γ families into those that intersect T-axis
and those that don't.

A similar analysis shows that the $\Gamma(2)$ curves are asymptotic
to the T-axis to the right of T_R and that in this range the axis
is a $\Gamma(2)$. The curve $\Gamma_c(2)$ separates the asymptotic from the
intersecting $\Gamma(2)$'s.

The point T_R is a point of adsorptivity reversal. Helfferich
and Klein call this the watershed point in their book on
Multicomponent Chromatography. Pan and Basmadjian (1971) noted the
same type of point for the adsorption of CO_2-He on 5A molecular
sieves and on carbon. Such a watershed point is a common feature
to adiabatic adsorption systems and has a large influence on the
way the adsorption columns behave.

Another important feature of this system is that for any
point on the hodograph plot, the velocity of propagation of a wave
is always faster on the $\Gamma(1)$ than on the $\Gamma(2)$. In other words at
each point in the c-T plane.

$$\frac{1}{\sigma(1)} > \frac{1}{\sigma(2)}$$

From physical reasoning, the wave associated with the initial state must move through the bed faster than the wave associated with the feed. If this were not the case the feed could overtake the initial wave, and the feed might appear in the breakthrough curve before the initial bed fluid. For this reason, when the path between the feed and the initial states in the hodograph plane is to be determined, <u>the path through the initial state always follows $\Gamma(1)$ and the path through the feed state always follows $\Gamma(2)$</u> unless a shock forms and the Σ's have to be substituted. The only exception to this rule occurs when the initial and feed states both lie on the same Γ. Then only one transition will occur and the path in the hodograph plane will be along that single Γ.

Examples of Adiabatic Adsorption Using the Hodograph Plot

These examples are taken from Rhee, et al (1979, 1972), for the benzene-nitrogen-charcoal system. The state is represented in the form

A: (c in millimoles/liter, n in moles/liter, T in K)

<u>Example 1: Elution by a pure solvent (Figure 3)</u>

Initial state A: (10,5.267,350)
Feed state B: (0,0,350)

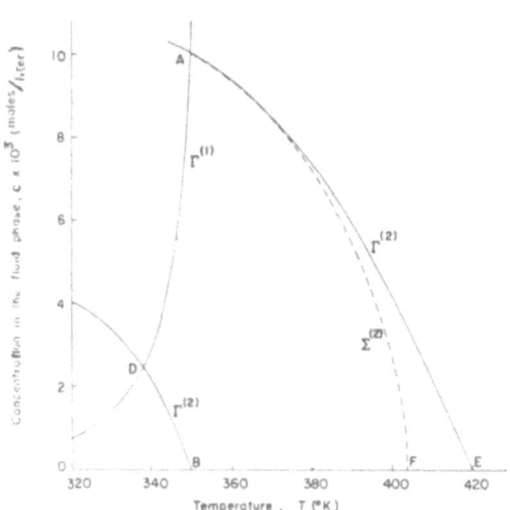

Figure 3. Hodograph plot for Examples 1 and 2--elution by pure solvent and saturation of a clean bed.

The path through the initial state A must be $\Gamma(1)$ and through the feed state B must be $\Gamma(2)$. The $\Gamma(1)$ and $\Gamma(2)$ curves intersect at D, and this is the plateau state D: (2.467,4.964,333.1). The feed and initial states are both.at 350K yet the intermediate state is cooler because sensible heat was taken from the bed to provide the heat of desorption. If the wave velocities are calculated from Equation 21a along the path, it is found that they decrease steadily from A to D along $\Gamma(1)$, then decrease abruptly as they change to $\Gamma(2)$, then decrease steadily again from D to B. Figures 4 and 5 show the characteristics with slopes of σ, and the concentration profile at $\tau=16$ respectively. Both transitions are simple waves.

Example 2: Saturation of a clean bed   (Figure 3)

    Initial state B: (0,0,350)
    Feed state A: (10.,5.267,350)

Although these are the same two states used in Example 1, the path in the hodograph plane is different. The feed must have a $\Gamma(2)$ through it (i.e., AE) and the initial state a $\Gamma(1)$ through it (i.e., EB) with E as the intermediate. Calculating the wave velocities along the path shows that the velocity is greatest at B, and with a value of 1/151, and is constant along BE. At E the velocity drops abruptly as the change is made to $\Gamma(2)$. Along $\Gamma(2)$ from E to A the wave velocity increases. But this is physically impossible since the feed would be moving faster than the intermediate plateau. Therefore the curve AE cannot be used and the $\Sigma(2)$ through A (i.e., AF) should replace it. Along AF there is only one wave velocity--the shockwave velocity given by Equation 22 to be 1/527.

Figures 6 and 7 show the characteristics and bed profiles. The constant velocity noted along BE is the speed of a purely thermal wave; there is no concentration change associated with it. This is not surprising since the initial and intermediate states are both solute free. This thermal wave moves through the bed at the same speed any pure thermal wave would move. It is uncoupled from the adsorption part of the problem. The reader may now realize that this example was the one sketched at the beginning of the chapter. The speed of pure thermal waves is always

$$1/\left(1 + \frac{C_{ps}}{C_{pf}}\right)$$

For gas-solid systems C_{ps} is usually much greater than C_{pf} and the thermal wave moves much slower than the carrier gas. For liquid-solid systems $C_{ps} \approx C_{pf}$ and the thermal wave moves at about half the speed of the solvent.

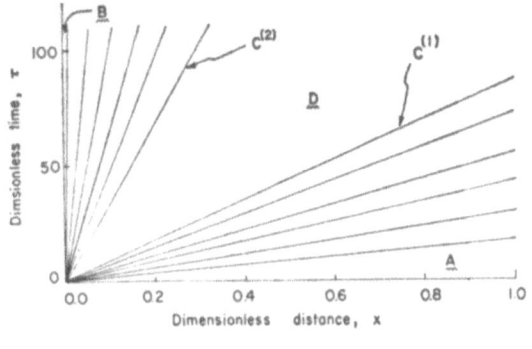

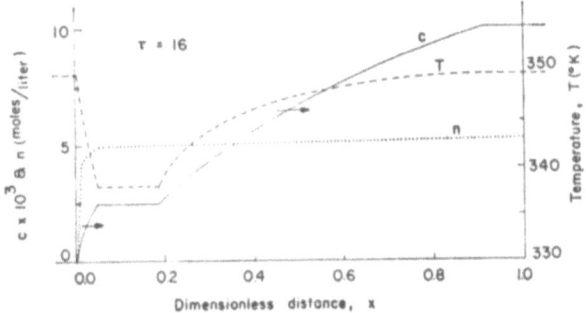

Figure 4:   Characteristics for Example 1

Figure 5:   Bed profiles for Example 1 at τ = 16

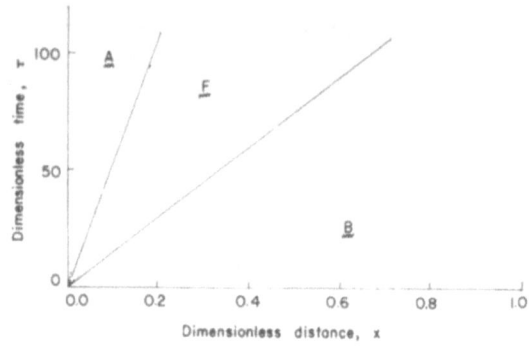

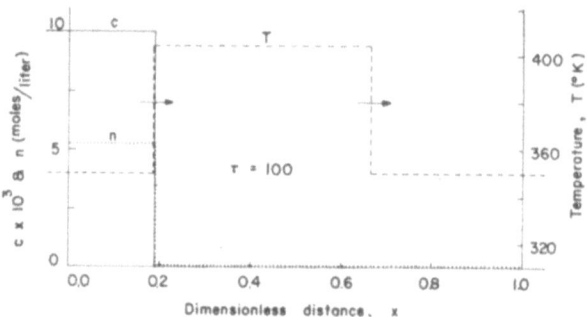

Figure 6: Characteristics for Example 2.

Figure 7: Bed profiles for Example 2.

Pure thermal waves can occur only if the feed state lies on a $\Gamma(2)$ to the left of $\Gamma_c(2)$, as in this problem; or if the initial state lies on a $\Gamma(1)$ to the right of $\Gamma_c(1)$. Only in these cases do the hodograph lines intersect the T-axis, and only along the T-axis do purely thermal waves exist. Not all systems with feeds on a $\Gamma(2)$ to the left of $\Gamma_c(2)$ give pure thermal waves however. The next example shows this.

Example 3: Further saturation of a partially saturated bed
(Figure 8)

Initial state B':   (0.1,1.015,350)
Feed state A: (10.0,5.267,350)

The path through the initial state should be a $\Gamma(1)$ (B' G) and the path through the feed should be a $\Gamma(2)$ (AG). However, since the wave velocities are in the reverse order along both Γ's the corresponding Σ's are used instead (i.e., B'H and HA). This gives rise to two shocks. Since B'H is not part of the T-axis, both shocks have temperature and concentration jumps.

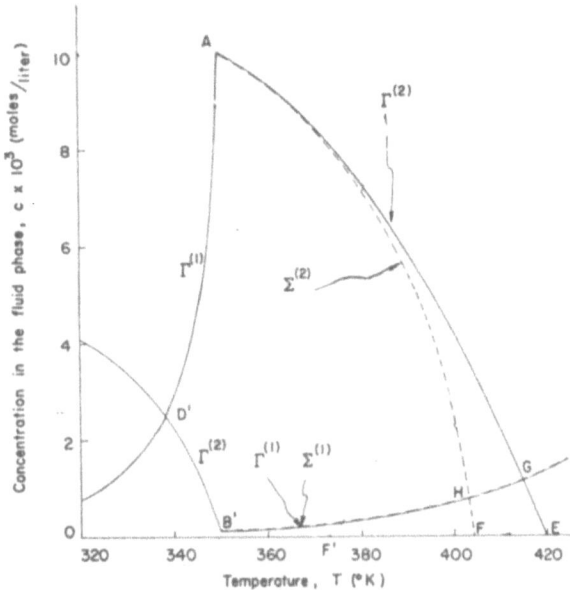

Figure 8. Hodograph plot for Example 3--further saturation
           of a partially saturated bed.

Figures 9 and 10 show the characteristics and column profiles.

In these three examples there have been simple waves, shock
waves, and a purely thermal wave, but the approach to eliciting
the bed profiles has been the same--the path from initial to
intermediate to feed states is drawn along $\Gamma(1)$ (or $\Sigma(1)$) then
$\Gamma(2)$ (or $\Sigma(2)$).

Now let us consider a few examples where the watershed point
is involved. The analysis is slightly more complicated.

Example 4: Elution in the vicinity of
              the watershed point   (Figure 11)

     Initial state A":   (5,0.320,550)
     Feed state B: (0,0,480)

Again the path through the initial state is a $\Gamma(1)$ (A" H).
From point H to point B, however, no $\Gamma(2)$ exists since the $\Gamma(2)$
through H only goes as far as T_R. The only path from T_R to B
lies on a $\Gamma(1)$. The problem is solved easily if we allow T_R, the
watershed point, to divide the problem into two problems: getting
from A" to T_R then from T_R to B. The path is A" H ($\Gamma(1)$),
HT_R ($\Gamma(2)$), and then T_R B ($\Gamma(1)$).

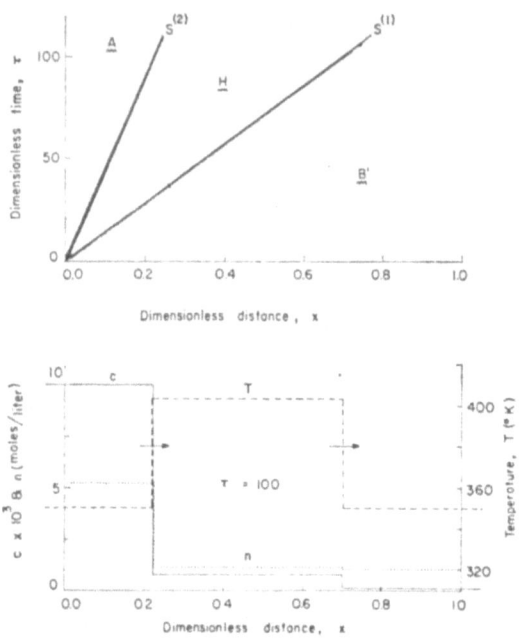

Figure 9.   Characteristics for Example 3.

Figure 10.   Bed profiles for Example 3

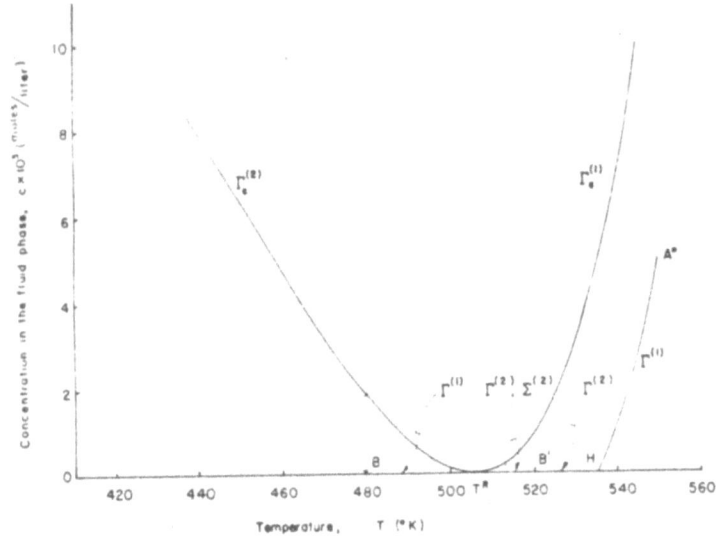

Figure 11.   Hodograph plot for Example 4--elution in the
vicinity of the watershed point.

Previously it was said that the feed was always on a $\Gamma(2)$ curve, unless the feed and initial states lie on the same $\Gamma(1)$ or $\Gamma(2)$. In this example points T_R and B lie on a single $\Gamma(1)$, and since they form a problem in themselves, the path through the feed point B can be on a $\Gamma(1)$. Figures 12 and 13 show the profiles for this example. The pure thermal wave, which always moves with velocity of 1/151, now lags behind the concentration wave. In example 2 the thermal was faster than the concentration wave. Note that the initial state A" lies on a $\Gamma(1)$ to the right of $\Gamma_c(1)$, and that a pure thermal wave appears.

Example 5: Further Saturation of a partially saturated bed   (Figure 14)

   Initial state A":   (5,0.320,550)

   Feed state A°:   (5,2.054,440)

The hodograph plane path through the initial state is along the $\Gamma(1)$ (i.e., A" H) and then along the $\Gamma(2)$ (HT_R) to the watershed point. This completes half the solution. The next half

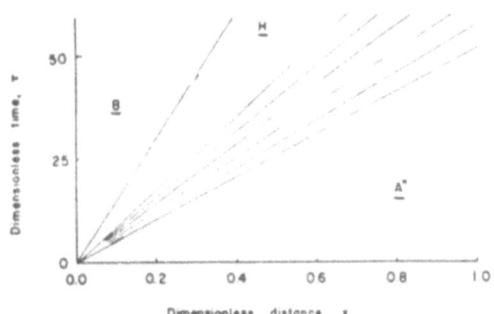

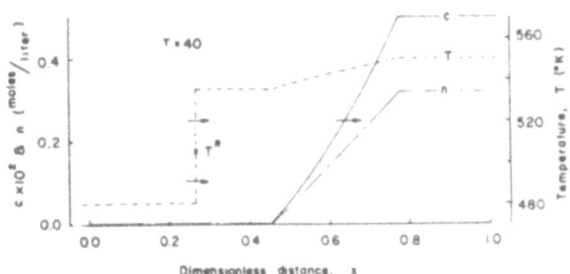

Figure 12.   Characteristics for Example 4.

Figure 13.   Bed profiles for Example 4

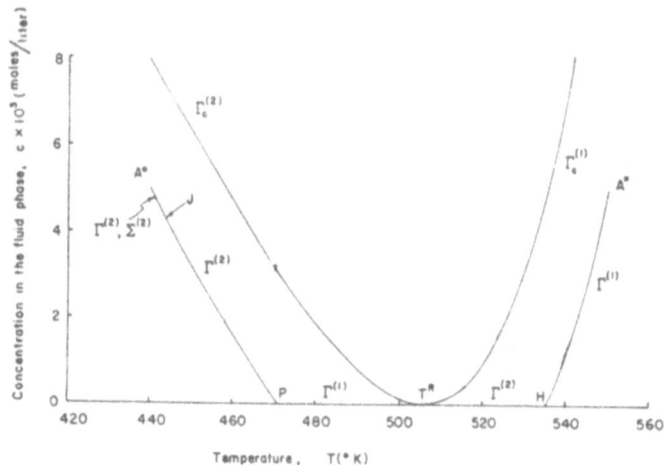

Figure 14. Hodograph plot for Example 5--further
saturation of a partially saturated bed

of the solution has T_R as the initial state and $A°$ as the feed.
The path is $T_R P$ ($\Gamma(1)$), PJ ($\Gamma(2)$), then $JA°$ ($\Sigma(2)$). The wave
velocities along $\Gamma(2)$ from P to $A°$ decreased from P to J but
increased from J to $A°$ requiring the switch to the $\Sigma(2)$.

Since in this example the watershed point divides the problem
into two problems, it should be expected that more than two
transitions exist. Figures 15 and 16 show the characteristics and
bed profile. The initial state is effectively eluted from the bed
by a cooler feed, even though both initial and feed states have the
identical concentration. Following the complete elution, the bed
saturates to the new level. Note that the purely thermal wave,
still with the speed 1/151, now moves at a speed between that of
the high and low temperature waves.

These examples illustrate how even very complex breakthrough
and column profiles can be calculated with reasonable ease once the
adsorption isotherm and thermal properties of the bed are known.
Computer calculations are needed only to map out the hodograph
plot; and these calculations, the solution of simple initial value
ordinary differential equations, is straightforward with any of the
numerous computer packages that are readily available. No finite
difference solutions of partial differential equations are involved,
yet shock front locations can be predicted easily.

The approach used in this chapter has followed very closely
the work of Rhee and Amundson and their coworkers at the
University of Minnesota. A similar analysis was also presented
by Pan and Basmadjian of the University of Toronto. These workers

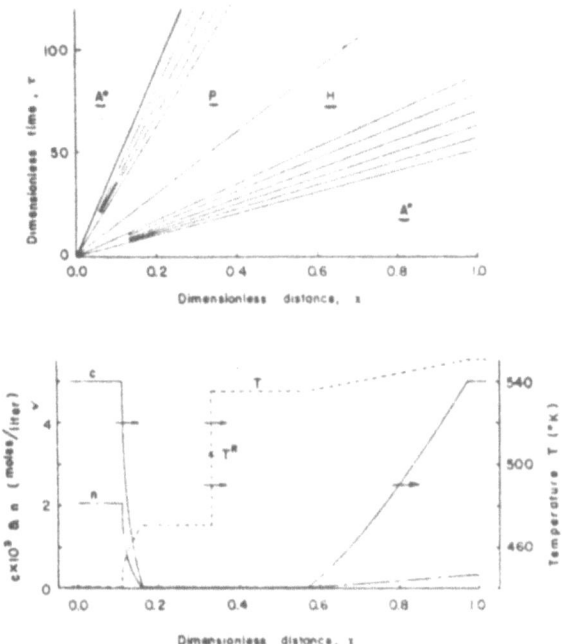

Figure 15.   Characteristics for Example 5.

Figure 16.   Bed profiles for Example 5.

also carried out experiments to see how well the equilibrium
theory predicts real column behavior.   Figure 17 shows their data
for CO_2-He on 5A molecular sieves.   The qualitative agreement
regarding the breakthrough times and concentrations is quite good.
The profile shapes, however, are far broader than the shock
transitions predicted by theory.   The equilibrium theory thus
gives the primary features of the breakthrough curves--breakthrough
times, plateau levels, and numbers of transitions.   A
nonequilibrium model is needed to give the secondary features--
profile spread and actual shape.

It should be pointed out here that Rhee, Heerdt, and Amundson
have also carried out studies on adiabatic adsorption involving
two solutes.   The theory developed from Equations 1 to 16 still
applies, but the hodograph space is no longer a plane; it requires
two dimensions.   However projections to two dimensions allow
practical use of the approach.   For two solutes and adiabatic
adsorption, there usually will be four constant (plateau) states
and three transitions unless the watershed point is crossed.   If
it is, additional plateaus and transitions can occur.   These
authors also treated countercurrent adiabatic adsorption.

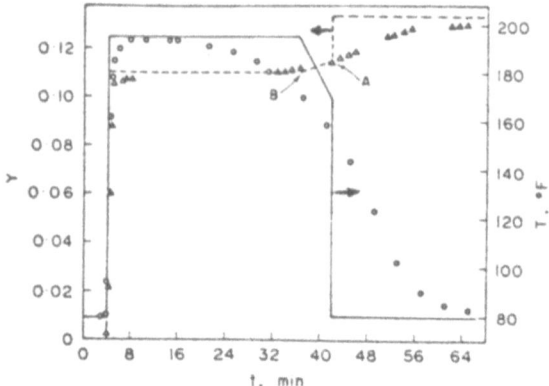

Figure 17.   Breakthrough curves for CO_2-He on 5A molecular
sieves.   Experimental points and equilibrium
theory prediction.

PARAMETRIC PUMPING

Parametric pumping is one of several cyclic separation
processes that Professor Wankat will discuss in a later chapter.
However, since parametric pumps often use nonisothermal fixed beds
of adsorbent, a very brief discussion of the process will be
presented here.

Two forms of parametric pumping have been described in the
literature: the recuperative thermal mode and the direct thermal
mode. In both of these a packed bed of adsorbent is subjected to
two periodic actions. First the direction of fluid flow is caused
to change periodically. Then, synchronously with the changes in
flow direction, the bed temperature is caused to change.

In the recuperative mode, first described by Wilhelm, et al
(1966), the temperature variations in the bed result from keeping
one end of the bed hot and the other cold so that flowing fluid
carries heat into and out of the bed periodically. Figure 18
shows the recuperative mode arrangement.

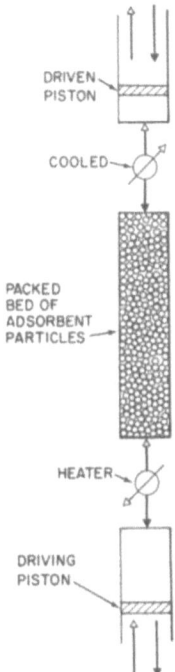

Figure 18.   Recuperative thermal parametric pumping

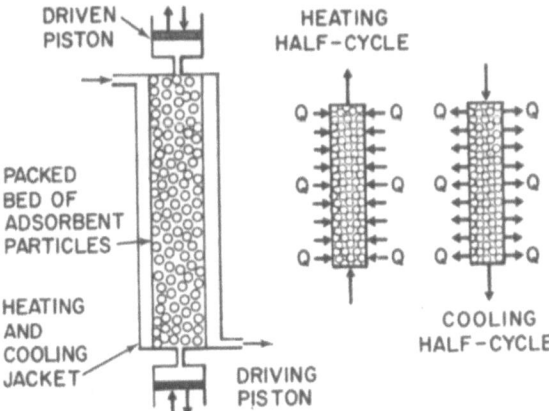

Figure 19.   Direct thermal parametric pumping

In the direct mode (Wilhelm and Sweed, 1968) the temperature is imposed directly through the walls of the bed by means of a jacket (Figure 19). When flow is in one direction the bed is hot; in the other direction it is cold.

The separation in parametric pumps results from the coupling of the alternating flow with the alternating adsorption and desorption of solute from the adsorbent caused by temperature changes. Clearly thermal parametric pumping is inherently a nonisothermal fixed bed sorption process, and so should be amenable to the equilibrium theory analysis.

In the recuperative mode, thermal waves start to move through the bed every time the flow direction changes. They do not necessarily leave the bed before the flow direction changes so that cycle after cycle thermal waves can accumulate in the bed. The early analyses of recuperative parametric pumping ignored the heat of adsorption's effect on temperature since this effect was small compared to the imposed heating and cooling. Sweed and Rigaudeau (1975) used an equilibrium theory analysis for linear adsorption (i.e., $f_1 = A(T)c_1$) with the recuperative mode, also ignoring the heat of adsorption effect.

They showed that waves build up within the bed cycle after cycle and many transitions and plateaus can exist at the same time. At steady periodic state, the equilibrium theory predicts a very large extent of separation provided that the thermal wave breaks through the bed on each half cycle. Real separations are poorer than the theory predicts because of axial dispersion of heat and mass and finite rates of mass and heat transfer. For gaseous systems the heat of adsorption cannot be ignored, and the equilibrium theory should be of great value. However, to date no one has attempted to apply the theory to the gaseous system.

The direct mode is nonisothermal, but obviously it is not adiabatic either. Each half cycle acts like an isothermal system. The multisolute equilibrium theory of Rhee applies on each half cycle.

NONEQUILIBRIUM CHROMATOGRAPHY

In the first half of this chapter we assumed that the fluid and solid in the bed were in compositional and thermal equilibrium with each other. Such a situation never exists in real chromatography! Dispersive forces, which are always present in real systems, cause shock transitions to spread out or broaden into the commonly observed "S" shape breakthrough curves. The same forces at work in gas chromatography make sharp, narrow pulses spread into broad gaussian ones.

The spreading effect has two primary causes. First, mass
and heat transfer resistances within the sorbent particles and
between the sorbent and fluid phases slow the approach to
equilibrium operation. Second, axial dispersion (molecular and
eddy) tend to diminish concentration gradients in the bed. In
this half of the chapter we shall examine how nonequilibrium and
other dispersive effects alter column behavior from what the
equilibrium theory predicts.

Nonequilibrium effects in single solute chromatography have
been studied extensively. In 1944, Thomas presented an elegant
analysis of nonequilibrium adsorption involving Langmuir isotherms
in the absence of axial dispersion. Devault (1943) showed that
if the adsorption equilibrium isotherm is of the "favorable"
type (i.e., $\partial^2 f/\partial c^2 < 0$), the profile on saturation approached a
"constant pattern" asymptotically with long time. On elution the
same isotherm gives "proportionate pattern" behavior in which the
wave width increases proportionately with time. In 1956
van Deemter, et al, showed for linear adsorption isotherms that
the variance of a pulse, i.e., its spread due to axial dispersion
and finite mass transfer, was equal to the sum of the variances
caused by the two mechanisms separately.

This allows us to treat the spread as though it were due to
mass transfer alone by adjusting the mass transfer coefficient to
give the addition spread due to axial dispersion. This lumping
of two dispersive mechanisms into one was expanded to slightly
nonlinear systems by Acrivos (1960). Rhee and Amundson (1972b)
also studied this lumping for nonlinear systems.

The spreading of waves in multisolute systems has received
less attention. Cooney and Lightfoot (1965) proved the existence
of constant pattern behavior for multisolute sorption when mass
transfer and/or axial dispersion are present. Vermeulen and
Clazie (1968) and Omatete (1972) modelled nonequilibrium
multicomponent chromatographs without axial dispersion. They
used a numerical integration method called the method of
characteristics to calculate bed profiles for real systems.
Cooney (1974) used numerical methods to compute profiles in
adiabatic adsorption where axial dispersion was ignored. Bradley
and Sweed (1975) analyzed the constant pattern profiles that arise
with both axial dispersion and the highly coupled mass transfer
that occurs in ion exchange. They developed a simple computational
scheme that allows calculation of profile shapes using ordinary--
not partial--differential equations. The lumping of axial
dispersion and finite mass transfer effects was used in this case.
The profiles are within a few percent of those calculated using

the partial differential equation model.

Model of Nonequilibrium Chromatography with Axial Dispersion

The material balance for nonequilibrium chromatography with axial dispersion is Equation 7 plus an additional term involving the Peclet number:

$$-\frac{1}{Pe}\frac{\partial^2 c_j}{\partial z^2} + \frac{\partial c_j}{\partial z} + \frac{\partial c_j}{\partial \tau} + \nu\frac{\partial n_j}{\partial \tau} = 0 \qquad (26)$$

where $Pe = D_{ax}/Lv$
= axial diffusivity/(bed length) (interstitial fluid velocity).

The rate of mass transfer between the phases is assumed to be proportional to "distances" from equilibrium

$$\frac{\partial n_j}{\partial \tau} = \sum_{k=1}^{M} s_{jk} \ (n_k{}^* - n_k) \qquad (27)$$

where

$$n_k{}^* = f_k \ (c_1, c_2, \ldots c_m), \qquad (28)$$

that is, $n_k{}^*$ is the solid phase concentration in equilibrium with the fluid. The Stanton numbers, s_{jk} are dimensionless mass transfer coeffieients.

Equation 27 states that the rate of accumulation of species j in the solid phase depends on more than the driving force of just species j. Species j can migrate to or from the solid because of driving force in other species. Vermeulen and Clazie (1968) used irreversible thermodynamics to show that Equation 27 is well suited to ion exchange where electrical forces, in addition to chemical potential forces, act to move ions. Bird, et al (1960) point out that heat and mass transfer are also coupled in this way. Hence Equation 27 is useful for adiabatic adsorption too.

Constant Pattern Behavior

As mentioned earlier, constant pattern waves are the result of dispersive forces acting on shock waves. The tendency of a wave to become sharp (i.e., become a shock) is balanced in real systems by the dispersive actions resulting in the common "S"-shaped wave.

Using a moving coordinate system, where

$$\xi = z - \frac{\tau}{\sigma s} \tag{29}$$

Bradley and Sweed (1975) lumped axial dispersion into mass transfer coefficient and reduced Equations 26 and 27 to

$$\frac{\partial n_j}{\partial \xi} \equiv - \sigma_s \sum_{k=1}^{M} s_{jk}^* \ (n_k^* - n_k) \tag{30}$$

where s_{jk}^*, the modified Stanton numbers combine the s_{jk}'s and the Peclet number. Their paper should be referred to for details. Suffice it to say, Equation 30 can be solved to give constant pattern profiles. For the highly coupled ion exchange system of H^+, Ag^+, and Na^+ on Duolite C-25, Bradley and Sweed (1975) computed such profiles using Equation 30 and also using the full partial differential equation model. The s_{jk}'s were taken from the report of Vermuelen and Clazie (1968). Figure 20 shows these profiles. Clearly, lumping of dispersive forces is reasonable even for quite nonlinear systems.

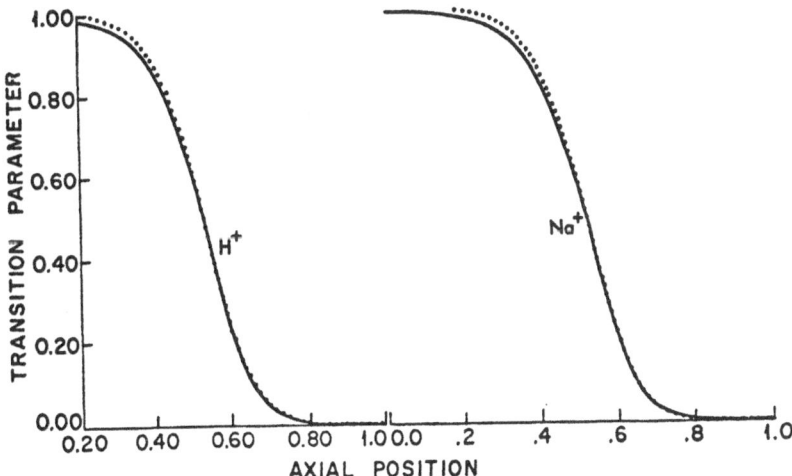

Figure 20. Calculated profiles for constant pattern ion exchange of H^+, Ag^+, Na^+ on Duolite C-25. Solid line are ODE solution; dotted lines are PDE solution.

Adiabatic Adsorption in Real Systems

We might reasonably ask whether experimental adiabatic chromatographs show the features found in the equilibrium theory (e.g., plateaus) and in nonequilibrium theories (e.g., constant pattern behavior). The answer is yes and no. Pan and Basmadjian (1970) carried out a number of experiments using CO_2-He adsorption on 5A molecular sieves. Figure 21 shows the breakthrough profiles for different length beds. Constant pattern behavior and plateaus are obvious here. However, at lower feed concentrations with the same system, the plateaus are not clearly evident (Figure 22) and the transitions are very spread. We cannot tell whether constant patterns have been reached yet.

It seems that the two transitions in Figure 22 have spread out so much they have met each other. The plateau has been eroded from both sides, and the shock fronts do not look like shocks anymore. So, real adiabatic adsorption can have close-to-equilibrium profiles or far-from-equilibrium ones.

Cooney (1974) investigated adiabatic adsorption using a partial differential equation model without axial dispersion and

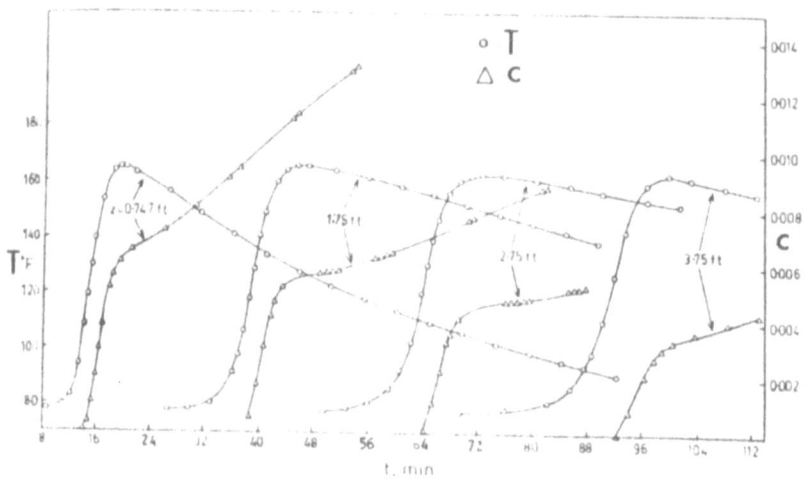

Figure 21.   Breakthrough profiles for CO_2 in He 5A molecular sieve. High feed concentration.

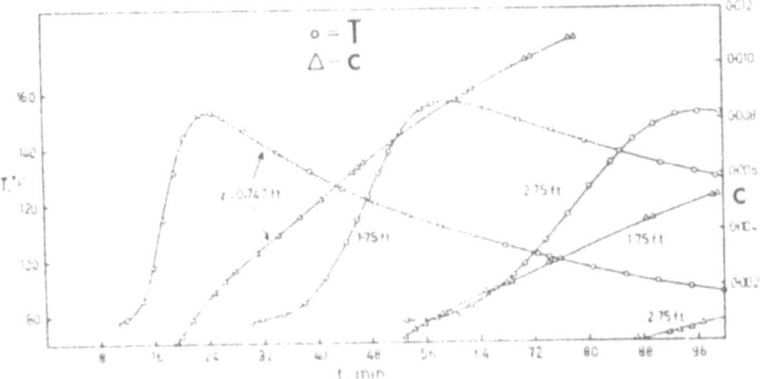

Figure 22.   Breakthrough profiles for CO_2 in He on 5A molecular
             sieve.   Low feed concentration.

showed that observations like those of Pan and Basmadjian could
be simulated numerically.

USE OF EQUILIBRIUM THEORY FOR REAL SYSTEMS

     The equilibrium theory provides a relatively easy calculation
of the best separation that an adsorption bed can give.   The
effects of feed and initial loadings and temperatures on
separation and breakthrough curves can be examined quickly.   The
theory allows us to know the range of conditions under which a
pure thermal wave can or cannot appear   Or which transition will
come first.   Or how many transitions there will be.   The theory
tells us whether a transition will be a shock.   If so we can
expect a constant pattern transition in the real system.

     The equilibrium theory predicts when concentration enhancements
can occur or when plateau temperatures can get too high or too low
for good operation.

     The equilibrium theory cannot predict how good a separation
really will be.   Dispersive forces, if large enough, can wipe out
much of the separation.   Even constant pattern analysis is of
limited value since the profiles may spread so much that the
plateaus and the constant pattern itself are destroyed.   Only
numerical methods can reliably treat real systems.

Real systems can be made to approach equilibrium operation, and in modern high performance liquid chromatographs (HPLC) they do. In HPLC the sorbent particle size is made very small to minimize the intraparticle mass transfer resistance. The small particles also have a very large total external surface area which also speeds the attainment of equilibrium. Commercial columns are very carefully packed and so have low axial dispersion. The difficulty in using such columns is the very high pressure drop they can develop (hundreds or thousands of pounds per square inch).

The close-to-equilibrium operation of these HPLC systems makes their behavior amenable to equilibrium theory analysis.

List of Symbols

c_j = concentration of species j in fluid, mol/m^3 of fluid

C_{pf} = heat capacity of fluid, J/m^3K, a constant

C_{pj} = heat capacity of species j in fluid phase, J/m^3K

C_{ps} = heat capacity of solid support, J/m^3K

$\dfrac{\mathcal{D}f_j}{\mathcal{D}c_j}$ = defined by Equation 13

f_j = equilibrium solid concentration of species j in solid, mol/m^3 of solid

$h_j{}^*$ = partial molar enthalpy of species j in solid, J/mol

h_j = partial molar enthalpy of species j in fluid, J/mol

H_s = enthalpy of solid phase including adsorbed molecules, J/m^3 of solid

H = enthalpy of fluid, J/m^3 of fluid

ΔH_j = heat of adsorption of species j, J/mol of j

K = constant in adsorption isotherm

L = bed length, m

M = number of species in the fluid phase plus one

n_j = concentration of solute j in the sorbent phase, mol/m^3 of solid

N = constant in adsorption isotherm

Pe = Peclet number, D_{ax}/Lv

$s_{jk}{}^*$ = modified mass transfer coefficient (dimensionless)

s_{jk} = mass transfer coefficient (dimensionless) for species j due to a driving force in species k

t = time, s

T = fluid temperature, K

T_s = temperature of solid phase, K

T_r = arbitrary reference temperature, K

v = interstitial velocity of fluid, m/s

x = distance from column inlet, m

z = dimensionless bed length

Γ_c = the Γ curve through the watershed point

Γ = a curve or family of curves in the C-T plane

Δ = (in Equations 22 & 23) difference operator

ϵ = void fraction in the bed, m^3 of fluid/m^3 of bed

ν = $(1-\epsilon)/\epsilon$

σ = reciprocal of concentration wave speed, dimensionless

τ = time, dimensionless

REFERENCES

1. Acrivos, A., *Chemical Engineering Science* 13, 1 (1960).
2. Aris, R., and Amundson, N. R., *Mathematical Methods in Chemical Engineering Volume 2: First Order Partial Differential Equations with Applications*, Prentice-Hall, Englewood Cliffs, New Jersey, 1973.
3. Bird, R. B., Stewart, W. E., and Lightfoot, E. N., *Transport Phenomena*, Wiley, New York (1962).
4. Bradley, W. G., and Sweed, N. H. *Rate Controlled Constant Pattern Fixed-Bed Sorption with Axial Dispersion and Nonlinear Multicomponent Equilibria*, AIChE Symposium Series No. 152, 59 (1975), Zwiebel, I. and Sweed, N. H., eds.
5. Cooney, D. O., *Numerical Investigation of Adiabatic Fixed Bed Adsorption*, Industrial & Engineering Chemistry Process Design and Development 13, 368 (1974)
6. Cooney, D. O., and Lightfoot, E. N., *Existence of Asymptotic Solutions to Fixed Bed Separations and Exchange Equations*, Industrial & Engineering Chemistry Fundamentals 4 233 (1965).
7. Devault, D., *The Theory of Chromatography*, Journal American Chemical Society 65, 532 (1943)
8. Helfferich, F. and Klein, G., *Multicomponent Chromatography: Theory of Interference*, Marcel Dekker, New York 1970.
9. Omatete, O. O., PhD Thesis, University of California, Berkeley (1972)
10. Pan, C. Y., and Basmadjian, D., *An Analysis of Adiabatic Sorption of Single Solutes in Fixed Beds: Equilibrium Theory*, Chemical Engineering Science 26, 45-57 (1971).
11. Pan, C. Y., and Basmadjian, D. *An Analysis of Adiabatic Adsorption of Single Solutes in Fixed Beds: Pure Thermal Wave Formation and Its Practical Implications*, Chemical Engineering Science 25, 1653-1664 (1970).
12. Perry's Chemical Engineers Handbook, 5th Edition, McGraw-Hill, New York (197).
13. Rhee, H. K., and Amundson, N. R., *An Analysis of an Adiabatic Adsorption Column: Part I. Theoretical Development*, Chemical Engineering Journal 1, 241-254 (1970).
14. Rhee, H. K., and Amundson, N. R., *An Analysis of an Adiabatic Adsorption Column: Part IV. Adsorption in the High Temperature Range*, Chemical Engineering Journal 3, 122-135 (1972a).
15. Rhee, H. K., and Amundson, N. R., *A Study of the Shock Layer in Nonequilibrium Exchange Systems*, Chemical Engineering Science 27, 199 (1972b).
16. Rhee, H. K., Heerdt, E. D., and Amundson, N. R., *An Analysis of an Adiabatic Adsorption Column: Part II. Adsorption of a Single Solute*, Chemical Engineering Journal 1, 279-290 (1970).

REFERENCES (continued)

17. Rhee, H. K., Heerdt, E. D., and Amundson, N. R., *An Analysis of an Adiabatic Adsorption Column: Part III. Adiabatic Adsorption of Two Solutes*, Chemical Engineering Journal 3 22-34 (1972).

18. Sweed, N. H., and Rigaudeau, J. M. M., *Equilibrium Theory and Scale Up of Parametric Pumps* in Adsorption and Ion Exchange AIChE Symposium Series No. 152, Zwiebel, I., and Sweed, N. H. eds., American Institute of Chemical Engineers, New York 1975.

19. Thomas, H. C., *Heterogeneous ion exchange in a flowing system*, Journal American Chemical Society 66, 1664 (1944).

20. van Deemter, J. J., Zuiderweg, F. J., and Klinkenberg, A., *Longitudinal diffusion and resistance to mass transfer as causes of nonideality in chromatography*, Chemical Engineering Science 5, 271 (1956).

21. Vermeulen, T., and Clazie, R. N., *Office of Saline Water Report No. 326* (1968).

22. Wilhelm, R. H., Rice, A. W., and Bendelius, A. R., *Parametric Pumping: a dynamic principle for separating fluid mixtures*, Industrial & Engineering Chemistry Fundamentals 5, 141 (1966)

23. Wilhelm, R. H., and Sweed, N. H., *Parametric Pumping: Separation of a Mixture of Toluene and n-Heptane*, Science 159, 522 (1968).

ION EXCHANGE AND CHEMICAL REACTION IN FIXED BEDS

Gerhard Klein

University of California, Sea Water Conversion Laboratory, Richmond, California, U.S.A.

ABSTRACT. Local-equilibrium theory of simple fixed-bed ion exchange is extended to systems in which chemical reaction accompanies ion exchange. Performance prediction for mono- and bivariant systems is discussed, and examples are given for cases with and without phase change.

1. INTRODUCTION

Chemical reactions that accompany ion exchange are an extremely common phenomenon. The neutralization reaction on which deionization processes are based, and complexing utilized to enhance the separation of trace counterion species with similar selectivities are among the best-known examples (Ketelle and Boyd, 1974). Of more recent origin is ion-exchange softening of waters containing bicarbonate alkalinity, with weak-acid resin, where hydrogen ion is doing double duty as regenerant and neutralizing agent for bicarbonate ion (Bresler et al., 1974; Klein et al., 1978A). George et al. (1967) have developed a process in which sulfate ion is removed from brines by precipitation with barium ion displaced from a cation exchanger by cations present in the feed brine. Precipitation has also been invoked in other saline-water conversion processes. In the Popper process, saline water is subjected to partial deionization in a bed containing a mixture of cation exchanger in the calcium form and an anion exchanger in the hydroxide form. As calcium and hydroxyl ions are being displaced from the resin, they precipitate, thus aiding the ion-exchange reactions by which the ions of the saline water are taken up by the resins. The product of this exhaustion step is primarily a saturated solution of lime water, which has a much lower total concentration

than the raw saline water. Regeneration is accomplished by means
of a suspension of calcium hydroxide, thus again involving a chemical reaction, this time that of dissolution (Popper et al.,
1963). Precipitation of silver chloride in briquettes of a cation
exchanger in the silver form has been used for emergency deionization of water during World War II (Tiger et al., 1964) and proposed for fixed-bed application by Glueck (1968).

In the foregoing examples, chemical reactions modify simple ion
exchange in a desired manner. But they may also have to be addres-
sed as problems. For example, precipitation of calcium sulfate can
occur during regeneration of a softening resin with sulfuric acid
or a sulfate brine (Cherney, 1966; Applebaum, 1968). The presence
of precipitate can then cause reduction in operating capacity and
clogging of the bed. In one application, however, this phenomenon
has been turned into a benefit by encouraging precipitation and
regeneration in the upflow direction to allow precipitate to be
separated from the resin by hydraulic classification. The precipitation reaction thus not only increases the concentration gradient
for regeneration, buy the precipitate also forms a sink for calcium
ion, so that the regenerant brine can be recirculated if desired
(Klein et al., 1979).

From these examples, it can be seen that chemical reactions accompanying ion exchange may cause problems, but can also be powerful
tools in modifying selectivity beneficially by operating on the
fluid medium. The alternative, of modifying the chemical nature
of the exchanger, is often technologically or economically infea-
sible.

Finally, chemical reactions accompanying ion exchange are likely
to assume increasing importance in the understanding of geochemical
and chemical minerals-leaching processes. Additional applications
will be discussed further below.

The theory of fixed ion-exchange beds in with chemical reactions
occur has been advanced by Shiloh (1966), Golden (1972), Golden
et al. (1974), Page (1968, 1971), and Page et al. (1975). This
work has been based on the premise that equilibrium exists everywhere locally in the bed. The dispersive effect of mass-transfer
inefficiencies can be estimated by the general methods indicated by Vermeulen et al. (1973). Schweich and Villermaux (1978) have
provided models to predict the kinetic behavior of trace-chromatographic pulses modified by chemical reaction.

This earlier, idealized theoretical development relied on a combination of point relations (such as electroneutrality, chemical and
ion-exchange equilibria, and conservation equations) and material
balances to predict column response. On this basis it also classified various systems by their number of "degrees of freedom".

The present discussion utilizes this work, but greatly simplifies
the treatment by strictly separating the point relations from material balances, which depend on operating conditions. In this man
ner, except for a modification required when stationary precipitates are involved, the problem becomes completely analogous to any

other local-equilibrium, fixed-bed sorption problem. The number of independent point relations minus the number of independent variables involved becomes the variance; the combination of the point relations defines isotherms, and they with the coherence condition, permit solution of an eigenvalue problem that leads to composition paths and composition-velocity contours. Finally, material-balance relations are used to determine the column response corresponding to given feed and presaturation compositions. In the present section, only cases of uniform presaturation of the bed and of constant feed composition are discussed, but adaptation of the method developed to arbitrary boundary conditions is possible with the technique indicated by Helfferich and Klein (1970). Modification of the ion-exchange process by chemical reaction tends to magnify certain abnormalities that do not appear at all in constant-separation factor systems and that can usually be observed only to a slight extent in other pure ion-exchange systems. They include selectivity reversals, discontinuities in composition-velocity eigenvalues, and modification of the topology of the composition-path grid. Systems involving ion exchange plus chemical reaction can thus constitute an interesting study subject for the fundamental aspects of fixed-bed theory, particularly also with respect to the practical problem of finding the location of the abrupt portions of "compound" transitions, that is, of transitions composed of abrupt as well as of gradual elements.

The representation of systems having a variance greater than 2 becomes difficult, and their numerical treatment, even with the aid of high-speed computers, tends to be prohibitively time-consuming. Except for the subject of variance and the general formulation of the eigenvalue problem, the present discussion has therefore been limited to mono- and bivariant systems, whose visualization can still be readily facilitated with graphical means. However, many higher systems are reducible to systems of lower variance because some of their components are present only in trace amounts or have selectivities sufficiently similar so that they can be treated as a group. The utility of the methods indicated here thus can often be extended to higher systems of interest.

The presentation is arranged in the following manner. The general relations leading from a given problem to its solution are presented in the section titled "Basic Concepts". This is followed by two sections of specific applications, with examples developed to varying degrees of detail. The first of these sections deals with reactions without creation or disappearance of one or more phases, and the second, with reactions causing a change in the number of phases; particularly with precipitation. The methods employed are summarized in a final section.

2. BASIC CONCEPTS

2.1 Introduction

Many of the salient features of the response of a fixed sorption bed of given presaturation-concentration profiles to a given feed--composition history can be predicted with the aid of theory based on the premise that the compositions of the phases are locally at equilibrium throughout the bed. Thus, for the relatively simple but practically important case of uniform presaturation and constant feed composition, this theory indicates that zones of constant composition (plateau zones) will appear in the column. It moreover will predict their number and in most cases, to a good approximation, their quantitative composition. The concentration profiles in the transition zones intervening between the plateau zones are also obtainable as functions of time.

In actual operation, fixed-bed behavior based on local equilibrium is modified by mass-transfer inefficiencies. Equilibrium theory under certain conditions predicts the appearance of concentration discontinuities or steps. In these cases, in reality, relatively sharp but slightly diffuse transitions arise instead and the concentration profiles in them tend to become invariant with respect to time (constant-pattern transitions). Other types of transition zones, according to local-equilibrium theory, broaden with time (proportionate-pattern transitions). For the latter, the equilibrium theory provides a close approximation to the profiles obtained when mass-transfer inefficiencies are significant. For both types of transition, the equilibrium theory predicts the average position of a concentration profile in a transition quantitatively as a function of time.

Local-equilibrium theory thus is seen to provide a valuable characterization of ideal column response, of which the actual response may be regarded to a qualitatively predictable aberration. Since the rigorous mathematical treatment of fixed-bed operations, with inclusion of kinetic considerations, not only tends to become exceedingly complicated, but also often unreliable because of the uncertain validity of the underlying premises and of the values of the parameters involved, the approach adopted here relies primarily on the local-equilibrim premise. Kinetic effects are dealt with only briefly and qualitatively. Further refinement in this direction can be attempted along the lines of the methods summarized by Vermeulen et al. (1973). A theoretical analysis of the kinetics of various ion-exchange processes involving reactions has been given by Helfferich (1965).

The key concepts involved in the theory of equilibrium fixed-bed sorption accompanied by reaction are those of variance, coherence, and composition velocity. Variance is the minimum number of concentration variables that determine the composition of the system at equilibrium. Coherence is the tendency of the con

centrations constituting a given composition to travel with the
same velocity, thus keeping the composition intact. Composition
velocity finally is the velocity of a coherent composition.
The most logical order of treating these concepts is probably that
given above. However, in the belief that this will make the presen
tation clearer, the concept of composition velocity will be develo
ped first for a simple monovariant system. Examples of the utiliza
tion of this concept for predicting clumn performance will be given.
Variance will be discussed next, as a basis of the need for the
concept of coherence. An understanding of coherence will then lead
to solution of the eigenvalue problem required for multivariant
systems.
Alternatives to the treatment presented here are stagewise calcula
tions, and calculations based on the method of characteristics.
The respective general framework of such calculations has been given by Pandya et al. (1965), and by Acrivos (1956). Into such a
framework, the relations valid locally for the systems under consideration can then be fitted. These approaches have the merit of
being applicable to monuniform initial and boundary conditions and
to systems of variance greater than 2, and perhaps of requiring
less theoretical background of the person employing them than the
method treated here. On the other hand, the present method, once
the required diagrams are produced, can lead to rapid performance
prediction for many initial and boundary conditions, provided that
they are uniform. Use of this method may also lead to a better
understanding of the basic characteristics of each system. Moreover,
stagewise and characteristics calculations can consume considerable
computer time, which, especially for systems of higher variance,
would tend to become excessive.
On balance, then, the composition-path method has been selected
for exclusive presentation here but without prejudice to other
methods, whose development may also prove useful, at least for cor-
roborative purposes.

2.2 Composition velocity in monovariant systems

As a composition wave moves through a sorption bed, each constituent concentration moves with a certain instantaneous velocity,
called "concentration velocity". In the particularly simple case
of a single-component adsorption system in which the fluid and
sorbent phases locally are everywhere at equilibrium, this concen
tration velocity coincides with the composition velocity, the composition being the set of fluid- and sorbent-phase conentrations
that are uniquely related by equilibrium. As will be seen further
below, single-component adsorption and binary ion exchange fall
into the class of monovariant systems. For such systems, we now
establish the relation which DeVault (1943) has demonstrated to
exist between the composition velocity and the slope of the isotherm.

To do this, we consider a continuous portion of an arbitrary concen
tration profile for one of the phases, i.e., a plot of concentration
against distance from the upstream end of the sorption bed. The cor
responding profile for the concentration in the other phase is then
uniquely determined by the sorption isotherm and the local-equilibrium premise.

Next, we imagine the region under consideration to be divided into
an arbitrary number of segments, and represent the continuous con
centration profiles approximately by the average value of the con
centrations in each segment. The resulting step functions can be
made to approximate the continuous concentration profiles as closely as desired by chosing the segments sufficiently small.

Fig. 1 shows one such segment schematically. The left-hand diagram
applies to the arbitrary time t, and the right-hand diagram to the
time t + Δt. The segment extends from bed-level z to bed-level
z + Δz, Δz being so chosen that it will be traversed by the concen
tration step shown in the time interval Δt.

The upper (dashed) lines in the diagrams represent concentrations
in the fluid phase, and the lower (solid) lines, concentrations in
the sorbent phase. Both concentrations are given in gram atoms,
moles, or equivalents per liter of sorbent bed. The fluid-phase
concentrations of Species i, dsignated by $\tilde{c}_i$, are based on unit
solution volume, and ε is the void fraction of the sorbent bed.
The sorbent-phase concentrations $\tilde{q}_i$ are based on unit weight of
sorbent, and ρ_B is the bulk density of the latter. The tildes of
over the concentration variables designate combinations of concentrations where indicated, as discussed following Eq. 2.

In the left-hand diagram, corresponding to time t, the concentration steps are at the upstream of the segment; in the right-hand
diagram, they have reached its downstream end, z + Δz. A material
balance can now be written stating that the accumulation of Sorba-
te i in the segment during time interval Δt must equal the amount
of i that has entered the segment, minus the amount that has left
it:

$$\Delta z (\rho_B \Delta \tilde{q}_i + \varepsilon \Delta \tilde{c}_i) = u \varepsilon \Delta \tilde{c}_i \Delta t \tag{1}$$

Here, u is the linear velocity of the fluid phase in the bed.
In the limit as the finite increments of the variables under consideration approach zero, Eq. 1 assumes the form

$$u_{\tilde{c}_i} = u_{\tilde{q}_i} = u / \left[(\rho_B d\tilde{q}_i / \varepsilon d\tilde{c}_i) + 1 \right] \tag{2}$$

where $u_{\tilde{c}_i}$ or $u_{\tilde{q}_i}$ equals $\lim\limits_{\Delta z \to 0, \Delta t \to 0}$ dz/dt. This is the desired relatio
between the composition velocity ($u_{\tilde{c}_i}$ or $u_{\tilde{q}_i}$) and the slope of
the isotherm $d\tilde{q}_i / d\tilde{c}_i$ at a composition point ($\tilde{c}_i, \tilde{q}_i$).

In the present discussion, the material balance represented by Eq.
2 sometimes logically must be applied not to a single molecular or
ionic species, such as perhaps bicarbonate ion, but to an atom or

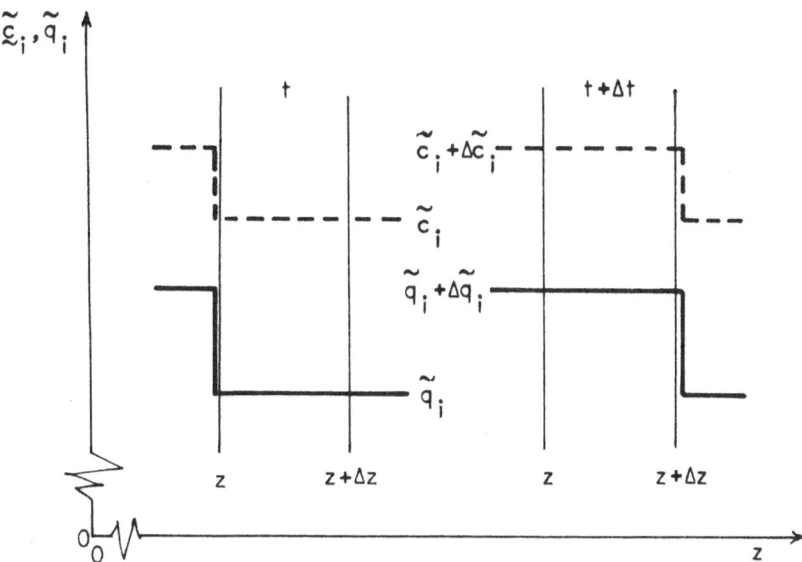

Fig. I.   Relation of composition velocity to isotherm slope in
single-component equilibrium adsorption systems. Pro-
gression of small concentration steps $\Delta \tilde{c}_i$ and $\Delta \tilde{q}_i$ through
bed segment of length Δz in time Δt

chemical group in all its forms, such as carbon in carbonate ion, bicarbonate ion, and carbonic acid. This accounts for the need of introducing a special symbol (here the tilde) to designate a combination of concentrations rather than the concentration of a single species. In material balances, the concentrations with the tilde are appropriate even when there exists only a single species containing the atom or group of interest. Nevertheless, where both the concentration of a single species and of a combination of species exist, separate symbols are still required for the concentrations of such single-species to permit their correct use in certain point relations, such as equilibrium expressions and electroneutrality equations. Examples of single and combined concentrations will be found further below.

Eq. 2 can be solved for $\varepsilon d\tilde{c}_i / \rho_B d\tilde{q}_i$ to give

$$\varepsilon d\tilde{c}_i / \rho_B d\tilde{q}_i = u_{\tilde{c}_i} / (u - u_{\tilde{c}_i}) \tag{3}$$

The right-hand member of this equation is seen to be the composition velocity as fraction of the apparent fluid velocity, the latter being the fluid velocity that would be measured by an observer traveling with the composition velocity. This relative velocity is analogous to the velocity of a sailboat running dead before the wind, expressed as a fraction of the velocity of the apparent wind, the latter being the velocity of the true wind minus the velocity of the boat. In either case, the moving entity under consideration (composition or boat) responds to the apparent velocity of the moving medium by which it is being propelled. It is therefore not surprising that most of the mathematical relations involving composition velocity take on a simpler form when the latter is expressed in terms of (dimensionless) relative velocity rather than (dimensional) linear velocity. Thus, with the definition

$$\underline{u} = u_{\tilde{c}_i} / (u - u_{\tilde{c}_i}) = u_{\tilde{q}_i} / (u - u_{\tilde{q}_i}) \tag{4}$$

Eq. 3 becomes

$$\underline{u} = \varepsilon d\tilde{c}_i / \rho_B d\tilde{q}_i \tag{5}$$

that is, the relative composition velocity, in the present case, simply equals ε / ρ_B times the reciprocal of the local isotherm slope, in terms of combined concentrations, if necessary. Eq. 5 will be referred to as "differential material balance".

For simple ion exchange, Helfferich and Klein (1973) have defined an "adjusted" composition velocity, which is the present relative composition velocity times $(\rho_B Q / \varepsilon C_0)$, where Q is the total exchange capacity and C_0 the total solution normality. Where reactions cause C_0 to vary, this definition is no longer useful. For cases

in which each composition velocity is constant throughout the
operation, the throughput ratio T, used by Vermeulen et al. (1973)
is the reciprocal of the adjusted composition velocity. However,
except for systems for which the local-equilibrium premise applies
closely, this relation no longer holds. Nevertheless, the throughput ratio may then still be defined as the number of equivalents
of product solution per equivalent of sorbent. This relates the
throughput ratio inversely to a mean rather than an instantaneous
velocity, a circumstance that, when overlooked, tends to give rise
to confusion.

We shall find further below that Eq. 5, giving the composition velocity relative to the apparent velocity of the fluid phase, also
applies to more complicated systems. For the moment, we proceed by
examining how this equation may be utilized to predict the behavior of the simple systems for which it was derived.

2.3 Fixed-bed response for monovariant system

The discussion of the present section will be confined to the pratically important case of uniform presaturation and constant feed
composition.

We shall first examine some formal consequences of the differential material balance (Eq. 5) and identify instances when this relation must be replaced by an "integral material balance". The
effect of mass-transfer inefficiencies will then be discussed qua
litatively, and constant- and proportionate-pattern breakthrough
will be related to abrupt and gradual breakthrough as predicted by
local-equilibrium theory.

Consider first a bed of a sorbent initially completely free of the
feed component. The isotherm of this component is arbitrary but
assumed to be concave upward throughout, as in the example of Fig.
2, where equilibrium sorbent-phase concentrations $\tilde{q}_i$ are plotted
against concentrations $\tilde{c}_i$ in the fluid phase. Such an isotherm indicates a distribution favoring the presence of the sorbent in the
fluid phase and is therefore unfavorable for the uptake of the
feed component. For short, this type of isotherm is referred to as
"unfavorable". It must be remembered, however, that the same isotherm is favorable for desorption.

Since Eq. 5 gives the relative velocity of each composition $(\tilde{c}_i, \tilde{q}_i)$,
we can readily construct a graph of the concentration in either
phase against ρ_B/ε times the relative composition velocity $\underline{u}$ by
measuring or calculating the slopes of the isotherm at various
points and plotting the abscissae $(\tilde{c}_i)$ and ordinates $(\tilde{q}_i)$ of these
points against the reciprocals of the corresponding slopes. The
result of this procedure as based on the isotherm of Fig. 2 is
shown in Fig. 3. The concentrations are seen to travel the faster
the lower they are, so that each slightly higher concentration
is travelling at the same velocity as it would had the bed been
presaturated to a slightly lower concentration. The concentration

velocities, being noninterfering, are therefore constant, so that instead of instantaneous velocities dz/dt, ratios of finite distan ces and times can be used. Eq. 4 then may be written in the form:

$$\underline{u} = (z_{\tilde{c}_i}/t)/\left[(z/t) - (z_{\tilde{c}_i}/t)\right] \qquad (6)$$

$z_{\tilde{c}_i}$ and z being the respective distances traversed in time t by the composition $(\tilde{c}_i, \tilde{q}_i)$ and by the fluid front, t may be cancelled, yielding

$$\underline{u} = z_{\tilde{c}_i}/(z - z_{\tilde{c}_i}) \qquad (7)$$

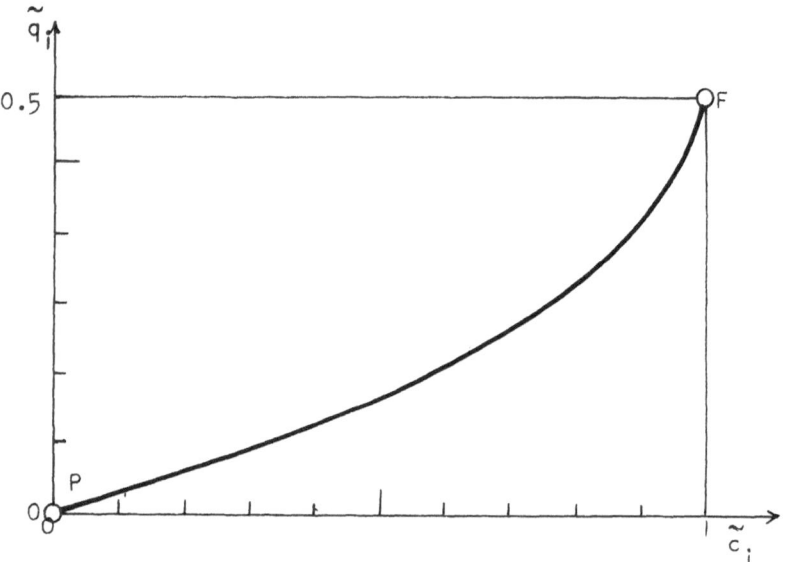

Fig. 2. Example of sorption isotherm unfavorable in terms of feed-component concentrations $\tilde{c}_i$ and $\tilde{q}_i$. F and P are the respective feed- and presaturation-composition points

Plots of the type of Fig. 3 may thus be viewed as concentration profiles showing ρ_B/ε times the distances to which concentrations have penetrated the bed as fractions of the distance that the fluid front has travelled beyond the respective concentrations. A given isotherm thus yields a master set of profiles, from which profiles corresponding to particular times can readily be derived. Eq. 7 can be simplified conceptually by neglecting $z_{\tilde{c}_i}$ with respect to z in the denominator. The error introduced by this approximation will often be negligible.

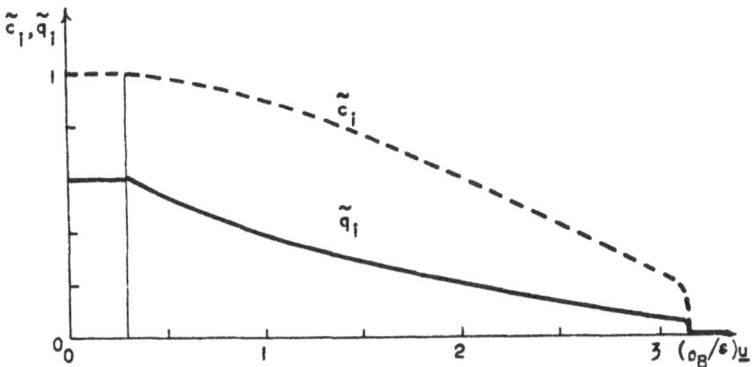

Fig. 3. Plot of concentrations $\tilde{c}_i$ and $\tilde{q}_i$ against ρ_B/ϵ times relative composition velocity $\underline{u}$, constructed from isotherm of Fig. 2

The concentration profiles of Fig. 3 show that there exists an upstream zone of uniform composition (plateau zone) in which the sorbent has come to equilibrium with the feed composition, and a downstream plateau zone, in which, in general, the effluent has come to equilibrium with the initial composition of the sorbent. These plateau zones are separated by a zone of varying composition, referred to as boundary or transition. The limits of the transition zone are determined by the limiting slopes of the isotherm.
All of these zones will of course only be present in the exchanger bed if, for the given time, the bed is long enough to contain them. In diagrams of the type of Fig. 3, this is supposed to be the case. In practice, the bed is of finite length and, after a certain time, the leading part of the profile will disappear from it. If feed is continued to be supplied, "breakthrough" in the form of increasing concentrations of the feed component in the effluent will occur. The corresponding effluent-concentration history can be obtained from the master profile with the aid of Eq. 6, but this time by holding the distance $z_{\tilde{c}_i}$ constant at the bed depth

of interest, and varying time t. Through $\underline{u}$ and its dependence on composition, c_i can thus be plotted against t, or effluent volume $\underline{u}St$, where S is the cross-sectional area of the bed.
What if the isotherm in terms of the concentrations of the feed component is "favorable", i.e., convex upward, or, what amounts to the same thing if Points F and P in Fig.2 are interchanged? Concentrations in the vicinity of the feed concentration would now travel faster, instead of more slowly, than concentrations in the vicinity of the presaturation concentration. Construction by the same method as used for Fig. 3 would lead to the master profiles of Fig.4 but these profiles are absurd because they show the simultaneous existence of three values of $\tilde{c}_i$ and of three values of $\tilde{q}_i$ at any $\underline{u}$

value in the zone in which the concentrations change! The performan ce prediction based on the material balance of Eq. 5 must therefo re be invalid.

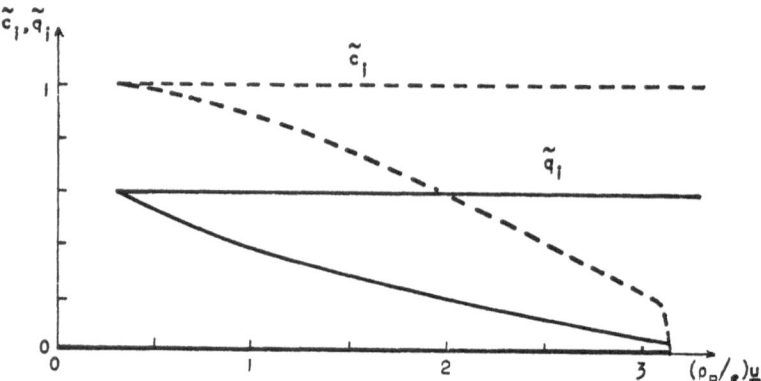

Fig. 4. Hypothetical plot of concentrations $\tilde{c}_i$ and $\tilde{q}_i$ against ρ_B/ϵ times relative composition velocity $\underline{u}$, constructed from Fig. 2, but with Points F and P interchanged. Each bed level in the transi tion zone exhibits three concentration values for either phase

If we observed an actual adsorption column governed by a convex-up ward isotherm in terms of feed-component concentrations, we would find a concentration wave in which the concentrations declined in the flow direction as sorption took place. Leading this wave would be a region of low concentration, which, if they tended to follow Eq. 5, would travel slowly. The high concentrations in the trailing part of the wave, on the other hand, would tend to move faster. The result would be a steepening tendency of the profiles, which, in the limit as mass-transfer inefficiencies became negligible, would approach abrupt concentration steps.

The material balance for such steps is given by Eq. 1, its deriva tion remaining valid in spite of the fact that, for the present purpose, the concentration steps are considered to be finite ins tead of small. With such steps retained in the deriavtion, the in tegral material balance corresponding to Eq. 5 becomes

$$\underline{u}_\Delta = \epsilon \Delta \tilde{c}_i / \rho_B \Delta \tilde{q}_i \tag{8}$$

Here, $\underline{u}_\Delta$ is the relative velocity of the concentration steps, and $\Delta \tilde{c}_i$ and $\Delta \tilde{q}_i$ are given by the equations

$$\Delta \tilde{c}_i = \tilde{c}_i'' - \tilde{c}_i' \tag{9}$$

and

$$\Delta\tilde{q}_i = \tilde{q}_i'' - \tilde{q}_i' \tag{10}$$

where $(\tilde{c}_i', \tilde{q}_i')$ is the composition of the adjacent downstream plateau zone, and $(\tilde{c}_i'', \tilde{q}_i'')$ the composition of the adjacent upstream
plateau zone. Applied to the isotherm of Fig. 2, with F and P interchanged, Eq. 8 now yields the master profiles of Fig. 5.

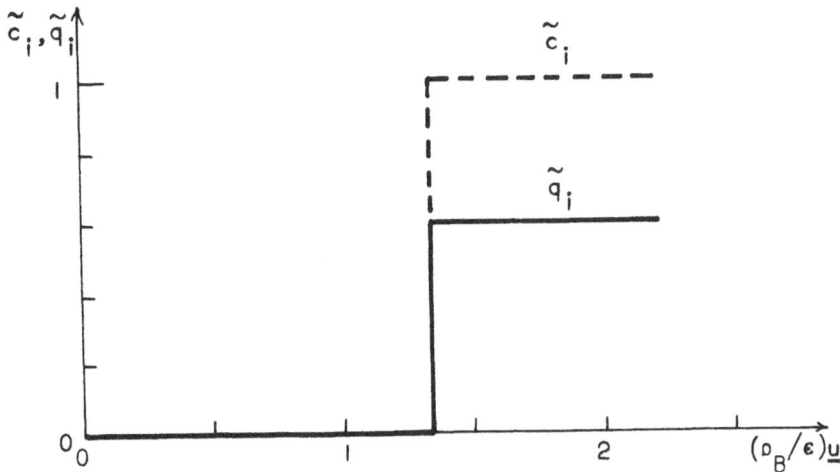

Fig. 5. Meaningful plots of concentrations $\tilde{c}_i$ and $\tilde{q}_i$ against ρ_B/ϵ
        times relative composition velocity $\underline{u}$, constructed from
        Fig. 2, but with Points F and P interchanged. Velocity of
        concentration steps evaluated with Eq. 8

To summarize the discussion so far, the differential material balance (Eq. 5) applies for unfavorable feed-component isotherms and
the integral material balance (Eq. 8) for favorable isotherms.
While derivations to this point apply directly to single-component adsorption, they are readily seen to be valid also for binary
ion exchange of ion species i and j, where

$$c_i + c_j = C_o \tag{11}$$

and

$$q_i + q_j = Q \tag{12}$$

Here, C_o is the total solution normality, and Q, the exchange capacity, in milli-equivalents per gram of exchanger. Since we are
dealing with simple concentration variables, the tildes have been
omitted. It follows from these relations that

$$dc_i = -dc_j \tag{13}$$

and

$$dq_i = -dq_j \qquad (14)$$

and hence that

$$dc_i/dq_i = dc_j/dq_j \qquad (15)$$

or

$$\Delta c_i/\Delta q_i = \Delta c_j/\Delta q_j \qquad (16)$$

so that the material balance for one of the counterion species implies that of the other and only one of these relations is independent. The situation therefore has not been complicated by the presence of two components instead of one and it suffices to base the calculations on the concentrations of either.

Most simple ion-exchange systems are governed by entirely favorable or unfavorable isotherms, the latter being typified by Fig. 2. However, when chemical reaction advenes, highly irregular isotherm shapes can arise. It is then of interest to examine the fixed-bed behavior of systems governed by such isotherms.

The effect of feed and presaturation compositions on the response of systems having isotherms with inflection points has been analyzed theoretically by Tudge (1961), and Tondeur (1967). Their results, and the effects of other abnormalities, such as discontinuities in slope of the isotherm, have been summarized by Golden (1972), who has formulated the following rule.

To deduce the equilibrium response of a system governed by an isotherm for which only one concentration can be varied independently, imagine a string stretched around the part of the isotherm that lies between the feed and presaturation-composition points. Only the side of the string corresponding to clockwise progression from the feed-composition to the presaturation-composition point determines the response, the other side is ignored. Composition regions for which the string hugs the isotherm correspond to gradual transitions or part of a transition, which are governed by the differential material balance (Eq. 5), and composition intervals corresponding to portions of the string that are not tangent to the isotherm correspond to concentration steps whose velocity is given by the integral material balance (Eq. 8). An example of the application of this rule is shown in Figs. 6 and 7.

Fig. 6 represents a hypothetical isotherm with two inflection points. A string stretched around the isotherm in clockwise direction from the feed-composition point F to the presaturation-composition point P would bridge the isotherm along the chords indicated as dashed lines, and hug it along the middle portion of the isotherm, which is convex downward and lies between the two points of tangency of these chords. The resulting concentration profiles,

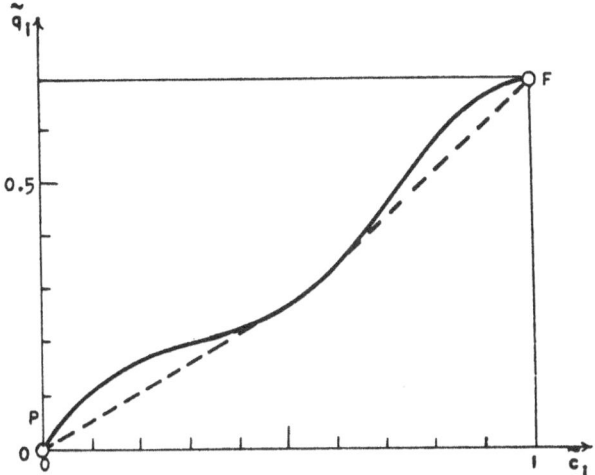

Fig. 6. Hypothetical isotherm with inflection points to illustrate application of Golden's Rule. F and P are the
feed- and presaturation-composition points. The reciprocals of the slopes of the dashed chords are proportional to relative step velocities. Cf. Fig. 7

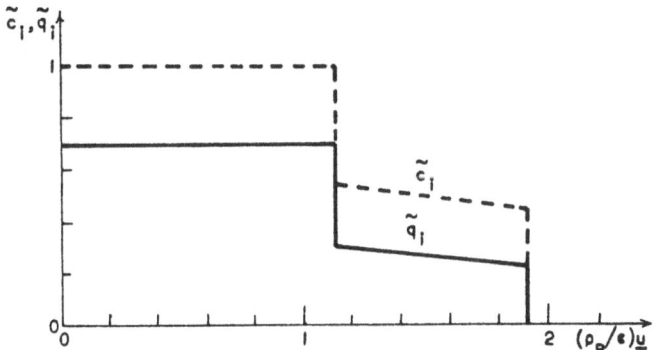

Fig. 7. Master concentration profiles derived from isotherm of Fig. 6
with Golden's Rule. Bridging of the chords drawn tangent to
the isotherm gives rise to two abrupt composition changes. In
an intermediate zone, application of the differential material
balance yields a gradual transition

shown in Fig. 7, exhibit the two usual terminal plateau zones, two abrupt composition changes, and an intermediate gradual transition. A discontinuity in isotherm slope (not illustrated here) between two portions of the isotherm that give rise to gradual transitions will cause the appearance of an additional plateau zone.

The quantitative prediction of column response when mass-transfer inefficiencies are taken into account is a difficult matter. However, the qualitative characterization of such response is of fundamental utility and will now be discussed.

The mechanism underlying the sharpening tendency of transitions for which local-equilibrium theory predicts concentration steps has already been explained. As the transition penetrates into the bed, this tendency increasingly continues to balance the spreading tendency due to mass-transfer inefficiencies until profile shapes invariant with respect to time (constant pattern) are reached. The profiles move through the bed with constant velocity. Consequently, each constituent concentration and composition also moves with this velocity, which may be obtained from material balance for a component i with the same reasoning as employed in the derivation of Eq. 1. Because the transition is diffues, this material balance takes the form

$$\Delta z \ \rho_B \int_{\tilde{q}_i'}^{\tilde{q}_i''} d\tilde{q}_i + \varepsilon \int_{\tilde{c}_i'}^{\tilde{c}_i''} d\tilde{c}_i = u(\tilde{c}_i'' - \tilde{c}_i')\Delta t \tag{17}$$

Here, Δz is the variable distance that the concentrations $\tilde{c}_i$ and $\tilde{q}_i$ have traveled in time Δt. The integration limits have been defined below Eq. 10. If we now let $\overline{\Delta z}$ equal the average distance that $\tilde{c}_i$ and $\tilde{q}_i$ travel during Δt, and define $\Delta\tilde{c}_i$ and $\Delta\tilde{q}_i$ as in Eqs. 9 and 10, we obtain the relation

$$\overline{\Delta z}(\rho_B\Delta\tilde{q}_i + \varepsilon\Delta\tilde{c}_i) = u\varepsilon\Delta\tilde{c}_i\Delta t \tag{18}$$

which is analogous, and formally almost identical, to Eq. 1. Derivation analogous to that of Eq. 5 then yields

$$\overline{u} = \varepsilon\Delta\tilde{c}_i/\rho_B\Delta\tilde{q}_i \tag{19}$$

$\overline{u}$ being the average velocity of the transition profiles. Comparison of this equation with Eq. 8, which gives the step velocity based on the local-equilibrium premise, shows the average velocity of the constant-pattern transition, and therefore the velocity of each of its constituent concentrations, to equal the step velocity.

Since constant-pattern transitions are relatively sharp, the step velocity as based on the local-equilibrium premise may be used to obtain a fair approximation to the breakthrough time. Because of the diffuseness of the transition, actual breakthrough will occur

somewhat earlier than so predicted, but the difference can be
evaluated by measuring a breakthrough curve with a column of any
length and diameter, but under otherwise representative conditions. This is the basis of the LUB (length-of unused bed) design
method (Lukchis, 1973) and equivalent methods (Michaels, 1952).
Even without this refinement, however, where local-equilibrium
theory predicts an abrupt transition, its position as calculated
by that theory usually gives a valuable preliminary indication
about the feasibility of a proposed process.

An even closer similarity to the behavior predicted by local-equi
librium theory is exhibited by proportionate-pattern transitions.
Here, provided the concentration profiles are sufficiently flat,
the phases have enough time to equilibrate, so that the agreement
between ideal and actual behavior is quantitative to a high degree of approximation, except toward the extremes of the transition. Unfortunately, this ready predictability of proportionate-
-pattern breakthrough is of relatively little practical value because the existence of proportionate-pattern transitions implies
a large sorbent inventory ant therefore makes processes in which
such transitions occur economically unattractive.

The formation of constant-pattern transitions is favored by high
selectivity for the component being taken up by the sorbent, and
by high mass-transfer rates; that of proportionate-pattern transi
tions, by low selectivity. The latter transitions can be formed
with considerable lower mass-tranfer rates. The approach to
either pattern will be the closer the greater the bed depth,
other things being equal.

Both constant- and proportionate-pattern transitions are reflec-
tions of transient phenomena in that the local composition in the
bed is a function of both distance and time. However, at least
formally, appropriate normalization of the distance or time variables can reduce the concentrations in such transitions to functions of a single variable. A different normalization procedure
applies to each of these transition types.

In many cases, the combination of equilibrium and kinetics is
such that neither of these limiting types of transitions arises.
The truly transient behavior which then ensues can be predicted,
at least in principle, with the aid of sophisticeted models as
summarized by Vermeulen et al. (1973).

2.4 Variance

When one attempts to extend the concept of the differential material balance and its consequences to systems with more components,
one faces the problem of determining the direction along which
the total derivatives $d\tilde{c}_i/d\tilde{q}_i$, on which the composition velocity
depends, must be evaluated. In a 2-component adsorption system,
for example, the isotherm may be represented by two surfaces (a
q_1- and a q_2-surface) over a c_1, c_2-base plane. The value of the

ratio dc_1/dq_1 (= dc_2/dq_2; cf. Section 2.5), and therefore of the composition velocity, now will depend not only on the coordinates (c_1, c_2) of the point at which it is determined, but also on the direction dc_1/dc_2 in the base plane along which this is done. But before we approach the eigenvalue problem of finding this direction, we are faced with another problem, namely that of defining what is a meaningful concentration variable, a question that can become nontrivial in systems in which ion exchange or adsorption is accompanied by one or more chemical reactions, especially when the latter give rise to a phase change.

In addition to the need for defining the system variables, it is also desirable to determine the level of complexity of the problem to permit a preliminary assessment of the practical feasibility of a solution. Pursuit of both of these objectives leads to the concept of variance, or equivalent concepts, which is of fundamental importance in the classification of sorption systems.

We here define "variance" as the minimum number of concentration variables required to determine the composition of a system at equilibrium. In the case of an adsorption system, the variance then simply is the number of adsorbate species involved that are present in larger than trace concentrations. In a monovariant system, for example, setting the concentration of the single adsorbate species present will also determine its equilibrium concentration in the sorbent. With two adsorbate species, two fluid-phase concentrations can be set independently and will determine the concentrations of both species in the sorbent; and so on for any number of components.

The analogy of simple ion exchange to adorpdion systems is still readily apparent. Because the sums of the normalities in the fluid and exchanger phases are constant, the concentration of any one exchangeable component in either phase is determined by the concentrations of the remaining components. In a simple n-component ion-exchange system, it is therefore only necessary to set $n - 1$ counterion concentrations in either phase to determine the equilibrium composition of the system. An n-component adsorption system is thus equivalent, in therms of variance, to an ideal (n + 1) -component ion-exchange system.

The situation becomes less obvious when chemical reaction accompanies ion exchange. Consider, for axample, the treatment of a water containing sodium bicarbonate by a softener consisting of an ion-exchange bed in the acid form. Here, the exchange of sodium and hydrogen ions is accompanied by the formation of carbonic acid.

In addition to the concentrations of these ions, the concentrations of bicarbonate ion and carbonic acid in the fluid phase now may also have to be taken into account. (Uptake of these species by the ion exchanger in neglected). Moreover, the total electrolyte concentration changes due to formation of carbonic acid. The question then arises how these changes affect the variance of the system.

In a pure cation-exchange system in which sodium and hydrogen

ions exchange on a strong-acid resin, the equations governing e-
quilibrium are the electroneutrality equations for each phase

$$c_{Na} + c_H = C_o \qquad (20)$$

$$q_{Na} + q_H = Q \qquad (21)$$

and one ion-exchange equilibrium relation, for instance of the
ideal mass-action form,

$$(q_H c_{Na})/(c_H q_{Na}) = \alpha_{HNa} \qquad (22)$$

where the separation factor α_{HNa} is a constant. The 4 concentration variables involved permit solution of these 3 equations if
one of the variables is set, and the system is therefore monovariant.

In the bicarbonate system, the electroneutrality relation for the
fluid phase, with the additional assumption that the concentration of carbonate ion is negligible, becomes

$$c_{Na} + c_H = c_{HCO_3} \qquad (23)$$

The first dissociation equilibrium of carbonic acid is also involved:

$$c_H c_{HCO_3}/c_{H_2CO_3} = K_1 \qquad (24)$$

where K_1 is the apparent molar dissociation constant. With Eqs.
21 and 22 unchanged, these equations, plus Eqs. 23 and 24, now
constitute a 4-equation system. However, while the number of equations has increased by one, two new concentration variables, namely c_{HCO_3} and $c_{H_2CO_3}$, have been introduced. But if it can be assumed that species containing carbon are excluded from the exchanger phase, and that carbonic acid either remains entirely in solution or escapes from it in only negligible quantities, the concentration C_C of carbon atoms in solution will be conserved. Hence
we have the additional relation

$$c_{HCO_3} + c_{H_2CO_3} = C_C \qquad (25)$$

where the concentrations are taken to be molalities. In spite of
the reaction which complicates the system, the latter is therefore still monovariant.

If exchangeable cation species other than hydrogen and sodium
ions are present, there will be one additional independent ion-exchange equilibrium relation and one fluid- and one exchanger-phase concentration for each, so that, with one additional equation
and two new variables, each such counterion species will increase

the variance by one. Additional coion species, on the other hand, which neither enter the exchanger phase nor participate in a reaction, do not affect the variance because their concentrations are constant.

As an example of variance when ion exchange in accompanied by complexing, we may consider zinc ion and its chloride complexes in the presence of another cation species and a cation exchanger, extending Helfferich's treatment (1962) by inclusion of the assmption that the electrically neutral complex can be present in the exchanger in significant concentration. The negatively charged complexes and chloride ion are assumed to be practically fully excluded from the exchanger by the Donnan relation.

We then have 4 relations for complexing equilibria in the fluid phase, corresponding to the reactions

$$Zn^{++} + Cl^- = ZnCl^+$$

$$ZnCl^+ + Cl^- = ZnCl_2$$

$$ZnCl_2 + Cl^- = ZnCl_3^-$$

$$ZnCl_3^- + Cl^- = ZnCl_4^{--}$$

There will be 2 independent relations for the ion-exchange equilibria involving the 3 cation species. Also, there will be 1 distribution equilibrium for the neutral complex. With the 2 usual electroneutrality relations, there will then be a total of 9 equations, involving the 11 variables c_{Cl}, c_M, c_{Zn}, c_{ZnCl}, c_{ZnCl_2}, c_{ZnCl_3}, c_{ZnCl_4}, q_M, q_{Zn}, q_{ZnCl} and q_{ZnCl_2}, where M stands for the cation species other than zinc. The variance thus is $11 - 9 = 2$. For zinc and its chloride complexes in the presence of an anion exchanger, we have the same 4 complexing equilibria, and the same number (2) of independent ion-exchange equilibria, involving 3 anion species, 1 distribution equilibrium, and 2 electroneutrality relations, as before, totalling 9; and 10 variables, namely c_{Cl}, c_{Zn}, c_{ZnCl}, c_{ZnCl_2}, c_{ZnCl_3}, c_{ZnCl_4}, q_{Cl}, q_{ZnCl_2}, q_{ZnCl_3}, and q_{ZnCl_4}. The variance is $10 - 9 = 1$. Here again, coion species have been assumed to be virtually excluded from the exchanger.

An interesting example of invariant systems is that of trace chromatography, where traces of several counterion species accompany one counterion species present in gross concentrations. Here, the concentrations of the trace components are not variable in the sense that they do not affect the rest of the system. Since the concentrations of the gross counterion are constant, the system is unvariant

Other examples of determining variance will be given in Sections 3 and 4.

The concept of variance as treated here differs from similar concepts as developed by Shiloh (1966) and Golden (1972), and is sim

pler. In the first place, in the present discussion variance is
considered an equilibrium property of the system and is thus independent of operating conditions (feed- and presaturation compositions). This viewpoint is a consequence of also separating the
equilibrium-dependent composition-path grid (cf. Section 2.6)
from operating conditions in the evaluation of column response.
In the second place, a system may exhibit regions of different va
riance. Here, the variance of such systems is taken as the highest
variance possible for it. In the earlier treatments, the variance
has been evaluated for each region and the latter have then been
combined to permit performance prediction. The present treatment
of variance in systems involving chemical reaction is entirely
analogous to that of systems in which such reaction is absent and
has the additional advantage that its relation to the Phase Rule
is more readily apparent.

2.5 Coherence

Coherence is defined as the invariance of given compositions in a
transition zone and therefore implies that each constituent concentra
tion of a composition continues to travel with the same (constant)
velocity as the composition. Coherence is a common phenomenon,
constant- and proportionate-pattern transitions being familiar
examples. In fixed-bed sorption beds operating at equilibrium,
and with constant boundary conditions, coherence always exists in
each gradual transition. It has been shown that coherence will
also be attained in initially noncoherent systems, provided that
a downstream zone of uniform presaturation and sufficient length
is present and that the feed composition remains constant long
enough (Helfferich and Klein, 1970).
For equilibrium systems of variance greater that 1, coherence
must be invoked to make the composition velocity determinate. For
gradual transitions, coherence implies equality of the concentration velocities in each composition Hence, for a given composition,

$$d\tilde{c}_i/d\tilde{q}_i = d\tilde{c}_j/d\tilde{q}_j \qquad (26)$$

for any two components i and j of the system. This relation is
termed the "differential coherence condition". The application of
this equation to systems in which chemical reaction plays a role
will be discussed in later sections.
Concentration steps in equilibrium systems also travel together.
For them, the "integral coherence condition" applies:

$$\Delta\tilde{c}_i/\Delta\tilde{q}_i = \Delta\tilde{c}_j/\Delta\tilde{q}_j \qquad (27)$$

2.6 The composition-velocity-eigenvalue problem

As mentioned before, in systems of variance greater than 1, a com position velocity can only be determined when the direction in composition space along which this velocity is evaluated is defined. This evaluation, together with the evaluation of the composition velocity itself, amounts to an eigenvalue problem and is done with the aid of the differential coherence condition (Eq. 26).

The solution of this problem for systems of any variance can be presented in matrix form (Mangelsdorf, 1966). However, as many systems of variance greather than 2 are not readily accessible to calculation even under the local-equilibrium premise, only the so lution for bivariant systems is presented here.

The equilibrium relations can be represented by the fully general relations

$$\tilde{q}_i = f_i(\tilde{c}_i, \tilde{c}_j) \tag{28}$$

$$\tilde{q}_j = f_j(\tilde{c}_i, \tilde{c}_j) \tag{29}$$

where i and j denote the components in terms of the concentrations of which it was chosen to represent the system, and f_i and f_j are appropriate functions. Differentiation yields

$$d\tilde{q}_i = f_{ii}d\tilde{c}_i + f_{ij}d\tilde{c}_j \tag{30}$$

$$d\tilde{q}_j = f_{ji}d\tilde{c}_i + f_{jj}d\tilde{c}_j \tag{31}$$

where

$$f_{kl} = \partial f_k/\partial \tilde{c}_l \tag{32}$$

Solution of Eq. 30 for $d\tilde{q}_i/d\tilde{c}_i$ and of Eq. 31 for $d\tilde{q}_j/d\tilde{c}_j$ yields

$$d\tilde{q}_i/d\tilde{c}_i = f_{ii} + f_{ij}d\tilde{c}_j/d\tilde{c}_i \tag{33}$$

$$d\tilde{q}_i/d\tilde{c}_j = f_{ji}d\tilde{c}_i/d\tilde{c}_j + f_{jj} \tag{34}$$

and application of the differential coherence condition of Eq. 29 then gives

$$f_{ii} + f_{ij}d\tilde{c}_j/d\tilde{c}_i = f_{ji}d\tilde{c}_i/d\tilde{c}_j + f_{jj} \tag{35}$$

This is a quadratic equation in $d\tilde{c}_j/d\tilde{c}_i$, with solution

$$(d\tilde{c}_j/d\tilde{c}_i)_{s,f} = \{(f_{jj}-f_{ii}) \pm [(f_{jj}-f_{ii})^2 + 4f_{ij}f_{ji}]^{\frac{1}{2}}\}/(2f_{ij}) \tag{36}$$

Except when the discriminant vanishes, two values for $d\tilde{c}_j/d\tilde{c}_i$ are
obtained. They correspond to directions along which two composition-velocity eigenvalues will be evaluated below. The subscripts
s and f refer to "slow" (lower velocity) and "fast" (higher velocity). In general, the number of direction values obtained by the
analogous procedure for a system of any variance equals the latter.
The expressions for $d\tilde{c}_j/d\tilde{c}_i$ as given by Eq. 36 can be substituted
into Eqs. 33 or 34, to yield, with Eq. 5, the following expres-
sion for the composition-velocity eigenvalues:

$$(d\tilde{c}_i/d\tilde{q}_i)_{s,f} = (\rho_B/\varepsilon)\underline{u}_{s,f} = 2/\{f_{ii}+f_{jj} \pm [(f_{jj}-f_{ii})^2 + 4f_{ij}f_{ji}]^{\frac{1}{2}}\} \tag{37}$$

Compositions for which the discriminant of Eq. 36 vanishes yield
a single direction and a single value for the composition velocity. Such composition points have been called "watershed points"
(Helfferich, 1967), and their topology and significance have been
discusssed by Tondeur (1970), Helfferich and Klein (1970) and
Golden et al. (1974).
The directions as given by Eq. 36 apply to composition points
$(\tilde{c}_i, \tilde{c}_j)$. With the aid of this equation, a direction field can be
constructed in the $\tilde{c}_i$, $\tilde{c}_j$-plane, in which the two directions, cor
responding to the plus and minus signs can be indicated by two
short line segments, each pair going through a number of selected
points. This direction field can then be integrated graphically
by drawing families of representative curves through it in such
a manner that their directions interpolate the directions of the
corresponding direction segments in their vicinity. The grid of
curves so obtained has been termed "composition-path" grid. An
example will be found further below.
It is apparent from Eq. 36 that an entire such path corresponds
either to the lower or higher composition-velocity eigenvalue indicated for each of its points. Therefore, the paths may be described as "slow" and "fast". They may be distinguished by representing them as solid and dashed curves, respectively.
Integration methods other than the direction-field method may of
course be used. Suitable numerical methods for computer calculations can readily be devised. In the case of equilibrium relations
of the constant-separation-factor or Langmuir type, the paths are
linear and a transformation to orthogonalize them has been given
(Helfferich and Klein, 1970; Rhee et al., 1970).
The composition-velocity eigenvalue of bivariant systems can also
be represented graphically, in the form of contour plots. Again,
such contours corresponding to the smaller eigenvalue for a given
point may be drawn as solid lines, and contours corresponding to
the larger engenvalue, as dashed lines.

Utilization of such plots for column-performance prediction will
be discussed briefly in the following section. Special problems
that arise when additional phases appear or disappear during the
reaction-exchange process will be illustrated in Section 4.

2.7 Fixed-bed response for bivariant systems

Because of the extreme irregularities in isotherm surfaces possible in fixed-bed systems in which chemical reaction occurs, pre
diction of column response can become challenging for bivariant
systems, and prohibitively difficult for systems of greater variance. Up to now, only ad-hoc solutions for specific complex bivariant problems have been developed. A unified procedure is yet
lacking. Examples are given in Sections 3 and 4. Here, only an
outline to the solution approach, covering some possible cases,
will be given.
The diagrams (or their computer-memory equivalents) usually requi
red to predict the column response to arbitrary feed and presaturation compositions comprise q_1- and q_2-contours, composition-
path, and composition-velocity-contour grids in the c_1, c_2-plane.
When it is possible to go from the feed-composition point F to
the presaturation-composition point P by first following a slow
path and then a fast path, and if the composition velocity increa
ses along these paths in the direction of progression, then all
that is required to construct a generalized profile is to plot
concentration against relative velocity for a sufficient number
of points along the paths. The point of intersection of the two
paths, where the composition velocity changes discontinuously,
corresponds to a middle plateau zone bounded by the two composi-
tion-velocities corresponding to the eingenvalues at the intersection point.
When the change of composition velocities along a slow and a fast
path is not monotonic, the resulting transition must be a least
partly abrupt. If both transitions are fully abrupt, then a combi
nation of the integral coherence condition (eq. 27) and of the
equilibrium relation (eqs. 28 and 29) permits two curves to be
constructed that pass through F and P, and whose intersection represents the composition of the intermediate plateau zone. Such
curves are tangent to the composition paths at their points of
origin ("anchor points"), but then in general deviate from the
composition path. They are not determined uniquely by the equilibrium relations, as are the composition paths, but depend as well
on their point of origin (Klein et al., 1967).
The existence of transitions that are partly gradual and partly
abrupt gives rise to a number of problems that exceed the scope
of the present discussion. For these cases, Golden's Rule, applied
separately to the slow and fast transition, can serve as criterion
of correctness of a hypothetical composition route.

The procedure of obtaining ideal effluent-concentration histories is analogous. Here, concentrations are plotted against the reciprocal of the relative composition velocity instead of against the composition velocity as such, to yield a universal effluent--concentration history.

3. REACTION WITHOUT PHASE CHANGE

3.1 Introduction

Chemical reactions accompanying ion exchange can cause the formation or disappearance of additional phases. The present section is limited to the discussion of systems in which such phase changes do not occur. Of numerous systems of this type that are of practical or theoretical interest, or both, only a few types are selcted here for analysis with various levels of detail. Three examples of monovariant systems include the neutralization of bicarbonates or carbonates by strong- or weak-acid resins in the hydrogen form, and binary ion exchange accompanied by complexing. Next, the variance of the trace-chromatographic separation of the complexed rare-earth ions is examined. The discussion of a three-component ion-exchange system with two complexing counterion species concludes the section as an example of a bivariant case.

3.2 Neutralization of bicarbonate with strong-acid exchangers

The treatment of waters containing bicarbonates with strong- or weak-acid cation exchangers in the hydrogen form is a practical example of ion exchange accompanied by reaction. Such treatment takes place in the first bed of a two-bed deionization train, and in simultaneous softening, bicarbonate removal, and total-dissolved-solids reduction, where the acid required does double duty as regenerant and neutralizing agent.
We first treat the case of solutions containing hydrogen ion, one other cation species, bicarbonate ion, and other anion species, and examine the variance of systems involving such solutions and either strong-acid or weak-acid exchangers. We shall then characterize the types of effective isotherms that can arise, and their consequences for column performance. Finally, we shall briefly discuss extension of the methods developed to certain systems of higher variance.
For the system involving strong-acid exchangers, with sodium ion as example of the second cation species, the point relations of primary interest comprise (1) the dissociation equilibrium of carbonic acid, (2) electroneutrality in the solution phase, (3) electroneutrality in the exchanger phase, (4) the ion-exchange equilibrium, (5) conservation of carbon in the solution phase (with the

premises that anionic species and carbonic acid are completely
excluded from the exchanger phase,and that the pressure in the
system is sufficiently high to keep carbonic acid in solution),
and (6) conservation of nonreacting anions. These relations contain the solution-phase concentrations of sodium, hydrogen, and
bicarbonate ions and carbonic acid, and the total concentration
of nonreacting anions; and the exchanger-phase concentrations of
sodium and hydrogen ions as variables, totaling seven in number.
Given the value of any one of these variables, the system can in
principle be solved for all the others, i.e., is monovariant. It
is thus possible, for example to establish the exchanger-phase
concentration of sodium as a function of its solution-phase concentration, and to obtain an effective isotherm which reflects
the combined effects of ion exchange and chemical reactions; in
this case the neutralization of bicarbonate ion or the dissociation of carbonic acid.

The discussion of the general properties of such an isotherm is
facilitated by considering a possible experiment involving addition of increments of exchanger in the hydrogen form to a given
volume of solution. After allowance of enough time for equilibrium
to be approached sufficiently closely, the solution, the exchanger, or both, would be analyzed for the constituents of interest.
If the total initial concentration of the solution is C_0, and its
concentration of species containing carbon is C_C, then, because
of hydrolysis, the solution will be slightly alkaline until an
amount of exchanger has been added that contains hydrogen ion e-
quivalent to the bicarbonate ion initially present in the solution. In this composition region, because of the strong tendency
of bicarbonate ion to form carbonic acid, the exchanger will lose
all of its hydrogen ion and be completely in the sodium form as
the concentration of sodium in solution becomes smaller in propor
tion to the amount of exchanger added. On the effective isotherm,
this process will be represented by the line $y_{Na} = q_{Na}/Q = 1$, extending between $c_{Na} = C_0 - CC$ and $c_{Na} = C_0$.

When more exchanger has been added than is stoichiometrically e-
quivalent to the amount of bicarbonate initially in solution,
further neutralization cannot take place and the total strong e-
lectrolyte concentration of the solution will remain constant at
the value $C_0 - C_C$, so that then, exchange of sodium and hydrogen
ions will take place unaffected by reaction and governed by the
isotherm applicable to the total strong-electrolyte concentration
prevailing. Finally, as the excess of exchanger approaches infini
ty, its sodium concentration, and the sodium concentration in solution will approach zero, at the lowest point of the isotherm.

For a given exchanger, the equilibrium relation for the exchange
of sodium and hydrogen, unaccompanied by reaction, can be characterized by the general relation

$$y_{Na} = f(c_{Na}, C_0) \tag{38}$$

where f is a suitable function. In the composition region with
reduced strong-electrolyte concentration, the latter is the total
concentration of nonreacting anions, $C_{an}=C_0-C_C$. Thus, to obtain
the isotherm relation for this region, C_0 in Eq. 38 must be replaced by $C_0-\dot{C}_C$. This yields

$$y_{Na} = f(c_{Na}, C_0 - C_C) \tag{39}$$

Constant-separation-factor isotherms as given by Eq. 22 represent
a particularly simple case. For them, if we let

$$x_{Na} = c_{Na}/C_0 \tag{40}$$

Eq. 38 becomes

$$y_{Na} = f(c_{Na}/C_0) = f(x_{Na}) \tag{41}$$

This means that the isotherm in terms of x_{Na} and y_{Na} is independent of the total solution normality. Given the value of x_{Na},
however, c_{Na} is dependent on the total solutions concentration,
so that, in the zone of interest.

$$y_{Na} = f(x_{Na}) = f\left[c_{Na}/(C_0 - C_C)\right] \tag{42}$$

The isotherm in this zone can therefore be obtained by multiplying c_{Na} in Eq. 41 by $C_0/(C_0-C_C)$, which is equivalent to compres-
sing the abscissae of the isotherm as given by Eq. 41 in the ratio $(C_0-C_C)/C_0$.
Various types of effective isotherm that can be obtained in the
manner indicated are shown in Fig. 8. Diagram a in this figure
shows the isotherm for a pure bicarbonate solution. The equilibrium here is "irreversible", practically regardless of the ion-
-exchange selectivity of the exchanger because of the strong tendency of bicarbonate ion to combine with hydrogen ion. Diagram b
is an effective isotherm for a solution of sodium bicarbonate and
other salts, and an exchanger selective for sodium. C_{an} is the
concentration of anions other than bicarbonate. Diagrams c and d
apply to similar solutions but with exchangers selective for hydrogen ion. Two cases are distinguished, according as the diagonal $y_{Na}=c_{Na}/C_0$ does not or does intersect the isotherm. For
isotherms without inflection points, intersection will ensue if

$$\lim_{c_{Na}, y_{Na} \to 0} dy_{Na}/dc_{Na} < 1/C_0 \tag{46}$$

where the derivative is the slope of the isotherm at the origin.
The role of this intersection in affecting breakthrough behavior
will be discussed below.
Effective isotherms based on ion exchange isotherms with one or
more inflection points have been excluded from consideration here,

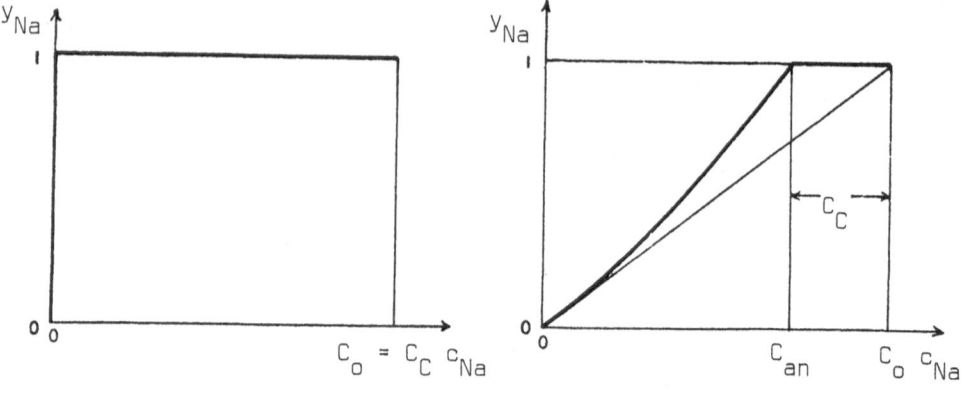

a. $C_o = C_C$. $MHCO_3$ solution

c. Exchanger selective for H. Non-intersecting diagonal. $C_C < C_o$

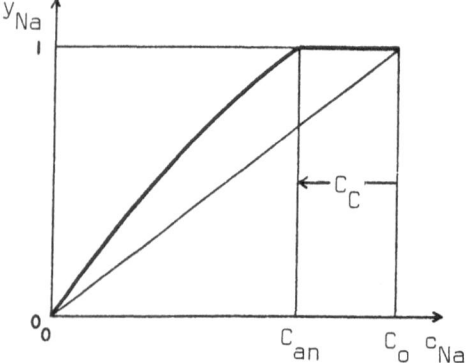

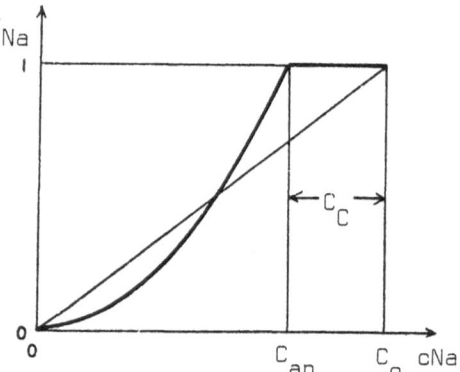

b. $C_C < C_o$. Exchanger selective for M

d. $C_C < C_o$. Exchanger selective for H. Intersecting diagonal

Fig. 8. Types of effective isotherms corresponding to the exchange of hydrogen and sodium ion on a strong-acid exchanger in the presence of bicarbonate ion

but their treatment, by analogy to the cases discussed, offers no
difficulty.
Starting with an effective isotherm, generalized sodium concentra
tion profiles or effluent-concentration histories may now be cons
tructed with the aid of Golden's Rule. The course of other concen
trations may then be established with the aid of the applicable
point relations.
Examples of the use of Golden's Rule, based on the isotherm of
Fig. 8, are shown in Fig. 9, where the slope of the dashed lines
determines the composition or step velocity according to Eq. 5 or
8. The feed-composition point in each case is the extreme right
upper point of the effective isotherm, $(C_0, 1)$. In Diagrams a, b
and c, the construction indicates a single abrupt transition of
velocity $\varepsilon C_0 / \rho_B Q$. In Diagram d, the curved and linear segments of
the route meet at a point such that the tangent to the isotherm
goes through the feed-composition point. The reciprocal of the
slope of the linear segment correspondes to the relative velocity
of the abrupt transition, according to the relation

$$\underline{u}_\Delta = \varepsilon(C_0 - c'_{Na})/\left[\rho_B Q(1 - y'_{Na})\right] \tag{47}$$

where the primes denote the coordinates of the point of tangency.
Similarly, the reciprocals of the slopes of the curved segment
correspond to the adjusted composition velocity, according the
relation

$$\underline{u} = \varepsilon dc_{Na}/(\rho_B Q dy_{Na}) \tag{48}$$

Fig. 10 shows the effluent-concentration histories derived from
Fig. 9. Only in Diagram d is sodium seen to leak ahead of the
abrupt transition. In two-bed deionization, this condition is undesirable because, if saturation is stopped at the breakthrough
of the abrupt transition, any sodium that has leaked out of the
cation-exchanger bed will appear in the final product, as it cannot be romoved in the second column, which contains only anion ex
changer. With the premise of equilibrium operation, sodium leakage
can be expected to be enhanced by the conbination of an exchanger
selective for hydrogen ion and a raw water of low bicarbonate con
tent. By the same token, a higher bicarbonate concentration of the
raw water may permit employment of an exchanger of intermediate
acid strength, and thus higher regenerant utilization.
The preceding discussion has been limited to raw waters containing
a monovalent metal ion (e.g. sodium ion) as the only cation species. In many cases, this restriction can be relaxed provided the
additional cations present are multivalent, that their concentrations are not too large, and that the equilibrium for the sodium-
-hydrogen exchange is not too unfavorable. A quantitative criterion, valid under the premise of equilibrium operation, is develo
ped below.
At the relatively low total raw-water concentrations of potential

392

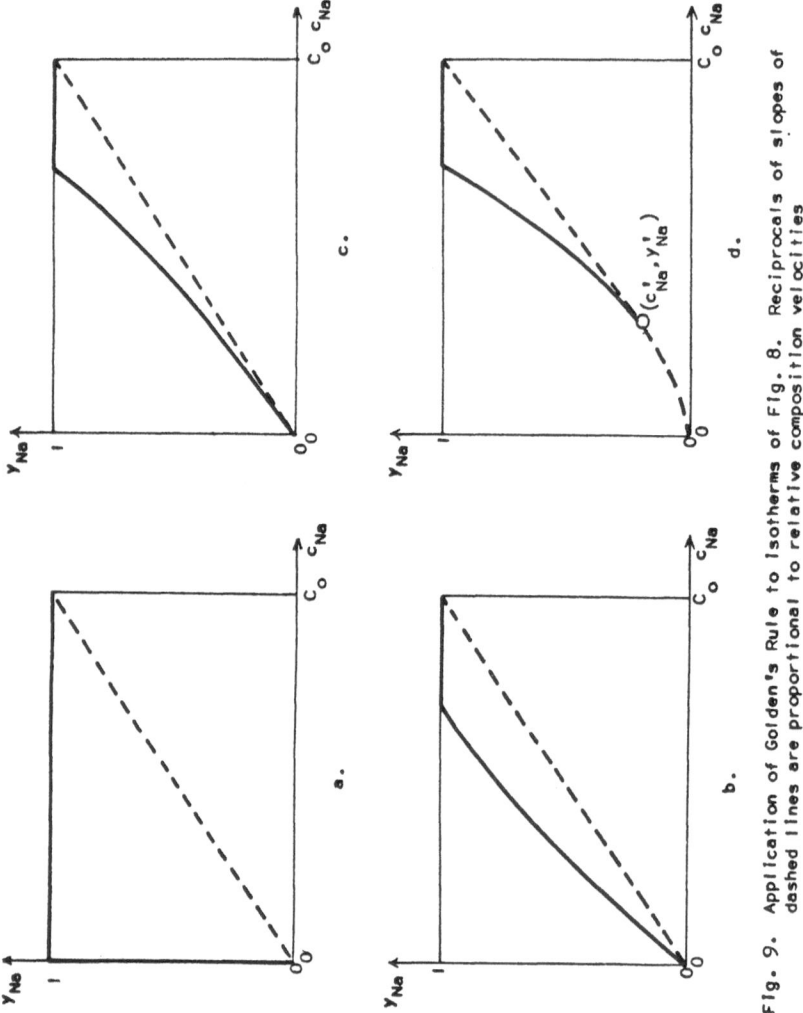

Fig. 9. Application of Golden's Rule to Isotherms of Fig. 8. Reciprocals of slopes of dashed lines are proportional to relative composition velocities

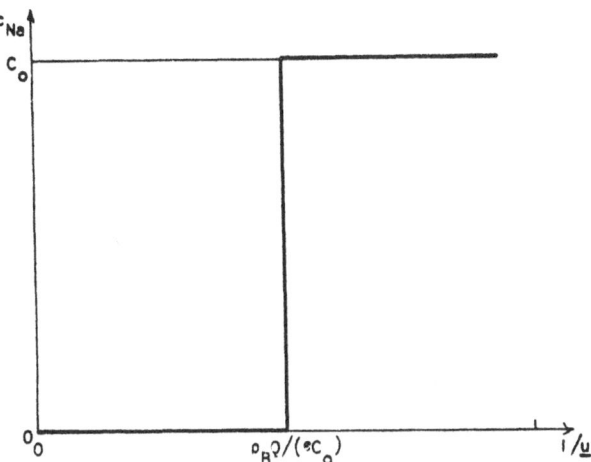

a. Effluent-concentration history for Figs. 9a, b, c

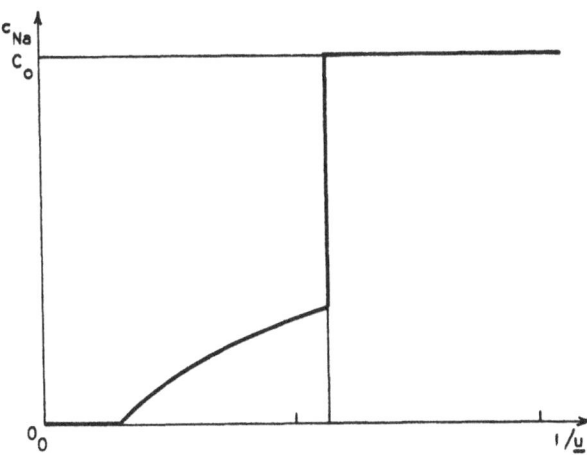

b. Effluent-concentration history for Fig. 9 d

Fig. 10. Effluent-concentration histories derived from Fig. 9

practical interest, the selectivity of the exchanger for multivalent cations may be expected to be so strong that these ions will
occupy virtually the entire exchange capacity in a zone near the
column inlet, of length such that its local capacity equals the
total number of equivalents of multivalent ions in the feed water
applied during a given interval of time. Within this zone, the
familiar frontal-analysis pattern will develop and only a trace
of sodium will be present. Downstream of the zone of multivalent
cations, there will be a zone containing sodium as the only cation; and further downstream, the column will contain exchanger
in the hydrogen form. A gradual transition may also appear, in
which the exchanger is partly in the hydrogen and partly in the
sodium form, as shown above.

The transition between the multivalent-cation and the sodium zones will tend to be very sharp, because of the high selectivity
of the exchanger for even the least strongly held of the multivalent cations relative to sodium. As long as the transition between
the sodium and the hydrogen zones involves only the exchange of
sodium and hydrogen ions, and the neutralization of bicarbonate
ion, its analysis need not include that of a multicomponent ion-
-exchange system. This will be the case whenever the diagonal
$y_{Na} = C_0$ does not intersect the compound isotherm for sodium (cf.
Figs. 8 and 9 $\underline{a}$, $\underline{b}$, $\underline{c}$; and 10 $\underline{a}$). When the diagonal $\underline{does}$ intersect the isotherm, interference of the least strongly held multi-
valent cation with the sodium-hydrogen exchange will be absent if
$\underline{u}_\Delta$, as determined from the compound sodium isotherm, equals or $\underline{ex}$
ceeds the adjusted velocity of the multivalent cation front, i.e.,
when

$$\left[\varepsilon C / (\rho_B Q) \right] (1 - c_{Na}^0 / C_0) \leqslant \underline{u}_\Delta .$$

In these expressions, c_{Na}^0 is the sodium concentration in the raw
water. If this criterion is not met, the problem assumes one or
more additional dimensions and its treatment exceeds the scope of
the present discussion.

3.3 Reaction and ion exchange involving carbonic acid,
      bicarbonates, carbonates, and weak-acid exchanger

Weak-acid ion-exchange resins exhibit a high degree of regenerant
utilization and high selectivity for calcium and magnesium ions.
For these reasons, they are finding increasing application in the
removal of mineral hardness and of carbonate and bicarbonate ions
from water (cf., e.g. Kunin and Vassiliou, 1963, and Bresler et
al., 1974). A preliminary theoretical and experimental study of
the response of weak-acid resins to carbonate and bicarbonate solutions has been made by Klein et al. (1978A).

In this work, the dissociation of bicarbonate ion can been taken into account, in addition to the dissociation of carbonic acid. This adds one relation to the six for strong-acid exchangers and bicarbonate. Carbonate ion is an additional species, however, so that this additional equilibrium relation does not affect the variance.

Another new feature in the present case is the weak-electrolyte nature of the exchanger, which contributes the concentrations of dissociated and undissociated ion-exchange groups as two additional variables. However, the equivalence of the sum of these variables to the total exchange capacity and the dissociation equilibrium for the exchanger phase compensate for this increase in the number of variables, so that the variance is still unchanged.

The equilibria governing the reactions in the solution phase are sufficiently familiar not to require further discussion here. As a specific example of the reactions in the exchanger phase, we consider cation exchange involving sodium and hydrogen ions, as being of direct practical interest. The derivation for other monovalent metal ions and for weak-base anion exchangers is analogous, and extension to multivalent counterions requires only obvious modifications.

The dissociation of the weak-acid exchanger may be represented by the reaction

$$HR = H^+ + R^-$$

where R stands for the (usually resinous) exchanger and R^- designates a dissociated ion-exchange group. The corresponding ideal mass-action equilibrium expression is

$$\frac{q_H q_R}{q_{HR}} = \frac{1 - \epsilon}{\rho_B} K = K_{resin} \tag{50}$$

Here, K is the molal dissociation constant in mols per liter of exchanger phase, and K_{resin} the dissociation constant in milliequivalents per gram of resin.

The ionized resin groups are occupied by sodium and hydrogen ions, so that

$$q_R = q_{Na} + q_H \tag{51}$$

Moreover, the total exchange capacity Q, in milliequivalents per gram of exchanger, is occupied by individual species as indicated by the following relation:

$$q_H + q_{Na} + q_{HR} = Q \tag{52}$$

With Eqs. 51 and 52, Eq. 50 now becomes

$$\frac{q_H(q_{Na} + q_H)}{Q - q_H - q_{Na}} = K_{resin} \tag{53}$$

On neglecting q_H as compared to q_{Na}, one obtains the approximate relation

$$\frac{q_H q_{Na}}{Q - q_{Na}} \cong K_{resin} \tag{54}$$

We now extend Helfferich's assumption of a constant-separation--factor relation for the exchange of dissociated hydrogen and sodium ions (Helfferich, 1962) to the case where the separation fac tor is not necessary equal to 1, to obtain

$$q_H = \alpha_{HNa}(q_{Na}/c_{Na})c_H \tag{55}$$

where α_{HNa} is the dimensionless separation factor.
When, instead of sodium, there is a metal ion M of valence ν_M ins tead of sodium ion, the apparent mass-action law becomes

$$K_{HM} = (q_H/c_H)^{\nu_M}/(c_M/q_M) \tag{56}$$

where K_{HM} is a constant selectivity coefficient. Eq. 55 can then be replaced by the more general relation

$$q_H = \left[K_{HM}(q_H/c_M)\right]^{1/\nu_M} c_H \tag{57}$$

For the constant-separation-factor case, substitution of the expression of Eq. 55 for q_H into Eq. 54 and solution of the resulting equation yields the following expression for q_{Na}-contours on a c_{Na}-vs-c_H plot:

$$c_{Na} = (\alpha_{HNa}/K_{resin})\left[q_{Na}^2/(Q - q_{Na})\right] c_H \tag{58}$$

In terms of the equivalent fraction y_{Na} of sodium ion on the exchanger, this equation assumes the form

$$c_{Na} = (\alpha_{HNa}Q/K_{resin})\left[y_{Na}^2/(1 - y_{Na})\right] c_H \tag{59}$$

The y_{Na}-contours, each representing a constant value of y_{Na}, are seen to be straight lines radiating from the origin. By virtue of Eq. 8, Q being constant, they are also contours of constant values of $q_H + q_{HR}$ or $y_H + y_{HR}$, the y's again being the corresponding equiva lent fractions. The linearity of the y-contours results from the constant-separation-factor equilibrium as expressed in Eq. 54.

For mass-action-type equilibria between heterovalent ions, the contours are curved. Specifically, for exchange between mono- and divalent ions, they are parabolic.

It follows from Eq. 59 that a unique grid of representative y_{Na}- contours is obtained if c_{Na} is plotted against $(\alpha_{HNa}Q/K_{resin})c_H$, as has been done in Fig. 11.

The compositions corresponding to the interaction of a weak acid and its salt are constrained to a curve in the c_H, c_{Na}-plane. For the simple case of acetic acid, this may be demonstrated as follows.

The ideal mass-action relation reflecting the dissociation equilib rium is

$$c_H c_{Ac}/c_{HAc} = K_{HAc} \tag{60}$$

HAc stands for acetic acid. Conservation of the species containing a carbon atom in the liquid phase is expressed by the equation

$$C_C = c_{HAc} + c_{Ac} \tag{61}$$

where C_C is the constant sum of the molalities of species containing carbon. The electroneutrality relation is

$$c_{Na} + c_H = c_{Ac} + c_{OH} + C_{an} \tag{62}$$

where C_{an} is the sum of the normalities of the anions of any strong acids present. This term is included here to allow for the possibility of the presence of neutral salts.

Together with the ideal dissociation equilibrium for water, where the dissociation constant is designated by K_w, Eqs. 60 through 62 may now be solved to relate c_{Na} to c_H, the constants C_C and C_{an} being taken to be know. The result is

$$c_{Na} = -c_H + \frac{C_C K_{HAc}}{c_H + K_{HAc}} + \frac{K_w}{c_H} + C_{an} \tag{63}$$

The analogous expression for the carbonic-acid system is

$$c_{Na} = c_H + \frac{C_C(c_H + 2K_2)}{c_H + (c_H^2/K_1) + K_2} + \frac{K_w}{c_H} + C_{an} \tag{64}$$

Here, K_1 and K_2 are the dissociation constants of carbonic acid and bicarbonate ion. Corresponding expressions for tribasic acids such as phosphoric acid are readily derivable.

In c_H, c_{Na}-coordinates, Eq. 63, 64 and analogous expressions describe a family of C_C-contours for a given value of C_{an}. A change in the value of C_{an} engenders a parallel translation of this grid. For any problems of practical interest, the first and third

398

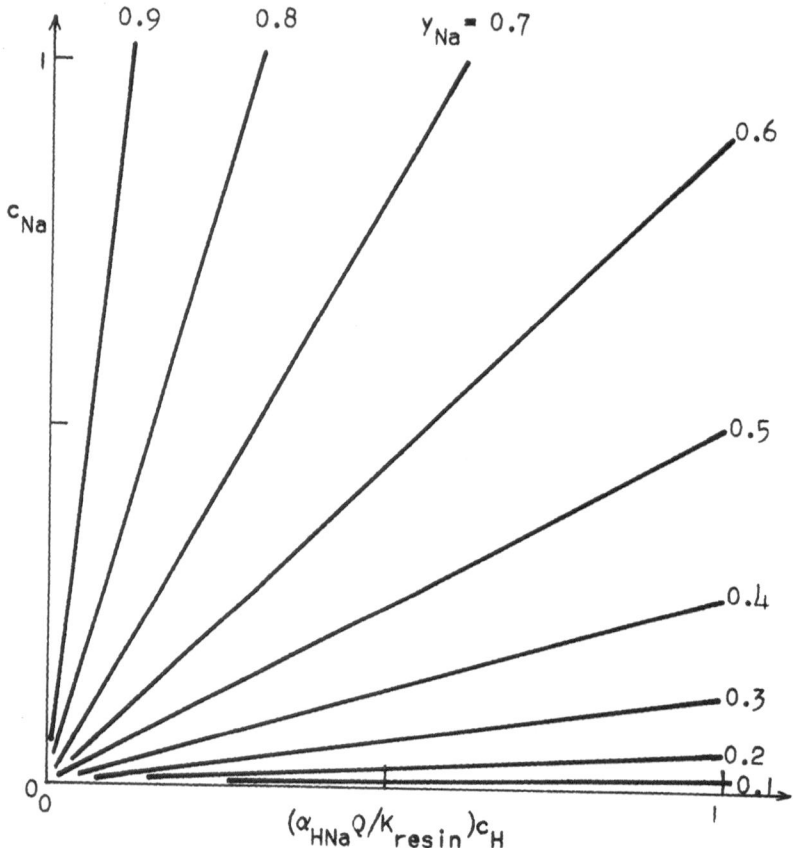

Fig. 11.   Ideal Na-isotherm surface for Na-H exchange on weak-
acid resin.   Slopes of y_{Na}-contours calculated with
Eq. 59

terms of the right-hand sides of these equations are negligible
as compared to the second term, so that for any particular system
$(c_{Na} - c_{an})/c_C$ then virtually is a unique function of c_H. Each of
these contours represents a part or all of a composition path. An
example for carbonic acid is shown as the dashed line in Fig. 12.
To predict the equilibrium performance of a column with respect
to a given feed solution containing a weak electrolyte, its salt,
or both, the value of the group $(\alpha_{HNa}Q/K_{resin})$ must first be established through appropriate equilibrium measurements. The
isotherm surface specific for the resin of interest in then represented by selected contours obtained with Eq. 59 and the composition path for the desired acid and total solution concentration
is constructed, as was done in Fig. 13 for an acrylic acid resin
and 0.126 formal carbonic acid. From the isotherm-surface plot, a
path isotherm can then be constructed by plotting solid-phase concentrations against the fluid-phase concentrations corresponding
to them along the path. In terms of hydrogen-ion concentrations,
this was done in Fig. 13, which was based on Fig. 12. From the
path isotherm, the column response is finally obtained with the
aid of the differential and integral material balances of Eqs. 5
and 8, and Golden's Rule. The effluent-concentration history so
derived from Fig. 13 is shown in Fig. 14.

3.4 Binary ion exchange with complexing

Examination of binary ion exchange in which one or both counterion
species complex with one or more coion species is of interest because it reveals the effect of complexing on isotherm configura-
tion, and therefore on the fixed-bed performance corresponding to
various combinations of feed and presaturation compositions.
Equations applicable to the exchange of two counterion species,
one or both of which form an electrically neutral complex with
the sole coion species present, have been given by Shiloh (1966).
The coion species alone or as part of a complex has been assumed
to be completely excluded from the exchanger.
The 5 (or 6) point relations involved in such systems are expressions of electroneutrality for both phases, expressions for one
(or two) complexing equilibria, one ion-exchange equilibrium, and
for conservation of the total concentrations of the complexed and
uncomplexed coion species. The variables involved are c_M, c_N, c_X,
c_{NX} (and c_{MX} if two complexes are formed), q_M, and q_N, totalling
6 or 7, as the case may be, M and N being the counterions, and X
the coion. These systems are therefore monovariant.
We consider here in detail the first type of system, in which
only one complex is formed. Moreover, we assume all ions to be monovalent and their concentrations to obey equilibria of the ideal
mass-action type. We also include the possibility that a second,
noncomplexing, coion species Y be present in addition to the
complexing one.

400

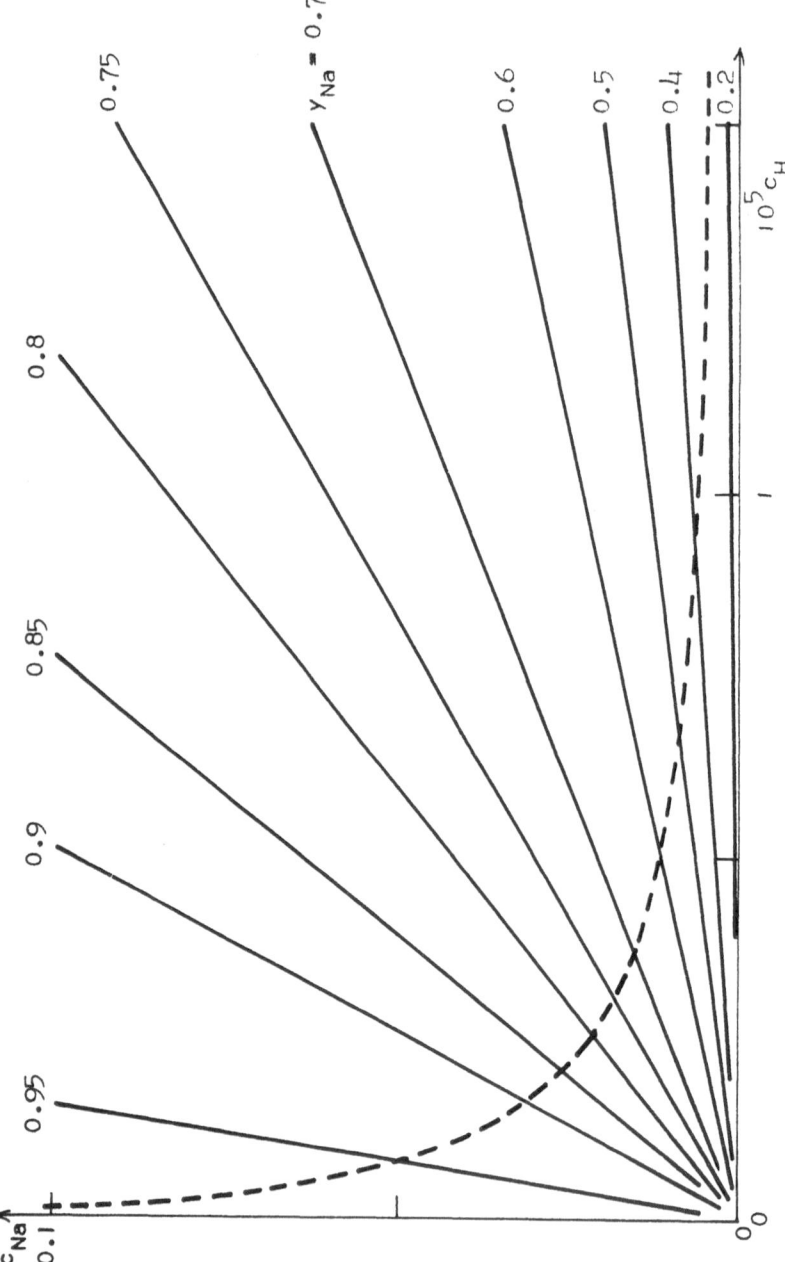

Fig. 12. y-contours, and composition path for 0.126 formal carbonic acid

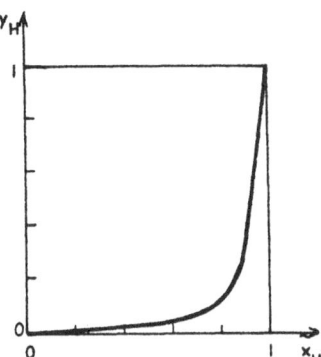

Fig. 13. Path isotherm for H-Na exchange with 0.126 formal carbonic acid feed solution and acrylic-acid resin. $(x = 1 - c_{Na}/c_C;$ $y_H = 1 - y_{Na})$

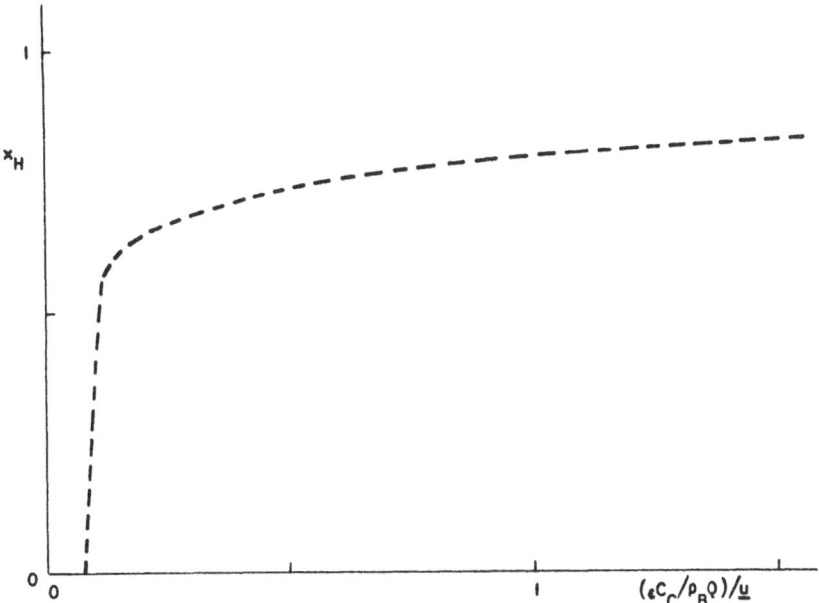

Fig. 14. Ideal response of exchanger bed in Na-form to 0.126 formal carbonic acid feed. Constructed from path isotherm of Fig. 13. Abscissa is proportional to reciprocal of composition velocity

We write the ion-exchange equilibrium in the form

$$q_M c_N / (q_N c_M) = \alpha_{MN} \tag{65}$$

The complexing-equilibrium relation may be represented in the usual manner by

$$c_{NX} / (c_N c_X) = K_{NX} \tag{66}$$

α_{MN} and K_{NX} are assumed to be constant. Conservation of X in the mobile medium corresponds to

$$c_X + c_{NX} = \bar{C}_X \tag{67}$$

$\bar{C}_X$ is a constant, representing the total concentration of X (complexed or noncomplexed) in the column feed. Electroneutrality of the solution is expressed by

$$c_M + c_N = c_X + C_Y \tag{68}$$

where C_Y is the combined constant concentration of noncomplexing coions present. In the exchanger phase, electroneutrality corresponds to

$$q_M + q_N = Q \tag{69}$$

Eq. 65 through 69 may be combined to yield

$$2c_X = - (C_Y + a) + \left[(C_Y + a)^2 + 4a\bar{C}_X \right]^{1/2} \tag{70}$$

where

$$a = \left[1 + \alpha_{NM} q_M / (Q - q_M) \right] / K_{NX} \tag{71}$$

α_{NM} being the reciprocal of α_{MN}.
To obtain pairs of values of c_M and q_M for plotting, values of c_X are calculated for arbitrarily selected values of q_M. Successive use of Eqs. 67, 66 and 68 then yields values of c_{NX}, c_N, and c_M. To show the effect of complexing of a counterion N for which the exchanger is selective ($\alpha_{MN}=0.2$), this method has been used to calculate the two intermediate curves of Fig. 15. For this figure, Q has been taken to be 2.0 milliequivalents per gram of exchanger, and the values of $\bar{C}_X$ and C_Y each to be 0.1 moles per liter. The lower of these curves corresponds to slight complexing ($K_{NX}=25$); the upper to a moderate stability constant ($K_{NX}=200$). The uppermost curve finally is for nearly complete complexing ($K_{NX}=10^5$). While this curve was computed by the method indicated, it could also have been constructed on the basis of the following considerations.
Near the corner $c_M=\bar{C}_X+C_Y$, $q_M=Q$ of the diagram, there is only a

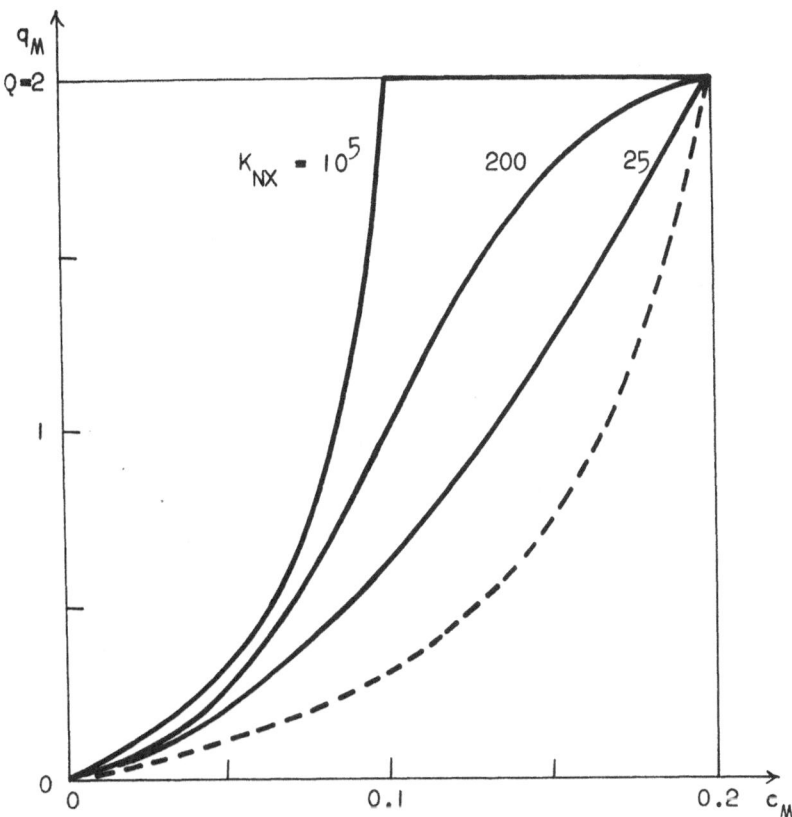

Fig. 15. Effect of complexing of counterion N on effective
isotherm for counterion M. $\alpha_{MN} = 0.2$, $Q = 2$ meq/g,
$\tilde{C}_X = C_Y = 0.1$ molar. Dashed curve is isotherm
for ion exchange without complexing

trace of N present. Most of this will be complexed because the
complexing ion X⁻ is present in excess and equilibrium strongly
favors complexing. As we proceed in the direction of decreasing
c_M, c_{NX} increases, because any N present is still complexed comple
tely. In this region, N thus does not compete with M for exchange
sites, and $q_M=Q$. At c_M-values below C_Y, however, all X has been
consumed as ligand and N is now free to participate in simple ion
exchange with M. In the present case, this exchange follows a
constant-separation-factor equilibrium, given by

$$q_m(C_y - c_M)/\left[c_M(Q - q_M)\right] = \alpha_{MN} \tag{72}$$

When the complexing ion is the only coion species present, the
isotherm for M would hug the q_M-axis and then follow the line
$q_M=Q$, thus corresponding to "irreversible" equilibrium.
Also shown on Fig. 15, as the dashed curve, is the constant-separation-factor isotherm for M,N-exchange in the absence of complexing, i.e., for the case $C_Y=c_M+c_N$.
Viewed together, this set of curves shows that suitable complexing can modify an ion-exchange isotherm significantly, making
it less unfavorable for ligands of moderate complexing power, and
making it favorable when the complexing tendency is sufficiently
large. The proportion of noncomplexing coions also affects the
isotherm in a predictable manner. Column-performance prediction
for any of these isotherm types and any combination of feed- and
presaturation compositions can readily be made with the aid of
Golden's Rule.

3.5 Multicomponent trace chromatography with complexing

Elution chromatography is a separation technique based on the selectivity of the sorbent for the components of a mixture. A volume of a solution of these components, small enough to occupy only
a relatively small fraction of the bed, is introduced first. Next,
the chromatogram is "developed". This means that a solution of a
component is applied for which the sorbent has less affinity than
for the components in the initial charge. The passage of the deve
loping solution has the effect of resolving the original mixture
into bands containing the individual components in pure form,
which travel at rates related inversely to their affinity for the
sorbent.
To make a separation feasible, the bed depth and time (or the
eluent volume) required must remain within practical limits. This
means that the ratios of the effective distribution coefficients
for any two components must not be too close to unity, and that
none of the coefficients themselves must be excessively large.
Neither of these conditions is met for the separation of the noncomplexed chlorides of the trivalent rare-earth cations on conven
tional strong-acid cation exchangers. However, Ketelle and Boyd

(1947) have solved this challenging problem by elution with a solution of a citric acid - ammonium citrate buffer. In forming
complexes with yttrium and the rare-earth cations, citrate exhibits greater specific selectivity between the ions of adjacent ra
re-earths than does the ion exchanger. This reduces uptake of the
uncomplexed ions by different factors, and thereby increases the
ratio of their velocities. In addition, the reduction of all the
distribution coefficients increases all the band velocities.
Inasmuch as 15 rare-earth ions, yttrium, their (probably neutral)
complexes, hydrogen, citrate, and ammonium ions, and citric acid
are involved, the system appears to be multivariant. In fact,
however, this is not so, as the following discussion will show.
In Ketelle and Boyd's experiments, the initial charge of the rare-
-earths and yttrium mixture was introduced into the top of a cation-exchanger bed in the hydrogen form. The only cation species
present in gross concentration in the eluent was ammonium ion,
which as can be shown from the experimental results, displaces hy
drogen ion from the exchanger ahead of any of the rare-earth and
yttrium bands. The latter therefore travel as traces in an ammo-
nium-ion environment.
The experimental effluent-concentration history for the elution
step exhibits a single peak for each component of the mixture,
suggesting the absence of charged complexes. Moreover, if adorp-
tion of any neutral complexes is neglected, the distribution coef
ficients are then given by $\lim\limits_{\tilde{c}_i, \tilde{q}_i \to 0} d\tilde{q}_i/d\tilde{c}_i$ (which, with Eq. 5, determine the band velocities). $\tilde{c}_i$ is the sum of the concentrations
of the uncomplexed and complexed forms, but the former is practically negligible as compared to the latter.
The significant species present in solution are ammonium, hydrogen, and dihydrogen citrate ions, citric acid, and yttrium and
the 15 rare-earth-ion species with their 16K complexes, K being
the number of complexes formed by yttrium and each of the rare-
-earth ions. Only ammonium and dihydrogen citrate ions, and citric
acid are present in major concentrations, and these concentra
tions, as well as the concentration of hydrogen ion, can be set
by proper formulation of the composition of the eluent solution.
There remain 16+16K trace-concentration variables associated with
the solution phase.
As already mentioned, ammonium ion will saturate the bed ahead of
any yttrium and rare-earth bands, so that the concentration of
ammonium ion in the exchanger virtually equals the known and cons
tant exchange capacity of the latter. There thus remain the 16
concentrations of yttrium and the rare-earth ions in the exchanger as variables related to this phase.
The relevant equations connecting these variables are 16K complexing-equilibrium relations and 16 equilibrium relations for the
exchange of yttrium and the rare-earth ions with ammonium ion.
The excess of the number of variables over the number of relations
connecting them is 16, and one might be tempted on superficial

examination to consider the variance of the system to be 16. However, in the trace region, with which we are concerned here, if one exchangeable ion gives rise to one complex, the value of the distribution coefficients $\tilde{q}_i/\tilde{c}_i$ will be unaffected by the value of $\tilde{c}_i$. With Eq. 5, the velocities of the individual yttrium and rare -earth bands then are independent of the initial quantities of these components and in spite of its apparent complexity, the sys tem may be considered as invariant.

If more than one exchangeable ion is present in the complex, $\tilde{c}_i$ will not be proportional to c_i, and, since in the trace region, $\tilde{q}_i$ is proportional to $\tilde{c}_i$, $\tilde{q}_i$ will not be proportional to c_i. The individual distribution coefficients then will depend on the initial amounts of the exchangeable ions present. In this case, the system may be regarded as a combination of 16 uncoupled monovariant systems.

3.6 Bivariant ion exchange with complexing

The effect of coion complexing on fixed-bed ion exchange involving gross concentrations of three counterion species has been studied theoretically by Shiloh (1966), and by Golden et al. (1974). In addition to illustrating the solution of such bivariant problems with specific models, this work demonstrated the kinds of effects that chemical reaction accompanying ion exchange can have on the diagrams characterizing a system, and on its fixed-bed performance.

In the cases examined, there were assumed to be present three monovalent exchangeable cation species, M, N and O, of which the first two form a neutral complex with the single, monovalent, coion species X. Complete exclusion of X and of its complexes from the exchanger phase was also assumed.

In preliminary calculations with various values of the separation factors for ion exchange and of the complex stability constants, Golden et al. (1974) found that up to three selectivity reversal loci could arise. For illustrative purposes, they selected a set of values that would give the maximum number of such reversals, namely $\alpha_{MO}=300$, $\alpha_{NO}=10$, $K_{MX}=800$, and $K_{NX}=20$.

In solving this problem, the concentrations of the counterion species M and N in the exchanger phase were selected as independent variables, and the solution-phase concentrations of each of these ions and of their complexes were added to yield effective combined fluid-phase concentrations for use in the coherence condition and in material balances yielding concentration and step velocities. The eigenvalue problem was then solved as indicated by Eq. 37, but with the c's and q's interchanged. Representative composition-paths, and contour curves of constant combined fluid--phase concentrations and of composition velocities were plotted against exchanger-phase concentrations, using the results of computer calculations.

The composition-path plot is shown in the triangular diagram of Fig. 16. Some of the paths are seen to be strongly curved. Since, for simple three-component ion-exchange systems governed by constant-separation-factor equilibrium, the composition paths are linear, this curvature is attributable to the effect of the complexing reactions.

For any intersection of paths (shown or interpolated), the composition velocity corresponding to the path shown as solid curve is lower that the velocity corresponding to the path shown as dashed curve. This has led to the short respective designations of "slow" and "fast" paths for these curves.

The arrows shown on the paths indicate the direction of increasing composition velocity. On the slow paths, including all three borders of the diagram, velocity extrema occur at the points marked by small crosses. At such points, the isotherms along the path have inflection points, and this is important in performance prediction, because, in practically all cases, any transition spanning an inflection point must be at least partly abrupt. On the border of the diagram opposite Corner N, there lie the points W_{N1} and W_{N2}, and on the border opposite Corner M, there lies the point W_M. These points are "watershed points" (Helfferich and Klein, 1970), at which the composition-velocity eigenvalues corresponding to either family of paths, and the slopes of the paths, coincide.

All fast paths are observed to start at points on the border opposite Corner N, between W_{N1} and W_{N2}. Paths do not penetrate into the triangle from Points W_{N1} and W_{N2}, but penetrating paths exist through Point W_M, connecting it to Points W'_{N2} and W'_{N1} on the border opposite Corner N.

The topology of bivariant composition-path grids has been examined by Tondeur (1971), who used conceptually orthogonalized diagrams to facilitate comparison of various systems. Fig. 17 represents the grid of Fig. 16 in this manner.

Another system, based on the constants $\alpha_{MO}=150$, $\alpha_{NO}=10$, $K_{MX}=10,000$ and $K_{NX}=150$, has been calculated by Shiloh (1965). The isothermal surfaces for this system exhibit two selectivity-reversal lines (where $\tilde{c}_i q_j = \tilde{c}_j q_i$) instead of the three of the system just discussed. The composition-path topology in both systems in identical, except that, in the second system, Component O takes place of Component M, and vice versa, so that the Points M, W_M, and O in Fig. 17 are replaced by the respective Points O, W_O, and M.

The existence of three watershed points in either system is at variance with the number of such points in many simple ion-exchange systems. Three-component ion-exchange systems governed by constant-separation-factor equilibrium, for example, have only one watershed point.

408

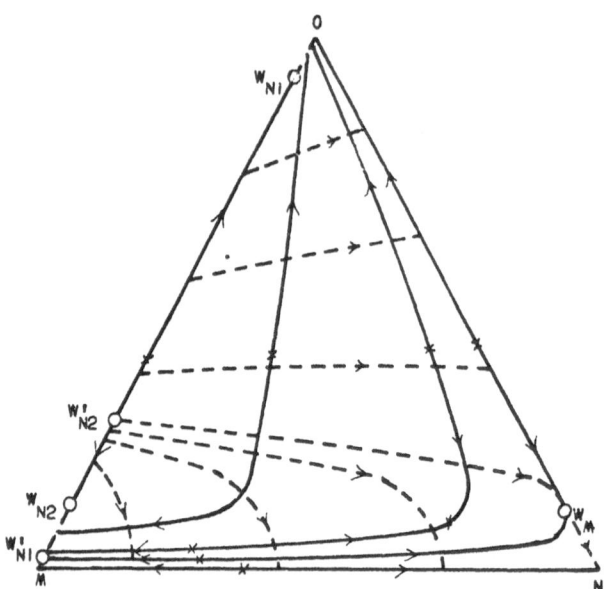

Fig. 16. Representative composition paths on exchanger-phase composition grid, for exchange of monovalent counterions M, N, and O, of which M and N form neutral complexes with monovalent coions X. Solid lines are "slow"; dashed lines, "fast" paths. Arrows indicate direction of increasing composition velocity. W_M, W_{N1}, and W_{N2} are watershed points. (After Golden et al., 1974)

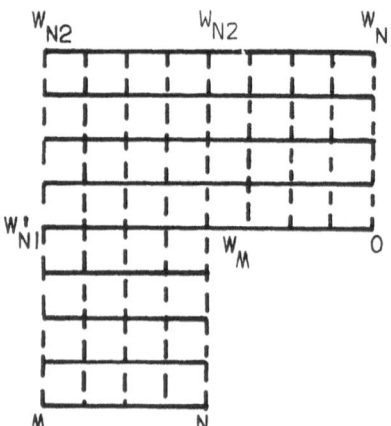

Fig. 17. Schematic composition-path topology corresponding to Fig. (after Golden et al., 1974)

4. REACTION WITH PHASE CHANGE

4.1 Introduction

Reactions accompanying ion exchange can result in phase changes, such as the formation and dissolution of precipitates and gases. Desirable and undesirable examples of precipitate formation have been given in Section 1. A familiar example of the formation of a gaseous phase as a result of a chemical reaction is the neutralization of bicarbonate ion by a cation exchanger in the hydrogen form.

The presence of one or more phases in addition to the exchanger and solution phases requires modification of the differential or integral material-balance and coherence conditions (Eqs. 5 or 8 and 15 or 16), and the modification depends on the velocities with which the additional phases move. Examples of types of movement are precipitates that cling to the exchanger particles, or trapped gases, that are stationary, precipitates fully entrained by the fluid phase, precipitates with a lower average velocity that that of the fluid phase, gases in upflow operation that move faster than the fluid phase, and gases that tend to rise in downflow operation, that move slowly than the fluid phase but could in certain cases have negative velocity. The case of arbitrary velocities has been treated in principle by Golden (1972); the discussion here will be limited to additional phases that are either stationary or move with the same axial velocity as the fluid phase.

As in ion exchange accompanied by reactions without phase change, the systems of interest here may be classified according to their variance. In the present section, we shall examine a monovariant system with moving precipitate and a bivariant system with both moving and stationary precipitates.

4.2 Monovariant ion exchange with formation or dissolution of mobile precipitate

We first consider a binary ion-exchange system in which a monovalent counterion species N may form a precipitate with the sole, monovalent, coion species, X. We assume the precipitate to be fully mobile and all equilibrium relations in terms of concentrations to be of the ideal mass-action type. The modifications required to extend this treatment to heterovalent ions are obvious. The case of precipitates formed by the counterion and several coions will be discussed further below.

Cases can be imagined in which precipitate is present throughout the entire bed during the half cycle of interest. However, such cases are not likely to be of practical interest. Moreover, they can be considered as a subcase of the more general systems in which part of the bed is undersaturated with respect to precipitate. In this type of system, the coion concentration c_X will be cons-

tant in the undersaturated region and decrease as the concentration c_N of N increases beyond the value at which c_X and c_N satisfy
the solubility-product relation. Thus, an effective isotherm in
terms of the concentrations of the nonprecipitating counterion
species may be expected to consist of a saturation branch at low,
and a pure-ion-exchange branch at high, concentrations of M in the
solution and the exchanger phases, c_X increasing with c_M along the
saturation branch, and remainning constant along the pure-ion-
-exchange branch.
The point relations governing the latter are the constant-separation-factor relation of Eq. 65, and the electroneutrality relations of Eqs. 68 (with $c_Y=0$) and 69; and the variable concentrations are c_M, c_N, q_M, and q_N. c_X has a constant value determined
by the operating conditions. Along the saturation of the effective isotherm, the additional solubility-product relation applies,
adding one equation, but also one variable, namely c_X, which now
is no longer constant. For both branches of the isotherm, the sys
tem is thus monovariant.
The effective isotherm is constructed by first plotting both the
saturation and the pure-ion-exchange isotherms. In the present ca
se, the formulation of the material-balance equations required
for performance prediction is simplest if this is done in terms
of the concentrations of the nonprecipitating counterion species,
M. Since the saturation branch must lie to the left, and the pure-
-ion-exchange branch to the right, of the intersection of the two
isotherms, the effective isotherm is obtained by ignoring the
nonapplicable parts of the component isotherms. An example of the
resulting effective isotherm is shown in Fig. 18, for
$\alpha_{NM}(=1/\alpha_{MN})=20$, and a maximum value of c_M equal to $10\sqrt{K_{spNX}}$,
where K_{spNX} is the solubility product of the precipitate ·NX.
Solution of the pure-ion-exchange branch of the isotherm offers
no difficulty. For the saturation branch, the point relations can
be combined to yield

$$y_m = 1/\left[1 - \alpha_{NM}/2 + (\alpha_{NM}/2)\sqrt{1 + 1/\xi^2}\right] \tag{73}$$

where

$$\xi = c_M/(2\sqrt{K_{spNX}}) \tag{74}$$

Provided $\alpha_{NM} > 2$, isotherms as given by Eq. 73 have an inflection
point at

$$\xi_\sim = 1/\sqrt{A^2 - 1} \tag{75}$$

where $\xi_\sim$ is the abcissa of the inflection point, and

$$A \equiv \left[3 + \sqrt{9 - 8(2/\alpha_{NM} - 1)^2}\right]/\left|4(1 - 2/\alpha_{NM})\right| \tag{76}$$

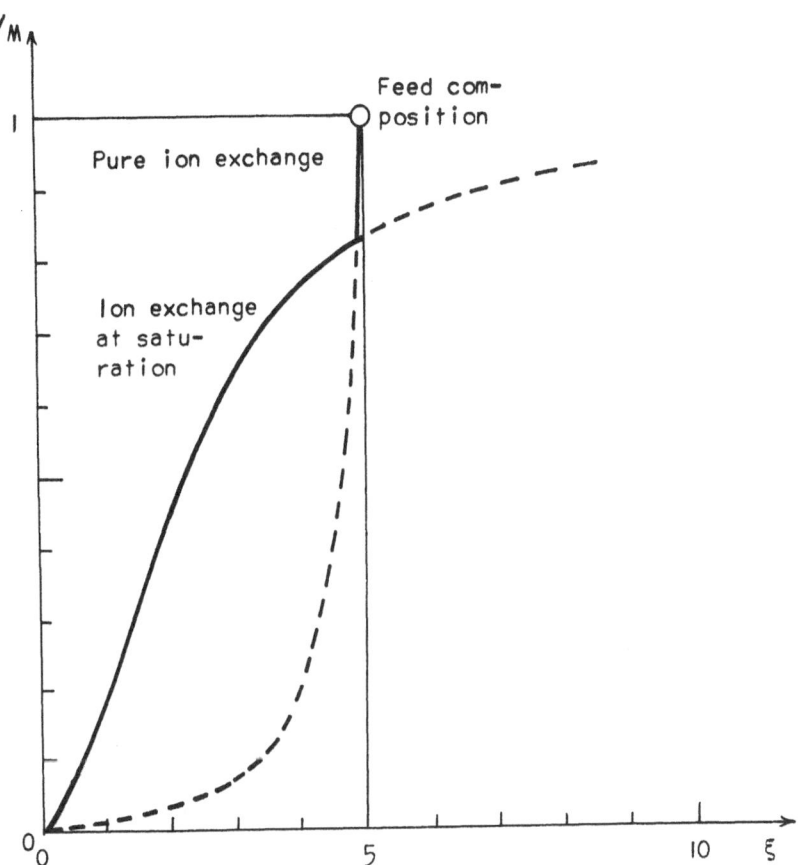

Fig. 18. Compound isotherm for ion exchange and precipitation. $\alpha_{NM} = 20$

As is to be expected, for precipitates of low solubility, the isotherm in terms of the concentrations of the nonprecipitating counterion species tends to become irreversible, i.e., of the form

$$0 < y_M < 1 \qquad \text{for } c_M \cong 0 \qquad\qquad (77)$$

$$y_M \cong 1 \qquad \text{for } c_M > 0 \qquad\qquad (78)$$

As the effective isotherm approaches irreversibility, its pure-ion-exchange branch will tend to represent a less and less in significant segment, and this branch can then be represented to a close approximation by the tangent to it at the maximum concentrations of the nonprecipitating counterion species that occur in the system.

Effective isotherms for binary systems in which one of the counterion species forms precipitates with more than one coion species have been examined by Golden (1972). If precipitates NX and NY are formed, for example, where Y is the second coion species, the variance in the region in which both precipitates exist is unchanged because the additional concentration variable, c_Y, is balanced by an additional solubility-product expression. In regions in which only one of the precipitates or no precipitate exist, the system is also monovariant, as before.

To obtain an effective isotherm, individual isotherms are constructed for pure ion exchange, for ion exchange in the presence of NX, and of NY, and of both precipitates. The appropriate segments of each isotherm are then combined to form the effective isotherm. Performance prediction for all these cases may be achieved with the aid of Golden's Rule, based on the types of effective isotherms discussed, which were constructed in terms of the concentrations of a nonprecipitating species. This gives rise to the question as to how cases could be handled in which all the counterion species of the system can precipitate.

Here, combined fluid-phase concentrations of the type

$$\tilde{c}_M = c_M + c_{MX} \qquad\qquad (79)$$

where M is the counterion, and X the coion forming the precipitate MX, would be used in the material-balance relations. The isotherm of interest, to which Golden's Rule can be applied, then would be plotted in terms of $\tilde{c}_M$ and q_M.

To consider the simplest case of this type, we examine a binary ion-exchange system in which both counterion species form precipitates with one coion species. In this case, the electroneutrality equation for the mobile-phase medium, and the two solubility-product relations alone suffice to determine the fluid-phase concentrations of both counterion species and of the coion species. The system thus is invariant, and a process cannot occur and therefore cannot be predicted.

4.3 Bivariant ion exchange with formation or dissolution of precipitate

An example of a bivariant ion-exchange system that in principle can involve either a stationary or a moving precipitate is incorporated in the intriguing desalination process proposed by Popper et al. (1963). We discuss this process here primarily for the new concepts which it illustrates. The process utilizes a mixed bed of a cation and an anion exchanger, regenerated, without previous separation of the resins, with the aid of a lime slurry.
In the exhaustion step, here idialized for simplicity, the solution to be desalted contains sodium chloride as the only solute. As the solution passes through the exchanger bed, sodium ion displaces calcium ion from the cation exchanger, and chloride ion displaces hydroxide ion from the anion exchanger. The equilibria governing these exchanges are made greatly more favorable for partial deionization through the precipitation of calcium hydroxide, which occurs simultaneously.
The product is a lime slurry in a saturated solution of calcium hydroxide, which, because of the relatively low solubility of the latter, can have a markedly lower electrolyte concentration than a not too dilute feed solution. The solids present in the product stream can be removed by mechanical means, and the dissolved solids can be either removed by deionization, or exchanged for innocuous ions. Such a secondary treatment would require smaller equipment and resin inventory, and less regenerant, thant the main step.
The practical problem to which this process addresses itself is that of making the desalting of fairly concentrated brines economically possible by using a regenerant which is not only inexpensive but also recyclable. It was thus proposed to regenerate with a saturated suspension of calcium hydroxide in a brine.
On the surface, this seemed thermodynamically impossible, for, if at the end of the exhaustion step, the resins were at equilibrium with solid calcium hydroxide, no further reaction could be expected to occur when a regenerant consisting of a suspension of calcium hydroxide in the feed solution was contacted with them.
Four factors, however, appeared to justify a theoretical analysis. The first was the proponents' claim that they had produced net desalted water in sustained cyclic operation. Second, at the end of the exhaustion step, not the entire bed, but only one of several zones in the bed, is at equilibrium with the feed as well as with solid calcium hydroxide. Third, the solubility of calcium hydroxide varies with the total electrolyte concentration. Finally, instead of a single regeneration step, the thermodynamically more plausible possibility was considered of carrying the regeneration out in two steps. In the first of these, a slurry of calcium hy-

droxide in feed brine would saturate the resins partially with
calcium and hydroxide ions. In the second step, regeneration
would be carried further by use of a slurry in fresh water.
Calculations based on the local-equilibrium premise and on both
the fixed- and moving-precipitate assumptions, carried out by
Page (1968, 1971), and reported by Page et al. (1975), appeared
to indicate that, under certain conditions, the amount of fresh
water produced could exceed the amount required for regeneration.
Here, however, we are not so much concerned with the potential
practical utility of this process as with the principles underlying it because these may find application in the analysis or
design of other processes with analogous features.

The exhaustion step of the calcium-hydroxide process may be regar
ded as an example of the more general case of any ion-exchange
step in which precipitate is being formed. Quite generally, column
zones will then arise in which the solution is either saturated
or undersaturated with respect to precipitate, and the boundaries
of these zones will be different for the cases of stationary and
mobile precipitate. We examine first the limiting ideal case in
which the precipitate is stationary and completely insoluble.

Here, precipitation will begin near the column inlet and then con
tinue in a front advancing toward the outlet end of the bed. The
exchange sites freed by the precipitating ions will take up salt
ions from the solution, so that, for the example of the calcium-
-hydroxide process, if equivalent amounts of the cation and anion
exchangers are employed, the solution issuing from the downstream
front of the zone containing precipitate will be deionized. At
this front, ion exchange and precipitation occur simultaneously.
Because precipitation is stoichiometric in this case, ion-exchange
equilibria do not affect the process.

If the precipitate is not completely insoluble, it will redissol-
ve in the undersaturated feed solution in an upstream region of
the bed, and the early effluent (after displacement of the column
voids), instead of being completely deionized, will be a saturated
solution of the precipitating ions. Moreover, when the precipitate
is somewhat soluble, the process is no longer governed exclusively
by stoichiometry, and the ion-exchange equilibrium relations will
affect the overall equilibrium. In an attempt to balance this
asymmetric effect, non-equivalent amounts of the exchangers are
used, and this, in turn, introduces a number of theoretical compli
cations. These will not be further considered here, as they do not
affect the concept of a precipitate zone between an upstream redis
solution front and a downstream precipitation and ion-exchange
front. A schematic representation of the stationary-precipitate
case with somewhat soluble precipitate is given in the left-hand
diagram of Fig. 19.

In the moving-precipitate case, again with the idealizing assumption of complete insolubility, precipitate is present downstream
of the ion-exchange front, since the precipitate, as soon as it is
formed, is swept on with the velocity of the mobile medium, which

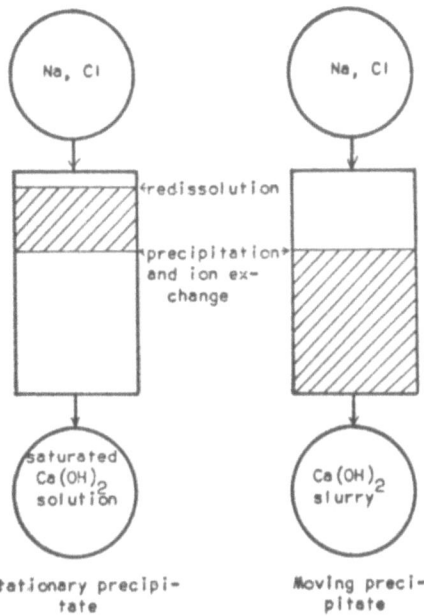

Fig. 19. Idealized behavior of stationary and moving precipitates in calcium hydroxide process. Equivalent amounts of exchangers, initially in calcium and hydroxide forms. Shaded zones contain precipitate

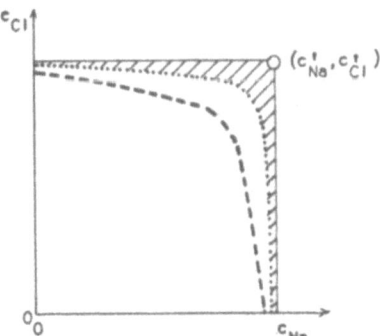

Fig. 20. Domain bound (dotted curve) and regime bound (dashed curve) in calcium hydroxide deionization process (schematic). $c'_{Na} = c'_{Cl}$ is the maximum electrolyte normality of the system

exceeds that of the ion-exchange and precipitation front. The effluent is an aqueous slurry of calcium hydroxide. A schematic
diagram of this process is shown as the right-hand part of Fig.
19.
In the stationary-precipitate case, undersaturated solution exists
upstream of the re-dissolution front; in the moving-precipitate
case, upstream of the precipitation front. In these undersaturated
zones, cation and anion exchange proceed independently, and the
total solution normality is constant. These zones thus are equivalent to two uncouple monovariant systems existing side by side.
In the saturated zones, the applicable point relations may be
exemplified by the solubility-product expression,

$$c_{Ca} c_{OH}^2 = 2K_{sp} \tag{80}$$

the ion-exchange equilibrium relations,

$$(q_{Ca}/c_{Ca})/(c_{Na}/q_{Na})^2 = K_{CaNa} \tag{81}$$

$$(q_{OH}/c_{OH})/(c_{Cl}/q_{Cl}) = \alpha_{OHCl} \tag{82}$$

and the electroneutrality relations,

$$q_{Ca} + q_{Na} = Q \tag{83}$$

$$q_{OH} + q_{Cl} = Q' \tag{84}$$

$$c_{Ca} + c_{Na} = c_{OH} + c_{Cl} \tag{85}$$

Here, the quantity of any ion entering into a concentration variable is measured in equivalent, and K_{sp} is the solubility product, Q, the capacity of the cation exchanger, and Q', that of the
anion exchanger. As there are 6 equations, involving 8 concentration variables, the system is bivariant. It is convenient to
choose the concentrations of the nonprecipitating species Na and
Cl to represent the system.
We consider a c_{Na}, c_{Cl}-plane of a system in which the concentrations of these ions in a certain pure NaCl solution are $c_{Na} = c'_{Na}$
and $c_{Cl} = c'_{Cl} = c'_{Na}$. If ion exchange with calcium or hydroxide ions
occurs, c_{Na}, c_{Cl}, or both, can only become smaller. For suitable
combinations of the extent to which cation and anion exchange
occur, the product of the molalities of the calcium concentration, and of the square of the hydroxide-ion concentration will
exceed K_{sp}, so that one may expect the diagram to consist of two
domains; one in which the solubility product relation is satisfied,
and one in which it is not. Such domains are shown schematically

in Fig. 20, where their boundary is represented by a dotted curve. This curve must simultaneously obey the electroneutrality relation for simple ion exchange, i.e.,

$$c_{Na} + c_{Ca} = c_{Cl} + c_{OH} = C_o = c'_{Na} = c'_{Cl} \qquad (86)$$

and the solubility-product relation (Eq. 80); hence, the relation

$$(C_0 - c_{Na})(C_0 - c_{Cl})^2 = 2K_{sp} \qquad (87)$$

which is linear in c_{Na} for any set value of c_{Cl}. Points on the curve can thus readily be obtained by calculating the values of c_{Na} corresponding to a selected set of values of c_{Cl}.

The area bounded by the resulting curve and the axes corresponds to the saturated domain and the (shaded) area bounded by the curve and the lines $c_{Na}=c'_{Na}$ and $c_{Cl}=c'_{Cl}$, to the undersaturated domain. As has already been mentioned, the saturated domain is bivariant, and the undersaturated domain may be considered to be composed of two independent monovariant systems.

For regeneration with a lime slurry, all points of interest on the c_{Na}, c_{Cl}-plot are in the bivariant zone, and performance predictions can be carried out most simply with the aid of material balances for sodium and chloride ions, which are not present in the precipitate. For the exhaustion step, however, the valid composition route in the c_{Na}, c_{Cl}-plane somehow has to bridge the domain bound, and this gives rise to an abrupt transition leading to compositions that lie on a curve which is distinct from the domain bound and which has been called "regime-bound locus" (Page, 1968). Such a curve, again schematically, is shown as the dashed line of Fig. 20.

The relations defining the regime-bound locus include the integral coherence condition (Eq. 27), written for sodium and chloride ions, and the 6 point relations (Eqs. 80 through 85). If the composition upstream of the transition corresponding to the regime--bound locus is known, this amounts to a system of 7 equations in 8 variables and therefore defines a curve in the c_{Na}, c_{Cl}-plane. Points on this curve can be obtained by adopting successive arbitrary values for the relative velocity of the composition step in which precipitation occurs, determining the corresponding upstream compositions, and calculating the downstream composition with the aid of the equations mentioned, and Eq. 8. u_Δ thus serves as a parameter in the construction of the regime-bound locus.

The performance calculations undertaken to date have been of the ad-hoc type and a complete composition-path grid has not been constructed. Such a grid would be required as basis for exhaustive composition-route search.

It is of practical interest to establish precipitate inventories. The amount of solid calcium hydroxide present as a function of time and distance can be calculated with the material balances of Eqs. 5 or 8, where, for moving precipitate,

$$\tilde{c}_{Ca} = c_{Ca} + c_{Ca(OH)_2}; \quad \tilde{q}_{Cl} = q_{Ca} \tag{88}$$

and, for stationary precipitate,

$$\tilde{c}_{Ca} = c_{Ca}; \quad \tilde{q}_{Ca} = q_{Ca} + q_{Ca(OH)_2} \tag{89}$$

Here, $c_{Ca(OH)_2}$ and $q_{Ca(OH)_2}$ are the respective numbers of equivalents of calcium in the precipitate, per liter of fluid medium, or
per kilogram of exchanger, ρ_B is the bulk density of the cation
exchanger in the packed bed. Analogous balances can be made for
hydroxide ion.

For the moving-precipitate case, any dissolution or precipitate
will release calcium and hydroxide ions into the solution phase,
and these ions can then either remain in solution, thus increasing
its concentration, or exchange for sodium or chloride ions,
without additional change in total solution normality. The drop in
"precipitate normality" thus balances the increase in solution
normality. The reverse relation applies to forming precipitate.
In general, then, in the case under consideration.

$$\Delta C = -\Delta c_{Ca(OH)_2} \tag{90}$$

where C is the total (variable) solution normality.
For the stationary-precipitate case and abrupt transitions, Eq. 8,
written in turn for calcium (using Eq. 89) and sodium, can be com
bined with Eq. 11 and 12 to yield

$$\underline{u}_\Delta = \varepsilon \Delta C / \left[\rho_B \, q_{Ca(OH)_2} \right] \tag{91}$$

Here, C_0 has been replaced by C to indicate that we are again
dealing with a variable total solution normality. Differential ma
terial balances that account for precipitate are governed by analogous considerations.

It is of interest to examine the results of the specific calculations mentioned above. They covered various proportions of anion
and cation exchangers, a range of total feed concentrations, and
the cases of moving and stationary precipitate. To conserve the
fresh water required for the preparation of the slurry used in the
second regeneration step, this step was carried only far enough to
result in a product water having a total-dissolved-solids level of
about 750 ppm (0.013 normal).

An anion-exchanger capacity about 50 percent in excess of the capacity of the cation exchanger was found best, and operation based
on the moving-precipitate assumption yielded about 15 percent more
net product than operation based on the stationary-precipitate as
sumption. For moving precipitate, the calculations indicated that
about 6.9 and 2.4 bed volumes of net water could be produced per

cycle for 0.15 and 0.20 normal feed brine, respectively. At about
0.30 normal feed brine, the calculated net production level fell
to zero.

5. SUMMARY

Chemical reactions accompany ion exchange in many actual and proposed processes of practical interest. Such reactions include neu
tralization and dissociation, complex formation, and precipitation.
The treatment of ion exchange with chemical reaction becomes analogous to that of other sorption system if certain concentration
variables are appropriately combined for use in material balances.
The systems of interest may be classsified according to variance,
i.e., the minimum number of concentration variables that must be
set to define the system at any point, when the system is at equi
librium.
For fixed-bed operation with local equilibrium, the velocity of a
given composition is found to have a number of eigenvalues equal
to the variance. If the bed is presaturated uniformly and receives
a feed stream of constant composition, these eigenvalues corresponds to successive zones of gradual composition changes, or
transitions. The system of equations permitting the solution of
the eigenvalue problem includes point relations (such as ion-
-exchange equilibria, and stoichiometry and electroneutrality equa
tions) and the differential coherence condition. The treatment of
abrupt composition changes or transitions is analogous. Here, step
velocities instead of composition velocities are found, and an in
tegral coherence condition is invoked.
The velocity of a concentration or of a conentration step is obtai
ned with the aid of a material balance. If such velocities are re
lated to the apparent velocity of the mobile medium passing the
composition or step of interest, relative velocities are obtained,
from which ideal concentration profiles and effluent-concentration
histories can be obtained readily. Such performance prediction is
particularly simple for monovariant systems, of which ordinary binary ion exchange is an example. Here, the relative velocities of
interest are inversely proporcional to the slopes of tangents or
chords of the applicable isotherm. For bivariant systems, solution
of the eigenvalue problem yields two families of composition
paths, which define the sequences of compositions possible in gra
dual transitions. In general, two paths pass through each point
in the composition field, each corresponding to a different compo
sition velocity. The exception is watershed points, at which these
velocities coincide.
Three examples have been presented to monovariant systems in which
phase changes do not occur. These include binary ion exchange accompanied by complexing, and the familiar and practically important neutralization of carbonates and bicarbonates by exchangers
in the hydrogen form. The variance of the trace-chromatographic

separation of the complexed rare-earth ions has also examined. An
example of a bivariant system without phase change was afforded
by three-component ion exchange accompanied by the formation of
complexes involving one coion- and two counterion species.
Examples of reaction with phase change are the formation and dissolution of precipitates and gases. Here, performance prediction
will depend on the mobility of the additional phase. Only the cases or fully mobile or fully stationary such phases have been considered in detail.
Examples have been presented of monovariant ion exchange with formation or dissolution of mobile precipitate. They include binary
ion exchange in which one of the counterion species forms a precipitate with one or two coion species. Finally, a proposed mixed-
-bed calcium-hydroxide deionization process has been discussed as
an example of bivariant ion exchange with formation or dissolution of precipitate. New concepts in this type of systems are the
use of mobile- and stationary-precipitate concentrations in material balances, and domain-bound and regime-bound loci. The domain
bound separates the regions in the composition field in which the
solubility product relation does and does not hold. The regime-
-bound locus defines the spectrum of compositions possible in the
plateau zone in which precipitate exists, as based on the feed
composition and the integral coherence condition.

ACKNOWLEDGMENT

Thanks are expressed to the National Science Foundation for partial support of this work, and to Eileen Smith for the preparation
of the typescript.

SYMBOLS

c_i	concentration of Species i in fluid medium (per unit volume of medium)
C	total (variable) solution normality
C_i	constant mobile-phase concentration of Species i
C_0	total solution concentration
K, K_{HAc}, K_1, K_2,...	apparent molar dissociation constants
K_{ij}	selectivity coefficient for counterion species i and j
K_{MX}, K_{NX}	complex stability constants

K_{resin}	dissociation constant of a weak-electrolyte ion-exchange resin (milliequivalents per gram of resin)
K_{sp}	molar solubility product
q_i	concentration of Species i in sorbent phase
Q, Q'	total exchange capacity (per unit weight of sorbent). Q' refers to an anion exchanger
S	cross-sectional area of bed
t	time
u	average axial component of linear velocity of fluid medium in sorbent bed
u_{c_i}, u_{q_i}	velocity of concentrations c_i, q_i
$\underline{u}$	relative composition velocity
$\underline{u}_\Delta$	relative velocity of a concentration step
x_i	equivalent fraction of Species i in solution
y_i	equivalent fraction of Species i in stationary medium
z	variable distance along axis of bed

Greek letters

α_{ij}	separation factor (cf. Eq. 22)
ϵ	void fraction of sorbent bed
ν_i	valence of Ion Species i
ρ_B	bulk density

Other symbols

f, s	(subscripts) fast or slow, as related to higher or lower of two composition-velocity eigenvalues
o	related to feed composition

refer to the respective composition at downstream and upstream and of a transition

indicates combined concentration

REFERENCES

Acrivos, A. (1956), "Method of Characteristics Technique", Ind.Eng.
Chemistry,48,703-1).

Applebaum, S.B. (1968), "Demineralization by Ion Exchange", Acade
mic Press, New York and London, 1968.

Bresler, S.A., F. Hussein , and E. Kreusch (1974), "Field Testing
of improved Ion-Exchange Techniques", Office of Saline Water Report INT-OSW-RDPR-74-970.

Cherney, S. (1966), "Scale-Prevention Studies and Feasibility
Analysis of Sea-Water Softening by Fixed-Bed Ion Exchange", M.S.
thesis in Chemical Engineering, University of California,
Berkeley.

DeVault, D. (1943), "The Theory of Chromatography", J.Am.Chem.Soc.,
65,532-40.

George, D.R., J.M.Riley, and L.Crocker (1967), "Preliminary
Process Development Studies for Desulfating Great Salt Lake Brines
and Sea Water", Bureau of Mines Report of Investigations 6928.
U.S.Dept. of the Interior, Washington, D.C..

Glueck, A.R. (1968), "Desalination by an Ion-Exchange-Precipita-
tion-Complex Process", Desalination,4,32-7.

Golden, F.M. (1972), "Theory of Fixed-Bed Performance for Ion
Exchange Accompanied by Chemical Reaction", Ph.D.dissertation,
University of California, Berkeley.

Golden, F.M., K.I.Shiloh, G.Klein, and T.Vermeulen (1974), "Theory
of Ion-Complexing Effects in Ion-Exchange Column Performance",
J.Physical Chemistry,78,926-35.

Helfferich, F. (1962), "Ion Exchange", McGraw-Hill, New York.

Helfferich, F. (1965), "Ion-Exchange Kinetics. V. Ion Exchange
Accompanied by Reactions", J.Phys.Chemistry,69,1178-87.

Helfferich, F. (1967), "Multicomponent Ion Exchange in Fixed
Beds", Ind.Eng.Chemistry Funds.,6,362-4.

Helfferich, F., and G.Klein (1970), "Multicomponent Chromatography"
Marcel Dekker, New York.

Ketelle, B.H., and G.E.Boyd (1947), "The Exchange Adsorption of
the Yttrium Group Rare Earths", J.Am.Chem.Soc.,69,2769-881.

Klein, G., D.Tondeur, and T.Vermeulen (1967), "Multicomponent Ion
Exchange in Fixed Beds", Ind.Eng.Chemistry Funds.,6,339-51

Klein, G., N.J.Norem, and T.Vermeulen (1978A), "Studies on the
Behavior of Carbonic Acid and Its Salts in Fixed Beds of Weak-
-Acid Exchangers", Symposium on Ion Exchange, Bhavnagar, India.

Klein, G., T.J.Jarvis, and T.Vermeulen (1979), "Fluidized-Bed
Ion Exchange with Precipitation-Principles and Bench-Scale Development", pp. 185-98 in Recent Developments in Separation Science,

N.N.Li, ed., vol. V. CRC Press, Inc., Cleveland, Ohio.

Kunin, R., and B.Vassiliou (1964), "New Deionization Techniques Based upon Weak-Electrolyte Ion-Exchange Resins", Ind.Eng.Chemistry,3,404-9.

Lukchis, G.M. (1973), "Adsorption Systems. Part I: Design by Mass-Transfer-Zone Concept", Chem.Engineering, 111-6.

Mangelsdorf, P.C., Jr. (1966), "Difference Chromatography", Analyt.Chemistry,38,1540-4.

Michaels, A.S. (1952), "Simplified Method of Interpreting Kinetic Data in Fixed-Bed Ion Exchange", Ind.Eng.Chemistry,44,1922-30.

Page, B.W. (1968), "Mixed-Bed Ion-Exchange Desalting by the Calcium Hydroxide Process: Theoretical Evaluation", M.S.thesis in Chemical Engineering, University of California, Berkeley.

Page, B.W. (1971), "Theory of Precipitation Ion Exchange with Applications to Calcium Hydroxide Desalting", Ph.D.dissertation in Chemical Engineering, University of California, Berkeley.

Page, B.W., G.Klein, F.M.Golden, and T.Vermeulen (1975), "Mixed-Bed Ion-Exchange Desalting by the Calcium Hydroxide Process", Am.Inst.Chem.Engrs.Sympos.Ser.,152,71,121-7.

Pandya, P.J., G.Klein, and T.Vermeulen (1965), "Saline Water Softening by Fixed-Bed Ion Exchange. Equilibrium-Stage Computation Method for Multicomponent Systems", University of California Sea Water Conversion Report 65-2.

Popper, K., R.J.Bouthilet, and V.Slamecka (1963), "Ion-Exchange Removal of Sodium Chloride with Calcium Hydroxide as Recoverable Regenerant", Science,141,1038-9.

Rhee, H-K, R.Aris, and N.R.Amundson (1970), "On the Theory of Multicomponent Chromatography", Phil.Trans.Roy.Soc.,267A,419-55

Schweich, D., and J.Villermaux (1978), "The Chromatographic Reactor. A New Theoretical Approach", Ind.Eng.Chem.Fundtls.,17,1.

Shiloh, K.I. (1966), "Ion Exchange Accompanied by Side Reactions: Column Performance", M.S.thesis in Chemical Engineering, University of California, Berkeley.

Tiger, H.L., S.Sussman, M.Lane, and V.J.Calise (1946), "Desalting Sea Water", Ind.Eng.Chemistry,38,1130-3.

Tondeur, D. (1967), private communication.

Tondeur, D. (1970), "Theory of Ion-Exchange Columns", Chem.Eng.J., 1,337-46.

Tondeur, D. (1971), "Dynamique des Transferts en Lit Fixe", J.Chimie Physique,68,No. 2,311-23.

Tudge, A.P. (1961), "Studies in Chromatographic Transport. II. The Effect of Adsorption Isotherm Shape", Can.J.Phys.,39,1611-8.

Vermeulen, T., G.Klein, and N.K.Hiester (1973), "Adsorption and Ion Exchange", Section 16 in 'Chemical Engineers Handbook', R.H.Perry and C.H.Chilton, eds., 5th ed., McGraw-Hill, New York.

Part 3

CYCLIC PERCOLATION OPERATIONS AND
NOVEL APPLICATIONS

DESIGN AND DEVELOPMENT OF CYCLIC OPERATIONS

Gerhard Klein

University of California, Seawater Conversion
Laboratory, Richmond, California

ABSTRACT. After introduction of the basic concepts governing cyc
lic sorption operations, superposition methods of design are sur-
veyed. Application of the local-equilibrium theory to the design
of fixed beds is discussed that consist of two sorbent layers which
exhibit favorable and unfavorable equilibria for the feed-sorbate
species. The development of a cyclic softening process with fluidized-bed regeneration accompanied by precipitation is outlined.

1. INTRODUCTION

Sorbent materials have a limited capacity, so that, as soon as
the latter is exhausted to the extent pratically possible, they
must be either discarded or regenerated. The cost of these materials is sufficiently high to make cyclic operation and reuse man
datory in most applications, and an analysis of cyclic operation
is therefore required as basis for rational process design and
optimization. Because of the number of independent operating variables involved, purely empirical process development is usually
out of the question.
Inasmuch as we are dealing with truly transient phenomena, governed by nonlinear differential equations, even the problem of describing the response of a uniformly presaturated bed to a feed of
constant composition ("single-step operation") is difficult. A de
gree of complexity is added in cyclic operation, when interference between the exhaustion and regeneration fronts may occur,
which is equivalent to nonuniform presaturation. Finally, a direct switch from exhaustant to regenerant is rarely made. Rather,
the usual cycle comprises exhaustion, backwashing actual regeneration in one or two steps, displacement, and rinse (Applebaum,

428

1968), with accompanying mixing, fluidization, and changes in total fluid-phase concentration and composition, in equilibrium,
and in kinetics. A full analysis of these phenomena appears yet
to be lacking.
Single-step calculation have been summarized by Vermeulen et al.
(1973).
In cyclic operation, neither the exhaustion nor the regeneration
step are usually carried to completion. It is then of interest to
predict the effect of varying the length of these steps on product purity and cost.
The approaches to the solution of this problem include semiempiri
cal superposition methods based on single-step experimental
effluent-concentration histories (Pancharatnam, 1968, Pancharatnam et al., 1969, Dodds and Tondeur, 1972A, B, 1974) and the much
less realistic calculations based on local-equilibrium theory.
However, this type of calculations is useful for the analysis of
special problems, such as reverse-flow regeneration in layered
beds (Klein and Vermeulen, 1975), and cyclic three-component ion
exchange (Grévillot and Tondeur, 1974). Finally, the method of
characteristics (Acrivos, 1956), equilibrium-stage calculations
(Pandya et al., 1965), and other numerical methods could be
adapted for the present purpose.
The present discussion outlines the superposition methods, the
local-equilibrium analysis of layered beds with reverse-flow regeneration, and a softening application in which regeneration is
performed in the upflow direction in order to cope with calcium
sulfate precipitation due to sulfates present in the regenerant.

2. BASIC CONCEPTS

The description of fixed-bed sorption processes involves a bewil
dering number of variables, including sorbent capacity, particle
size, bed length and cross-sectional area, void fraction, rate,
direction, and duration of exhaustion and regeneration flows, com
position and total concentrations of exhaustant and regenerant,
temperature, equilibrium and rate parameters, and overall process
arrangement. Additional parameters are required to define subsidi
ary operations, such as backwashing, displacement, and rinsing.
For industrial operations, the overall objective of optimization
usually is to minimize the cost of unit quantity of the product,
and then such factors as the cost of the regenerant, equipment,
operation, amortization, sorbent replacement, pumping, off-time,
controls and automation, tanks, and piping, must be taken into
account.
A number of the design variables can be combined into dimensionless groups, with the aid of which scaling up is greatly facilitated. Fixed-bed sorption operation are said to lend themselves
to scaleup by considerably larger factors than other chemical-en-
gineering operations.

Of particular interest here is the throughput ratio T, which, for ion exchange is the number of equivalents of product per equivalent of exchanger:

$$T = C_o(V - \varepsilon v)/\rho_B Qv \qquad (1)$$

Here, C_o is the total solution normality, V the influent volume, ε the void fraction, v the bed volume, ρ_B the bulk density of the exchanger, and Q the exchange capacity in milliequivalents per gram. Eq. 1, specifically applicable to the exhaustion step, may be adapted for the regeneration step by affixing a dagger to the symbols C_o, V, and T, to indicate that these now refer to regeneration.

For an exchanger that is uniformly contaminated with the feed component at the beginning of the exhaustion step, for feed contaminated with regenerant, or both, and for other sorption processes, analogous definitions can be provided (Vermeulen et al., 1973). The concept of the throughput ratio can also be extended to multicomponent (and other multivariant) systems. To characterize cyclic operations, throughput-ratio intervals ΔT and $\Delta T^\dagger$ are defined.

It is also convenient to normalize the concentration variables. For the fluid-phase concentration c of the exhaustant, during exhaustion,

$$x = c/C_o \qquad (2)$$

or, during regeneration

$$x^\dagger = c/C_o^\dagger \qquad (3)$$

For the sorbent-phase concentration q of the exchanger,

$$y = q/Q \qquad (4)$$

x and y are dimensionless equivalent fractions.

Operation of a column gives rise to a continuous sequence of concentration profiles, which may be plotted in terms of y against 1/T, the average y at any feed volume V being denoted by $\bar{y}$. On the basis of one equivalent of sorbent, $\bar{y}$ then is the number of equivalents of exhaustant present in the sorbed state.

The effluent concentration x is a function of T (effluent-concentration history). An effluent-concentration history in which x increases with T is called a breakthrough curve. Exhaustion is interrupted at breakthrough, i.e., when the concentration of the exhaustant in the effluent has reached a specified value. This may be taken as a point value, x, which may be termed instantaneous leakage, or as the average, or integral, leakage:

$$\bar{x} = \int_0^{\Delta T} x\, dT/T \qquad (5)$$

If we designate the average of the solid-phase concentration of the exhaustant at the end of the exhaustion step by $\bar{y}$, and at the end of the regeneration step by $\bar{y}^\dagger$ then $Q\Delta\bar{y}$, given by

$$Q\Delta\bar{y} = Q(\bar{y} - \bar{y}^\dagger) \qquad (6)$$

is the operating capacity of the sorbent. Material balance for the exhaustion step with one equivalent of sorbent yields the relation

$$\Delta\bar{y} = (1 - \bar{x})\Delta T \qquad (7)$$

The regenerant utilization is given by $\Delta\bar{y}/\Delta T^\dagger$.
Finally, the kinetic treatment of fixed-bed behavior is greatly facilitated by use of the number-of-transfer-units (N) concept, defined so that $(\partial y/\partial(NT))_N$, which equals $- (\partial x/\partial N)_{NT}$ represents a dimensionless driving force.
The concept of composition velocity and other topics have been elaborated in detail in the section titled "Ion Exchange and Chemical Reaction in Fixed Beds".

3. SUPERPOSITION METHODS

The mathematical analysis of cyclic fixed-bed sorption operations with unidirectional flow provided by LeMaguer (1967) is useful as a standard of comparison with other models, but it is of limited practical utility because of the restrictive premises on which it is based. These include the assumption of a constant-separation-factor exhaustion equilibrium that is exactly reversed during regeneration, and of the same number of transfer units for both the exhaustion and the regeneration steps. These assumptions stem from the underlying second-order kinetic model, which, when applied to an individual step, can offen simulate actual column performance quite satisfactorily. Moreover, the advanced mathematical derivations on which this treatment is based add little to conceptualization, and computer calculations or extensive tables are required for an evaluation of individual cases.
Less rigorous but less restricted in its application, and easy to apply is the semiempirical method proposed by Pancharatnam (1968) and reported by Pancharatnam et al. (1969). Here, the effluent history under cyclic operation is obtained by appropriate superposition of the histories of the individual operating steps, based on a uniformly presaturated bed. The procedure for a single cycle is illustrated in Fig.1. Here, Diagrams a and b show the single-

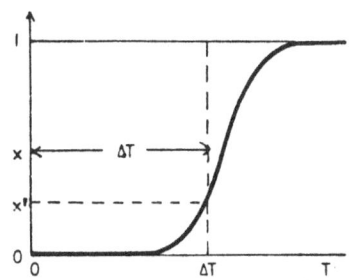

a. Single exhaustion. x' = maximum exhaustant concentration

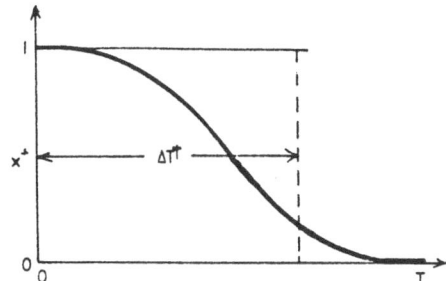

b. Single regeneration

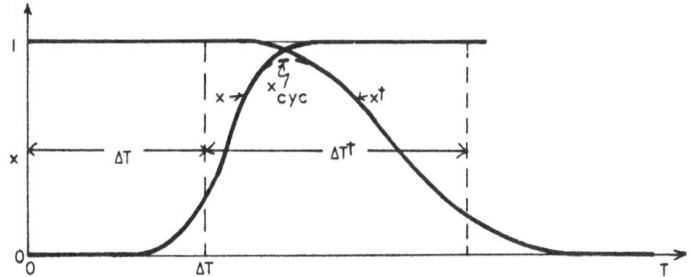

c. Diagram b translated horizontally, so that $\Delta T^{\dagger}$begins at $T = \Delta T$. $x_{cyc} = 1 - x - x^{\dagger}$

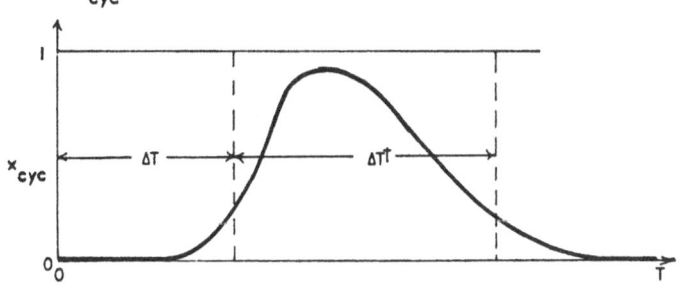

d. Cyclic effluent-concentration history, from Diagram c

Fig.1. Simple superposition method. T = equivalents of effluent solution per equivalent of exchanger. x, $x^{\dagger}$, x_{cyc} are the respective equivalent fractions of exhaustant in single-step exhaustion, regeneration, and in cyclic operation

-step histories of the exhaustant concentration x as functions of
the throughput ratio T. To combine these histories for a cycle (beginning with an exhaustion step), the single-step exhaustion history is copied without change from Diagram a, and shown as the
curve marked x in Diagram c. If x' is the maximum point value of
x permitted in the effluent, then the throughput-ratio increment
ΔT corresponding to the exhaustion step is given by the abcissa
of the point on the exhaustion history that has the ordinate x'.
We now combine Diagrame b with Diagram a by shifting the origin
of Diagram b to the point (ΔT,o) of Diagram c to obtain the trans
lated regeneration history, marked $x^{\dagger}$. The cyclic history in the
vicinity of the point at which the curves x and $x^{\dagger}$ cross is now
obtained by adding the concentrations of the regenerant ion,
which here approach trace values. Thus, in this region, the cyclic concentration of the exhaustant, x_{cyc}, is given approximately
by $1 - c_{cyc} = 1 - x + 1 - x^{\dagger}$, or

$$x_{cyc} = x + x^{\dagger} - 1 \qquad (8)$$

x_{cyc} in this neighborhood is shown as a dashed curve. The same
expression applies to other regions of the cyclic history in Diagram c, in which either x or $x^{\dagger}$ approach 1, so that x_{cyc} then
equals either x or $x^{\dagger}$,respectively. The resulting cyclic history
is shown as Diagram d. The interval during which spent regenerant
appears in the effluent is noted as $\Delta T^{\dagger}$.
The construction for the next half cycle is analogous. Here, the
problem is that of finding the cyclic effluent-concentration history in the vicinity of the intersection of the single-cycle rege
neration history of Cycle 1, and of the exhaustion history of
Cycle 2. If we designate the throughput-ratio intervals for the
first cycle by ΔT_1 and $\Delta T_1^{\dagger}$, then the origin of the second-cycle
exhaustion history will be translated to $T = \Delta T_1 + \Delta T^{\dagger}$. (The symbol $\dagger$ refers to regeneration). The vertical superposition of the
exhaustant concentrations becomes even simpler than before, becau
se, near the intersection of the two histories, it is now the
exhaustant concentrations directly that approach trace values, so
that

$$x_{cyc} = x_1^{\dagger} + x_2 \qquad (9)$$

where the subscripts refer to the cycle numbers.
This method was tested experimentally for several binary ion-
-exchange systems with Dowex 50 resin, in some of which widely
differing exhaustion and regeneration flowrates and total solution concentrations were used, and some of which were governed by
strongly nonlinear isotherms.
For the exchange of sodium and hidrogen ions,which is governed by
a nearly linear isotherm, and where the flowrates, total concen-

trations, and number of transfer units for both the exhaustion and regeneration steps were the same, the superposition method predicted the cyclic effluent-concentration history almost exactly, even at small values of the throughput-ratio increments ΔT and $\Delta T^{\dagger}$. For heterovalent exchange (Na-Mg or Na-Ca), when widely differing exhaustion and regeneration flowrates and total solution concentrations were used, and when the isotherms were strongly nonlinear, larger values of ΔT and $\Delta T^{\dagger}$ were required to achieve comparable accuracy, probably largely because of the non-linearity of the isotherms involved.

The superposition method was also compared to calculations performed by LeMaguer (1967), which were based on the reaction-kinetic model (Thomas, 1948) and subject to the restrictions mentioned earlier. In the equations underlying these calculations, a memory term appears, which measures the difference of the actual (integral) leakage from the value predicted by superposition. Although this term is defined only for cases in which the separation factors of the exhaustion and regeneration steps are reciprocals, and in which the number of transfer units for one of these steps equals the ratio of the number of transfer units and the separation factor for the other step, it was possible, by choice of suitable experimental conditions underlying the superposition method, to show that relatively large values of this term did not cause serious errors.

Pancharatnam's simple superposition method has been modified and improved by Dodds and Tondeur (1972A), who account for residual exhaustant at the end of the regeneration step, and for residual regenerant at the end of the exhaustion step. This develoment is based on the additional assumption that the fractional removals of exhaustant during regeneration, and of regenerant during exhaustion, are the same functions of the throughput ratio as in single-step operation.

The reproducible cyclic effluent-concentration history is synthesized from single-step exhaustion and regeneration effluent-concentration histories by the following procedure.

1. Given ΔT and $\Delta T^{\dagger}$, the fractional removals of regenerant during exhaustion (F_s) and of exhaustant during regeneration (F_r) are calculated by graphic integration of the respective single--step effluent-concentration histories, using the following relations:

$$F_s = (1 - \bar{x})\Delta T \tag{10}$$

$$F_r = \bar{x}^{\dagger}\Delta T^{\dagger} \tag{11}$$

2. The residual stoichiometric capacity of the bed for the regeneration step is calculated by the expression

$$\Delta \bar{T}^\dagger = F_s / (F_s + F_r - F_s F_r) \tag{12}$$

and for the exhaustion step by

$$\Delta \bar{T} = (1 - F_r) \Delta \bar{T}^\dagger \tag{13}$$

The single-step effluent-concentration histories are now transfer
red to the cyclic plot in such a way that the average T-value, $\bar{T}$
of the single-step exhaustion history is at $T = \bar{T}$, and the average
$T^\dagger$-value, $\bar{T}^\dagger$, of the single-step regeneration history at $T = \Delta T + \bar{T}^\dagger$,
and so forth for other cycles. Inapplicable parts of the construc
tion are finally erased to produce the effective cyclic effluent-
-concentration history.
In a later paper, Dodds and Tondeur (1972B) have analyzed the
effects of changes in total solution concentration. The cyclic
histories constructed as indicated in these studies are in good
agreement with corresponding experimental effluent-concentration
histories over a wide range of conditions, but the limits of vali
dity of these results remain to be explored.

4. BEDS CONTAINING TWO LAYERS OF DIFFERENT TYPES OF CATION OR ANION EXCHANGERS

An interesting example of design involving the reverse-flow regeneration of fixed beds is that of the layered-bed column. This re
cently described sorption technique improves regenerant utilization by using a layer of an exchanger having unfavorable equilibrium for the exhaustant near the exhaustant inlet, and a second
layer, of an exchanger havin favorable equilibrium, at the other
end of the column. When counterflow regeneration is employed, the
equilibrium of the second layer becomes unfavorable for regeneration, leading to a dispersion of the concentration contours in
the column. However, when these concentrations traverse the first
layer, they again become focussed, so that the overall effect is
that of high regenerant utilization. Similarly, during exhaustion
dispersion occurring in the first layer is corrected in the second
with resulting adequate product purity.
The following graphical design method is based on the premise of
local equilibrium throughout. In it, the exchanger layer traversed first by the exhaustant solution is designated Layer 1; the
other layer, Layer 2. Although for hydrodynamic reasons, upflow
may occasionally be used for exhaustion, exhaustion is assumed to
be downflow and regeneration upflow.
Dimensionless quantities are used for simplicity. Distance is mea
sured from the exhaustant inlet, in terms of total number of equi
valents of exchanger, Z; and time T in terms of the cumulative
number of equivalents of solution per equivalent of exchanger,

supplied to or put through the column, regardless of flow direction. Concentrations are given as equivalent fractions. Since the
sum of the equivalent fractions of both components in either phase is one, it suffices to use the concentration of only the
exhaustant. β is the relative number of equivalents in Layer 1.
In general, there will be four isotherms of interest; two for
exhaustion and two for regeneration.

The basis of the construction are concentration countours, whose
slopes in a T,Z-diagram, are related to the isotherm slopes. In
the diagrams used, the cumulative throughtput, in equivalents of
solution, and regardless of flow direction, T, is the abcissa,
and the equivalents of exchanger, Z, between the top of Layer 1
and the bed level under consideration, is the ordinate. The latter is directed downward to symbolize the customarily downward
flow in fixed-bed columns. The scales on both axes are identical.
β marks the Z-value corresponding to the boundary between the
exchanger layers. Effluent-concentration histories and cyclic per
formance are based on a total of one equivalent of exchanger capa
city, so that the construction normally extends to the line $Z=1$.
In Fig. 2, representing an exhaustion step starting with completely regenerated exchangers, a bundle of contours emanates from
the origin. Their direction is obtained from the exhaustion isotherm for Layer 1. Where they intersect the line $Z=\beta$ (the boundary between the exchangers), they assume directions taken from the
isotherm for Layer 2.

The leading contours in Layer 2 intersect, giving rise to a parti
ally abrupt transition (concentration step), whose velocity is
obtained from the isotherm as the slope of the chord between the
limiting concentrations of the step. When additional contours mer
ge with the trajectory, the span of the step involved, and thus
the slope of the trajectory, change. The entire step trajectory
can thus be approximated by a series of line segments intervening
between intersections of the step trajectory and the characteristics merging with it.

An effluent-concentration history for the exhaustion step can be
construced by plotting the values of the x contours that intersect the line $Z=1$ against the T values of the intersections. The
result is an abrupt rise in exhaustant concentration, followed by
a gradual transition. To achieve perfect product purity, one
would, in this case, end the exhaustion step where the abrupt
transition occurs.

A concentration profile at this T value is obtained by plotting
the contour values of the x contours intersecting the line $T=\Delta T$
against the Z values of these intersections. The corresponding y-
-profile can be obtained by reference to the isotherms.

In proceeding to the regeneration step, the question arises where
a given x contour that has become discontinuous as the result of
an isotherm shift accompanying a change in total solution concentration continues on the "new" side of the discontinuity (at a T

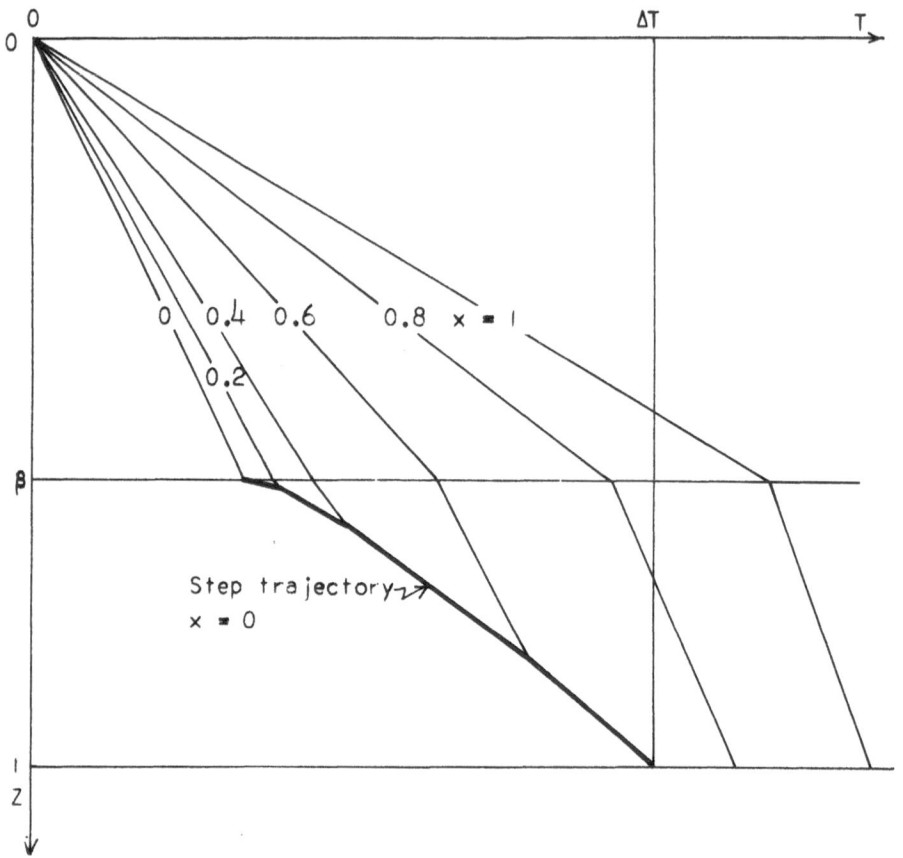

Fig. 2. T,Z-Diagram for exhaustion of fully regenerated exchangers.
Exhaustant begins to break through abruptly at T = ΔT

value infinitesimally larger than ΔT). The solution to this problem is based on the invariance of the y profile with total solution concentration and on the one-to-one correspondences between
a given y value and the x values related to it by the "old" and
"new" isotherms (valid at $T \leqslant \Delta T$ and $T \geqslant \Delta T$, respectively).
In practice, a construction like that of Fig. 3 may be used;
where the isotherms for the exchanger layer of interest are plotted above the x profiles, in which the Z coordinate is oriented
downward.
Let P_1 (x_{old}, Z_{old}) be a point on the old x profile, shown as solid curve. P_2 then is the point on the old isotherm having the
coordinates (x_{old}, y_0), and P_3 the point on the new isotherm having the coordinates (x_{new}, y_0). x_{new} thus is the fluid-phase concentration at the bed level Z_{old} immediately after the change in

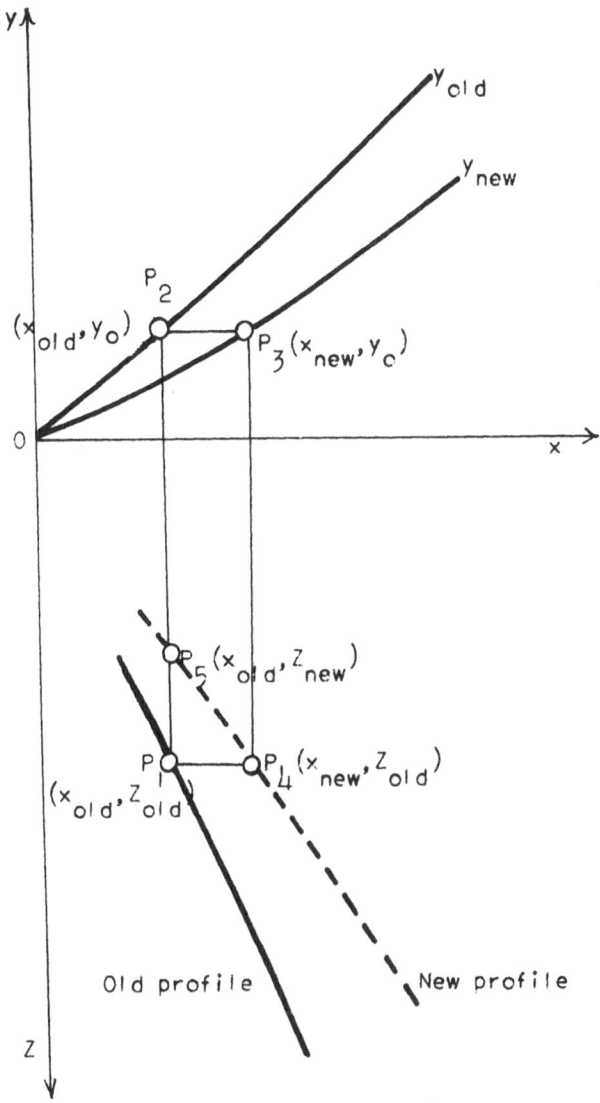

Fig. 3. Switch of x-profile accompanying isotherm shift

total solution concentration, and the point P_4 (x_{new}, z_{old}) is a
point on the new x-profile. This construction illustrates the procedure for generating the entire new profile, shown as the dashed
curve. From it, the point of departure of the new x-contour having the same contour value as an old x-contour is readily established as the ordinate of the point on the new profile which has
this contour value as abcissa (for example, for the contour value

x_{old}, the ordinate z_{new} of point P_5 in the figure).
The procedure for a switch from regeneration to exhaustion is i-
dentical.
Provided cyclic reproducibility has been attained, material balan
ce shows that, when no exhaustant is allowed to appear in the
exhaustion effluent, the exchanger utilization equals the amount
of product per cycle:

$$\Delta \bar{y} = \Delta T \tag{14}$$

Moreover, when no regenerant is lost in the regeneration effluent,

$$\Delta T = \Delta T^{\dagger} \tag{15}$$

5. FLUIDIZED-BED REGENERATION WITH PRECIPITATION

In the south-western United States and Mexico, enormous amounts
of saline agricultural waste water accumulate in tile-drainage
systems. The use of this water has been envisaged for power-plant
cooling by Sephton and Klein (1976). However, because of high calcium, and very high sulfate levels, the water has a high scaling
potential with respect to calcium sulfate and must be subjected
to softening pretreatment, for which ion exchange in principle
was found to be an attractive candidate method. The softened
water would be concentrated in cooling towers and, if necessary,
in an evaporator, the blowdown from which would serve as cost-
-free regenerant.
The difficulty to be expected with this scheme as carried out in
fully fixed-bed operation is that during regeneration the sulfate-
rich blowdown can be expected to precipitate calcium sulfate as
calcium ion is being displaced from the exchanger by sodium ion.
Such precipitation would then clog the bed, as shown by Cherney
(1966).
In the process developed to overcome this difficulty in the case
of agricultural waste water (Klein et al.,1976, Jarvis, 1976, and
Haugseth and Beitelshees, 1974), precipitation during regeneration
was allowed to occur, thus enhancing the regeneration process,
and clogging was prevented by upflow operation with a rate sufficient to cause fluidization. It was possible to find a combination
of bed depth (about 3 ft) and flowrate (about 10 gal per min-ft^2)
such that the calcium sulfate crystals formed remained small and,
in spite of their having a specific gravity greather than that of
the resin, were carried out of the column through overflow open-
ings, while the resin remained in the column, with negligible
carryover of only the smallest particles.
The principal features of the overall flowsheet of the process
are shown in Fig. 4. Here, a line leading from the top of the ion-
exchange columns is seen to carry the calcium-sulfate effluent
slurry into a stirring-and settling tank. Stirring encourages the

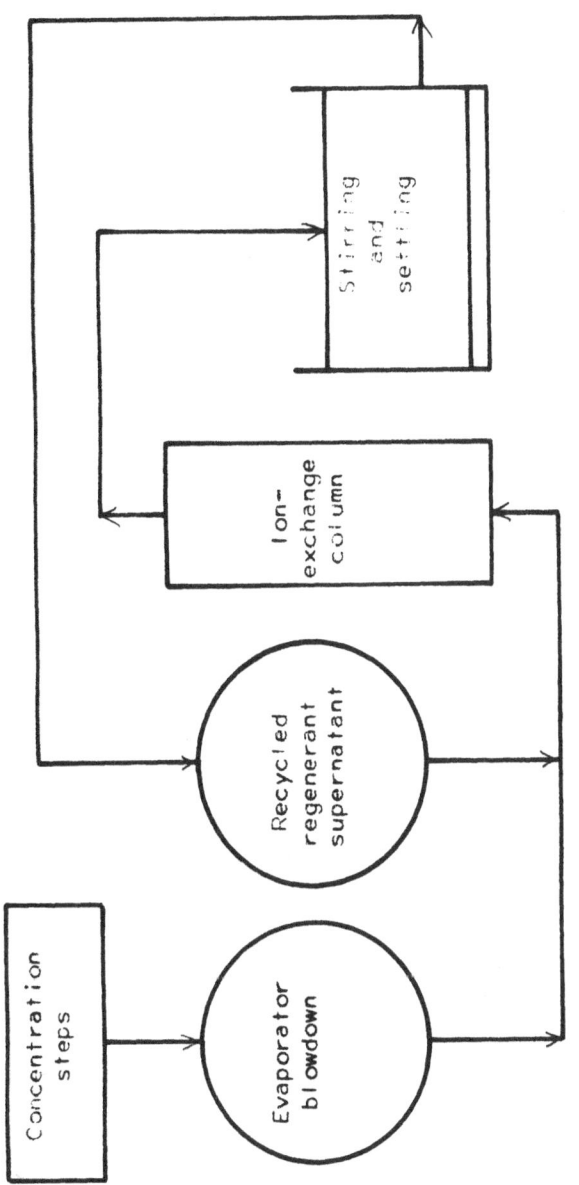

Fig. 4. Regeneration of fluidized bed (schematic)

growth of larger crystals at the expense of smaller ones and faci
litates subsequent settling. A considerable inventory of resulting
regeneration-effluent solution in accumulated and recycled as the
first portion of the regenerant of the next cycle. This portion
removes most of the calcium from exchanger. Most of the remainder,
and magnesium, is then purged from the resin with a second,
smaller, regenerant portion, which consists of softened blowdown
brine. This recycling scheme permits adequate regeneration with
the strictly limited amount of blowdown brine available, at a
flowrate so high that normally this brine would not suffice. The
sulfate in the recycled brine serves as a sink for calcium. Excess
settled precipitate is drawn off and wasted from the stirring and
settling tank from time to time. No chemicals are added to incre
ase the waste-disposal burden of the process.
The development of this process is an example of scaleup from
runs with one-inch laboratory columns and showed that laboratory
experiments could be very successfully translated to the 2000-gal-
lon-per day pilot-plant scale by keeping the flowrates and bed
depth the same.

REFERENCES

1. A. Acrivos, Ind. Eng. Chemistry, 48, 703-10 (1956).
2. S. B. Applebaum, Demineralization by Ion Exchange, Academic
    Press, New York and London, 1968.
3. S. Cherney, M.S. thesis in Chemical Engineering, University
    of California, Berkeley, 1966.
4. J. A. Dodds and D. Tondeur, Chem. Eng. Science, 27, 1267-81,
    1972A.
5. J. A. Dodds and D. Tondeur, Chem. Eng. Science, 27, 1291-8,
    1972B.
6. J. A. Dodds and D. Tondeur, Chem. Eng. Science, 29, 611-9,
    1974.
7. G. Grévillot, D. Tondeur and J. A. Dodds, J. Chromatography,
    102, 421-8, 1974.
8. L. A. Haugseth and C. D. Beitelshees, Evaluation of Ion Ex-
    change Pretreatment for Membrane Desalting Processes, U.S.
    Dept. Int. Bureau of Reclamation Rept. REC-ERC-74-26, Denver,
    Colorado, 1974.
9. T. J. Jarvis, M.S. thesis in Chemical Engineering, University
    of California, Berkeley, 1976
10. G. Klein and T.Vermeulen, Am. Inst. Chem. Engrs. Sympos. Ser.,
    71, No. 152, 69-76, 1975.
11. G. Klein, T. J. Jarvis and T. Vermeulen, Fluidized-Bed Ion
    Exchange with Precipitation - Principles and Bench-Scale De-
    velopment". P. 185 in Vol. V of Recent Developments in Sepa-
    ration Science, CRC Press, N.N.Li, ed., West Palm Beach, Florida
12. M. LeMaguer, M.S. thesis in Chemical Engineering, University
    of California, Berkeley, 1967.

13. S. Pancharatnam, M.S. thesis in Chemical Engineering, University of California, Berkeley, 1968.

14. S. Pancharatnam, G. Klein and T. Vermeulen, Study of Design-Optimization Procedure for Ion-Exchange and Adsorption Columns, U.S. Dept. Int., OSW Res. Developt. Prog. Rept. 477, Washington D.C., 1969.

15. P. J. Pandya, G. Klein and T. Vermeulen, Saline-Water Softening by Fixed-Bed Ion Exchange. Equilibrium Stage Computation Method for Multicomponent Systems, University of California Sea Water Conversion Laboratory Rept. 65-2, 1965.

16. H. H. Sephton and G. Klein, A Method of Using Irrigation Drainage Water for Power-Plant Cooling, Proc. First Desalination Congress of the American Continent, Mexico City, 1976.

17. H. C. Thomas, Annals New York Acad. Sciences, 44, Art. 2, 161-82, 1948.

18. T. Vermeulen, G. Klein and N. K. Hiester, Adsorption and Ion Exchange, Sect. 16 in Chemical Engineers Handbook, R. H. Perry and C. H. Chilton, eds., 5th ed., McGraw-Hill, New York, 1973.

CYCLIC SEPARATION TECHNIQUES*

Phillip C. Wankat

School of Chemical Engineering, Purdue University,
West Lafayette, Indiana, USA

ABSTRACT. The cyclic separation techniques, parametric pumping, pressure swing adsorption and cycling zone adsorption, are considered in detail. The equipment configurations and experimental data obtained are reviewed, and then explained using a retention argument theory. This theory is then related to the local equilibrium theory and compared with other theories. The differences and similarities of these methods are discussed and the relationship to two-dimensional separators is developed. An extensive bibliography and a glossary are provided.

INTRODUCTION

Adsorption and chromatographic processes can be operated with countercurrent or crossflow of the sorbent bed, or in a simulated countercurrent fashion. However, the most common method of operation is to use a fixed bed with some type of time dependent separation. For example, in preparative chromatography a pulse of feed is developed with solvent or carrier gas which is followed by a second feed pulse. In common industrial adsorbers the feed pulse becomes quite large and desorption is usually accomplished by heating the bed, decreasing the pressure, or using some type of desorbent. The desorption is often in the opposite direction of the adsorption step. After desorption and an optional cooling/drying step, the operation is repeated. These common operational methods can thus be considered to be cyclic processes since the same steps are repeated over and over.

---

*This work was partially supported by the National Science Foundation - USA under grant ENG 74-02002A01

The cyclic processes (parametric puming, pressure swing adsorption and cycling zone adsorption) which will be discussed in this paper differ from the more common methods in some of the details of the operation. However, there is no sharp dividing line where the operation suddenly becomes a cyclic operation. Probably the most important difference is in the philosophy used in the cyclic processes. This philosophy is to use the cyclic nature of the process to obtain the separation instead of operating in a cyclic fashion because one is forced to do so. The cyclic processes use part of the fluid being separated to achieve the desired desorption, and the separation is based on the differences in sorption caused by changing a thermodynamic variable. These points should become clear when the processes are discussed in detail, and all terms are defined.

This paper will start with a description of batch thermal parametric pumping equipment and the operational cycle used. Then typical separations obtained will be presented and a simple chromatographic retention argument will be used to qualitatively explain the separation and provide a physical feel for why the method works. Then the important extensions of parametric pumping to open systems, recuperative mode operation and use of cyclic variables other than temperature will be discussed. These extensions will be discussed in terms of the equipment arrangement, data obtained, and a retention argument explanation of the separation.

The discussion will then shift to the related pressure swing adsorption process which is commercially significant for gas separations. The equipment arrangements, separation data and a physical explanation of why the technique works will be presented.

In parametric pumping and pressure swing adsorption the flow is reversed during operation. In cycling zone adsorption flow reversal is not employed. The differences and simularities between this mode of operation and parametric pumping will be explored. To close this mainly qualitative material, the large variety of sorbents and chemical systems which have been studied will be summarized in tabular form.

A quantitative understanding of the separations will be developed through the local equilibrium model of adsorption. This will first be done for a half cycle of any process and then be made specific for the various operation modes of parametric pumping and cycling zone adsorption. The theoretical predictions will be compared to experimental results and the large number of theoretical papers utilizing the local equilibrium model will be summarized. Other theoretical models will be briefly discussed.

The final section of the paper will first consider extensions of the cyclic processes to two-dimensional operation. Then related operational methods will be discussed to provide a perspective for where the cyclic processes fit into the spectrum of separation techniques.

Two distinctive features of this paper are provided to aid the reader. A glossary of terms is provided at the end of the paper to explain the jargon which has developed. Also the literature is covered extensively in a series of tables which briefly outline the data and theoretical models which have been published.

PARAMETRIC PUMPING

Basic Ideas and Physical Arguments

The basic ideas for parametric pumping were first discussed by Wilhelm and his coworkers [123-125]. Although the first experiments involved an open, recuperative mode system (entering fluid heated or cooled), until recently most of the experimental and theoretical work has involved direct mode systems (whole column heated or cooled). As an introduction to parametric pumping we will first consider the batch, direct mode system shown in Figure 1. In this equipment a jacketed column is packed with a suitable adsorbent or ion exchange resin. During the first half of a cycle fluid flows from the bottom reservoir up through the hot column into the top reservoir. In the second half of the cycle the fluid flows from the top reservoir downward through the cold column into the bottom reservoir. This cycle is repeated until the experiment is ended and the reservoirs are emptied. Most laboratory systems have employed syringe pumps for the two reservoirs. The experiments are usually started with the sorbent and fluid in equilibrium at the hot temperature and the bottom reservoir contains this fluid. Since the sorbent can sorb less solute at the hot temperature than at the cold temperature, we would expect the concentration of solute on the solid to increase when the column temperature drops. At the same time the fluid concentration decreases. Thus during downflow a less concentrated fluid is pumped into the bottom reservoir. During upflow a more concentrated fluid is pumped into the top reservoir and a partial separation is obtained. After many cycles most of the solute will be "pumped" into the top reservoir and the top section of the column.

The separation which can be obtained in this type of device is well illustrated by the classical results obtained by Wilhelm and Sweed [124]. These authors removed toluene from n-heptane using silica gel as the adsorbent and cycling the temperature

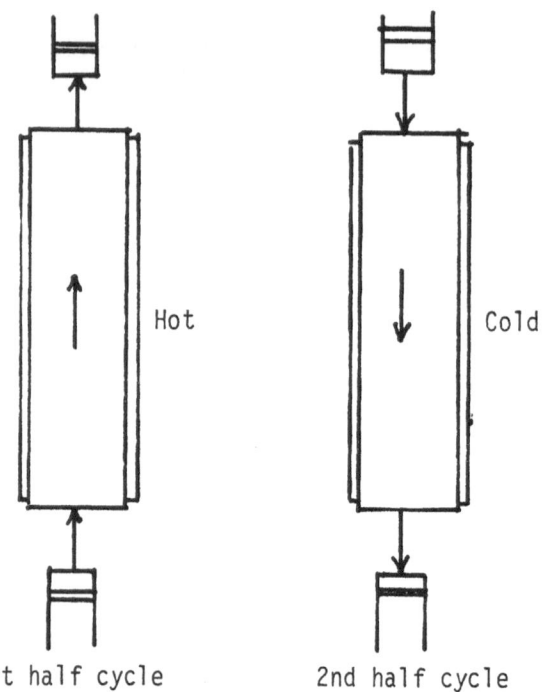

1st half cycle   2nd half cycle

Figure 1. Batch, Direct Mode Parametric Pump

between 4° and 70°C or between 15° and 70°C in a direct mode
parametric pump. Their results are shown in Figure 2. These
results were obtained with the fluid and adsorbent initially in
equilibrium at ambient temperature. Clearly the toluene has been
"pumped" to the top reservoir and removed from the bottom reservoir.
As expected, more separation is obtained when the temperature is
cycled over a wider temperature range since the difference between
the amount adsorbed at the two temperatures is greater. More
separation is also obtained with longer cycle times (slower
fluid velocities) since mass transfer rate limitations have less
effect on the separation.

These experimental results were responsible for a great
surge of interest in parametric pumping and in cyclic separations
in general. From the vantage point of ten years later we can
develop a simple physical picture of how parametric pumping works.
This physical picture will be developed using the retention
argument common in chromatography.

The retention argument looks at the rate of movement of
solutes through a packed bed. We will temporarily restrict

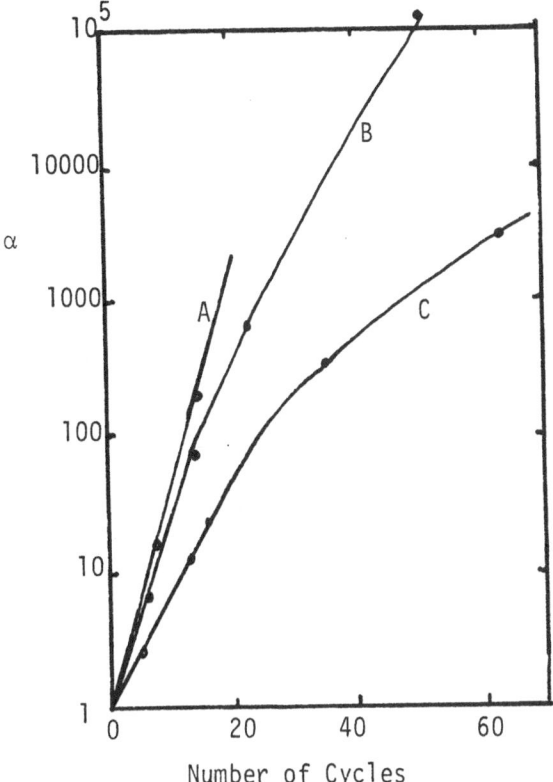

Figure 2. Experimental results for batch direct mode parametric pumping separation of toluene from heptane on silica gel. [124]

A   140 min cycle 70^0 and 4^0
B   30 min cycle 70^0 and 4^0
C   8.5 min cycle 70^0 and 15^0

$$\alpha = \frac{\text{Concentration in top reservoir}}{\text{Concentration in bottom reservoir}}$$

ourselves to the simple linear equilibria of the form

$$q = A(T)c* \tag{1}$$

For common adsorbents the amount of material adsorbed decreases as temperature is increased. Thus A(T) is a monotonically decreasing function of temperature. The amount of solute in the mobile phase compared to the total amount of solute in the column is easily calculated as:

$$\frac{\text{Amt in mobile phase}}{\text{total amount}} = \frac{(LA_c)\varepsilon c}{(LA_c)\varepsilon c + (LA_c)(1-\varepsilon)q\rho_B} \tag{2}$$

If we assume that the solid and fluid are everywhere in local equilibrium and substitute eq. (1) into (2), we obtain after algebraic manipulation

$$\frac{\text{Amt in mobile phase}}{\text{total amt solute in column}} = \frac{1}{1 + \left(\frac{1-\varepsilon}{\varepsilon}\right) A(T)\rho_B} \tag{3}$$

When the temperature increases this ratio also increases.

If the fluid has a constant velocity of v, then the solute velocity is just vx (relative amount of time the solute is in the mobile phase). Assuming a random process of adsorption and desorption, the average solute velocity becomes

$$u_{solute} = v\left(\frac{\text{amount solute in mobile phase}}{\text{total amount solute in column}}\right) \tag{4}$$

or

$$u_{solute}(T) = \frac{v}{1 + \frac{1-\varepsilon}{\varepsilon} A(T)\rho_B} \tag{5}$$

Eq. (5) allows us to explore the behavior of solute in the column as the temperature and flow directions are changed. This will obviously result in a crude, first order picture, but it does include the major effects of convective flow and equilibrium.

On a plot of distance up the column versus time eq. (5) tells us that the average solute molecule moves at a slope equal to $u_{solute}(T)$. This slope can be positive or negative depending upon v, and depends upon the column temperature. For the results shown in Figure 2 the column was hot during upward flow and cold during downward flow. A schematic of the paths of average solute molecules is shown in Figure 3. Since the solute velocity is greatest during upflow when the column is hot, solute which enters the column during the early part of a cycle will eventually exit into the top reservoir. Once in the top reservoir the solute

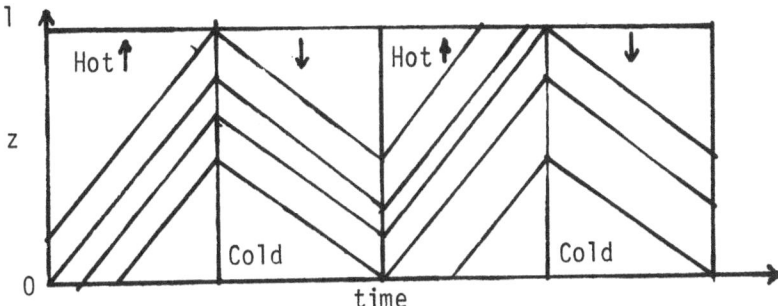

Figure 3. Schematic of Paths in Direct Mode Parametric Pump

can be refluxed to the column but will not exit into the bottom
reservoir since its downward velocity is low. As the number of
cycles increases the bottom reservoir will become increasingly
diluted and the top reservoir more concentrated (Obviously, the
reverse phenomenon occurs if the column is cold during upflow
and hot during downflow).

This action of "pumping" solute up to the top reservoir
is aided by the changes in concentration which occur everytime
the column temperature is changed. When the temperature increases,
solute is desorbed and the fluid concentration must increase.
Conversely, when the temperature drops the solute is adsorbed and
the fluid concentration also drops. This behavior can be quanti-
fied by writing a mass balance on a segment of column of length
ΔX when the column temperature changes from T_1 to T_2. The mass
in this segment of column must be the same before and after the
temperature change.

$$(\Delta X A_c) \, \varepsilon(c_2 - c_1) + (\Delta X A_c)(1-\varepsilon)\rho_s(q_2 - q_1) = 0 \tag{6}$$

If we assume that the equilibrium is linear and that solid and
fluid are locally in equilibrium, then equation (1) can be
substituted into eq. (6). Upon rearrangement this becomes,

$$\frac{c_1}{c_2} = \frac{\varepsilon + (1-\varepsilon)\rho_s \, A(T_2)}{\varepsilon + (1-\varepsilon)\rho_s \, A(T_1)} = \frac{u_{solute_1}}{u_{solute_2}} \tag{7}$$

If $T_2 > T_1$ eq. (7) shows that $c_2 > c_1$ since $A(T_2)$ will be greater than $A(T)_1$. Thus everytime the column temperature is increased the fluid concentration increases by the amount shown in eq. (7). The upward flowing fluid is thus more concentrated and the solute moves faster. These two effects combine to "pump" the solute into the top reservoir.

The picture of the simple batch, direct mode parametric pump can be completed by considering what happens in the two reservoirs. This requires the external mass balances on the system and knowledge of the reservoir behavior. Wilhelm and Sweed [124] arranged their apparatus so that the reservoirs were reasonably well mixed and there was negligible reservoir dead volume. Under these conditions the fluid concentration exiting a reservoir and being pumped into the column, is constant during each half cycle. This concentration can be determined as the weighted average of fluid concentrations entering the reservoir during the previous half cycle.

Some of the conditions which will affect separation can be explored with simple physical arguments. If the column is cold during upflow and hot during downflow solute will be pumped towards the bottom of the column. If the heating and cooling of the column is out of phase with the alternating flow then some of the material entering the top reservoir will have undergone a transition from hot to cold (see Fig. 3). As shown in eq. (7) this transition decreases the fluid concentration and thus will decrease the concentration in the top reservoir. In a similar fashion the bottom reservoir concentration will be increased since some of the fluid entering this reservoir will undergo a temperature change from cold to hot. Thus operating the column out of phase will decrease the separation.

The effect of cycle period can also be qualitatively explored. If the cycle period is too long some of the solute from the bottom reservoir will pass directly into the top reservoir without undergoing a temperature change. Since this temperature change creates the separation, this solute will not be concentrated and less separation will be obtained. This direct breakthrough will not occur if

$$[L/u_{solute}(T_H)] < t_u \qquad (8)$$

where t_u is the time for the upward flow half cycle. On physical grounds cycling too fast is also detrimental since the entire bed will not be heated or cooled during much of the cycle. Since the temperature difference observed by the adsorbent is less than the maximum, less separation will occur.

This simple model predicts that the separation is controlled by differences in equilibrium adsorption at the two temperatures. Thus solutes which have large changes in adsorption will be easy to separate. Also factors such as very rapid adsorption, small particle size and slow fluid velocity which make operation closer to equilibrium should also increase the separation obtained. The exact type of cycling used (square wave versus sine wave) should be relatively unimportant since this affects only fluid velocity. The total reservoir displacement is important since it controls the fraction of the bed storage capacity which is used. If total reservoir displacement is too large (or cycle time is too long) breakthrough occurs and separation decreases.

Excess volume in the reservoirs which is not pumped into the column (dead volume) represents material which undergoes no separation. Thus more cycles will be required to achieve the desired separation if dead volume is present. Also on physical grounds reservoirs which are unmixed and thus retain the partial separations which occur in the fluid entering the reservoir should produce more separation. Philosophically, separated streams should not be remixed.

All of the predictions which can be made based on physical arguments are qualitatively in agreement with experimental results and with the quantitative theories which will be discussed later. We wish to emphasize that this simple explanation of parametric pumping shows that the separation results because the solute is stored on the adsorbent during downflow and released into the fluid during upflow. This alternate storage and release upon command provides the mechanism for separation. Complex nonlinear couplings undoubtedly do occur, but they are not the primary mechanisms for separation. With very rapid cycling a separation based on the rates of adsorption and desorption could be developed; however, this is not the case for the usual cyclic systems. The excellent qualitative agreement between this simple theory and the experimental separations indicates that the correct reasons for separation have probably been elucidated.

Extension to open systems

For production purposes some type of open system with feed input and product withdrawal during part or all of each cycle is desired. The first open systems were recuperative mode systems [123,125], but open direct mode systems were studied shortly afterwards by Gregory and Sweed [39,40] and Chen and Hill [12]. Many configurations for open systems have been studied and the advantages of continuous versus semi-continuous feed has been hotly debated. For the purposes of this introduction we will discuss the system shown in Figure 4 [40]. The physical arrangement

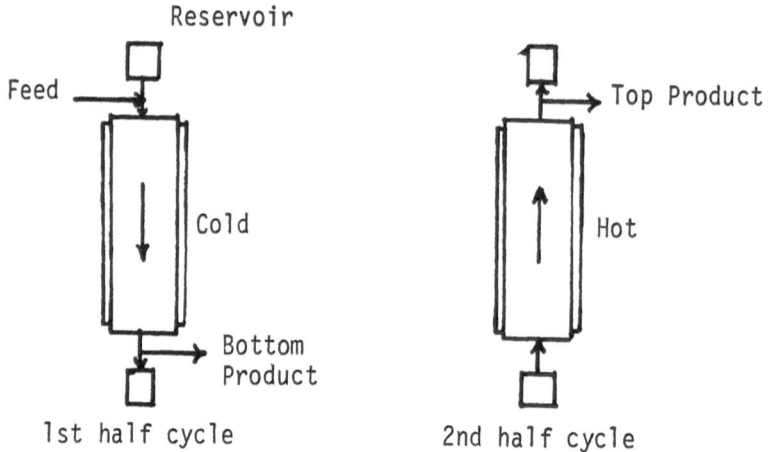

Figure 4.   Top feed, semicontinuous, direct mode
            parametric pump

for the other open systems are quite similar and the theoretical
analyses and experimental results are also similar. The physical arrangement shown in Figure 4 is the same as for the batch
systems shown in Figure 1 except that provision has been made
for feed addition and product withdrawals.

Typical experimental results for removal of NaCl from water
using an ion retardation resin which was thermally sensitive are
shown in Figure 5 [40]. The top product concentration increases
and the bottom product concentration decreases as the number of
cycles is increased until a limit cycle condition is reached
where each cycle is a repeat of the previous cycle. The experi-
menter obviously has control over the temperature difference used,
the fluid displacement per cycle, the cycle time and the reflux
ratios at top and bottom of the column. Since the batch system
can be considered to be the limit of an open system as we go to
total reflux, we would expect the same qualitative effects as in
a batch parametric pump. Increasing the temperature difference
increases the separation. Increasing the fluid displacement per
cycle has little effect until breakthrough occurs and then the
separation decreases rapidly. Increasing the cycle time, which
means decreasing fluid velocity at constant fluid displacement,
at first rapidly increases the separation obtainable because mass
transfer rate limitations become less controlling. The separation
then levels out to a relatively flat plateau. Gregory and Sweed
[40] found by numerical simulation that there is an optimum cycle
time. This optimum cycle time is also predicted by staged
parametric pumping models [43,109].

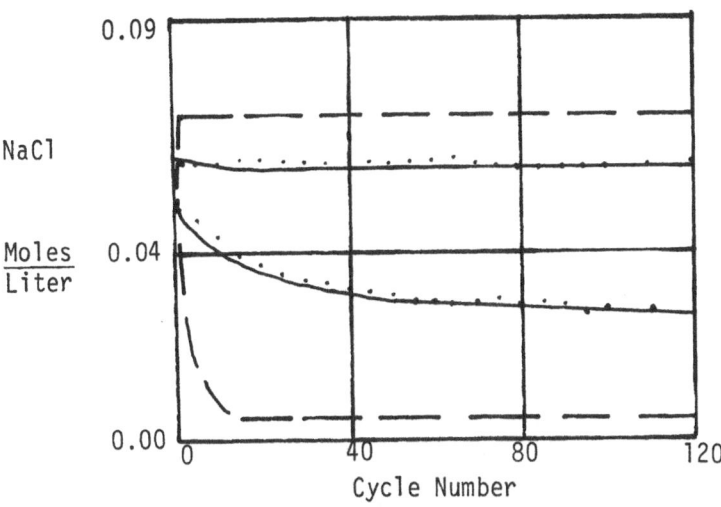

Figure 5.   Experimental results for open direct mode
parametric pump removing NaCl from water on
ion retardation resin [40].

·Experiment.   ———Numerical Simulation.
---Local Equilibrium Theory

The important new variables introduced with open systems are
the top and bottom reflux ratios. These ratios were defined
analogously to the external reflux ratio, L/D, in distillation.
That is the reflux ratio R at each end of the column is,

$$R = \frac{\text{fluid volume sent to reservoir (and thus back to column)}}{\text{fluid volume taken as product}}$$

(9)

The definition in eq. (9) is applied over a complete cycle.
Increasing the top reflux ratio increases the top product concentration, and increasing the bottom reflux ratio decreases the
bottom product concentration. Thus higher reflux ratios increase
the separation, but at the price of throughput. A complicating
factor is that an optimum bottom reflux ratio exists (other
variables held constant) since the fluid velocity in the column
will increase [40]. This prediction requires a model more
sophisticated than the local equilibrium model.

The effect of these variables can be understood utilizing
the retention argument discussed previously for the batch pump.
Again the solute is "pumped" towards the top reservoir, but now
there is an additional downward flow which must be overcome.
Since the pump in Figure 4 has a top feed, there must be greater

downward flow than upward flow. In Figure 3 this would either require a greater downward velocity and hence steeper slope during the cold portions of the cycle, or a longer downflow period than upflow period. To obtain a separation the net solute movement on Figure 3 still must be upwards. The condition to obtain separation can easily be determined using eq. (5) for the solute velocity. Since upward solute movement must be greater than downward movement separation will occur if,

$$[u_{solute}(T_H)]t_u > [u_{solute}(T_c)]t_D \qquad (10)$$

In addition, eq. (8) should be satisfied.

Figure 5 shows both the experimental and theoretical predictions. The local equilibrium model is shown to overpredict the separation. Since the assumptions are essentially the same, the retention arguments presented here will also overpredict the separation. This is probably caused by relatively slow mass transfer rates. To complete the analysis of open systems eq. (6) or (7) would have to be employed to determine the concentration changes when the temperature is changed, and the mass balances around each reservoir are required.

Many other experimental results with open direct mode systems using similar experimental arrangements have appeared in the literature [19,21,23]. Rice and Foo [77] used two columns to obtain continuous products in removing NaCl from water. A somewhat different semi-batch system for removing urea from water using activated carbon was reported by Sweed [92]. In the semi-batch system no top product is removed in Figure 4 and the solute is allowed to accumulate in the top reservoir for a large number of cycles until that reservoir is emptied. The commercial Sirotherm process [10,11] can be considered as a modification of an open parametric pumping system.

Recuperative Mode Parametric Pumping

In the recuperative mode the column is adiabatic while the entering fluid is either heated or cooled. Obviously, the recuperative mode can be operated as either a batch or an open system. One possible arrangement for an open recuperative mode system is shown in Figure 6 [119]. This figure does not show the reservoirs which are required if a single column is used. Two column arrangements with both columns reversing flow directions and having inlet fluid temperatures out of phase with each other are possible. If feed and both product streams are set to zero this system reduces to a batch recuperative mode pump.

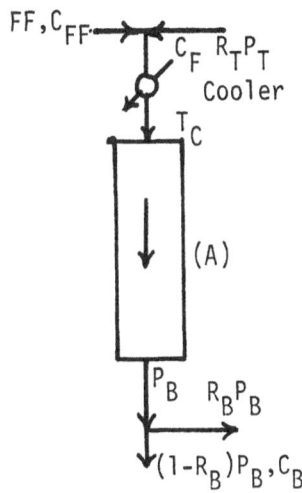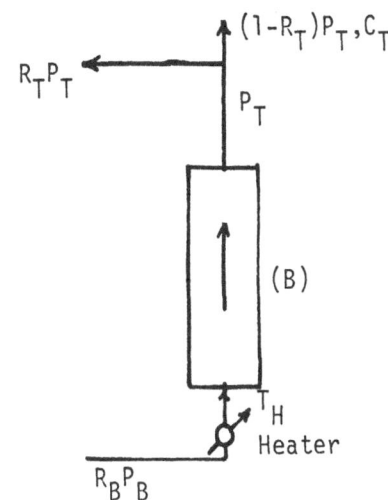

Figure 6. Open Recuperative Mode Parametric Pump [119]
A. Downflow Period t_D
B. Upflow Period t_u

The original recuperative mode separations obtained by Wilhelm et al [123,125] were quite small. After the dramatic direct mode separations were obtained [124,125], the recuperative mode fell into semi-oblivion. It has been used for separations using pH as the cyclic variable, but no large separations using temperature as the cyclic variable have been reported. This is surprising since large separations have been predicted in a batch recuperative mode system [95], and large energy savings can be obtained in open systems [119]. I am confident that large separations will be obtained in batch and open thermal recuperative mode parametric pumps when someone takes the time to try it. Recuperative separations using pH as the cyclic variable are shown in the next section.

The physical reasons why a recuperative mode parametric pump is capable of large separations and the conditions necessary to achieve these separations are easily delineated by the retention arguments developed previously. Since the entire column does not change temperature simultaneously we must follow the movement of the temperature wave in the column. If we can ignore the heat of adsorption and the heat of mixing and assume the column is adiabatic, then the energy in the mobile fluid compared to the total energy in fluid, solid and column wall in a segment of the column is,

$$\frac{\text{Energy in mobile phase}}{\text{Total energy in column segment}} = \frac{(\Delta XA_c)\epsilon\rho_f \, C_f(T_f - T_{ref})}{[(\Delta XA_c)\epsilon\rho C_f(T_f - T_{ref})}$$
$$+ (\Delta XA_c)(1-\epsilon)C_B(T_s - T_{ref})\rho_B$$
$$+ (\Delta X)(WC_W)(T_w - T_{ref})]$$

(11)

In eq. (11) W is the weight of column wall per length and T_{ref} is any convenient reference temperature. The velocity of the thermal wave in the column is just this ratio multiplied times the fluid velocity. After assuming local equilibrium so that $T_s = T_f = T_w$ and simplifying we have

$$u_{thermal} = \frac{v}{1 + \frac{1-\epsilon}{\epsilon}\frac{C_B\rho_B}{C_f\rho_f} + \frac{W}{A_c}\frac{C_W}{C_f\epsilon\rho_f}}$$

(12)

Note that with the simplifying assumptions made here $u_{thermal}$ is independent of temperature. Comparison of eqs. (12) and (5) show they have a similar form but there is an additional term in eq. (12) to account for thermal storage in column walls. Just as eq. (5) represented the movement of the average solute molecule, eq. (12) represents the average rate of movement of the thermal wave. A more exact analysis is needed to include dispersion and heat transfer rate effects.

Eqs. (5), (7) and (12) are sufficient for an approximate analysis of the recuperative parametric pump. Since the solute velocities depend upon the temperature level, the logical analysis procedure is to determine the temperature throughout the column and then solve for the solute movement. The thermal waves can easily be determined by plotting lines with a slope of $u_{thermal}$ on a graph of distance in column versus time. This is shown in Figure 7 for batch operation. The solute waves can now be superimposed on the thermal waves shown in Figure 7. Everytime a temperature boundary is reached u_{solute} will change. This is shown in Figure 8 for a system with linear isotherms. As in the direct mode the solute will be "pumped" to the top reservoir since the solute can move further during upflow then downflow.

To complete the analysis of the recuperative mode operation we need to consider the change in solute concentration when the temperature is changed, and the external mass balances around the

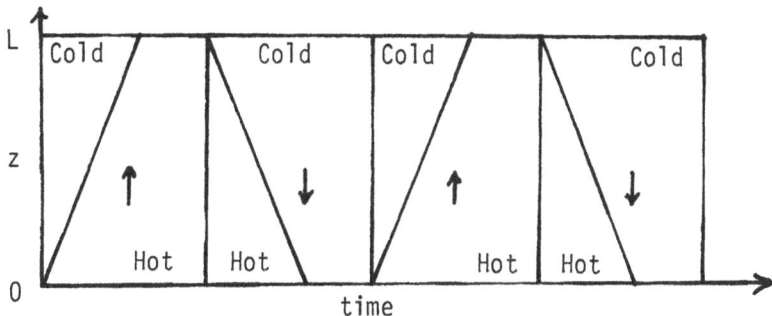

Figure 7. Thermal Waves in Batch, Recuperative Mode
Parametric Pump

reservoirs. The latter equations will be covered later. The
effect of temperature changes on solute concentration can be
determined by a mass balance on a differential section of column
Δz over which the temperature changes during a time interval Δt.
To ensure that all material in the differential section undergoes
a temperature change the control volume is selected so that
$\Delta t = \Delta z/u_{th}$. This mass balance then becomes,

$$[\varepsilon - \frac{\varepsilon v}{u_{th}}](c_2 - c_1) + (1-\varepsilon)\rho_s(q_2 - q_1) = 0 \qquad (13)$$

where 2 refers to conditions after the temperature change. If we
assume that solid and fluid are locally in equilibrium and that
the equilibrium isotherm is linear, then eq. (13) reduces to

$$\beta = \frac{c(T_H)}{c(T_c)} = \frac{\dfrac{1}{u_{solute}(T_C)} - \dfrac{1}{u_{thermal}}}{\dfrac{1}{u_{solute}(T_H)} - \dfrac{1}{u_{thermal}}} \qquad (14)$$

Comparison of eq. (13) to (6) and eq. (14) to (7) shows that the
recuperative mode system involves extra terms due to the movement of the temperature wave. In the limit of $u_{thermal} \to \infty$
eqs. (13) and (14) reduce to (6) and (7), respectively.

The recuperative mode separation can now be followed in
Figure 8. The solute which starts at the origin travels at a
velocity $u_{solute}(T_H)$ both during upflow and during downflow
until it is overtaken by the temperature wave (point A). At
point A the solute concentration is decreased according to eq. (14)

458

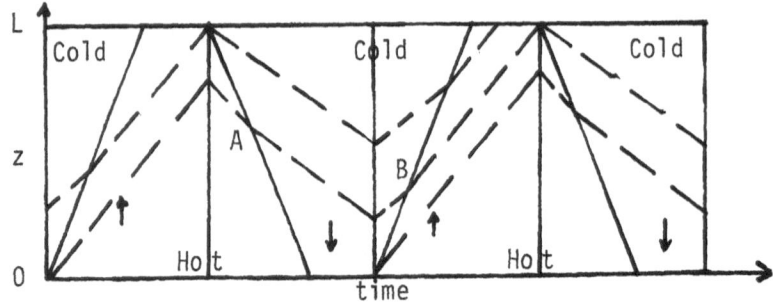

Figure 8. Solute and Thermal Waves in Batch Recuperative
Mode Pump

and the solute velocity decreases to $u_{solute}(T_c)$. The solute
velocity remains at this value until the solute is overtaken by a
hot temperature wave (point B). Here the concentration increases
and the solute velocity increases. Following the solute movement
in the column it is clear that the net movement is upwards and the
solute will eventually be "pumped" into the top reservoir.
Material that is in the top reservoir and is refluxed to the column
does not move down to the bottom reservoir but is returned to the
top reservoir. If sufficient cycles are available very large
separations can be obtained.

Comparison of Figures 8 and 3 shows that the net upward
velocity of the solute is greater in the direct mode. Thus in
batch operation the direct mode will require fewer cycles to
reach the desired degree of concentration. In open systems the
direct mode can operate with greater throughput than the recuperative mode [119]. These comparisons assume that the entire column
temperature can be changed rapidly in the direct mode. If a
considerable period of time is required for the temperature changes
the recuperative mode may be a better separator.

The recuperative mode does have a major advantage in that
energy requirements are considerably less than in the direct mode
and heat recovery is simpler. Wankat [119] predicted 15 times
less energy required for one example calculation when he compared
recuperative and direct modes. The recuperative mode will also
be easier to construct in large sizes. More research can be
expected on this technique.

Extensions and Modifications of Parametric Pumping

Parametric pumping has been extended and modified to separate
multicomponent mixtures, to utilize extraction parametric pumping

and to separate gases. The large variety of mixtures which have been separated will be outlined later in Tables 1, 2 and 3. In this section some of the extensions will be described.

Any thermodynamic variable which affects the adsorption of solute can be used as the cyclic variable. Sabadell and Sweed [80] were the first to use pH as the cyclic variable. They separated Na^+ and K^+ on a cation exchange resin in an open recuperative mode parametric pump. Chemicals were directly added to the solution being separated to obtain the desired pH. Thus this system utilizes chemical energy instead of thermal energy. pH has been used as the cyclic variable to separate proteins in aqueous solution, [27, 83] and some of the results are shown in Figure 9 [27]. Note that these are inverse separations since the concentrated solution appears at the top instead of the bottom as expected.

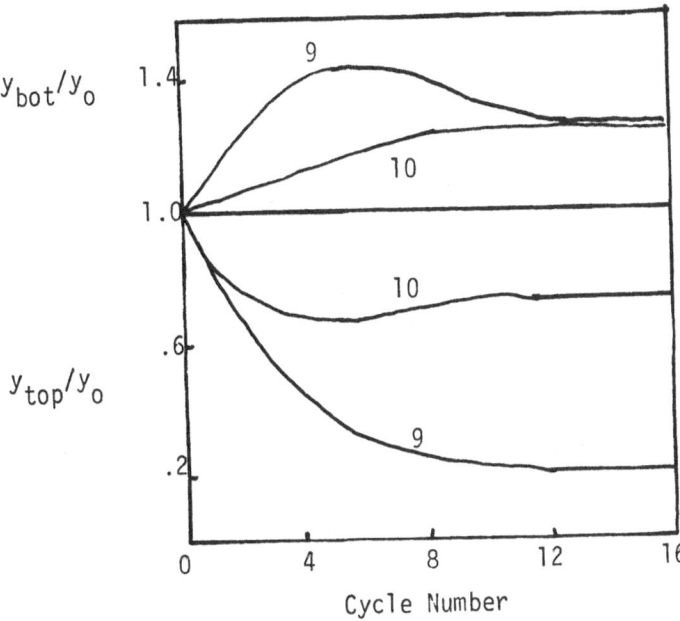

Figure 9. Separation of Haemoglobin in Haemoglobin Albumin-Water system by pH parametric pumping [27].

Run 9 pH_1 = 7.8 pH_2 = 5.5

Run 10 pH_1 = 7.6 pH_2 = 6.0

The reasons for these inverse separations still need to be delineated. The data are fairly scattered and are shown on the original figure [27].

Multicomponent separations refer to the separation of two or more solutes from each other, and has not been extensively studied. Theoretical methods for batch [14] and open system multicomponent separations [120] have been developed. Both of these methods apply to dilute, non-interacting solutes and produce pure solute in carrier. Experimental multicomponent separations of non-interacting solutes have involved removing one solute from one end of the column and leaving a mixture of solutes in diluent in the other reservoir. This straight-forward extension of the single solute separation system has been used to study fructose-glucose-water separation [1,20,25] separation of trypsin-α-chymotrypsin-water [83], and separation of albumin-haemoglobin-water [26,27].

When ions are being separated by ion exchange the ions compete for sites, and the concentration of one ion strongly affects the sorption of another ion. This condition is quali-tatively different from the non-interacting solute case. Multi-component separations with these ionic systems is easier than for non-interacting solute systems since the ion which is preferentially sorbed changes when the temperature is changed. Complete binary separations of K^+ and H^+ in water and ternary separation with K^+ at one end, H^+ at the other end, and a band of almost pure Na^+ in the middle were obtained [15]. The experi-mental results obtained for the ternary separation are shown in Figure 10. These separations were theoretically explained [16].

The major difficulty in extending parametric pumping to extraction is to keep the solvent phase stationary when the flow direction is reversed. This was achieved by two methods [109]. A helix was used for continuous flow separations. Alternating layers of water and diethyl ether were obtained in each loop of the helix. The second method was a series of centrifuge test tubes operated with discrete transfer steps followed by an equilibration period (counter-current distribution operation). Two equilibrium staged theories were developed for the direct mode and were compared to the experimental results obtained with the acetic acid-diethyl ether-water system. The experimental separa-tions were less than the theoretically predicted separations but followed the same trends.

Gas separations by thermal parametric pumping have not been as successful as the numerous liquid separations. The major difficulties are the increased dispersion in gas systems, the pressure increase (or bulk flow if pressure is equalized when the temperature is increased, and the low rates of heat transfer which require a waiting period to obtain heating and cooling. Perhaps the largest thermal parametric pumping separation of gases was obtained by Patrick et al [65] for the removal of SO_2

461

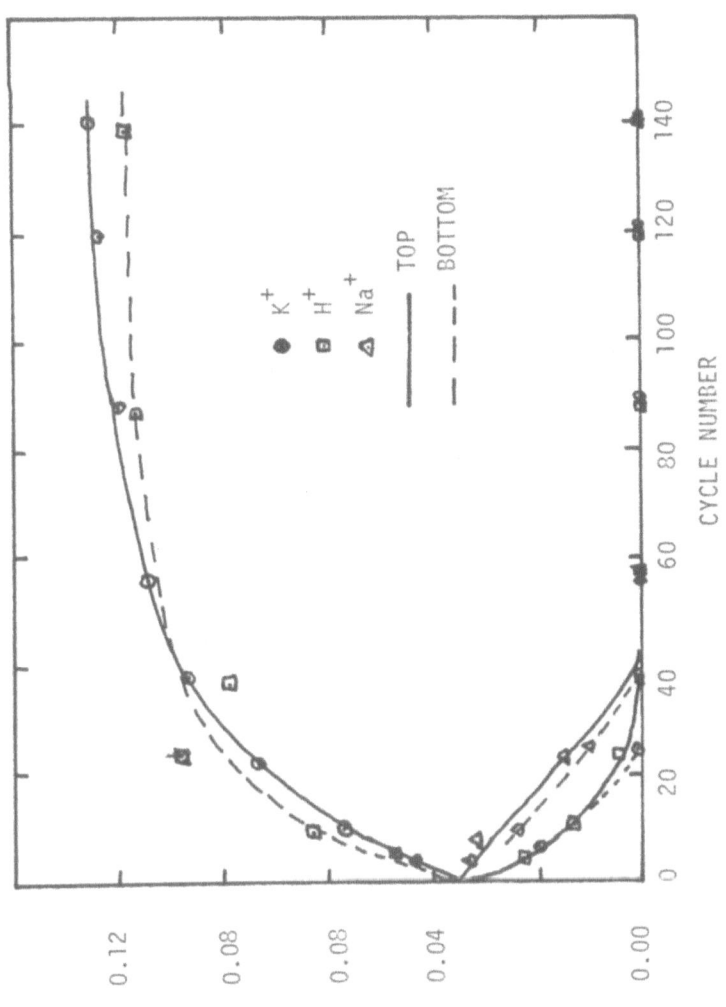

Figure 10. Separations of K^+, H^+, Na^+ in water on Dowex 50 x 8 by batch, direct mode thermal parametric pumping [15].

from air using silica gel as the adsorbent. This separation is
shown in Figure 11. Note that the separation is significantly
less than that predicted by the local equilibrium model. Other
gas separations are outlined in Table 3.

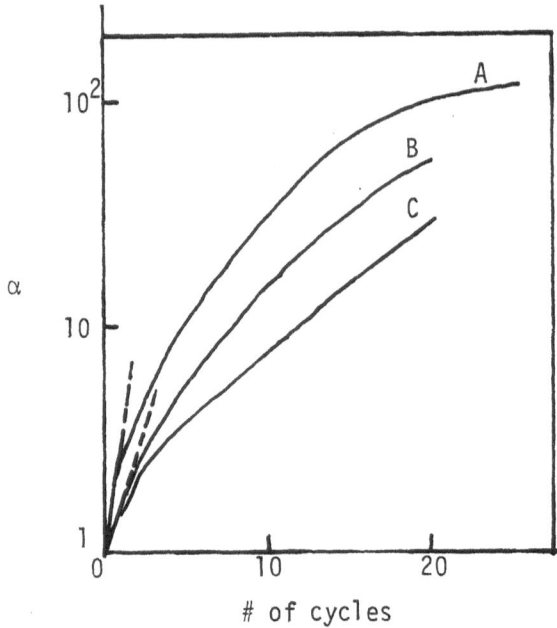

# of cycles

Figure 11.   Separation factors for separation
             of SO_2 from air on silica gel by
             direct mode thermal parametric
             pumping [65].

————Experiment A.   Temperature 400 to 200 and
400 to 100°F.   B.   400 to 300°F
                 C.   300 to 200°F
------Local equilibrium theory

    Since temperature is a difficult cyclic variable to use in
gas systems, it is natural to explore the use of other variables.
Pressure seems to be an obvious variable since it can be rapidly
changed and it has a large affect on gas adsorption. A process
similar to parametric pumping utilizing pressure cycling was
independently developed by Skarstrom [87,88]. This pressure swing
adsorption or heatless adsorption process will be discussed in
the next section.

    This completes the qualitative description of parametric
pumping. The purpose of this rather lengthly section has been
to explain physically how parametric pumping works, to describe
some of the experimental techniques which have been utilized, and

to illustrate some of the experimental separations which have been achieved. With this background we will move more rapidly in the qualitative descriptions of the other cyclic processes.

PRESSURE SWING ADSORPTION

Heatless adsorption was developed by Skarstrom [87,88] originally for drying compressed air and has since become a widely used industrial separation process known as pressure swing adsorption. The same principles used to understand parametric pumping can also be applied to pressure swing adsorption with the one addition that gas volumes increase when the pressure is dropped for purging.

A simple two column pressure swing apparatus for drying air is shown in Figure 12 [87]. Note that this is quite similar to

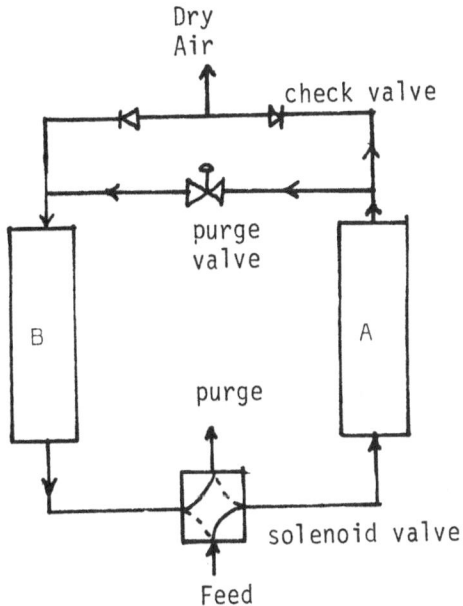

Figure 12. Two Column Pressure Swing Apparatus for Drying Compressed Air [87].

the open, recuperative mode parametric pump shown in Figure 6. The operation of the pressure swing dryer shown in Figure 12 is relatively simple. The feed gas at a high pressure ($\sim$100 psig) enters into column A. The water in this gas is adsorbed and a dry product gas exits out the top of the column. A portion of this dry product gas is expanded to a low pressure ($\sim$2psig) and used as a purge for column B. At low pressures less water is adsorbed and a wet, low pressure purge gas is removed from column B. Approximately every six minutes the solenoid valve is switched. The high pressure feed gas now goes into column B and pressurizes that column. At high pressure the water is now adsorbed in column B and a dry, high pressure product is removed. Simultaneously, column A is depressurized and purged with a portion of the high pressure product from column B. Three days were required for start-up with an initially wet desiccant to dry 4000 ppm feed gas to 2 ppm [87]. Eventually the product lines out at less than 1 ppm moisture. Operation is not isothermal; instead, the heat of adsorption is retained in the bed and then utilized for desorption. Any appropriate adsorbent can be used, and 13 years of use without changing the adsorbent have been reported [88].

Typical experimental results for this simple pressure swing apparatus are shown in Figure 13. These results show the removal of CO_2 from helium using silica gel as the adsorbent and cycling from 59 psia to 21 psia with approximately 9.1 bed volumes fed per half cycle of two minutes [84]. The variable γ is the volumetric ratio of purge to feed. As can be seen from Figure 13 this purge ratio controls both the rate at which solute concentration is decreased and the final steady state concentration obtained. The theoretical results shown in Figure 13 were for the local equilibrium theory which ignores mass transfer limitations and dispersion effects and thus is essentially the same as the retention argument theory presented earlier.

A wide variety of other gas systems have been separated and several of these such as drying, hydrogen purification, separation of oxygen from air, and separation of normal from branched paraffins are commercial operations. The references for these separations are outlined later in Table 3. Many of the commercial applications use cycles which are more complicated than the simple two column arrangement shown in Figure 12. The purposes of these more complicated processes are to utilize some of the blow-down gases for repressurizing beds and to decrease the amount of purge gas required. The basic reasons for separation remain the same as those which are delineated below for the simple pressure swing process. Since pressure swing adsorption is the subject of a separate paper, the commercial processes will not be further discussed here.

The retention arguments previously developed for parametric pumping can be applied to pressure swing adsorption to explain the separation. However, there is one major difference between pressure swing adsorption and parametric pumping. In the pressure swing adsorption system less solute is adsorbed at low pressure <u>and</u> the volume of gas increases greatly when the gas

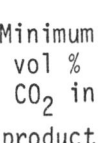

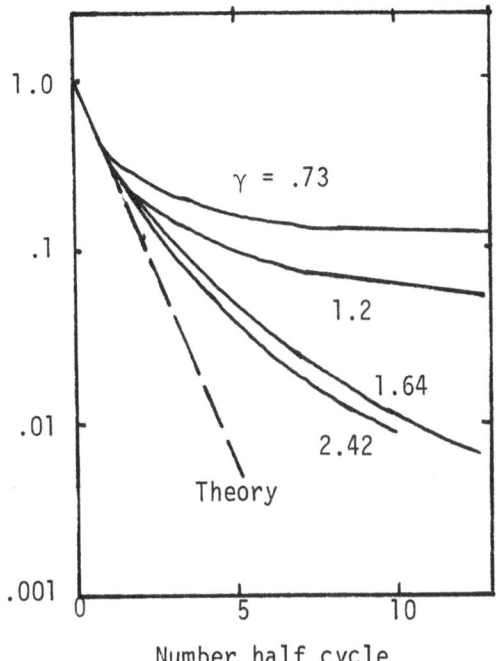

Minimum vol % CO_2 in product

Figure 13. Pressure swing adsorption removal of CO_2 from helium on silica gel [84].

is depressurized. Since the purging or solute removal step depends upon volumetric flow of gas and not mass flow, much less mass of gas is needed for purging than is run through the column during adsorption. Thus two physical phenomena aid the PSA separation while only one (the isotherm shift) is useful in parametric pumping. This difference may explain why pressure swing adsorption is a well accepted commercial process and parametric pumping is not.

The retention argument theory parallels the parametric pumping development. We assume that the adsorption isotherm is represented by a simple linear equilibria of the form

$$q = B(P)c^* \tag{15}$$

which is similar to eq. (1) but the constant is now a monotonically increasing function of pressure. If we assume that solid and fluid are everywhere in local equilibrium we obtain,

$$\frac{\text{amount solute in gas phase}}{\text{total amount solute in column}} = \frac{1}{1 + (\frac{1-\varepsilon}{\varepsilon})B(P)\rho_B} \tag{16}$$

Following the arguments presented earlier, the average solute velocity is,

$$u_{solute} = \frac{v}{1 + \frac{1-\varepsilon}{\varepsilon} B(P)\rho_B} \qquad (17)$$

If equal mass flow rates are used during adsorption and purge, the gas velocity, v, during purge will be much greater. Since u_{solute} will be greater during purge because of the pressure effect on the isotherm even if the gas velocity is constant, the purge step can use roughly the same gas volume as the adsorption step. Assuming that the ideal gas law, PV = nRT, is followed, the number of moles of purge gas required is,

$$\frac{n_{purge}}{n_{feed}} = \frac{P_{low}}{P_{high}} \frac{V_{purge}}{V_{feed}} = \gamma \frac{P_{low}}{P_{high}} \qquad (18)$$

In practice the volumetric purge to feed ratio, $\gamma = V_{purge}/V_{feed}$, should be from 1.1 to 1.5 to obtain complete cleanup of the feed [88]. The extra value of γ is required because of dispersion and diffusion of the gases. Since the half cycle times are equal, the gas velocity during purge will be equal to γ times the gas velocity during adsorption.

Pressure swing adsorption is analogous to the recuperative mode of parametric pumping since the pressure of the gas stream is changed at the ends of the column. However, the velocity of the pressure waves is very high (equivalent to a very steep slope of the thermal waves in Figures 7 and 8) so that as a first approximation the pressure changes can be assumed to be instantaneous. Thus the average molecular paths for either column A or B in Figure 12 can be drawn as having a slope

$$u_{solute(ads)} = \frac{v}{1 + \frac{1-\varepsilon}{\varepsilon} B(P_{high})\rho_B} \qquad (19)$$

during adsorption and

$$u_{solute(purge)} = \frac{\gamma v}{1 + \frac{1-\varepsilon}{\varepsilon} B(P_{low})\rho_B} \qquad (20)$$

during purge. In these equations v is the gas velocity during the adsorption step. The schematic of the paths of the average solute molecules is shown in Figure 14 for one of the two columns. Note that this is for continuous operation with $(n_{feed}-n_{purge})$ moles of gas removed as high pressure product every cycle and n_{purge} moles of low pressure waste removed every cycle, where n_{purge} and n_{feed} are related by eqn. (18).

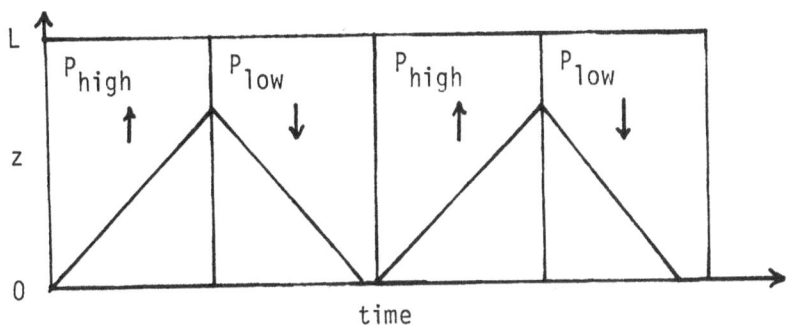

Figure 14.   Average paths in two column pressure swing adsorption

This retention argument shows that the separation is basically due to two effects: the isotherm change when the pressure is dropped and the volume expansion which occurs.   The latter effect is probably more important.   This model is vastly oversimplified and ignores mass transfer rates and dispersion except to lump them into γ.   It also ignores the complex phenomena which occur during blowdown and repressurization (see Mitchell and Shendalman [62] for more details).   The model does provide a qualitative explanation for the observed separations, and relates pressure swing adsorption to parametric pumping.

CYCLING ZONE ADSORPTION

Basic principles and experimental results

In parametric pumping and pressure swing adsorption the flow is reversed (reflux) during the purge step to regenerate the adsorbent and remove the solute.   Pigford and his coworkers [8,68] developed an alternate process called cycling zone adsorption which does not utilize flow reversal.   Instead unidirectional flow is employed and fractions of the outlet stream are collected as product.   This process is similar to parametric pumping and pressure swing adsorption in that changes in a thermodynamic variable such as temperature are used as a signal to the

the solute to adsorb or desorb. Increased concentration can be achieved by using several zones in series or by utilizing a moving thermodynamic wave in the column. Operation can be in the "standing wave" or "traveling wave" modes which are equivalent to the direct and recuperative modes of parametric pumping.

The basic cycling zone adsorption schemes were presented by Pigford and his coworkers [8,68] and were reviewed by Wankat [113]. The basic single column arrangements are shown in Figure 15. In the direct mode the entire bed is heated or cooled during each half cycle. Since the solute is adsorbed

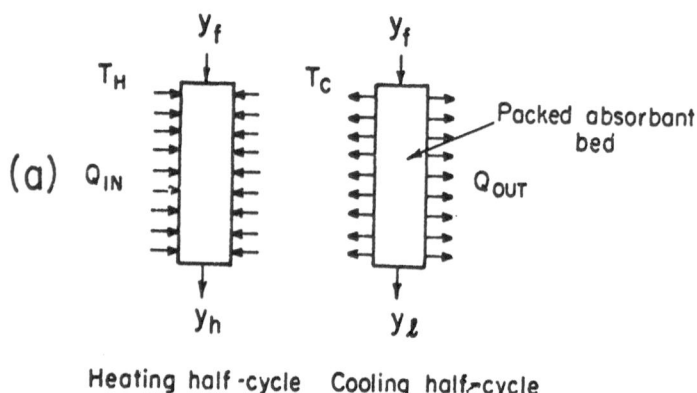

Heating half-cycle   Cooling half-cycle

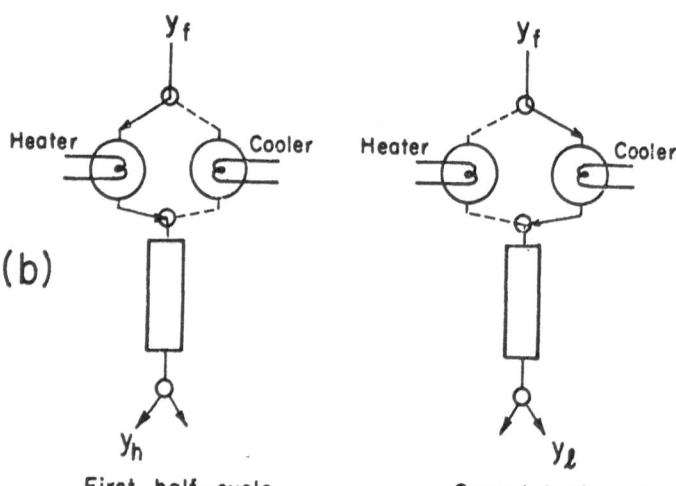

First half-cycle   Second half-cycle

Figure 15.   Single zone operation of cycling zone adsorption.
a) Direct mode.   b) Traveling wave mode [113].

during the cold half cycle, a dilute product is expected during
this period. When the bed is heated, the stored solute is
desorbed and the fluid becomes concentrated producing a concentrated product. This simple arrangement is thus very much like
normal adsorption with cocurrent regeneration. Continuous
product can be obtained by using two columns in parallel.

In travelling wave mode shown in Figure 15b the entering
fluid is heated or cooled and the bed itself is adiabatic. When
the moving temperature wave reaches a point in the bed, the
adsorbent is heated or cooled and either releases or stores
solute. The concentrated and dilute solute waves then exit at
times which are controlled by the movement of the thermal wave.

The experimental results obtained for removal of acetic acid
from water on activated carbon using both direct and traveling
wave modes are shown in Figure 16 [8]. The theoretical curves

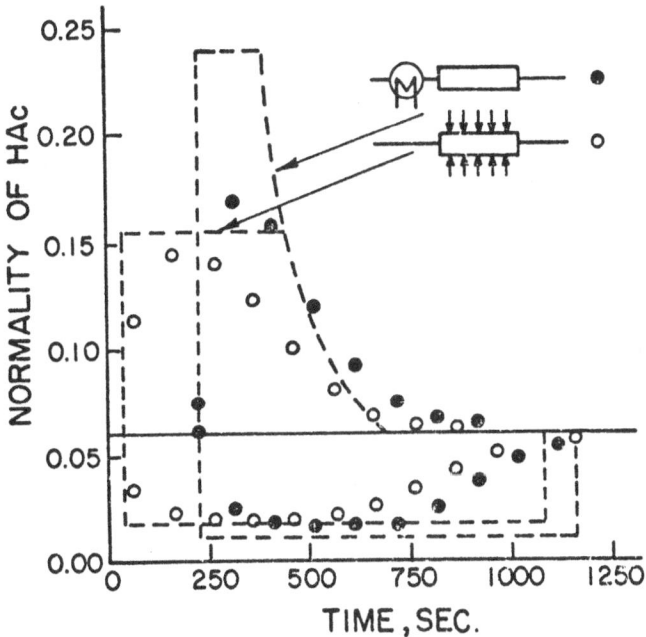

Figure 16. Cycling Zone Adsorption separation of acetic acid
from water on activated carbon in a single zone
[8].

shown in Figure 16 were obtained with the local equilibrium model.
The results shown in this figure have been shifted along the
time axis. The concentrations above the feed concentration were

obtained during the heating half cycle while those below were
during the cooling half cycle. Note that the travelling wave
results lag behind the temperature change because a finite
period is required for the thermal wave to traverse the column.
Also note that larger separations were obtained with the travelling wave mode. This result is in contrast with parametric pumping where the recuperative mode produces smaller separations and
will be explained later. As in parametric pumping more separation can be obtained if larger temperature differences are
employed or if a slower fluid velocity is used.

Two methods for increasing the cycling zone separation have
been developed. The first of these is to add more zones in
series while the second is to adjust the velocity of the moving
thermodynamic wave. The first method will be presented here but
the second will be presented after the retention argument reasons
for separation have been presented. The multiple zone systems
are shown in Figure 17 for the direct mode [8,68]. In this

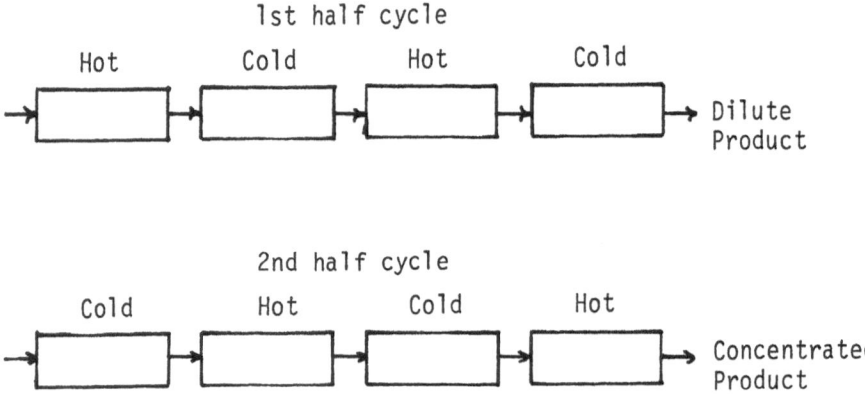

Figure 17. Multiple Zone Direct Mode Cycling Zone Adsorption

system alternate zones are at different temperatures and the
temperatures are changed totally out of phase with each other.
If the temperature switches are appropriately timed, a concentrated product leaving the first zone will undergo a temperature
change from cold to hot in the second zone and be further concentrated. In this way the separation obtained can be multiplied.
Multiple zone traveling wave systems have also been studied by
placing heat exchangers between the columns. In this case timing
becomes much more difficult because of the lag caused by the
traveling thermal wave.

An example of the experimental results obtained in two-zone
direct mode systems is shown in Figure 18 for a gas separation
[105]. The molecular sieve adsorbent preferentially adsorbed
nitrogen at low temperatures and desorbed nitrogen at high
temperatures; thus, gas leaving a cold region has a higher
concentration of oxygen. The temperatures of the two columns
were cycled as sine waves and were 180° out of phase. Thus

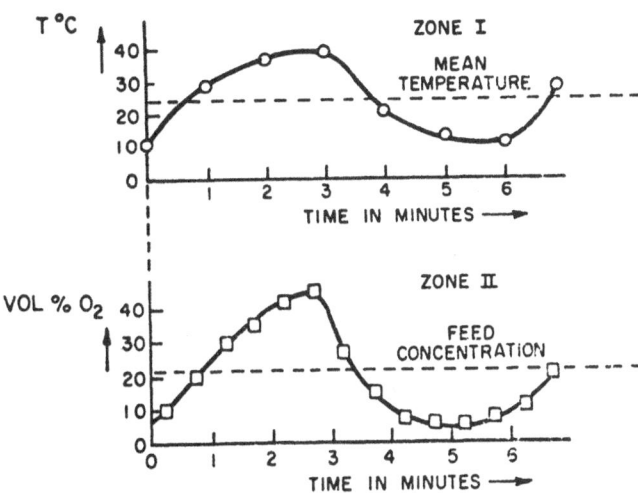

Figure 18. Cycling Zone Adsorption Separation of Oxygen and
Nitrogen on Molecular Sieve in Two Column Direct
Mode System [105].

when the first zone in Figure 18 is hot the second zone is cold,
and the product gas leaving the second zone is concentrated in
oxygen. One problem with this gas separation was the large
flow fluctuations caused by alternate adsorption and desorption
and the expansion and contraction of gas due to temperature
cycling. Two columns in series were used to partially minimize
this effect.

Retention Argument Explanation of Cycling Zone Adsorption

The retension argument used to explain parametric pumping
can easily be applied to cycling zone adsorption. Perusal of
the basic equations of that argument, eqs. (1) to (7) for direct
mode and eqs. (1) to (5) plus (11) to (14) for recuperative mode,
shows that none of these equations involved the idea that flow
direction changed. It was only in the application of these
equations that changes in flow direction were invoked. Thus
the velocity of an average solute molecule in cycling zone

472

adsorption will be described by eq. (5), the velocity of the
thermal wave by eq. (12), and the effect of temperature changes
by either eqs. (6) and (7) or eqs. (13) and (14). This means
that all of the basic ideas of the retention argument can be
carried over without modification to cycling zone adsorption as
long as the same assumptions are valid. This broad applicability
should not be surprising since the theory basically describes the
behavior of adsorption columns when a thermodynamic variable is
cycled, and both parametric pumping and cycling zone adsorption
utilize this phenomena to obtain separation.

The application of the retention argument to CZA is straightforward and is shown in Figure 19 for a two zone direct mode
system. The separation that is obtained can be observed by
following paths AA', BB' and DD'. The solute which is fed to

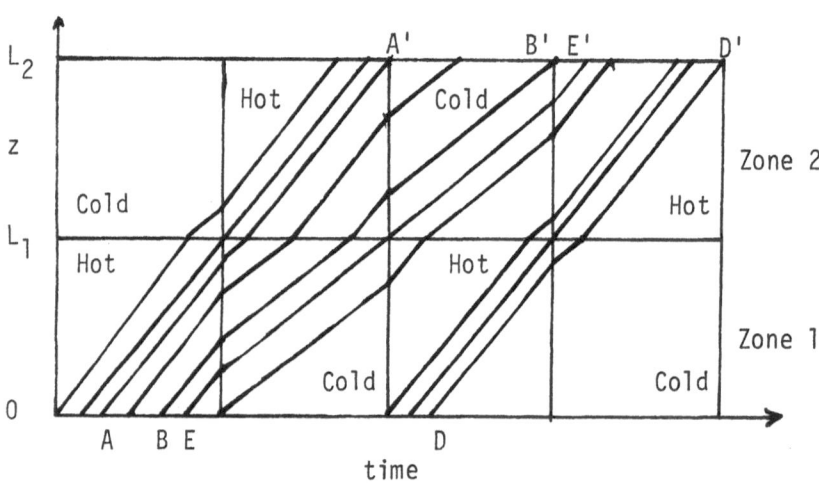

Figure 19.   Concentration Waves in Two Zone Direct Mode
Cycling Zone Adsorption

the first zone during time period AB exits the second zone
during period A'B'. Since this period is longer, the solute
exiting from the cold zone is obviously more dilute. This is
also indicated by the molecular paths being farther apart.
Solute exiting from the hot zone, B'D', is obviously concentrated
since BD > B'D'. A quantitative mass balance can also be done
by measuring the horizontal distances on Figure 19 and noting
that this distance is proportional to flow.

The effect of the temperature changes on the solute
concentration in the molecular paths can be deduced from eq. (7)
for linear isotherms. Eq. (7) applies when the adsorbent changes

temperature and thus refers to the vertical divisions in Figure 19. The horizontal division between zones 1 and 2 represents a feed to zone 2 and does not involve solute concentration changes (solute velocity changes because the temperature has changed). If we follow an arbitrary path between paths AA' and BB', we find that this solute undergoes two changes in adsorbent temperature and both are hot to cold. Eq. (7) shows that this hot to cold change decreases fluid concentration. Thus fluid exiting from A' to B' is dilute. In a similar fashion any path can be followed and the appropriate concentration changes determined. For example, path EE' goes from hot to cold and then cold to hot so it exits at the entering fluid concentration.

These paths can be followed to quantitatively predict the separation obtained. This was done by Baker and Pigford [8] for the very similar local equilibrium model. They found that the same separation is predicted for CZA as for PP if the same number of zones is used in CZA as cycles in PP. Thus the additional number of zones multiplies the separation and takes the place of the reflux and additional cycles in PP. This argument assumes no dispersion or mass transfer rate limitations. In practice, larger separations have been achieved with parametric pumps since the sharp concentration wave is not moved from zone to zone and less dispersion occurs. The multiplication of separation can be seen in Figure 19 by following different paths and determining whether the fluid concentration is increased or decreased.

It was previously mentioned that the other way separation can be increased in cycling zone adsorption is to use the traveling wave mode and adjust the wave velocity of the cyclic variable. This was first reported by Baker and Pigford [8] and has been exploited by Wankat and his coworkers [13,31,112-115]. With this procedure very large separations can be obtained with a single zone. The retention argument results for the recuperative mode parametric pump can be applied directly to cycling zone adsorption but now flow reversal is not used. This involves using eq. (12) to plot the thermal waves, eq. (5) to plot the concentration waves, and eq. (14) to determine the concentration changes when the adsorbent temperature changes. These plots are shown in Figure 20 for the three cases where $u_{thermal} > u_{solute}(T_H)$, $u_{solute}(T_H) > u_{thermal} > u_{solute}(T_c)$ and $u_{solute}(T_c) > u_{thermal}$. Different predictions are made in each of these cases.

The usual case for temperature waves in liquid is shown in Figure 20a where the thermal wave velocity is greater than the solute wave velocity. This corresponds to the traveling wave separation shown in Figure 16. Following the paths and using eq. (14) the outlet concentrations can easily be determined. A quantitative mass balance can be done to show that the solute input equals output.

474

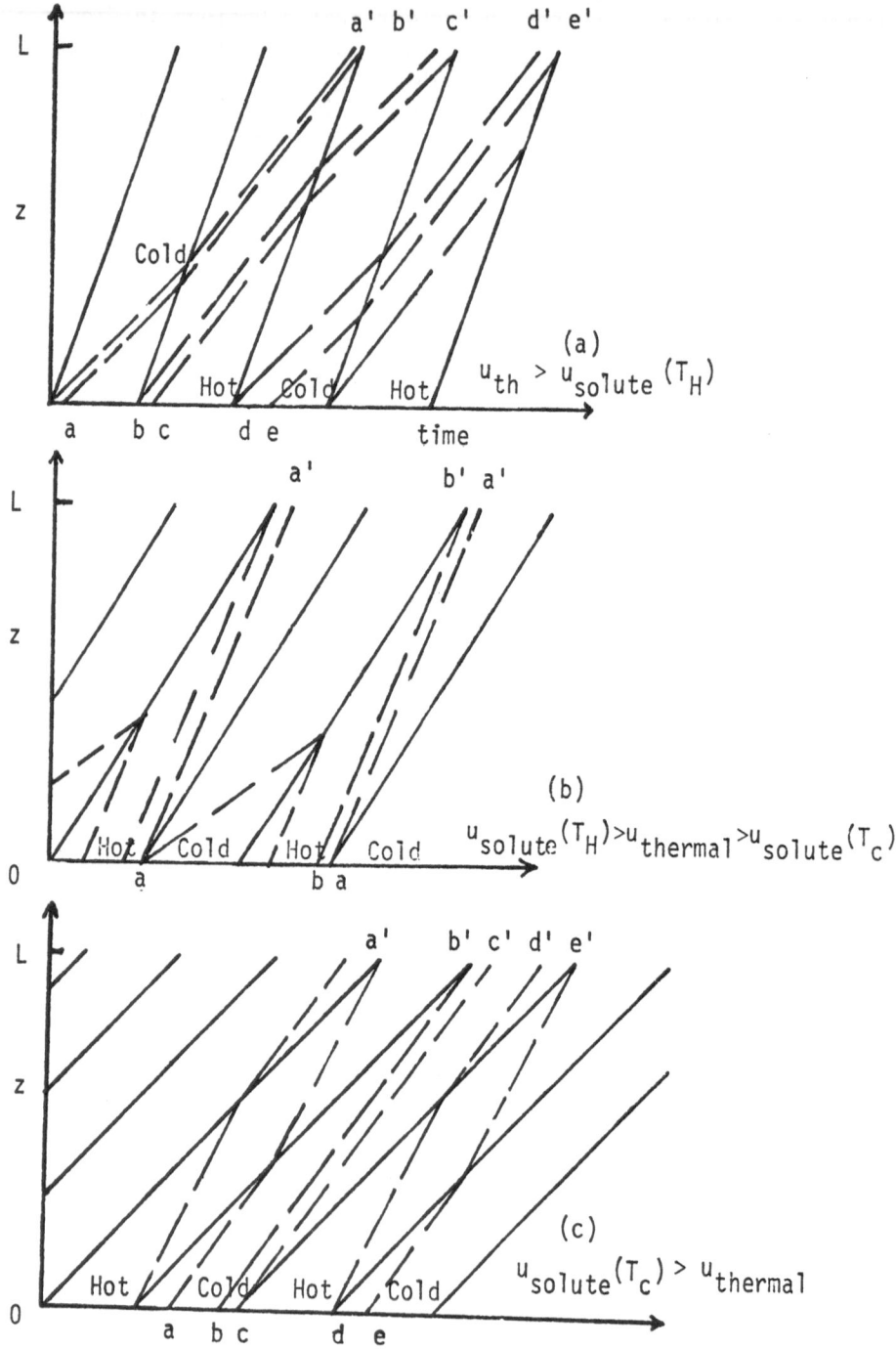

Figure 20.   Concentration and Thermal Waves in Traveling
            Wave Cycling Zone Adsorption

When the thermal wave velocity is decreased sufficiently, the results shown in Figure 20b are obtained. This corresponds to the traveling wave separations obtained with pH as the cyclic variable [13,31] which are shown later in Figure 21. Both Figure 20b and eqn. (14) predict physically impossible results in this case which means that a somewhat more sophistica- ted analysis is required (and will be done later). Eq. (14) predicts that C_1/C_2 is negative which is physically impossible. Figure 20b shows that all solute input between a and b will collect at the cold to hot thermal wave boundary and will all exit with the thermal wave at b' at infinite concentration. From b' to a' solute concentration is unchanged. Between a' and the point of infinite solute concentration no solute exits. These results are obviously physically unrealistic, but we can expect large separations when the velocity of the cyclic variable lies between the solute wave velocities. This is caused by "trapping" of the solute at the wave front of the cyclic variable and can be seen in Figure 21.

In Figure 20c the thermal wave velocity has been further decreased until it is less than both solute velocities. This circumstance will naturally occur in many gas systems. Eq. (14) predicts that solute undergoing a temperature shift from hot to cold is concentrated instead of diluted as is expected in liquid systems. Solute going from cold to hot is diluted. Thus reverse separations are predicted. Figure 20c agrees with these predictions since solute entering between a and b undergoes a temperature change from cold to hot and exits between a' and b' as a dilute material. Physically, this can occur because when it is hotter the solute velocity increases and the faster moving solute molecules are further apart and dilute. Similar results have been predicted in parametric pumping [119] and are explained by analogy. No experimental results corresponding to this case have been reported in the literature. Predicted separation for these gas systems are small.

Extensions of Cycling Zone Adsorption

Several extensions of cycling zone adsorption have been reported in the literature. These include the use of variables other than temperature as the cyclic variable, studies on extraction cycling zone separation and continuous multicomponent separation.

The separation of fructose-glucose-water mixtures using pH as the cyclic variable has been extensively studied [13,31]. In this system the pH wave velocity was naturally between the solute wave velocities at low and high pH. Thus the situation shown in Figure 20b was valid and large separations were expected.

476

The results of Busbice and Wankat [13] for separation of fructose
from water as plotted by Nelson et al [64] are shown in Figure 21.
pH was varied between 5.0 and 8.5. The theoretical results were
obtained with a continuous flow equilibrium stage model using
4 stages. Note that large separation factors are obtained since
very little fructose exits the column for long periods. The peak
concentration exits slightly before the high pH wave. This
phenomena will be physically explained later when the local
equilibrium model is presented.

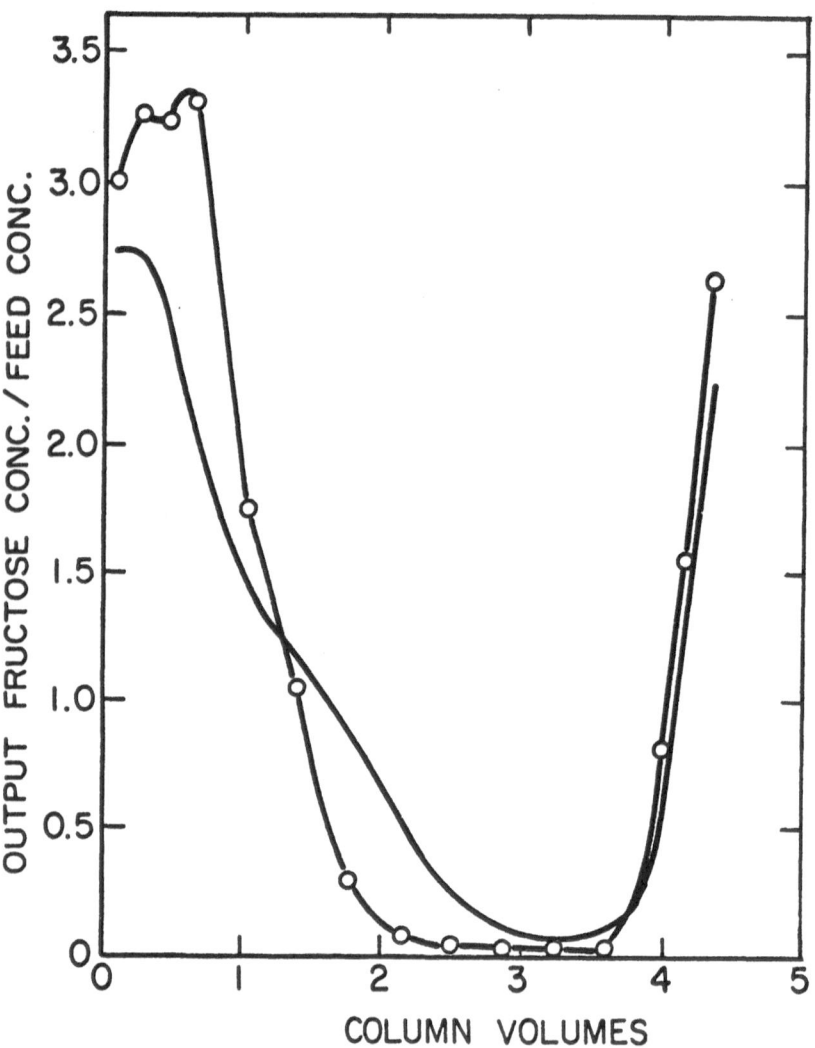

Figure 21. Separation of Fructose-Water by Cycling Zone
Adsorption Using pH as the Cyclic Variable.
O-O data. ——— theory. [64].

The extraction studies are referenced in Table 2. All of the cycling zone studies used test tube systems to keep the stationary phase in place, but column or mixer-settler systems could also be employed. The separation of bromocresol green and phenol red from water is shown in Figure 22 [115]. Na_2CO_3 concentration was used as the traveling cyclic variable. This separation introduces the idea of multicomponent separations which was first developed by Wankat [114].

The technique for multicomponent separations can be explained by using the retention arguments developed earlier and referring to Figure 22a. As the Na_2CO_3 concentration increases both dyes are more strongly attracted to the stationary 2-butanol phase. Ideally, the levels of the cyclic variable are chosen as follows: At the high level (5N) both dyes are strong attracted to the 2-butanol and hence both solute velocities are less than Na_2CO_3 velocity. At the intermediate level (0.1N) the phenol red is not strongly attracted to the stationary phase but the bromocresol green still is, and thus phenol red would ideally move faster than the cyclic variable but the bromocresol green moves slower. At the low Na_2CO_3 level (0.0005N) both dyes are more strongly attracted to the mobile aqueous phase and ideally both dyes would move faster than the cyclic variable. Now at the wave front changing from 5N to 0.1 N, phenol red is "trapped" at this wave boundary (corresponds to Figure 20b) resulting in a large concentration peak. The bromocresol green is not trapped since it's velocity is slower than Na_2CO_3 at both Na_2CO_3 levels (see Figure 20a). At the wave front change from 0.1 N to 0.0005 N phenol red would ideally move faster at both levels than the cyclic variable and would look like Figure 20c. At this wave front bromocresol green would ideally move faster than the cyclic variable at the 0.0005N level and slower at the 0.1N level. Bromocresol green would then be "trapped" at this wave front and exit in high concentration. Under ideal circumstances one component would concentrate at one wave front and the other at a different wave front. This separation can be extended to many components by adding more steps in the input wave shown in Figure 22a. With thermal waves an increasing series of steps would be used since the highest temperature gives the fastest solute velocities.

The results shown in Figure 22b are not ideal. Since the Na_2CO_3 remained almost entirely in the aqueous phase, its wave velocity was equal to the fluid velocity. Thus, it was impossible to have u_{solute} greater than the cyclic wave velocity. At 0.0005N the phenol red was barely soluble in the 2-butanol and it traveled at the fluid velocity also. Phenol red then stayed at its feed concentration and exits at its feed concentration when the bromocresol green peaks giving the "shoulder" shown

478

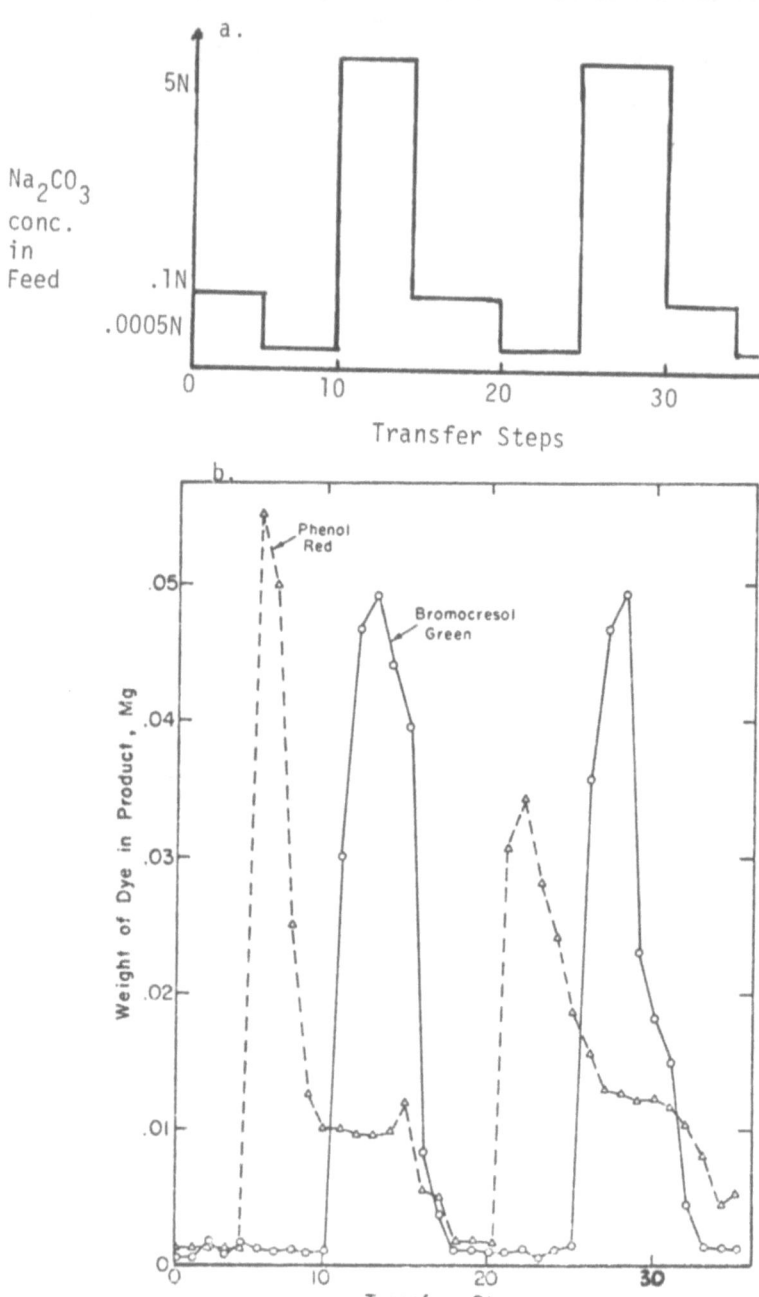

Figure 22. Separation of Two Dyes by Traveling Wave Cycling
            Zone Adsorption. a. Input variation of Cyclic
            variable. b. Output weights of dye. [115].

in Figure 22b. The results shown in Figure 22b are startup results with initially concentrated solutions in all tubes; thus, the phenol red peak concentration decreases as the limit cycle is approached.

Experimental results in columns [31] show that partial multicomponent separations are easy to obtain, but very clean separations are difficult to obtain because of solute-solute interactions and the effects of non-linear isotherms. Also, for many systems the natural velocity of the cyclic variable is too high and this method is difficult to apply. Research on various ways of reducing the cyclic variable velocity should prove fruitful in overcoming this problem. An alternate multicomponent technique has been developed [117] but has not been experimentally tested.

This completes the physical and qualitative description of the cyclic separations. In the next section the experimental literature will be summarized in annotated tables. Then we will proceed to the local equilibrium analyses of these separations.

CHEMICAL SYSTEMS STUDIED

In the previous sections the basic cyclic operations have been explained and some data has been presented. A wide variety of other chemical systems have been studied using many different sorbents. In this section these separations are briefly outlined in three tables. These tables list the sorbent (or solvent) used, the separation studied, the type of cyclic operation and the cyclic variable used, and give the references. Table 1 gives this information for liquid systems where a solid sorbent is used. Table 2 lists the liquid-liquid extraction systems which have been studied. Table 3 lists the cyclic gas studies which have been reported in the literature. Hopefully, these tables will serve to economically show the wide variety of systems which have been explored and will rapidly lead the interested reader to the appropriate references.

THEORY: LOCAL EQUILIBRIUM MODEL

The most popular theoretical description for the cyclic separations has been the local equilibrium model first developed by Pigford et al [67]. This model has also been used extensively for adsorption systems [5,85,106].

The mass and energy balances for sorption processes occurring in a non-isothermal fixed bed can be derived from differential balances on each phase. We will assume that radial gradients in

Table 1. Sorbents and Chemical Systems Used in Cyclic Liquid Systems

Sorbent	Chemical System	OPN	Variable	Reference
Activated carbon	acetic acid from water	CZ	T	[8,68]
		2-D CZ	T	[116]
	urea from water	PP	T	[92]
	oxalic acid from water	PP	T	[72]
	fructose-glucose-water	PP	T	[1]
Fullers earth	fructose-glucose-water	PP	T	[20]
Silica gel	toluene-n-heptane	PP	T	[19,21,23,124,125]
		CZ	T	[78]
	benzene-n-hexane	PP	T	[108]
	toluene-n-heptane-aniline	PP	T	[24]
Affinity adsorbent trypsin inhibitor on Sepharose 4B	trypsin-water	PP	pH	[83]
	trypsin, α-chymotrypsin water	PP	pH	[83]
Dihydroxyboryl succinamyl aminoethyl cellulose	glucose-fructose-water	CZ	pH,T	[13]
		CZ	pH	[31]
C-18 bonded phase HPLC adsorbent	dipetides from water	CZ	conc. dichloro-acetic acid	[63]

Sorbent	Chemical System	OPN	Variable	Reference
Solid BF_3-dimethyl sulfoxide complex	$B^{10}F_3$ from $B^{11}F_3$	PP	T	[81]
Ion Exchange Resins				
weak electrolyte resins-mixed bed	NaCl from water	PP	T	[79,123,125]
	(composite resin)	CZ	T	[35,54]
		CZ	T	[86]
(Composite resins)	Desalination brackish water (commercial)	"Sirotherm"	T	[10,11]
Ion retardation resin	NaCl from water	PP	T	[40,77,93]
	$NaNO_3$ from water	PP	T	[22]
Cation exchange resin	$K^+ - Na^+$ from water	PP	pH	[80]
	$K^+ - H$ – water	PP	T	[15]
	$K^+ - Na^+ - H^+$ – water	PP	T	[15]
	glucose-fructose-water	PP	T	[25]
Sephadex resin	albumin-haemoglobin-water	PP	T	[26]
	albumin-haemoglobin-water (inverse separation)	PP	pH	[27]

CZ = Cycling Zone Adsorption   PP = Parametric Pumping   T = Temperature

Table 2. Solvents and Chemical Systems Used in Extraction Cyclic Systems

Solvent	Chemical System	OPN	Variable	Reference
Diethyl ether	acetic acid from water	PP	T	[109]
Toluene	diethylamine from water	CZ	T	[110,112]
	2-D analog CZ		T	[116]
2-butanol	phenol red from water	CZ	conc. Na_2CO_3	[113]
	phenol red – bromocresol green–water	CZ		[115]

CZ = Cycling Zone   PP = Parametric Pumping   T = Temperature

Table 3. Adsorbents and Chemical Systems Used in Cyclic Gas Systems

Adsorbent	Chemical System	OPN	Variable	Reference
Alumina	removal of water vapor and oil vapors from air (commercial)	PSA	P	[87,88]
	air drying (commercial)	PSA	P	[127]
Activated Carbon	Methane from hydrogen (commercial)	PSA	P	[32,88]
	hydrocarbon vapors from air	PSA	P	[51]
	ethane-propane	PP	T	[48,49]
	argon-propane	PP	T	[49]
	propane-propylene (no sepn)	PP	T	[49]
	methane from helium	CZ	T	[9,68]
Fibrous Carbon	methane from nitrogen	PP	T	[60]
Molecular Sieves	water vapor from air (commercial)	PSA	P	[12,88]
	hydrogen purification (commercial)	PSA	P	[12,32,60,88,89]
	n-paraffin from branched (commercial)	PSA	P	[12,44,53,57,58]
		Pressure Variation	P	[102]

Adsorbent	Chemical System	OPN	Variable	Reference
Molecular Sieves (cont.)	oxygen from air (N_2 is adsorbed)(commercial)	PSA	P	[51,56,88]
	oxygen from air	CZ	P	[30]
	nitrogen from oxygen		T	[105]
	methane from nitrogen		P	[52,104]
Silica Gel	water vapor from air (commercial)	PSA	P	[87,88]
	methane from hydrogen (commercial)	PSA	P	[32]
	carbon dioxide from helium	PSA	P	[62,84]
	sulfur dioxide from air	PP similar to CZ	T	[65]
	ethane-propane-N_2		conc. feed in carrier	[97]
	ethane-propane-CO_2	"	"	[97]
	methane-ethane-He	"	"	[97]
	methane-ethane-CO_2	"	"	[97]
	ethane-ethylene-He	"	"	[97]
	ethane-ethylene-N_2	"	"	[97]
	ethane-N_2-He	"	"	[97]
Benzyl ether coated on firebrick	propylene-butene-1,He	"	conc. feed in carrier	[97]
	propane-propylene-He	"		[97]

Adsorbent	Chemical System	OPN	Variable	Reference
Palladium deposited on alumina	deuterium from hydrogen	PSA	P	[121]
Unspecified	CO$_2$ from air (commercial)	PSA	P	[69]
	water and hydrogen from air (commercial)	PSA	P	[69]
	air drying (commercial)	PSA	P	[28,122]
	air drying (commercial)	Combination Refrigeration Heating Pressure Swing	P,T	[28]
	hydrogen purification	PSA	P	[17]
	hydrogen purification	combined cryogenic and PSA	T,P	[17]

CZ = Cycling Zone Adsorption   PP = Parametric Pumping   PSA = Pressure Swing Adsorption
P = Pressure   T = Temperature

temperature, velocity and concentration are negligible, that thermal diffusion and pressure diffusion can be neglected, and that chemical reactions do not occur (reversible chemisorption is allowed). Then for one of the solutes the solute balance on both the fluid and solid phases is:

$$\varepsilon \frac{\partial c}{\partial t} + \rho_B(1-\varepsilon) \frac{\partial q}{\partial t} + \varepsilon v \frac{\partial c}{\partial z} + \varepsilon\, D\, \frac{\partial^2 c}{\partial z^2} = 0 \qquad (21)$$

and

$$\rho_B(1-\varepsilon) \frac{\partial q}{\partial t} = -\,k_M a_p (c^* - c) \qquad (22)$$

where ρ_B is the bulk density of the solid including any fluid contained in the pores, and c^* is the fluid concentration in equilibrium with the solid. When these forms of the solute mass balances are used, the equilibrium isotherm relating fluid concentration c^* to q must include the fluid in the pores. For multicomponent systems eqs. (21) and (22) must be written for each solute. In general, these equations will be coupled since the amount adsorbed depends on the concentration of all solutes.

With the same assumptions that were made previously plus the assumptions that no latent heat effects occur, no electrical or magnetic fields are present, no radiant heat transfer occurs, and viscous heating can be neglected the energy balance for both phases becomes

$$\rho_f C_f \varepsilon \frac{\partial T}{\partial t} + \rho_B C_B(1-\varepsilon) \frac{\partial T_S}{\partial t} + \rho_f C_f v \varepsilon \frac{\partial T}{\partial z}$$

$$- D_T\, \rho_f C_f\, \varepsilon \frac{\partial^2 T}{\partial z^2} = h_w A_w (T_w - T_{Amb}) - \frac{C_w W}{A_c} \frac{\partial T_w}{\partial t} \qquad (23)$$

where C_B is the bulk heat capacity of solid particles. The energy balance on the solid phase is,

$$\rho_B C_B(1-\varepsilon) \frac{\partial T_S}{\partial t} = -\,h_p a_p (T_s - T) + (1-\varepsilon)\,\rho_B \Delta H_{ads} \frac{\partial q}{\partial t} \qquad (24)$$

In addition to the mass and energy balances an equilibrium expression of the general form

$$q = q(T_s, c_1^*, c_2^* \ldots c_n^*) \qquad (25)$$

is required.

Despite the simplifications involved in the development, these equations are formidable and have resisted analytical solution, but they have been numerically integrated [79]. To obtain convenient analytical solutions further assumptions are usually invoked.

The mass and energy balances can be greatly simplified by making the set of assumptions known as the local equilibrium assumptions. The heat and mass transfer rates are assumed to be high so that the solid is locally in equilibrium with the fluid. Thus $c = c^*$ and $T = T_S$. In addition, the axial dispersion and diffusion terms will be neglected. The density and heat capacities are assumed to be constant during each portion of the cycle. The equilibrium is assumed to depend upon only one solute concentration for each solute. The heat of adsorption is assumed to be small and axial dispersion of the temperature wave is neglected. For direct mode operation the entire column is assumed to be at the set temperature and temperature changes are assumed to be rapid so that radial gradients can be ignored. Thus for the direct mode neither energy balance equation is required. For the traveling wave mode in addition to assuming that $T = T_S$, the column is assumed to be adiabatic and the wall temperature is assumed to be the same as the fluid temperature. The equations will be developed here for the case where temperature is the thermodynamic variable. However, the mathematical approach is general and can be modified for other cyclic variables such as pH, concentration or pressure.

The local equilibrium assumptions greatly simplify the mass and energy balances. Since the rates of heat and mass transfer are very high and the heat of adsorption can be neglected, neither eq. (22) nor (24) are needed. After suitable simplification and manipulation, the mass balance eq. (21) becomes

$$\frac{\partial c}{\partial t} + u_c \frac{\partial c}{\partial z} = - \frac{u_c}{v\varepsilon} (1-\varepsilon)\, \rho_B \frac{\partial q}{\partial T}\frac{\partial T}{\partial t} \tag{26}$$

where

$$u_c = \frac{v}{1 + \frac{(1-\varepsilon)}{\varepsilon}\rho_B \frac{\partial q}{\partial c}} \tag{27}$$

The energy balance equation (23) simplifies to,

$$\frac{\partial T}{\partial t} + u_{th} \frac{\partial T}{\partial z} = 0 \tag{28}$$

where

$$u_{th} = \frac{V}{1 + \dfrac{\rho_B C_B (1-\varepsilon)}{\varepsilon \, \rho_f C_f} + \dfrac{W C_w}{\varepsilon A_c \rho_f C_f}}$$ (29)

These equations will be solved by the method of characteristics for half a cycle of any process, and then the results will be made specific for the different processes. The energy balance eq. (28) can be solved independently of the mass balance. Appropriate general initial and boundary conditions are

$$T = T_0(z) \qquad t = 0, \; 0 \le z \le L$$ (30)

$$T = T_F(t) \qquad z = 0, \; t \ge 0$$ (31)

where $T_0(z)$ is a specified initial temperature distribution and $T_F(t)$ is a specified feed temperature dependence. The solution of eq. (28) subject to eqs. (30) and (31) by the method of characteristics is straightforward. Temperature will be constant along characteristics given by,

$$\frac{dz}{dt} = u_{th}$$ (32)

On a graph of axial distance, z, versus time, t, eq. (32) plots as a straight line of slope u_{th}. Along each of these "characteristic lines" temperature is constant. Thus this solution predicts that temperature waves propogate through the column at a velocity u_{th} and that the wave shape is unchanged as it passes through the column (i.e., square waves remain square waves, etc.). This means that the feed temperature function given in eq. (31) can be used as the solution if it is moved through the column at a velocity u_{th}. Thus the temperature within the column is given by the function

$$T = T_F(t - z/u_{th})$$ (33)

These distance versus time plots have been shown previously when the retention argument was developed. In that argument the thermal wave was shown to have a velocity of $u_{thermal}$. Comparison of eqs. (29) and (12) shows that u_{th} equals $u_{thermal}$ and the local equilibrium and retention argument predictions are exactly the same for the temperature wave. Thus the thermal wave solutions shown in Figures 7 and 8 for parametric pumping and in Figure 20 for cycling zone adsorption are also predicted by the local equilibrium model. Thus the retention argument is a physical description of the local equilibrium model.

Solution of the solute balance eq. (26) is somewhat more complex. For direct mode operation temperature is constant throughout each portion of the cycle, $\frac{\partial T}{\partial t} = 0$, and the RHS of eq. (26) is zero. This is also true for the traveling wave mode if square waves are used as the input function except where the temperature changes. Thus for the majority of the cycle eq. (26) simplifies to

$$\frac{\partial c}{\partial t} + u_c \frac{\partial c}{\partial z} = 0 \tag{34}$$

Solution of eq. (34) by the method of characteristics is simple since concentration will be constant along the characteristics and the technique used for finding the temperature can be applied.

General initial and boundary conditions similar to those used for temperature are also appropriate for concentration.

$$c = c_0(z), \quad t = 0, \quad 0 \le z \le 1 \tag{35}$$

$$c = c_F(t), \quad z = 0, \quad t > 0 \tag{36}$$

where $c_0(z)$ and $c_F(t)$ are the initial concentration profile in the column and the feed concentration profile, respectively.

The solution by the method of characteristics states that concentration is constant along characteristic lines given by

$$\frac{dz}{dt} = u_c \tag{37}$$

On a plot of distance versus time the characteristics are straight lines with a slope of u_c. This solution is only valid in the constant temperature portions of the cycle and when characteristics do not intersect each other. Eq.(27) shows that for non-linear isotherms u_c depends upon both concentration and temperature (in general, on the cyclic variable) through the term $\partial q/\partial c$. In the constant temperature portions of the cycle the characteristics will be straight.

When the temperature of the adsorbent changes a concentration discontinuity occurs. The concentration change can be determined by a mass balance over an element of the column at the temperature boundary. This mass balance was developed earlier and presented as eq. (13) which is also valid for the local equilibrium theory. For the direct mode the term $\varepsilon v/u_{th}$ drops out.

For the linear isotherm $q = A(T)c^*$, the solution for the concentration shift given by eq. (13) is particularly simple and is given by eq. (14).

Note that for linear isotherms the concentration wave velocity in eq. (27) simplifies to

$$u_c = \frac{v}{1 + (\frac{1-\varepsilon}{\varepsilon})\, \rho_B\, A(T)} \tag{38}$$

In this case u_c is concentration independent. Comparison of eqs. (38) and (5) shows that u_c equals u_{solute} for linear isotherms. Thus for linear isotherms the local equilibrium model solution and the retention argument solution are the same. Thus, the paths shown in Figures 3, 7 and 8 for parametric pumping, and in Figures 19 and 20 for cycling zone adsorption can also be considered as characteristics. Since concentration remains constant along these characteristics except at the temperature boundaries where the change in concentration is determined by Eq. (14) the characteristics can be used to predict the exiting concentrations. Thus for linear isotherms the local equilibrium solutions have already been presented in disguise.

The local equilibrium solutions using linear isotherms often overstate the separations achieved. This was shown in Figure 5 for an open parametric pump with a liquid system and in Figure 11 for a batch gas system. This model is very useful as a correlating equation for the data if the "equilibrium" parameters are fit to the data obtained from one run of the apparatus. References for this approach and other uses of the local equilibrium model are given in Table 4.

Solution for Non Linear Isotherms

For nonlinear isotherms eq. (27) shows that the concentration wave velocity depends upon both temperature and concentration. Now it is possible for characteristics to intersect each other if they are at different concentrations. The intersection of two characteristics of different concentration would imply that two different concentrations could occur at the same location simultaneously and this is clearly physically impossible. The result is a "shock" wave where the concentration jumps from one concentration to the other. The velocity of the shock wave is

Table 4. Applications of Local Equilibrium Model
in Cyclic Separations

Application	OPN	Ref.
Development of local equilibrium Model, graphical and analytical solution-batch fit batch separation data to model	PP	[67]
Development of model and graphical solution-batch	PP	[96]
Generalization of analytical solution-batch	PP	[4]
Effect of non-linear isotherms-batch	PP	[7,70]
Effect of reservoir dead volume-batch	PP	[18,39]
Effect of non-mixed reservoirs-batch	PP	[98]
Comparison with experiment, agreement poor	PP	[20,39,40, 65,81,83]
Method for batch separation of multi-component mixtures-theory	PP	[14]
Open system operation-theory	PP	[18,19,39]
Comparisons of continuous and semi-continuous systems	PP	[21,39]
Fit open system experimental data to linear-equilibrium theory--reasonable fit but semi-empirical approach	PP	[1,19,21-25]
Optimization of open systems	PP	[22]
Application to recuperative mode (traveling wave) Note: all other PP studies are direct mode.)	PP	[95,119]
Effect of Sine instead of square waves batch	PP	[75]
Non-linear and competing solutes-batch and open	PP	[16]
Analysis of energy use in direct and recuperative modes	PP	[119]

Application	OPN	Ref.
Two-dimensional analogs	PP,CZ, PSA	[116,118]
Reviews of parametric pumping	PP	[76,92,94]
Direct mode, linear equilibrium model development	CZ	[45]
Direct and traveling wave modes, linear and non-linear equilibria model development. Comparison of theory and experiment.	CZ	[8]
Traveling wave analysis and optimization with linear equilibrium	CZ	[61]
Multicomponent separations with linear equilibria - theory.	CZ	[114]
Review of cycling zone separation	CZ	[113]
Multicomponent separations with linear or nonlinear equilibria	CZ	[117]
Development of model and comparison with experiment. Non-linear equilibria	pressure cycle different from PSA	[104]
Model development and comparison with experiment. Linear equilibria	PSA	[84]

controlled by a macroscopic mass balance instead of the solute balance given by eqs. (21) and (26). For favorable isotherms such as Langmuir or Freundlich the concentration wave velocity increases as the concentration increases. The characteristics can intersect and a shock wave can occur if material of low concentration is being displaced by material of higher concentration (and hence a higher value of u_c).

The velocity of the shock wave can be calculated from a macroscopic mass balance similar to that used to calculate the change in concentration when the temperature was changed. Pick a column section of height Δz and do a mass balance on this section for a time period $\Delta t = \Delta z/u_{sh}$ during which time the shock wave passes through this section of the column. If c_{j+1} is the concentration after the shock wave has passed and c_j is the concentration before the shock wave,

$$u_{sh} = \frac{v}{1 + \rho_B \left(\frac{1-\varepsilon}{\varepsilon}\right) \frac{q_{i+1}-q_i}{c_{i+1}-c_i}} \tag{39}$$

The shock wave velocity thus depends upon the concentrations on both sides of the shock wave and on the equilibrium expression used.

The relationship between the shock velocity and the concentration wave velocities will be illustrated for the specific case of a Freundlich isotherm, $q = A(T)c^k$. The solute wave velocity is,

$$u_{c_j} = \frac{v}{1 + \left(\frac{1-\varepsilon}{\varepsilon}\right)\rho_B A(T) kc_j^{k-1}} \tag{40}$$

Since $0 \leq k \leq 1$ u_{c_j} will increase as the concentration increases if k is not zero or one. The shock wave velocity becomes

$$u_{sh} = \frac{v}{1 + \left(\frac{1-\varepsilon}{\varepsilon}\right)\rho_B A(T)\left(\frac{c_{j+1}^k - c_j^k}{c_{j+1} - c_j}\right)} \tag{41}$$

for the Freundlich isotherm. If $k = 1$ the Freundlich isotherm becomes a linear isotherm, eqs. (40) and (41) simplify to the linear isotherm concentration wave velocity, (38), and no separate shock wave occurs. If $0 < k < 1$ and $c_{j+1} > c_j$ (j+1 is after the shock, j before), then $u_{c_{J+1}} > u_{sh} > u_{c_j}$ and a shock wave does occur. Shock waves are plotted as lines of slope u_{sh}.

We have seen that shock waves occur for favorable isotherms if a fluid is displaced by a more concentrated fluid. What happens if a fluid is displaced by a less concentrated fluid? In this case since $c_j > c_{j+1}$ and $u_{c_j} > u_{c_{j+1}}$, the characteristics cannot intersect and a shock wave cannot occur. Instead a diffuse wave occurs where the concentration and hence the velocity varies. As the concentration decreases smoothly from c_j to c_{j+1} the solute velocity decreases from $u_c(c_j)$ to $u_c(c_{j+1})$.

With nonlinear isotherms the retention argument solutions
are no longer valid and must be modified by including shock and
diffuse waves. Use of these more realistic isotherms will
prevent the prediction of physically unrealistic infinite
concentrations as shown in Figure 20b for cycling zone adsorption.
To illustrate the use of non-linear isotherm solution this one
zone CZA system is shown in Figure 23. Since both hot and cold

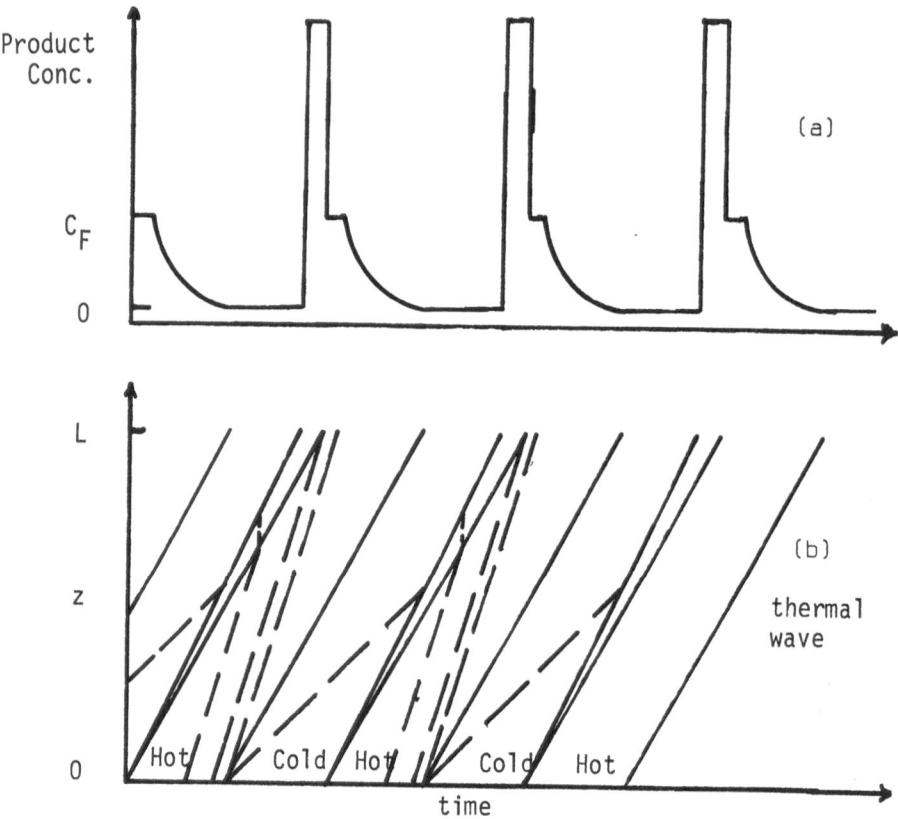

Figure 23. Non-Linear Isotherm Solution for Traveling Wave Cycling
Zone Adsorption with $u_c(T_H) \Big|_{C=C_F} > u_{th} > u_c(T_C) \Big|_{C=C_F}$

    A. Predicted Outlet Concentrations
    B. Characteristics

characteristics would intersect the thermal wave, a high concentration is built up. As the concentration increases the solute wave
velocity increases according to eqs. (38) and (40). The concentration increases until $u_c(T_C)$ becomes greater than u_{th} preventing an
infinite concentration. This concentration can be determined by

solving eq. (13). This high concentration now displaces a fluid
of lower concentration creating a shock wave. These characteristics and the shock wave are shown in Figure 23b. After the
shock wave, a small period at the feed concentration occurs because
of breakthrough of material fed during the hot cycle. Then a
diffuse wave occurs as material of feed concentration is displaced by less concentrated fluid. The resulting product
concentrations are shown in Figure 23a. If an additional zone
were used the feed to this zone would be the product concentrations shown in Figure 23a. Now shock and diffuse waves would
occur at the feed point because of the varying concentrations in
the feed.

Application of the non-linear theory to model cycling zone
adsorption separations is straightforward. This was done by
Baker and Pigford [8]   using the Freundlich isotherm and the
results were shown in Figure 16 for a situation similar to
Figure 20A. Note that dispersional processes spread out the
steep shock wave, but the fit between theory and experiment is
quite good considering the simplicity of the theory.

Application of the non-linear theory to parametric pumping
requires external mass balances to determine the concentrations
entering the column from the combination of feed and reservoirs.
In addition, the spread of diffuse waves while flow is in one
direction and the overtaking of the diffuse wave by a shock wave
when flow is in the other direction must be considered. The
start-up behavior can be determined by following each cycle,
using a procedure very similar to that shown for cycling zone
adsorption. Limit cycle behavior can also be directly calculated
[16].

External Balances

The solution is completed by considering the external mass
balances which include the effects of feed, products and reflux
from the reservoirs. These external balances were done for batch
direct mode parametric pumps with mixed [4,67] and unmixed [98]
reservoirs, and the recuperative mode was studied for mixed
reservoirs [95]. For open parametric pumps the external equations
for a variety of direct mode [18,39] and recuperative mode [119]
systems have been solved. Pressure swing adsorption [84] and
cycling zone adsorption [8] were also studied.

Since the exact details of the external balances depend
heavily on the system studied, the reader is referred to the
original literature. One example of the external balances will
be given here for the open recuperative mode parametric pump
shown in Figure 6 [119]. The reflux ratios, R_T and R_B, and the

496

appropriate product and reflux rates are shown on Figure 6. An overall mass balance at the top and at the bottom of the column can be written. These can then be solved for the overall flows,

$$P_T = \frac{R_B(FF)}{1-R_TR_B} \tag{42}$$

$$P_B = \frac{(FF)}{1-R_TR_B} \tag{43}$$

The bottom reflux ratio can also be related to the up and down-flow times shown in the characteristic diagrams such as Figure 24.

$$R_B = \frac{t_u}{t_d} \left| \frac{v_u}{v_d} \right| \tag{44}$$

Eq. (44) shows that we are not free to choose arbitrary values for cycle times, flow and bottom reflux ratio.

To determine the separation achieved a solute balance on the entire system for one cycle can be used.

$$c_{FF}\frac{(FF)}{\rho_f} = c_B\frac{P_B}{\rho_f}(1-R_B) + c_T\frac{P_T}{\rho_f}(1-R_T) \tag{45}$$

Usually, additional information will be required to determine the separation obtained. However, for the important special case where $c_B = 0$ the solution is independent of the type of reservoirs employed, and eq. (45) can be solved immediately for c_T.

$$c_T = \frac{(FF)c_{FF}}{(1-R_T)P_T} = \frac{c_{FF}(1-R_TR_B)}{(1-R_T)R_B} \quad \text{for } c_B = 0 \tag{46}$$

Note that as long as $c_B = 0$, the top product concentration is controlled by this overall mass balance. The top product concentration will be the same whether direct or recuperative modes are used and whether the isotherm is linear or nonlinear as long as $c_B = 0$. The characteristic solution is needed to determine that c_B is indeed zero.

When the bottom product concentration is not zero, all of the operating and physical parameters will affect the separation. To illustrate this the particular case of an open recuperative mode pump separating a gas will be considered. To simplify the illustration linear isotherms will be assumed, and we will limit our results to operation at the cyclic steady state where each cycle is a repeat of the previous cycle. For gas system $u_h > u_c > u_{th}$ and $c_B > 0$ regardless of the values of R_B and R_T. The characteristics for this particular case are shown in Figure 24. Note that for this gas system equations (13) and (14) predict reverse separation, $\beta < 1$, as indicated on Figure 24.

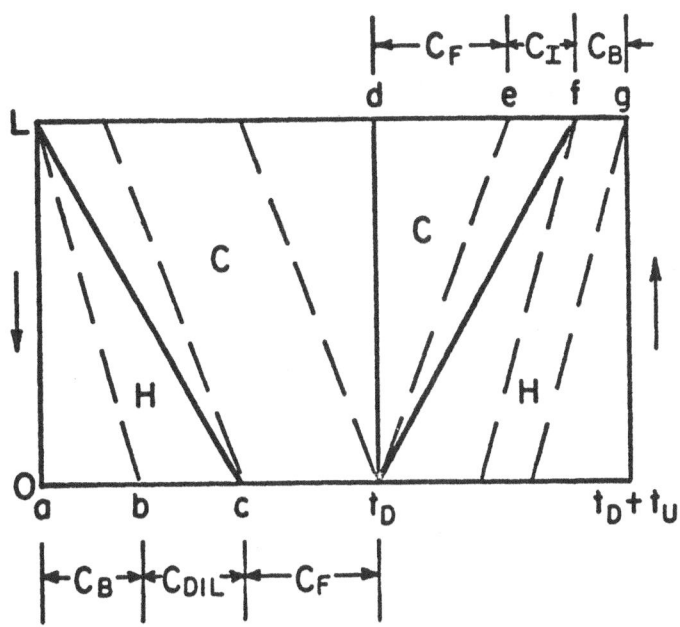

Figure 24. Characteristic solutions for recuperative mode parametric pump in cyclic steady state for case where $u_h > u_c > u_{th}$. [119]

Mass balances utilizing the characteristics can be done on the top and bottom of the column. For completely mixed reservoirs the top and bottom concentrations are an average of the exiting concentrations. Thus for the top of the column

$$c_T = \frac{de}{dg} \, c_F + \frac{ef}{dg} \, c_I + \frac{fg}{dg} \, c_B \qquad (47)$$

where $c_I = c_B/\beta$ and de, dg, etc. are distances shown on Figure 24. The actual feed to the column, c_F, is determined from a mass balance around the top mixing point as,

$$c_F = \frac{c_{FF}(FF) + R_T P_T c_T}{FF + R_T P_T} \qquad (48)$$

The ratio $\frac{de}{dg}$ represents the fraction of time that the top product leaves with a concentration of C_F. This and the other fractions can be calculated from the points of intersection of the characteristics.

$$\frac{de}{dg} = \frac{L}{u_c t_u}, \quad \frac{ef}{dg} = \frac{\frac{L}{u_{th}} - \frac{L}{u_c}}{t_u}, \quad \frac{fg}{dg} = \frac{t_u - \frac{L}{u_{th}}}{t_u} \tag{49}$$

Combining eqs. (45),(47) and (48) with eqs. (49), the definition of β, and the expressions for P_T and P_B, we obtain after algebraic manipulation,

$$C_B = \frac{C_{FF}[1- R_B \frac{L}{u_c t_u}](1 - R_B R_T)}{(1-R_B)(1 - \frac{L}{u_c t_u}R_B R_T) + R_B(1-R_T)(1 - \frac{L}{u_h t_u})} \tag{50}$$

Once C_B is known C_T can be found from the overall balance eq. (45) as

$$C_T = \frac{(1-R_B R_T)C_{FF}}{R_B(1-R_T)} - \frac{(1-R_B)C_B}{R_B(1-R_T)} \tag{51}$$

An alternative is to carefully construct the characteristic diagram and calculate the mass balance from actually measured ratios. For other cases such as where $u_{th} > u_H > u_c$ the analysis is very similar to that shown here.

One way to increase the separation when $C_B > 0$ is to use unmixed reservoirs and withdraw the most dilute material as the bottom product and the most concentrated material as the top product. Since this can be done in a variety of ways the reader should refer to the original literature [98,119].

The local equilibrium model has been widely used to study the cyclic separation techniques. The literature in this area is summarized in Table 4.

OTHER THEORIES

The local equilibrium model has been the most popular model to study the cyclic separations. However, as was shown previously, if linear equilibrium isotherms are used the theory makes physically impossible predictions. These physically impossible predictions do not occur when appropriate non-linear isotherms are employed in the theory, but the separation achieved is often overpredicted. Since the local equilibrium model does not provide good a priori predictions it should not be used for design.

A variety of other theoretical models have been developed and published in the literature. These models are outlined in Table 5. Due to space limitations it is not our intention to discuss any of these models in other than general terms.

Equilibrium stage analyses have been employed by several investigators to explain parametric pumping and cycling zone separations in staged systems [109,110,112-114,116] and in continuous contact columns [13,31,42,43,64,108,116]. In the staged systems the staged model is naturally appropriate. To model continuous contact columns an appropriate HETP must be defined. This HETP is considerably less than that obtained from pulse experiments [13,31,64] and best modeling results have been obtained by fitting cyclic experiments [31,64]. The fit between theory and experiment which can be obtained was illustrated in Figure 21 [64] for separation of fructose and glucose from water by CZA. The staged theory has also been employed to formally relate parametric pumping to distillation [42,43] and help explain the separations obtained. The equilibrium stage model is also related to the mixing cell models [46,124,125] and the STOP-GO model [40,41,49,93,95,96] neither of which assume equilibrium.

The STOP-GO model is a simplified numerical integration scheme [96] which can provide a very good fit between theory and experiment. This fit was illustrated for the parametric pumping separation of NaCl from water in Figure 5 [40]. Note that the STOP-GO model provided an excellent prediction of separation while the local equilibrium model greatly overpredicted the separation. To use this model considerable experimentation was required to determine the appropriate parameters. Complete numerical integration without the simplifications of the STOP-GO model has also been used to provide excellent simulations of data [79].

The effect of radial heat transfer into the bed in direct mode parametric pumping has been extensively studied utilizing frequency response solutions of the partial differential equations [33,34,71,73,74,76,77]. These results have shown the importance of radial heat transfer, have explained some of the reverse separations which were obtained experimentally [72], and have been used to predict the conditions for optimum separation.

Table 5. Models Other Than Local-Equilibrium Model

Model	OPN	Ref.
Overall balances	PSA	[107,121]
Overall balances with assumed mass transfer zone	PSA	[32]
Equilibrium stage analysis	PP	[42,43,108 109]
	CZ	[13,31,64, 110,112, 113,114
	2D	[116]
Equilibrium stage analysis with chemical reaction	PP	[37]
Mixing-cell model	PP	[46,124,125]
	General	[55]
Near equilibrium model	PP	[46]
Near equilibrium model with chemical rxn	PP	[2,3,36]
STOP-GO model--simplified numerical integration	PP	[40,41,49 93,95,96]
Frequency response solution of partial differential equations	PP	[33,34,71,73, 74,76,77]
General integral analysis	PP	[47]
Complete numerical integration	PP	[79]
Reduce equations to O.D.E. by method of characteristics and then numerically integrate	PSA	[62]
Relation between one and two-dimensional process for general model without dispersion	General	[118]

EXTENSION TO TWO-DIMENSIONAL SYSTEMS

Two-dimensional separation techniques are another area of separations research which has been extensively studied [90,91]. This area has developed independently of the cyclic studies. However, it has recently been shown that the cyclic separations and the steady state two-dimensional separations are mathematically analogous [116,118]. In this section this analogy will be demonstrated and some of the possibilities for new two-dimensional separation schemes will be discussed.

The prototype of a two-dimensional separator is the rotating annulus system shown in Figure 25. In this apparatus the annulus is packed with sorbent and fluid flows through the packing while the annulus rotates at angular velocity w. For the system shown in Figure 25 where feed is introduced from $\theta = 0$ to $\theta = \theta_F$, a steady state separation occurs where each solute traverses a helical path to the exit port. This basic apparatus can be changed so that it will mimic one-dimensional separations such as parametric pumping or cycling zone adsorption.

The analogy between the time-independent one-dimensional and the steady state two-dimensional processes is easily seen by writing the mass and energy balances for the two-dimensional apparatus. Assuming radial gradients are negligible, the mass balance in cylindrical coordinates is

$$\varepsilon w \frac{\partial c}{\partial \theta} + \varepsilon v \frac{\partial c}{\partial z} + (1-\varepsilon)\rho_B w \frac{\partial q}{\partial \theta} - D \frac{\partial^2 c}{\partial z^2} - D \frac{1}{r^2} \frac{\partial^2 c}{\partial \theta^2} = 0 \qquad (52)$$

The rotation of the annulus results in a solid body rotation of both phases. The energy balance is

$$\rho_f C_f \left[\varepsilon w \frac{\partial T}{\partial \theta} + \rho_B C_B (1-\varepsilon) w \frac{\partial T_s}{\partial \theta} + \rho_f C_f \varepsilon v \frac{\partial T}{\partial z} \right.$$

$$- D_T \frac{\partial^2 T}{\partial z^2} - D_T \frac{1}{r^2} \frac{\partial^2 T}{\partial \theta^2} = h_w A_w (T_w - T_{Amb}) - \frac{C_w W}{A_c} w \frac{\partial T_w}{\partial \theta} \qquad (53)$$

The solute balance on the solid phase for steady state rotating system is

$$(1-\varepsilon) \rho_B w \frac{\partial q}{\partial \theta} = - k_M a_p (c^* - c) \qquad (54)$$

The equilibrium expression is the same as Eq. (25).

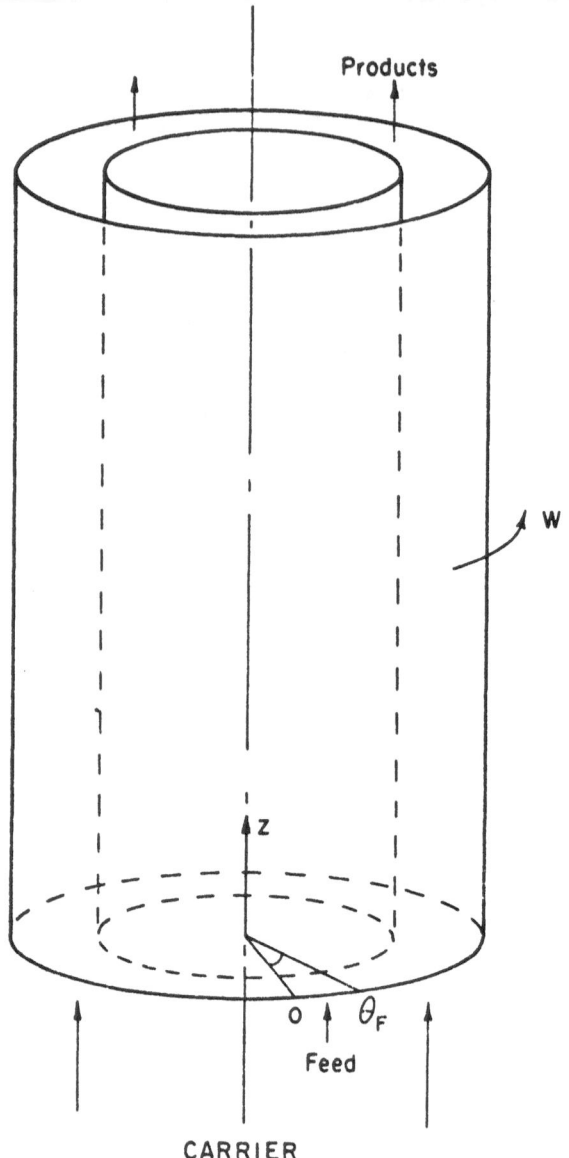

Figure 25. Rotating Annulus Apparatus for Two-Dimensional
Separation [118].

Comparison of eqs. (52) and (53) to eqs. (21) and (23), respectively, shows that there is an extra diffusion term in the equations for the two-dimensional system. Under conditions where mass or thermal diffusion are important, the one- and two-dimensional systems are not analogous. However, the diffusional terms can commonly be ignored at least for a first approximation. Under the conditions of negligible diffusion and dispersion, there is a term-by-term correspondence between eqs. (21) and (52), eqs. (23) and (53), and eqs. (22) and (54). If the simple transformation

$$t \rightarrow \theta/w \tag{55}$$

is made, then eqs. (21), (23), and (22) are transformed into eqs. (52), (53) and (54), respectively, for the case with negligible diffusion. As long as q is not a function of t or θ, this transformation is valid regardless of the equilibrium relationship which is used. In addition, the appropriate boundary conditions for the one-dimensional system will transform to the two-dimensional boundary conditions when diffusion can be neglected.

The analogy relates conditions for limit cycle operation of one-dimensional systems where each cycle is a repeat of previous cycles to steady state operation of the two-dimensional system. A limit cycle has been reached when

$$c(z,t')\Big|_{\text{cycle } n-1} = c(z,t')\Big|_{\text{cycle } n} = c(z,t')\Big|_{\text{cycle } n+1}$$

Since the partial differential equations and the boundary conditions can be transformed from the time dependent, one-dimensional form to the steady state, two-dimensional form, the systems are analogous, and the solutions obtained for the column system can also be employed for the two-dimensional system. This analogy requires that diffusion be negligible, but does not require the more restrictive local equilibrium assumptions. The analogy holds strictly for comparison of column operation to the rotating cylinder type of two-dimensional apparatus with axial flow shown in Figure 25.

Since the steady state two-dimensional system is analogous to the time-dependent one-dimensional column systems, we would expect that two-dimensional analogs of CZA, PP and PSA can be developed. The two-dimensional analogue of CZA has been studied experimentally and theoretically in both staged and continuous flow systems [116]. The staged two-dimensional systems are also analogous to the staged one-dimensional systems. A good fit between theory and experiment was obtained for the staged system [116]. The separations obtained in the rotating annulus were quite small because of channeling and non-uniform flow.

504

The general analogy between one and two-dimensional systems, and conceptual drawings and theoretical solutions for a variety of two-dimensional systems were developed by Wankat [118]. A two-dimensional, open direct mode parametric pump with mixed reservoirs is illustrated in Figure 26. The feed is input during downflow into the cold adsorbent bed, and regeneration is done during upflow in the hot bed. In Figure 26 the right side of the annulus has downward flow and the left side upward. A series of vertical dividers would allow for flow reversal in the annulus. Some of the dilute product is used as purge to regenerate the bed, and some of the concentrated product is refluxed to increase the concentration of the concentrated products.

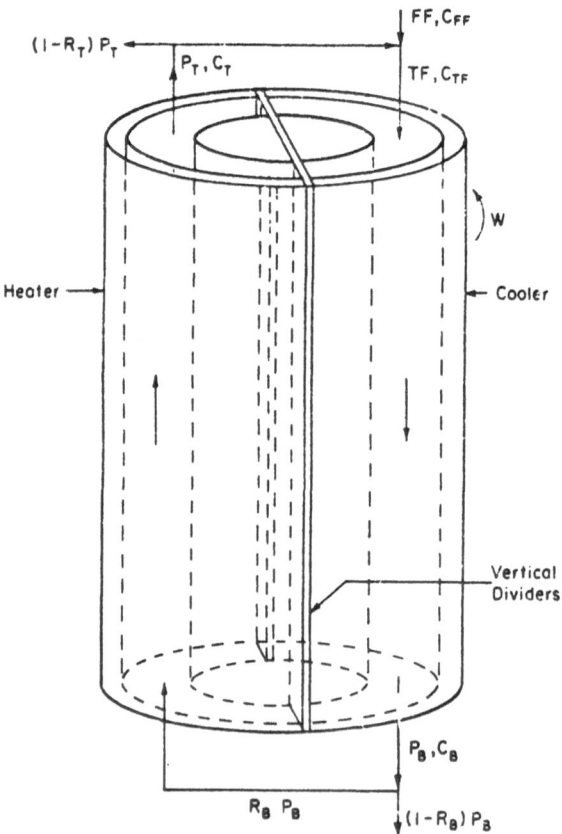

Figure 26. Two-dimensional direct mode parametric pump.
Top feed with completely mixed reservoirs [118].

The theoretical solutions obtained for parametric pumping are applicable to this system when the transformation shown in eq. (55) is used. Thus, to a first approximation, the physically very different one and two-dimensional systems will produce the same separations. This equivalence allows the designer latitude in how he wishes to achieve the separation.

RELATED TECHNIQUES

To finish this paper and to help show where these techniques fit into a broader spectrum of techniques, the cyclic separations will be briefly related to a variety of other methods. Several related papers are listed in Table 6.

Table 6. Related Papers

System	Reference
Cyclic electrodialysis for removal of NaCl from water	[99-101]
Cyclic Operation of a vapor-liquid separator for ethanol-water	[50]
Cyclic pressure variations in freeze drying for water removal	[38]
Chromathermography	[126]
Review of continuous methods of chromatography	[90,91]
Cyclic operation of membrane gas separators	[66]
Review of unsteady separation schemes where flow rates are varied	[29,82]
Review of cyclic reaction systems and cyclic operation of reactors	[6]
Mathematical analysis of heat and concentration waves	[103]

Investigations of other separation techniques such as electrodialysis have used the idea of cycling a thermodynamic variable to produce separation [38,50,99-101]. These examples are thus very closely related to PP, CZA or PSA, and other unit operations such as absorption could be utilized as parametric pumps. Chromathermography [126] is related to CZA except a moving heater changes the column temperature and produces the desired separation. The two-dimensional analogs of the cyclic methods have already been discussed.

The cyclic methods are also related to usual adsorption and chromatographic techniques. No clear dividing line exists between these methods, and processes such as the Sirotherm process [10,11] which is similar to parametric pumping was developed from usual sorption processes.

Somewhat more distant from the techniques discussed here are the unsteady separations which pulse flow rates in distillation, extraction or ion exchange [29,82]. Finally, it should be noted that there is an extensive literature on cyclic reactions and cyclic operation of reactors [6] which is at least formally related to the cyclic separations.

GLOSSARY

batch   a parametric pump with no feed input or product withdrawals

closed   same as batch

continuous   operation with feed during all parts of the cycle

concentration wave   rate at which concentration differences
  velocity   move through the column

cyclic variable   the thermodynamic variable such as temperature or pH which is cycled to force the separation

cycling zone adsorp-   a cyclic separation processes with
  tion   undirectional flow of fluid and one or more beds in series. The bed temperatures or the inlet fluid temperatures to each bed are varied periodically and are out of phase with the adjoining bed.

CZA   abbreviation for cycling zone adsorption

diffuse wave	decreasing concentration wave front predicted by the local equilibrium model when a concentrated solution is displaced by a dilute solution
direct mode	operation where the entire bed is heated or cooled simultaneously
heatless drying	pressure swing adsorption applied to drying of gases
limit cycle	condition after many cycles where each cycle is an exact repeat of the previous cycle. Analogous to steady state although conditions during a cycle are changing
local equilibrium model	theoretical model which assumes that fluid and solid are locally in equilibrium, that dispersion and diffusion are negligible, that heat of adsorption is negligible and that fluid velocity is constant
method of characteristics	mathematical method for solving certain types of partial differential equations
mixed mode	operation where both entire bed and entering fluid are heated and cooled
mixed reservoirs	all contents of reservoir are assumed to be well-mixed
non-mixed reservoirs	reservoir contents are totally nonmixed
open	a system with feed input and product withdrawals during all or part of each cycle
parametric pumping	a cyclic separation with periodic reversal of fluid flow direction utilizing fluid stored in reservoirs. The bed or fluid temperatures (or other cyclic variable) are varied periodically to force the separation
parapump, PP	abbreviations for parametric pumping

pressure swing adsorption	a cyclic separation process for gas separation which utilizes adsorption at high pressure and counter flow desorption at low pressure
PSA	abbreviation for pressure swing adsorption
recuperative mode	parametric pumping operation where the fluid flowing into the column is heated or cooled and the column itself is adiabatic. In general applies to operation where cyclic variable is changed in fluid flowing into the column
repeating state	same as limit cycle
reservoir dead volume	In parametric pumping this is volume in reservoir which remains in the reservoir at the end of each half cycle and thus is not pumped into the column.
semi-batch	operation of open parametric pump with product withdrawal at one end of column only. Concentration in reservoir at other end increases
separation factor,α	(Solute concentration in concentrated product)/(Solute concentration in dilute product)
semi-continuous	operation with feed during part of a cycle but not all of the cycle
shock wave	discontinuous concentration front predicted by the local equilibrium model when a dilute solution is displaced by a concentrated solution
standing wave mode	same as direct mode but applied to cycling zone adsorption
thermal wave velocity	rate at which temperature differences move through the column
traveling wave mode	same as recuperative mode but applied to cycling zone adsorption

two-dimensional | a process where the major separation occurs in two spatial dimensions (e.g. rotating annulus)

ultimate separation | the maximum separation which can be obtained in a batch parametric pump

NOMENCLATURE

a_p	surface area for mass transfer, eq. (22)
A_c	cross sectional area of column
$A(T), B(P)$	constants for linear adsorption isotherm
A_w	area of wall
c	fluid concentration
c^*	fluid concentration in equilibrium with solid
C_f	heat capacity of fluid
C_B	bulk heat capacity of solid
C_w	heat capacity of wall
D	mass diffusivity
D_T	thermal diffusivity
FF	fresh feed per cycle
$\Delta Hads$	heat of adsorption
h_w	wall heat transfer coefficient
k	exponent in Freundlich isotherm
k_M	mass transfer coefficient, eq. (22)
L	length of column
n	moles of gas
P, P_{high}, P_{low}	pressure
P_B, P_T	flow per half cycle at bottom and top of open parametric pump
q	amount of solute adsorbed
r	radial coordinate
R_B, R_T	bottom and top reflux ratios, eq. (9)
t	time

510

t'	time since start of cycle
t_d	period for downward flow
t_u	period for upward flow
T	fluid temperature
T_c, T_h	cold or hot temperature
$T_0(t)$	initial temperature distribution
$T_F(t)$	feed temperature
T_S	solid temperature
T_W	wall temperature
u_c, u_{solute}	concentration or solute velocity, eqs. (5), (27), (38)
u_{sh}	shock wave velocity, eq. (40)
$u_{th}, u_{thermal}$	thermal wave velocity, eqs. (12), (29)
v	fluid velocity
V	volume of gas
w	angular velocity
W	mass of wall
ΔX	segment of column
z	column axis

Greek Letters

α	separation factor
β	$C(T_H)/C(T_C)$ given in eq. (14)
γ	V_{purge}/V_{feed}
ε	porosity
θ	angular coordinate
ρ_B	bulk density of solid
ρ_f	fluid density

REFERENCES

1. Z. M. Ahmed, paper F2-2, AIChE-GVC joint meeting, Vol IV of preprints, Munich, Germany, Sept. 17-20, 1974.
2. G. P. Apostolopoulos, Ind. Eng. Chem. Fundam. 14, 11 (1975).
3. G. P. Apostolopoulos, Ind. Eng. Chem. Fundam., 15, 148 (1976).
4. R. Aris, Ind. Eng. Chem., Fundam., 8, 603 (1969).

5. R. Aris and N. R. Amundson, Mathematical Methods in Chemical Engineering, Vol. 2., First-Order Partial Differential Equations with Applications, Englewood Cliffs, N.J.: Prentice Hall, 1973.

6. J. E. Bailey, Chem. Eng. Communic., 1, 111 (1974).

7. B. Baker, Ind. Eng. Chem. Fundam., 9, 304 (1970).

8. B. Baker and R. L. Pigford, Ind. Eng. Chem. Fundam., 10, 283 (1971).

9. D. E. Blum, Ph.D. thesis, Univ. California-Berkeley, 1971.

10. B. A. Bolto, K. Eppinger, A. S. MacPherson, R. Sicidak, D. E. Weiss and D. Willis, Desalination, 13, 269 (1973).

11. B. A. Bolto, Chem. Tech, 5, 303 (1975).

12. D. W. Breck, Zeolite Molecular Sieves, Structure, Chemistry and Use, N.Y.: Wiley, 1974.

13. M. E. Busbice and P. C. Wankat, J. Chromatog, 114, 369 (1975).

14. T. J. Butts, R. Gupta and N. H. Sweed, Chem. Eng. Sci., 27, 855 (1972).

15. T. J. Butts, N. H. Sweed and A. A. Camero, Ind. Eng. Chem. Fundam., 12, 467 (1973).

16. A. A. Camero and N. H. Sweed, AIChE Journal, 22, 369 (1976).

17. Chemical Design, "Processes, Plants, Equipment," with fliers on "Heatless Adsorption for Hydrogen Upgrading," and "Hydrogen Purification Plants," Chemical Design, Inc., Lockport, N.Y. (No Date).

18. H. T. Chem and F. B. Hill, Separat. Sci., 6, 411 (1971).

19. H. T. Chen, J. L. Rak, J. D. Stokes and F. B. Hill, AIChE Journal, 19, 356 (1972a).

20. H. T. Chen, J. Jaferi, and J. D. Stokes, paper 9e presented at 73rd National Meeting AIChE, Minneapolis, MN., Aug. 28, (1972b).

21. H. T. Chen, E. R. Reiss, J. D. Stokes and F. B. Hill, AIChE Journal 19, 589 (1973).

22. H. T. Chen and J. A. Manganaro, AIChE Journal, 20, 1020 (1974).

23. H. T. Chen, J. A. Park, and J. L. Rak, Separat. Sci., 9, 35 (1974).

24. H. T. Chen, W. W. Lin, J. D. Stokes and W. R. Fabisiak, AIChE Journal, 20, 306 (1974b).

25. H. T. Chen and V. J. D'Emidio, AIChE Journal, 21, 813 (1975).

26. H.T. Chen, D.I. Cho, J. Del'Osso and P. Falcon, paper 34b at 82nd National Meeting AIChE, Atlantic City, N.J., Aug. 30, 1976.

27. H. T. Chen, T.K. Hsieh, H. C. Lee and F. B. Hill, AIChE Journal, 23, 695 (1977).

28. Deltech Engineering, "Deltech Air Dryer," Bulletin 300 C (1967); "Reactivation Air Requirements," Bulletin 308C (no date); "Deltech Heatless Air Dryer," Bulletin 312B (no date); "Heatless Air Dryer Dimensions," Bulletin 327E (no date). "Deltech Filter-Dryer Combination Assures Quality Film Processing,"

Bulletin 380 (no date). "Heat Pump Dryer," Bulletin 800C, "Suggested Specification for Compressed Gas Dryer," Bulletin 816B, (no date),Deltech Engineering, Inc., New Castle, Delaware

29. R. Dodds, P. I. Hudson, L. Kershenbaum and M. Streat, Chem. Eng. Sci., 28, 1233 (1973).

30. D. Domine and L. Hay, in Molecular Sieves, Society of Chemical Industry, London, 1968. pp. 204-216.

31. J. C. Dore and P. C. Wankat, Chem. Eng. Sci. 31, 921 (1976).

32. K. J. Doshi, C. H. Katira, H. A. Stewart, paper 38a presented at 68th National Meeting of AIChE, Houston, Texas (March 2, 1971).

33. S. C. Foo and R. G. Rice, AIChE Journal, 21, 1149 (1975).

34. S. C. Foo, and R. G. Rice, AIChE J., 23, 120 (1977).

35. V. R. Ginde and C. Chu, Desalination, 10, 309 (1972).

36. S. Goto and M. Matsubara, Ind. Eng. Chem. Fundam. 15, 148 (1976).

37. S. Goto and M. Matsubara, Ind. Eng. Chem. Fundam., 16, 193 (1977).

38. P. F. Greenfield, Chem. Eng. Sci., 29, 2115 (1974).

39. R. A. Gregory and N. H. Sweed, Chem. Eng. J., 1,207 (1970).

40. R. A. Gregory and N. H. Sweed, Chem. Eng. Journal, 4, 139 (1972).

41. R. A. Gregory, AIChE Journal, 20, 294 (1974).

42. G. Grevillot and D. Tondeur, AIChE Journal, 22, 1055 (1976).

43. G. Grevillot, and D. Tondeur, AIChE J., 23, 840 (1977).

44. G. J. Griesmer, W. F. Avery and M.N.Y. Lee, Hydrocarbon Processing, 44 (6), 147 (June 1965).

45. R. Gupta and N. H. Sweed, Ind. Eng. Chem. Fundam., 10, 280(1971)

46. R. Gupta and N. H. Sweed, Ind. Eng. Chem. Fundam., 12, 335 (1973).

47. F. J. M. Horn and C. H. Lin, Berichte der Bunsengessel-schaft fur Physikalische Chemie, 73, 575 (1969).

48. T. J. Jenczewski and A. L. Myers, AIChE Journal, 14, 509 (1968).

49. T. J. Jenczewski and A. L. Myers, Ind. Eng. Chem. Fundam., 9, 216 (1970).

50. L. M. Joseph and R. H. Kadlec, Ind. Eng. Chem. Proc. Des. Develop., 14, 187 (1975).

51. W. H. King, Jr., Anal. Chem., 43, 984 (1971).

52. D. E. Kowler and R. H. Kadlec, AIChE Journal, 18, 1207 (1972).

53. G. R. Landolt and G. T. Kerr, Separat. Purific. Methods, 2, 283 (1973).

54. J. A. Latty, Ph.D. thesis, Univ. California, Berkeley, 1974.

55. R. Lavie and M. J. Reilly, Chem. Eng. Sci., 27, 1835 (1972).

56. H. Lee and P. E. Stahl, in I. Zwiebel, D. B. Broughton, D. T. Camp (eds.), "Gas Purification by Adsorption," AIChE Symp. Series, 69 (#134), 1 (1973).

57. M.N.Y. Lee, in N.N. Li (ed.), Recent Developments in Separation Science, vol. I, Cleveland, CRC Press, pp. 75-112, 1972.

58. Linde Division - Union Carbide, Hydrocarbon Processing, 51 (9), 210 (Sept. 1972b); 53 (9), 199 (Sept. 1974).
59. Linde Division - Union Carbide, Hydrocarbon Processing, 51 (91), 221 (Sept. 1972a); 52 (4), 128 (April, 1973); 54 (4), 130 (April 1975).
60. M. A. McAndrew, Ph.D. thesis, Princeton University, (1967).
61. D. Meir and R. Lavie, Chem. Eng. Sci., 29, 1133 (1974).
62. J. E. Mitchell and L. H. Shendalman, AIChE Symp. Ser., 69 (134), 25 (1973).
63. W. C. Nelson and P. C. Wankat, J. Chromatography, 121, 205 (1976).
64. W. C. Nelson, D. F. Silarski and P. C. Wankat, Ind. Eng. Chem. Fundam., 17, 32 (1978).
65. R. R. Patrick, J. T. Schrodt and R. I. Kermode, Separat. Sci., 7, 331 (1972).
66. D. R. Paul, Ind. Eng. Chem. Proc. Des. Develop., 10, 375 (1971).
67. R. L. Pigford, B. Baker and D. E. Blum, Ind. Eng. Chem. Fundam., 8, 144 (1969a).
68. R. L. Pigford, B. Baker and D. E. Blum, Ind. Eng. Chem. Fundam., 8, 848 (1969b).
69. Puregas, Bulletins AP-976A, AP984-3, M1-73, AP770E3000-373, and AP963E: "Heatless Air Dryers for Ultra dry Air", "Carbon Dioxide Adsorbers", "HE 200 Series Heatless Dryers for Hydro-carbon Adsorption", "Series HF 3000 Heatless Dryers, Instal-lation, Operation and Maintenance Manual", and "Drying Com-pressed Air and Other Gases with Puregas Heatless Dryers (1976)", respectively, General Cable Apparatus Div., Westminister, Co. (no date).
70. H. K. Rhee and N. R. Amundson, Ind. Chem. Eng. Fundam., 9 303 (1970).
71. R. G. Rice, Ind. Eng. Chem. Fundam., 12, 406 (1973).
72. R. G. Rice and M. Mackenzie, Ind. Eng. Chem. Fundam., 12, 486 (1973).
73. R. G. Rice and S. C. Foo, Ind. Eng. Chem. Fundam., 13, 396 (1974).
74. R. G. Rice, Ind. Eng. Chem. Fundam., 14, 202 (1975a).
75. R. G. Rice, Ind. Eng. Chem. Fundam., 14, 362 (1975b).
76. R. G. Rice, Separation and Purification Methods, 5, 139 (1976).
77. R. G. Rice and S. C. Foo, Chemeca 77, Canberra, Australia, pp. 179-183, (14-16 Sept. 1977).
78. R. D. Rieke, Ph. D. thesis, Univ. California, Berkeley (1972).
79. R. W. Rolke and R. H. Wilhelm, Ind. Eng. Chem. Fundam., 8, 235 (1969).
80. J. E. Sabadell and N. H. Sweed, Separat. Sci., 5, 171 (1970).
81. H. G. Schroeder and C. E. Hamrin, AIChE Journal, 21, 807 (1975).
82. V. N. Schrodt, Ind. Eng. Chem., 59 (6), 58 (1967).
83. A. G. Shaffer and C. E. Hamrin, AIChE Journal, 21, 782 (1975).

84. L. H. Shendalman and J. E. Mitchell, Chem. Eng. Sci., 27, 1449 (1972).

85. T. K. Sherwood, R. L. Pigford and C. R. Wilke, Mass Transfer, N.Y.: McGraw-Hill (1975).

86. T. T. Shih, and R. L. Pigford, in N. N. Li (ed.)., Recent Developments in Separation Science, Vol. III, Part A, pp. 129-150, CRC Press, Cleveland 1977.

87. C. W. Skarstrom, Annals N.Y. Acad. Sci., 72, 75 (1959).

88. C. W. Skarstrom, in N.N. Li (ed.), Recent Developments in Separation Science, Vol. II, Cleveland, CRC Press, 1972, pp. 95-106.

89. H. A. Stewart and J. L. Heck, Chem. Eng. Prog., 65 (9), 78 (1969).

90. M. V. Sussman, and R. N. S. Rathore, Chromatographia, 8, 55 (1975).

91. M. V. Sussman, Chem.Tech., 6, 260 (1976).

92. N. H. Sweed, in E.S. Perry and C.J. Van Oss (eds.), Progress in Separation and Purification, vol. 4, Wiley-Interscience, NY, pp. 171-240 (1971).

93. N. H. Sweed and R. A. Gregory, AIChE Journal, 17, 171 (1971).

94. N. H. Sweed, in N.N. Li (ed.), Recent Developments in Separation Science, Vol. I, CRC Press, Cleveland, pp. 59-74, 1972.

95. N.H.Sweed, and J. M. Rigaudeau, in I. Zwiebel and N.H. Sweed, Adsorption and Ion Exchange, AIChE Symp. Ser., 71, (152), (1975).

96. N. H. Sweed and R. H. Wilhelm, Ind. Eng. Chem. Fundam., 8, 221 (1969).

97. D. W. Thompson, Trans. Instn. Chem. Engrs., 39, 289 (1961).

98. D. W. Thompson and B. D. Bowen, Ind. Eng. Chem. Fundam., 11 415 (1972).

99. D. W. Thompson and M. E. Abu-Goukh, paper presented at 24th Canadian Chemical Engineer's Conference, Ottawa, Canada, Oct. 20-23, 1974.

100. D. W. Thompson and D. Bass, Can. J. Chem. Eng., 52, 345 (1974).

101. D. W. Thompson, D. Bass and M. E. Abu-Goukh, Can. J. Chem. Eng., 52, 479 (1974).

102. B. P. Trading, Ltd., Hydrocarbon Processing, 53 (9), 204, (Sept. 1974).

103. G. A. Turner, Heat and Concentration Waves, Analysis and Application, N.Y.: Academic, 1972.

104. P. H. Turnock and R. H. Kadlec, AIChE Journal, 17, 335 (1971).

105. E. Van der Vlist, Separat. Sci., 6, 727 (1971).

106. T. G. Vermeulen, G. Klein and N. K. Hiester, in R.H. Perry and C.H. Chilton (eds.), Chemical Engineers Handbook, 5th et., N.Y.: McGraw-Hill, sec. 16, 1973.

107. J. L. Wagner and H. A. Stewart, paper presented at 3rd joint meeting I.I.QP.R and AIChE., San Juan, Puerto Rico, May 20, 1970.

108. N. Wakao, H. Matsumoto, K. Suzuki, and A. Kawahara, Kagaku Kogaku, 32, 169 (1968). (in Japanese).
109. P. C. Wankat, Ind. Eng. Chem. Fundam., 12, 372 (1973a).
110. P. C. Wankat, Separat.Sci., 8, 473 (1973b).
111. P. C. Wankat, Separat. Sci., 9, 85 (1974a).
112. P. C. Wankat, J. Chromatography, 88, 211 (1974b).
113. P. C. Wankat, J. C. Dore, and W. C. Nelson, Separat. Purific Methods, 4, 215 (1975).
114. P. C. Wankat, Ind. Eng. Chem. Fundam., 14, 96 (1975).
115. P.C. Wankat and J. W. Ross, Separat, Sci., 11, 207 (1976).
116. P. C. Wankat, A. R. Middleton, and B. L. Hudson, Ind. Eng. Chem. Fundam., 15, 309 (1976).
117. P. C. Wankat, Chem. Eng. Sci.. 32, 1283 (1977a).
118. P. C. Wankat, AIChE Journal, 23, 859 (1977b).
119. P. C. Wankat, Chem. Eng. Sci.,(1978).
120. P. C. Wankat, (in preparation, 1978).
121. K. Weaver and C. E. Hamrin, Chem. Eng. Sci., 29, %873 (%974).
122. A. L. Weiner, Chemical Engineering, p. 92 (Sept. 16, 1974).
123. R. H. Wilhelm, A. W. Rice, A. R. Bendelius, Ind. Eng. Chem. Fundam., 5, 141 (1966).
124. R. H. Wilhelm and N. H. Sweed, Science, 159, 522 (1968).
125. R. H. Wilhelm, A. W. Rice, R. W. Rolke and N. H. Sweed, Ind. Eng. Chem. Fundamentals, 7, 337 (1968).
126. A. A. Zhukhovitskii, in R. P. W. Scott (ed.), Gas Chromatography 1960, p. 293, Butterworths, London 1960.
127. -------------- "Heatless Dryer Solves Freeze-Up Problems", Actual Specifying Engineer, 28 (4), 126 (Oct. 1972).

DUAL-STEP COUNTERCURRENT PROCESSES

Daniel Tondeur

Maître de Recherche at the Centre National de La Recherche Scientifique.
Laboratoire des Sciences du Génie Chimique - Ecole Nationale Supérieure des Industries Chimiques, 1 , rue Grandville - 54042 NANCY-FRANCE

1. INTRODUCTION - TERMINOLOGY

Most chemical extraction or separation processes using an auxiliary, non-consumable phase, are "dual-step" or "multi-step" processes with extractant recycle. For example liquid-liquid extraction usually implies an extraction step, where the liquid extractant is contacted with the solution to be treated, and a regeneration step, where the extractant is stripped from the solute it contained, to allow its recycle to the extraction step. In what follows, we call "saturation" the step where the extractant picks up a substance to be removed from a flow, and "regeneration" the step where this solute is removed from the extractant.

There are basically two ways in which such a system can operate, In the first, (Mode 1) the two steps are performed in distinct vessels, and the extractant phase moves, more or less continuously from one vessel to the other. In the other, the extractant remains in a vessel, but it is contacted with the phase to be extracted and the regenerant alternatively. Figure 1 shows schematically these two modes, in their simplest versions.

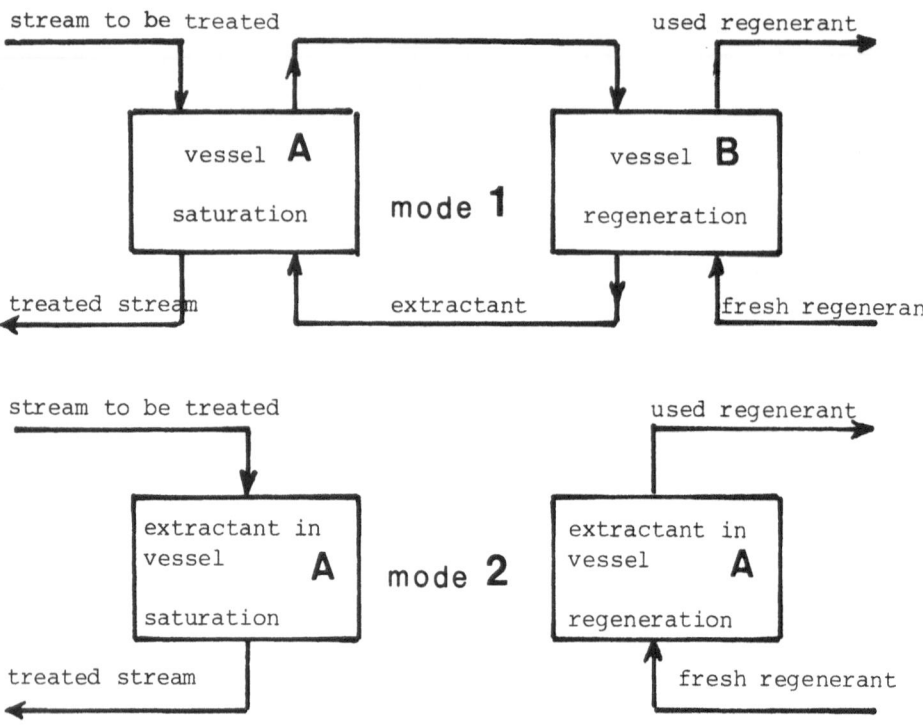

<u>Figure 1</u> : the two basic operating modes

The circulations materialized by the arrows are drawn here in such way that countercurrent operation is performed everywhere. Obviously, other combinations of flow directions can be used.

Finding an unambiguous terminology for these two modes is not obvious. One may be tempted by distinctions such as moving versus fixed extractant, or bed ; continuous versus discontinuous operation ; two-vessels versus one-vessel operation ; steady-state versus non-steady state ; space-cyclic versus time-cyclic process ; circulating versus non-circulating extractant ; non-confined versus confined extractant.... Notice that in mode 1, the movement of the extractant is sometimes done by discrete pulses, between which the operation is fixed-bed ; that in mode 2, the extractant may undergo some movement, for instance by fluidisation, that both modes have steady and unsteady states, as we shall see below ; that mode 2 may be operated with two vessels out of phase. With this in mind, we leave the reader think about the most appropriate term...

The specific problem of such two-step processes is the coupling between the steps ; clearly, each step depends on the preceding through the state of the extractant. This coupling desappears when a condition is applied to one step, such that the extractant comes out of this step always in the same state. For example, if the regeneration is always complete, whatever the state of saturation of the extractant is. We are concerned with the more general case where coupling exists.

The questions we propose to investigate here are three :

. The question of the dynamics : what is the unsteady behaviour, the response to a perturbation ?

. The question of the reflux : what is it, how to create it when is it possible ?

. The question of the cascade : what flow sheets are possible, other than those of figure 1 ?

2. OVERALL DYNAMICS OF TWO STEP PROCESSES

We are interested in knowing how a two-step process tends toward its limit regime after an initial perturbation, of step type. We are not interested by the details of the process inside the vessel, but merely by the compositions of the various effluents we thus only consider an "overall" behaviour. Let us examine successively the two modes of operation, for the extraction of a solute from a stream S.

2.1 A simple model for mode 1

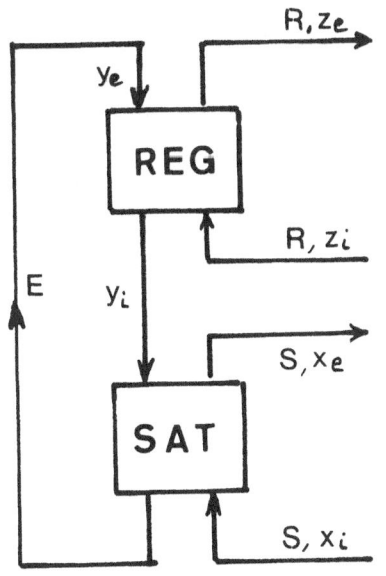

Figure 2 : Notations for Mode 1

S,E,R : flow rates of stream extractant, regenerant

x,y,z : concentrations of solute in stream, extractant, regenerant

i,e : subscripts for inlet and exit of a vessel

We start by writing material balances on each of the vessels.

Saturation vessel :

$$S (x_i - x_e) = E (y_e - y_i) + \frac{d}{dt} \left[\text{holdup} \right]_{sat} \qquad (1)$$

Regeneration vessel :

$$R (z_i - z_e) = E (y_i - y_e) + \frac{d}{dt} \left[\text{holdup} \right]_{Reg} \qquad (2)$$

Obviously, we shall not be able to get information on the unsteady behaviour unless something is known, or assumed, on the holdup in each vessel, thus on what happens "inside". First, we express the holdup in each vessel as the sum of holdup in extractant and holdup in stream :

$$\text{holdup in saturation vessel} = v_S(1-\varepsilon_S)y_S + v_S\varepsilon_S x_S \qquad (3)$$

$$\text{holdup in regeneration vessel} = v_R(1-\varepsilon_R)y_R + v_R\varepsilon_R z_E \qquad (4)$$

where v_S, v_R are the vessel volumes, ε_S, ε_R the fraction of vessel volume occupied by stream or regenerant ; y_S, x_S, y_R, z_R are average concentrations of the different phases in each vessel. (average over the whole phase contained in the vessel at a time t). We shall suppose here the following quantities constant and known: inlet concentrations x_i, z_i ; flow rates, S, E, R ; vessel volumes v_S and v_R ; fractional volumes of phases ε_S, ε_R.

Contactor Efficiencies. Let us first recall the classical definitions of efficiency, summarized in the table below. These efficiencies are defined here with reference to the extractant.

	HAUSEN efficiency	MURPHREE efficiency
saturation	$H_S = \dfrac{y_e - y_i}{y_i^* - y_i}$ (6) where y_i^* is in equilibrium with x_i	$M_S = \dfrac{y_e - y_i}{y_e^* - y_i}$ (8) where y_e^* is in equilibrium with x_e
regeneration	$H_R = \dfrac{y_e - y_i}{y_e - y_i^{**}}$ (7) where y_i^{**} is in equilibrium with z_i	$M_R = \dfrac{y_e - y_i}{y_e - y_e^{**}}$ (9) where y_e^{**} is in equilibrium with z_e

Hausen efficiencies are defined with respect to a maximal concentration change such that the <u>outgoing</u> extractant would be in equilibrium with the <u>incoming</u> stream. Thus H = 1 for an infinite number of theoretical stages. Murphree efficiencies are defined with respect to a maximum change such that the <u>outgoing</u> extractant would be in equilibrium with the <u>outgoing</u> stream. Thus M = 1 when the contactor is equivalent to one theoretical stage. Using Hausen efficiencies is more appropriate in percolation operations were usually a certain number of theoretical stages intervene.

<u>Constant Hausen efficiencies.</u> We suppose H_S and H_R are constant. At steady state, this is equivalent to assuming each vessel behaves like a staged equipment with a well defined and constant number of theoretical stages. Recall that in a linear system at steady state (system where phase equilibria, and steady-state operating lines are straight), the Hausen efficiency is simply related to the number of theoretical stages N by Kremser's equations

$$ H = \frac{r^{N+1} - r}{r^{N+1} - 1} \quad \text{with} \quad r = \frac{S}{KE} \tag{10} $$

In the unsteady-state, no such relation holds however.

With constant inlet concentrations x_i and z_i, y_i^* and y_i^{**} are constant (and supposed known). Equation 6 and 7 may then be solved for y_e and y_i, to give:

$$ y_e = (H_S y_i^* + H_R y_i^{**} - H_S H_R y_i^{**})/D \tag{11} $$

$$ y_i = (H_S y_i^* + H_R y_i^{**} - H_S H_R y_i^*)/D \tag{12} $$

$$ D = H_S + H_R - H_S H_R \tag{13} $$

we notice that y_e and y_i are constant, and depend only on the efficiencies of the contactors and of the inlet concentrations of the stream and regenerant. It may be convenient to introduce an <u>overall efficiency</u> Ho as the ratio of the actual change in y to that which would occur if both contactors had an infinite number of theoretical stages ($H_S = H_R = 1$). It is easy to show from equation 11, to 13 that

$$ \boxed{ Ho = \frac{y_e - y_i}{y_i^* - y_i^{**}} = \frac{H_S H_R}{H_S + H_R - H_S H_R} } \tag{14} $$

The overall efficiency should thus not be taken just as the product of the two efficiencies, although the difference introduced when doing so is small. Notice that H_c is closely related to what some people call the <u>operating capacity</u> or the "<u>useful capacity</u>", which would be defined here as $C_o = y_e - y_i$.

<u>Further assumptions ; unsteady-state solution</u>. Since y_e and y_i are time-independent, the extractant works at steady-state. It is then reasonable to assume that the average extractant concentrations y_S and y_R in the vessels are constant, and their time derivative vanish . We next assume that, for a given x_i, the exit concentration x_e can be expressed as an increasing function of the average stream concentration x_S in the vessel, and similarly in the regeneration vessel. Thus we write

$$x_s = f\ (x_e)$$

$$z_s = g\ (z_e) \tag{15}$$

The exact nature of such a relationships could only be deduced from a detailed examination of the concentration profiles inside the vessels. A linear relation is not unreasonable. Here, we make no particular assumption on the form. The material balance of Equation 1 is now rewritten :

$$S\ (x_i - x_e) = E\ C_o + \frac{d}{dt}\ (v_s\ \varepsilon_s\ x_s)$$

with $\dfrac{d}{dt}\ x_s = \dfrac{df}{dx_e}\ \dfrac{d\ x_e}{dt} = f'\ \dfrac{dx_e}{dt}$, from Equation 15, and $f' > 0$.

Then, x_s being eliminated ;

$$\frac{d\ x_e}{dt} + \frac{S}{v_s \varepsilon_s\ t'}\ x_e = x_i\ \frac{S}{v_s \varepsilon_s\ f'} - \frac{E\ C_o}{v_s \varepsilon_s\ f'}$$

Introducing the average residence time of the stream in the saturation vessel,

$$t_{ss} = v_s \varepsilon_s / S \tag{16}$$

we obtain

$$\frac{d\ x_e}{dt} + \frac{x_e}{f'\ t_{ss}} = \frac{x_i}{f'\ t_{ss}}\ \left[1 - B_S\right] \tag{17}$$

where $B_s = \dfrac{E\ C_o}{S\ x_i} = \dfrac{E\ Ho}{S\ x_i}\ (y_i{}^* - y_j{}^{**})$ (18)

is a relative capacity flux, the ratio of the flux of operating capacity to the flux of solute to be extracted. The solution of the differential equation 17 is :

$$x_e = x_i\ (1-B_s)\left[1 - \exp - \frac{t}{f' t_{ss}}\right]$$ (19)

The expression for z_e is identical, with x_i replaced by z_i, f' by g', t_{ss} by $t_{RR} = V_R \epsilon_R/R$, and $- B_s$ by $+ B_R = \dfrac{E\ C_o}{R\ z_i}$

This solution is the sum of a constant term x_i $(1-B_s)$ which represents the steady state solution, and of a transient term, which decreases exponentially (since f'> o), and the faster the smaller the residence time of the stream in the vessel. The solutions for x_e and z_e are decoupled here, (due to the y's staying constant) but not independent, since the efficiency of both vessels appear in the steady state solution through B_s, thus Ho ; but the other vessel has no influence on the transient exponential.

Simple analytical solutions are thus possible, with the two sets of assumptions used : constant Hausen efficiencies, and relation between holdup in vessel and outgoing concentration.

Constant Murphree efficiencies. Equations 6 and 7 may be solved in the same way as before, with M_s and M_R replacing H_s and H_R, and $y_e{}^*$ and $y_e{}^{**}$ replacing $y_i{}^*$ and $y_i{}^{**}$ respectively. Hower, $y_e{}^*$ and $y_e{}^{**}$ are not constant, since x_e and z_e are not constant. Thus y_e and y_i vary with time. In order to obtain an explicit solution, we shall assume linear equilibrium relations, so that

$$y_e{}^* = K_s\ x_e$$
$$y_e{}^{**} = K_R\ z_e$$ (20)

The material balance for the two vessels must be solved simultaneously this time. The average holdup concentrations y_s, y_R do not vanish. We shall assume, as before, that the holdup is a univoque function of the outgoing concentration, so that

$$y_s = f_1\ (y_e) \qquad y_R = f_3\ (y_i)$$
$$x_s = f_2\ (x_e) \qquad x_R = f_4\ (z_e)$$ (21)

Introducing an <u>overall Murphree efficiency</u>

$$Mo = \frac{y_e - y_i}{y_e^* - y_e^{**}} = \frac{M_s M_R}{M_s + M_R - M_s M_R} \tag{22}$$

letting in addition

$$X = x_e - x_i \qquad Co = y_e - y_i \qquad Z = z_e - z_i$$

and using (20) to eliminate y_e^* and y_e^{**}, we obtain the system

$$0 = \frac{d}{dt}\left[(1-\varepsilon_s)f'_1 Mo\ (K_s X - K_R Z) + \varepsilon_s f'_2 X\right] + \frac{\varepsilon_s X}{t_{ss}}$$

$$+ \frac{Mo\ (1-\varepsilon_s)}{t_{es}}\left[K_s X - K_R Z + K_s x_i - K_R z_i\right]$$

$$0 = \frac{d}{dt}\left[-(1-\varepsilon_R)f'_3 Mo(K_s X - K_R Z) + \varepsilon_R f'_4 X\right] - \frac{\varepsilon_R Z}{t_{rr}}$$

$$+ \frac{Mo\ (1-\varepsilon_R)}{t_{er}}\left[K_s X - K_R Z + K_s x_i - K_R z_i\right]$$

In operator form, this is

$$\begin{cases} (A_1 L + B_1)X + (C_1 L + D_1)Z = E_1 \\ (A_2 L + B_2)X + (C_2 L + D_2)Z = E_2 \end{cases} \tag{23}$$

where L designates the operator $\frac{d}{dt}$

The solution of such a system is written

$$X = x_e - x_i = F_1 e^{\lambda_1 t} + F_2 e^{\lambda_2 t} + x_e^{ss} \tag{24}$$

where λ_1 and λ_2 are the eigenvalues of the system, obtained by calculating the determinant of the system 23, and solving the 2nd degree equation obtained, with L as unknown. x_e^{ss} is the particular solution, corresponding to the steady state, which can be obtained directly from the material balances. We obtain

$$x_e^{ss} = \frac{S\ x_i\ (1+b_R) + R\ z_i\ b_R}{S\ (1 + b_s + b_R)} \tag{25}$$

with

$$b_s = \frac{E \ Mo \ K_S}{S} \qquad b_R = \frac{E \ Mo \ K_R}{R} \qquad (26)$$

This more complicated model does not bring much more insight into the problem, although it illustrates a mathematical approach that is more general. Eigenvalue problems arise very often in unteady state processes, and we know that their real or complex nature, and the sign of their real part determines the oscillatory behaviour and the stability of the response.

2.2 A simple model for mode 2

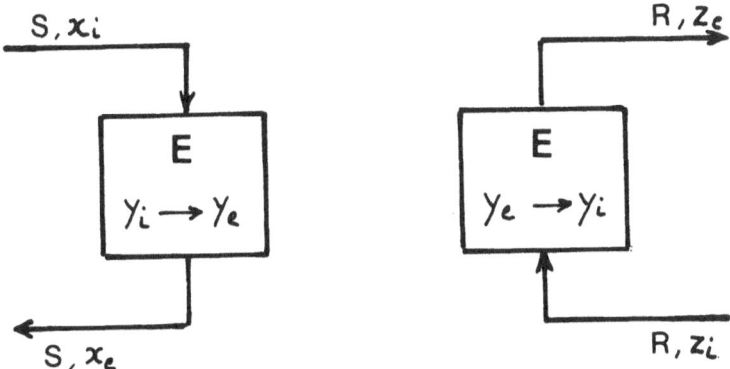

Figure 3 :Notations for Mode 2

    S,R : flow rates of stream and regenerant

    E : quantity of extractant confined in vessel

    x,y,z : concentration of solute in stream, extractant and regenerant

Here, we define y_i as the initial average concentration of the extractant at the start of the saturation step and y_e is the average concentration at the end of the saturation step. Beginning and end of saturation of course correspond to end and beginning of regeneration respectively.

Let us write a material balance over the whole saturation step, the duration of which we designate by t_s.

$$S \ t_s \ (x_i - \overline{x_e}) = E \ (y_e - y_i) \qquad (27)$$

where $\overline{x}_e$ is the concentration of the saturation effluent, averaged over the whole saturation step. Using again the concept of constant Hausen efficiencies, we can write

$$H_s = \frac{y_{e,j+1} - y_{ij}}{y_i^* - y_{ij}} \qquad H_R = \frac{y_{ej} - y_{is}}{y_{ej} - y_i^{**}} \qquad (28)$$

The subscript j designates the jth step after the start-up, which we suppose here to be a regeneration. Step j + 1 is thus a saturation. y_{ej} is thus the initial state of the extractant for regeneration j, y_{ij} the final state after that regeneration, which is also the initial state of saturation j + 1; $y_{e,j+1}$ is the final state after saturation j+1. Eliminating y_{ij} between the expressions 28, we obtain

$$y_{e,j+1} + (D-1) \ y_{ej} = H_s \ y_i^* + H_R \ y_i^{**} - H_s \ H_R \ y_i^{**} \qquad (29)$$

This is a linear, first order finite difference equation with constant coefficients, which is to be compared to the ordinary differential equation 17. The solution of 29 is the sum of the solution of the homogeneous solution, which corresponds to the transient regime, and of a particular solution of the complete equation, which corresponds to the "steady state", where $y_{e,j+1} = y_{ej}$. We obtain

$$y_{ej} = K \ (D-1)^j + \frac{H_s \ y_i^* + H_R \ y_i^{**} - H_s y_i^{**}}{D} \qquad (30)$$

transient term   stationnary term

and $y_{ij} = y_{ej} \ (1-H_R) + H_R \ y_i^{**}$

C_{oj} = operating capacity at step $j = y_{ej} - y_{ij}$ (31)

$C_{oj} = y_{ej} \ H_R \ (1 - y_i^{**})$

and the average effluent concentration of a saturation j+1

$$\overline{x}_{e,j+1} = x_i - \frac{E \ C_{o,j+1}}{S \ t_s} = x_i - \frac{E \ H_R}{S \ t_s} \ (1 - y_i^{**}) \ y_{e,j+1} \qquad (32)$$

We can see that the final solution is in many respects quite similar to that of Mode 1. In particular the steady state term in Equation 30 is identical to Equation 11. Equation 32 can be compared to Equation 10. By working on averaged quantities, the limit regime of a cyclic operation can thus be described in a manner similar to the steady state of a continuous process.

3. REFLUX

3.1 Operation without equilibrium shift

Let us consider again the flow sheet of Mode 1

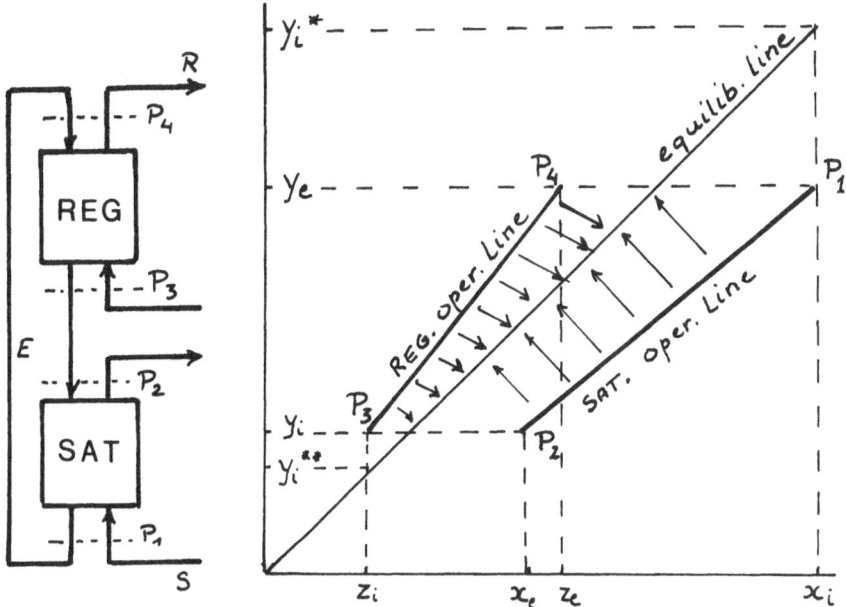

Figure 4 : Mc Cabe-Thiele diagram for the steady-state operation of Mode 1 without equilibrium shift

Figure 4 illustrates the operating conditions in a case where saturation and regeneration occur under the same thermodynamic conditions, that is, the equilibrium isotherm is the same for both saturation and regeneration. Since in saturation, we want to transfer solute from the stream S to the extractant (direction of arrows , increasing y, decreasing x), the operating line of the saturation section must lie below the equilibrium line. For the reverse reasons, the operating line of the regeneration section must lie above. This entails certain conditions on the regenerant one can use, in order to achieve a given quality of extraction, defined by x_e for example :

for the reasons mentioned, points P_2 and P_3 must lie on opposite sides of the equilibrium line, and this entails that z_i is necessarily smaller than x_e. In other words, the regenerant must be purer than the product. At the limit, one can imagine that point P_2 and P_3 almost coïncide, and lie close to the equilibrium line, then the two operating lines would have to diverge, which means that the flow rate R of regenerant must be larger than the flow-rate S of the stream. If they did not diverge, the two operating lines would then practically coïncide with the equilibrium line, which implies that an infinite number of stages is required in both vessels with the result that $x_i \sim z_e$, that is we have transferred solute to the regenerant to achieve just the same concentration.

Thus in general, we have to satisfy the following constraints, if we want to achieve some result with a finite equipment :

. use a regenerant that is purer than the product desired

. obtain a used regenerant which is less concentrated than the stream to be treated

. use a regenerant flow-rate higher thant the stream flow-rate

The question is then : can one improve over this situation ?

3.2 The equilibrium shift

The constraints mentionned above amount to giving the regenerant some potential energy (in this case, chemical) with respect to the saturant. Other ways of achieving this would be to modify some thermodynamic variable, in order to shift the equilibrium. This can be done by acting on temperature, pressure, or chemical potential, for example. Whatever the thermodynamic variable chosen, let us consider the effect of an equilibrium shift on the Mc Cabe-Thiele diagram. Figures 5 and 6 show various situations envisageable.

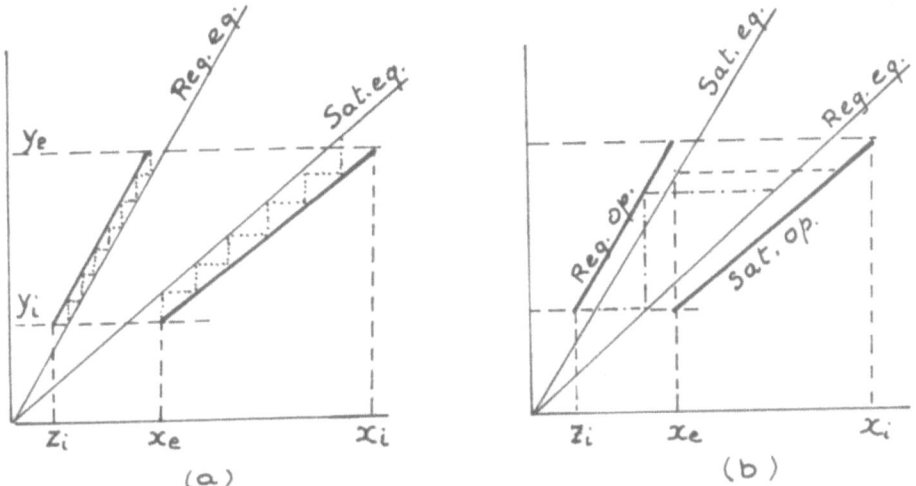

Figure 5 : Inefficient and efficient use of equilibrium shift

Figure 5a illustrates the wrong use of an equilibrium shift : the equilibrium line have become close to the operating lines ; the number of stages required to achieve the same result is increased with respect to the previous case. The thermodynamic conditions of the regeneration and saturation columns should be reversed (figure 5 b) : for the same result (same phase compositions) many less stages are needed. But then the regeneration operating line does not need to lie above the saturation equilibrium line, nor the saturation operating line to lie below the regeneration equilibrium line. At the cost of a few more stages, we can go to the situation of Figure 6 a : the operating lines lie between the equilibrium lines. There is no longer any obstacle to points P_2 and P_3 coinciding; in other words its is possible to use a regenerant having the same composition as the effluent stream. On Figure 6 a, the slope of the regeneration operating line is still larger than that of the saturation operating line. This means that we would still need more effluent stream than we have produced to regenerate the extractant. But very few stages were used. We can, at the cost of more stages, invert the slopes of the operating lines, and thus regenerate with a flow rate smaller than that of the effluent stream. This is illustrated on figure 6 b. The consequence is that we can use part of the effluent stream to regenerate the extractant, and do without any fresh regenerant : we have than introduced reflux. The flowsheet corresponding to the situation of figure 6 b is shown on figure 7.

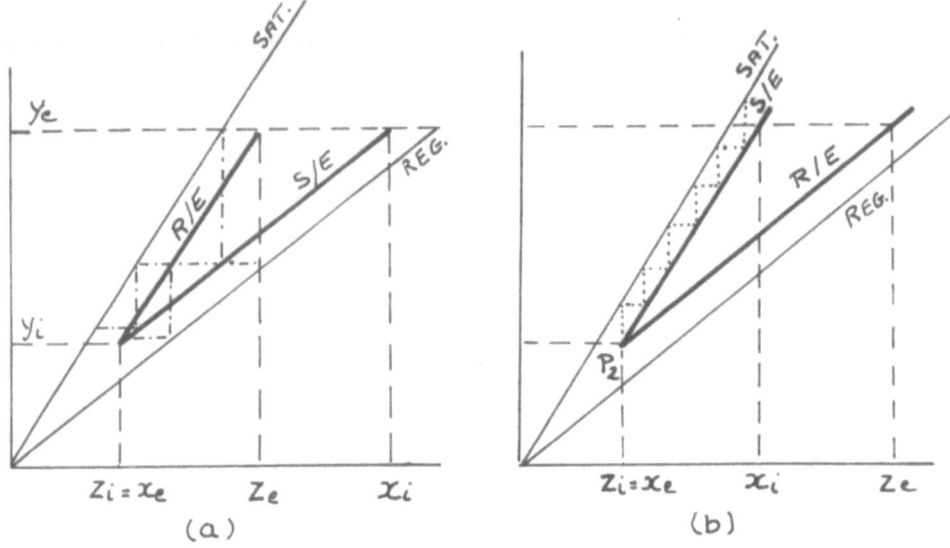

Figure 6 : Decreasing regenerant purity and flow rate

3.3 The stripping section with reflux

We refer to the flow-sheet of figure 7 and the diagram of figure 6 b and 7 b. We note the following :

. The interversion of the slopes of saturating and regeneration operating lines allows to regenerate with part only of the product stream.

. This entails that the used regenerant is more concentrated than the initial stream ($z_e > x_i$).

. In doing this, we have increased the number of stages required to achieve a given result.

. The composition x_e of point P_2 is at the intersection of the operating lines. By proper adjustment of the slopes, even with an imposed number of stages, it is possible to bring P_2 closer to the origin ; this is done by bringing the two operating lines closer together, in other words, by decreasing the production S-R.

. At the limit, by increasing the reflux and the number of stages, one can theoretically make x_e as small as one wants.

. The same is not true of z_e, which has an upper bound, as showns on figure 7 b. This bound corresponds to an infinity of stages on the rich side of the saturation column and of the regene-

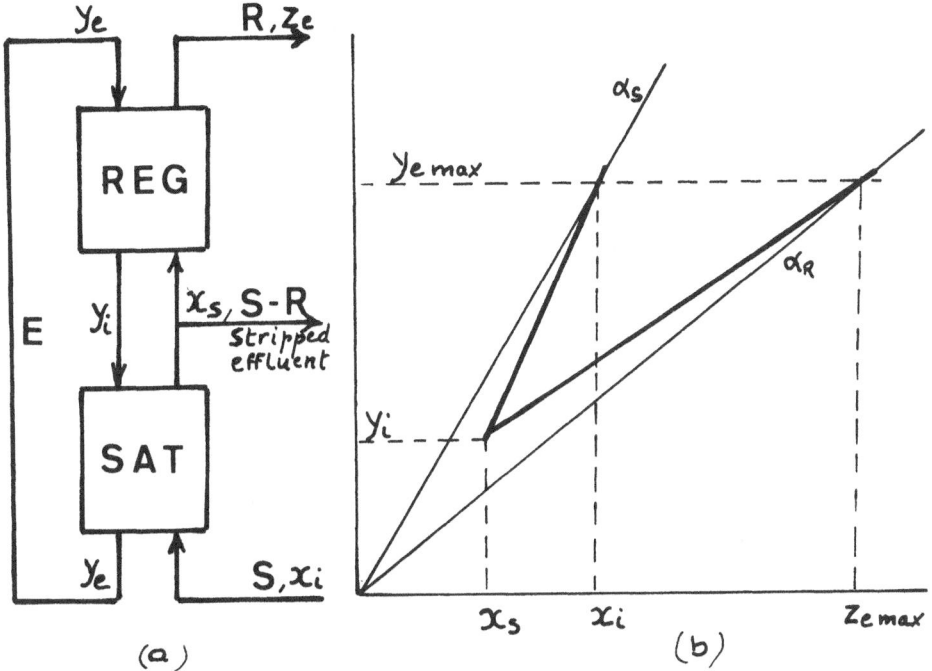

Figure 7 : The stripping section and its limiting enrichment

ration column. If α_S and α_R are the slopes of the straight equilibrium lines, the maximum enrichment is given by

$$z_e \ max = \frac{\alpha_S}{\alpha_R} \ x_i$$

. For these reasons : unlimited stripping, but limited enrichment, this flow arrangement will be called a <u>stripping section</u>.

. The reflux ratio is the ratio R/S-R of the flow used to regenerate and the production.

3.4 The enriching section with reflux

It is now easy to conceive an enriching section by interchanging the flow conditions of the regeneration and saturation vessels. The stream is introduced in the regeneration vessel and part of the effluent of this vessel is refluxed to the saturation vessel (Keep in mind that these two vessels are distinguished by the fact that a thermodynamic potentiel is different, for example temperature

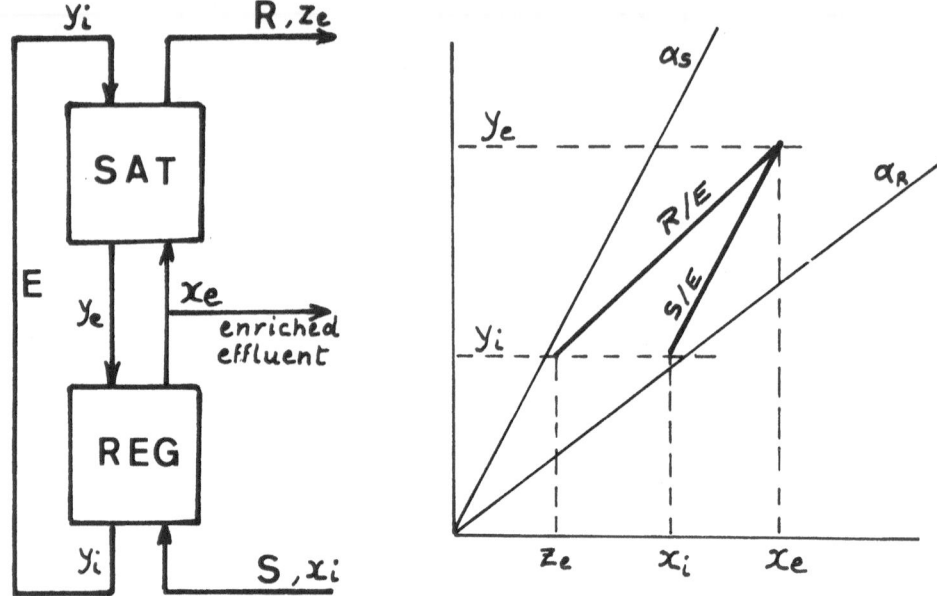

Figure 8 : The enriching section and its operating diagram

so that the equilibrium curves are such as shown). The Mc Cabe-Thiele diagram is that of figure 8. Similarly to the stripping section, the composition x_e of the rich product can be made indefinitely larger by increasing the reflux, and the number of stages. The composition z_e of the poor product on the other hand, is bounded at a value

$$z_e \, \min = \frac{\alpha_R}{\alpha_S} x_i$$

4. CASCADE ARRANGEMENTS

4.1 The complete fractionation process

A complete fractionation process is now built by combining a stripping section and an enriching section. This is done by connecting the (R, z_e) effluent of each section to the (S, x_i) inlet of the other. A decision must then be made on the feed location. Figure 9 shows such an arrangement, where the feed is introduced between the two saturation vessels. The steady state Mc Cabe-Thiele diagram of this system is shown. We see that it combines the opera-

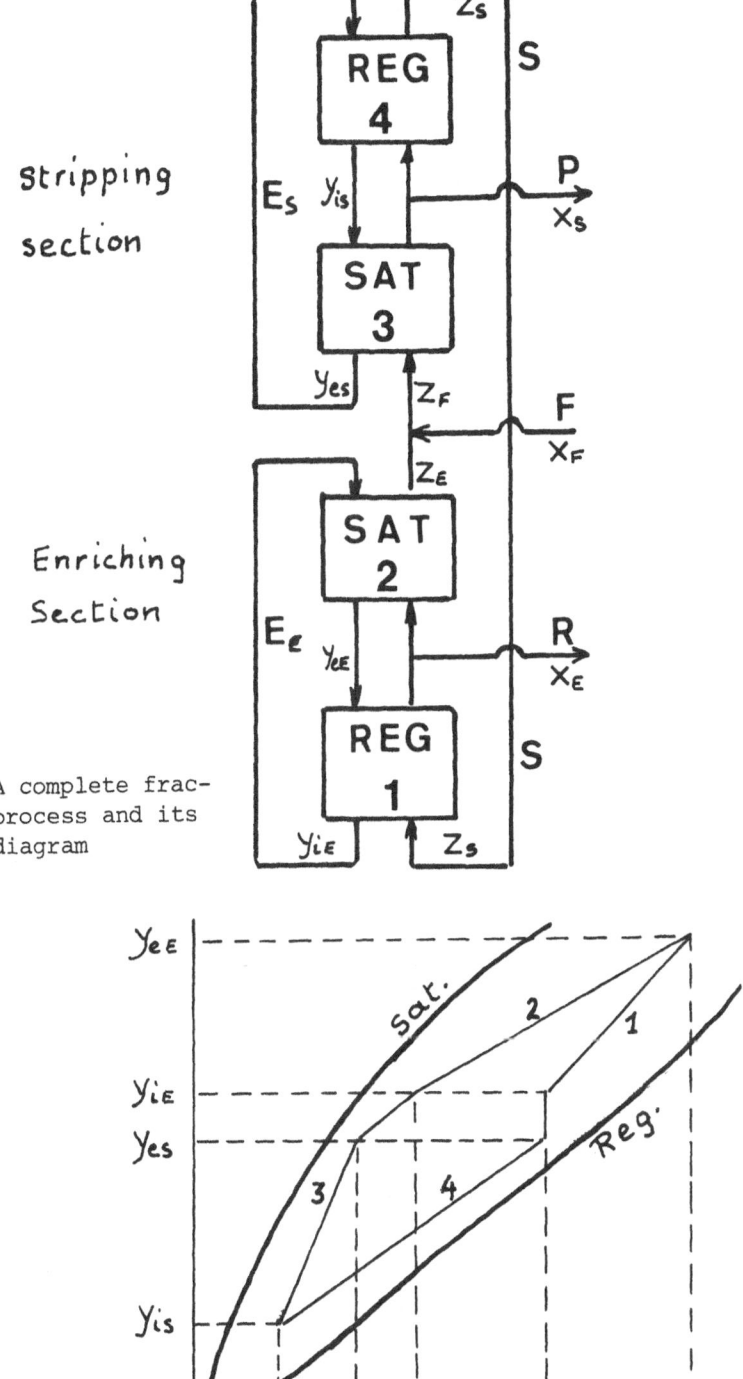

Figure 9: A complete fractionation process and its operating diagram

ting diagrams of both sections shown on figures 7 and 8. The operating diagram consists in four operating segments, numbered 1 to 4, each corresponding to one of the vessels. The slope of each segment (ratio of flow rates in the corresponding vessel) can be adjusted independently, since we can adjust independently for instance the extractant flow rate in each section, and the feed and one product flow-rate. At total reflux (no feed, no products), the operating lines in each section coincide, and the diagram reduces to two operating lines. The area of the closed polygonal formed by the operating lines can be considered as a measure of the production, which becomes zero at total reflux, and increases as the operating get closer to the corresponding equilibrium lines. Clearly, the production, and the product compositions, can be adjusted by adjusting the slopes of the operating lines, thus the flow-rates. An additionnal degree of freedom exists if the two sections are operated under a different set of thermodynamic conditions (say, temperatures). There are then two saturation and two regeneration equilibrium curves.

If one wanted to trace the analogy with distillation, one could say that the regeneration vessels play the role of the condenser and the reboiler ; there is a top and a bottom reflux ratio (S/P and $(S-R)/R$ respectively). When temperature is the thermodynamic variable used to shift the equilibrium, such processes have been called thermal extraction by Le Goff and Tondeur (1).

4.2. Simplest flowsheet

Simpler systems are obtained by merging the two extractant circuits into one. The two regeneration sections and the two saturation sections can then also be merged into one, and the resulting system is shown on figure 10

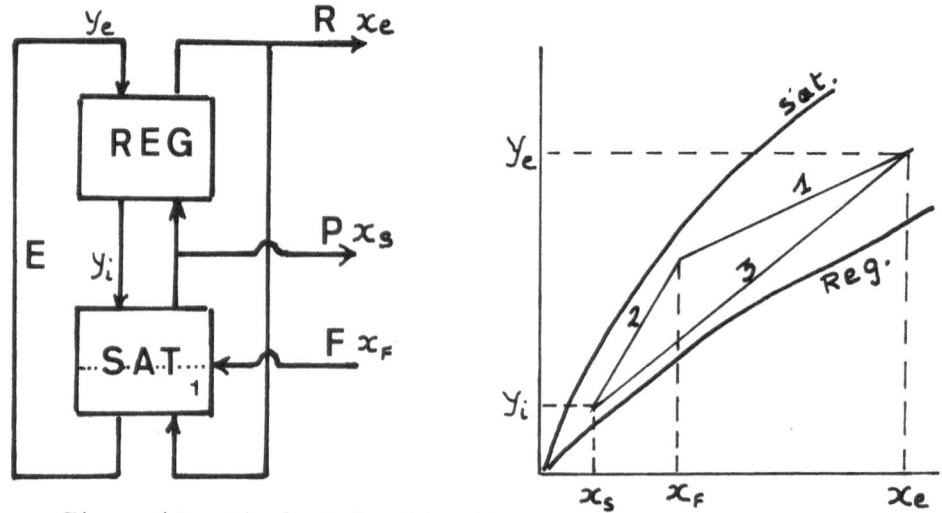

Figure 10 : Simplest fractionation process

One degree of freedom has been lost, since there is only one extractant flow rate left.

4.3 Minimum reflux

As in distillation, a specified composition of the products can be obtained under different conditions: less reflux can be used if more theoretical stages are used. But there is a value of the reflux in each section below which the specified separation becomes impossible. At that minimum reflux value, the separation requires an infinite number of stages, and this corresponds, on the Mc Cabe-Thiele diagram, to an operating line intersecting, or being tangent to an equilibrium line. Various situations where this occurs are illustrated on Figures 11. The notion of minimum reflux is not quite so simple to handle here as in binary distillation. One has to state carefully what is specified, what is kept constant, and thus as a function of which variable the reflux is calculated and its minimum sought.

The occurence of pinch points in the operating diagram, and their location at one extremity or the other of a contactor, are strongly dependent on the curvature of the equilibrium curves. In the frequent case where the equilibrium curves have the same concavity and are near each other, fitting the operating diagram in the space between these curves may be delicate. Figures 12 show possible flowsheet modifications to achieve a good production, while avoiding pinch points.

4.4 Analytical solutions

Analytical expressions giving the separation ratio z_e/x as a function of the equilibrium parameters α_{Sat}, α_{Reg}, of the flowrate ratios, and of the Hausen efficiencies of the contactors have been presented by Le Goff and Tondeur (1) in the case of linear equilibrium curves. When both regeneration and saturation contactors have an efficiency H equal to 1 (infinite number of theoretical stages) the expressions are :

Enriching section

Stripping section

$$\frac{x_e}{z_e} - 1 = \left(\frac{\alpha_S}{\alpha_R} - 1\right) \frac{\alpha_R \; E/R}{\alpha_R \; E/R + 1} \qquad \frac{z_e}{x_S} - 1 = \left(\frac{\alpha_S}{\alpha_R} - 1\right) \frac{\alpha_S E/R}{\alpha_S E/R - 1}$$

5. CLOSURE

Parametric pumping, ion-exchange cycles and other fixed-bed cyclic operations, examined in this course, are examples of Mode 2, which have been abundantly studied at steady and unsteady

536

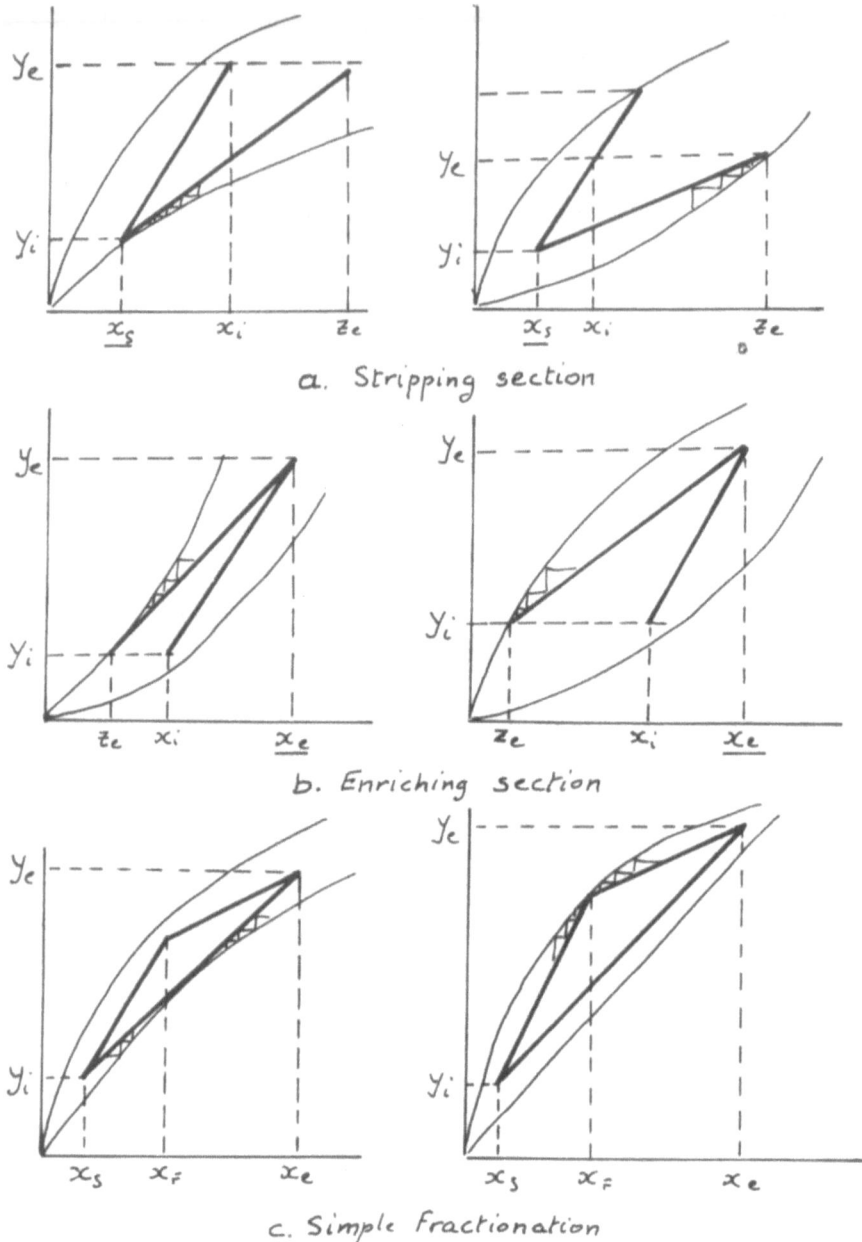

a. Stripping section

b. Enriching section

c. Simple Fractionation

Figure 11 : Illustration of possible reflux limitations

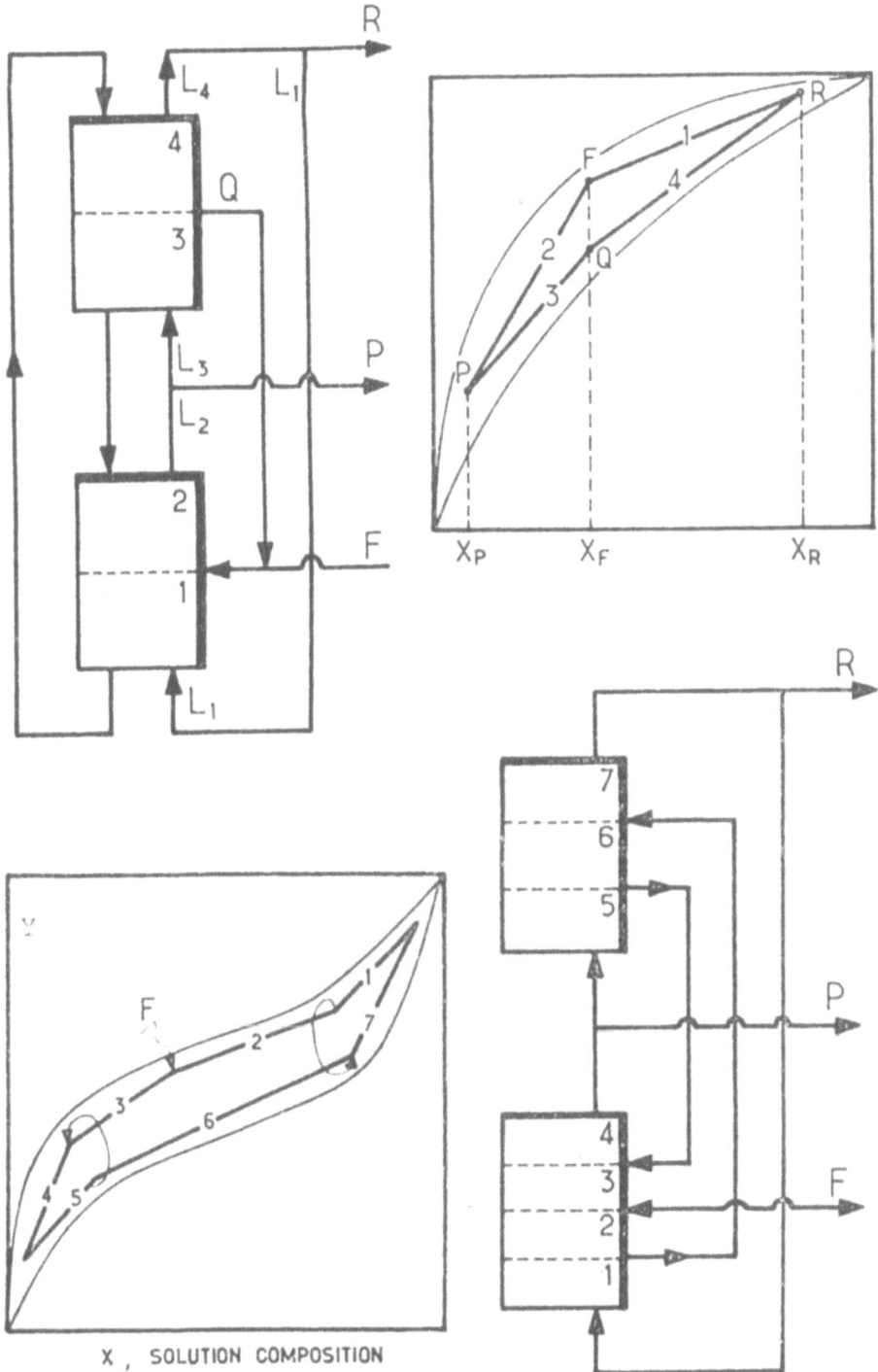

Figure 12: Flowsheet modifications avoiding pinch points.

state. Strangely, little attention seems to have been given to the dynamics of coupled columns in Mode 1, in particular with reflux, except in the case of isotopic separations. Dual temperature processes with chemical exchange have been in use for a long time, in particular for deuterium concentration. Pratt (2) gives a short account of such processes.

REFERENCES

1. LE GOFF P., TONDEUR D., La thermo-extraction, Entropie, n° 79, 5, 1978

2. PRATT H.R.C., Countercurrent Separation Processes, Chap. 6, Elsevier, 1967

THE CHROMATOGRAPHIC REACTOR

Jacques VILLERMAUX

Professor of Chemical Engineering at the Ecole Nationale Supérieure des Industries Chimiques.
Director of the Laboratoire des Sciences du Génie Chimique. CNRS-ENSIC. NANCY-FRANCE.

1. INTRODUCTION : WHAT IS A CHROMATOGRAPHIC REACTOR ? HISTORICAL BACKGROUND AND GENERAL CHARACTERISTICS.

The chromatographic reactor has been defined by LANGER et al [1] as "a chromatographic column in which a solute or several so-lutes are intentionally converted, either partially or totally, to products during their residence in the column. The solute reactant or reactant mixture is injected into the chromatographic reactor as a pulse. Both conversion to product and separation take place in the course of passage through the column ; the device is truly both reactor and chromatograph".

The reaction may take place either in the mobile phase (gas or liquid) or in the stationary phase (liquid supported or solid). In addition to its retention properties, the stationary phase may act as a catalyst for the reaction. The two functions (retention/ catalysis) may be performed by a single species or by a mixture of species, each playing a specific role. Under certain conditions, the coupling between reaction and separation has a favourable ef-fect on the course of the chemical reaction, by modifying the spa-tial distribution of products. In particular, one can try to drive reversible reactions to completion by continuously withdrawing one or several products from the reacting zone as it passes through the column.

The principle of the chromatographic reactor has been known for more than 15 years. It has been studied and used in two princi-pal directions.

- For analytical purposes, in order to study the kinetics and

mechanism of chemical reactions. These applications are well covered in two articles by Langer et al. [1] [2]. The method has been
especially well developed by C.S.G. Phillips and coworkers in
Oxford.

- For preparative purposes, in order to obtain products in
better conditions than in the conventional way.

In the present paper, we will essentially deal with this second aspect, although results derived in sections 2 and 3 below
might also be useful for analytical applications.

The first idea for a chromatographic reactor seems to have
been published in 1961 almost simultaneously in the Soviet Union
by Roginskii and coworkers [3] and in the United States by two re-
searchers of Esso Engineering Co, Magee [4] and Dinwiddie [5].
These authors consider a reversible reaction A $\rightleftharpoons$ B + C. Assume
that the reaction is carried out in a chromatographic column where
A, B and C are unequally retained : this is especially the case if
C is hydrogen which is not adsorbed. Owing to the escape of C from
the reacting zone, the equilibrium is continuously shifted to the
right and the reaction can be driven, if not to completion, at
least to an extent higher than that corresponding to steady-state
processing. This has been proven experimentally by Roginskii and
coworkers [6] [7] [8] [9] who obtained improvements in conversion
about 20 to 30 % over equilibrium in the case of dehydrogenation
of cyclohexane into benzene and butene into butadiene. The same
observation has been made by Murakami et al [10] for the dealky-
lation of cumene into benzene and propylene, by Phillips et al [11]
for the dehydrogenation of ethylbenzene into styrene and by Unger
et al [12] in the case of the synthesis of ammonia from nitrogen
and hydrogen. These authors used a mixture of catalyst (promoted
iron oxide) and adsorbant (molecular sieve). More recently, high
conversions of ethane into ethylene (80 %) have been obtained at
400-500°C over cadmium exchanged zeolites by Antonucci et al [13].
In our Laboratory, Schweich et al [14] have obtained the total dehydrogenation of cyclohexane into benzene over a Pt - Al$_2$O$_3$ catalyst at 200°C and have derived the experimental conditions leading
to total conversion of the reactant. Besides this laboratory work,
several patents have been taken out, concerning chromatographic
reactors, or, more generally improvement of chemical reactors per-
formances by cyclic pulse processing or by any device producing a
spatial separation of reacting species [4] [5] [21] [22]. However, to
our knowledge, it does not seem that these patents have been followed by the development of industrial applications in spite of the
fact that their possibility had been recognized early on [4] [5] [21]
[22] [23].

Simultaneously, mathematical models and simulations have been
published to study the performance of chromatographic reactors.

The major assumptions in these models are summarized in Table I
below :

TABLE I - MODELLING OF CHROMATOGRAPHIC REACTORS

Authors	Reactions	Adsorption	Propagation of species	Axial Dispersion	Rate of reaction	Input concentration signal	Experimental verification of the model
Magee [15]	A ⇌ B + C	linear instantaneous	$u_A = u_B$	negligible	instantaneous local equilibrium	pulse	no
Gore [16]	A ⇌ B + C	linear finite rate		negligible	finite, A and B adsorbed C gaseous	delayed sine wave gaussian	no
Hattori [17]	A ⇌ R + S A ⇌ 2 R {A + B → R R + B → S A → R 2R → S}	linear instantaneous	$u_A = u_S$ $u_A = u_B$	negligible	order n	rectangular triangular pulse gaussian	yes
Chieh [18]	A ⇌ B + C	Langmuir-Hinshelwood finite rate		yes	finite rate A, B adsorbed C gaseous	delayed sine wave	no
Wetherhold [19]	A ⇌ B + C	Freundlich instantaneous		negligible	finite, in the mobile phase	rectangular pulse	yes
Schweich [20]	$A_1 ⇌ A_2 + A_3$	linear instantaneous		yes	finite rate	rectangular pulse	yes

Most of these models rely on the assumptions of the "ideal chromatographic reactor" as defined by Langer et al [2] :

(1) A pulse of reactant reacts as it travels through the column and the reaction products are instantaneously separated from the reactant, and in many cases, also from each other

(2) the mobile phase is incompressible - i.e. its linear velocity is constant through the column

(3) axial dispersion and band spreading are negligible

(4) the rates of mass transfer, adsorption and absorption are fast and not limiting - i.e. the reaction rate is limiting

(5) the adsorption isotherms are linear

(6) the column operates isothermally and heat effects are negligible.

However, as can be seen on table I, some models use less conservative assumptions (especially on points (3) (4) and (5)).

Related theoretical studies deal with moving bed chromatographic reactors [24] [25] [26] [27]. We will exclude these more complex devices, from the scope of this study.

Now, being given a chemical reaction the extent (or the selectivity) of which is amenable to improvement in a chromatographic reactor, what kind of requirements may we try to fullfill in operating the reactor ?

Several objectives are possible, and sometimes contradictory. Let us consider first a reversible reaction A $\rightleftarrows$ B + C or A + E $\rightleftarrows$ R + S, and compare classical steady-state processing to pulse processing (single or periodic).

One may aim at :

(1) Obtaining a total conversion of reactants

(2) Improving the conversion over steady state processing for the same mean feedflowrate of reactants.

(3) Improving the conversion of reactants over steady state processing for the same amount of reactants processed.

(4) Obtaining the products in a concentrated and purified state.

In the case of a consecutive-competitive reaction system

$A + B \longrightarrow R$, $R + B \rightarrow S$, the goal may be

(5) Improving the productivity of R

(6) Improving the yield of R

Matsen et al [23] presented conditions which they felt to be advantageous for a chromatographic reactor :

(I) The reaction must be reversible, and the equilibrium constant must be small.

(II) The separation of products should limit the reaction - i.e. the reaction rates should be relatively high.

(III) At least two chromatographically separable products must be formed.

(IV) Reactants must not be separated in the reactor.

We will try to give partial answers to questions (1) —— (6) in sections 3 and 4 of this paper.

Before discussing these non-linear situations, it is interesting to investigate thoroughly the case of linear processes, for which most mathematical derivations can be performed to the very end, thanks to the availability of linear transform methods. This is the object of section 2 below, which may be considered as the continuation of the companion paper of this series "Theory of linear chromatography" [28], to which the reader is referred.

2. LINEAR CHROMATOGRAPHY IN PRESENCE OF FIRST ORDER REACTIONS

In this section, we assume that solutes may take part in first order reactions either in the mobile phase or in the stationary phase. We will consider successively the case of a first order irreversible reaction $A \xrightarrow{k} B$ and of a first order reversible isomerization $A \underset{k_2}{\overset{k_1}{\rightleftharpoons}} B$. The assumptions are those of linear chromatography (ref. [28], sec. 1), the reactions taking place in isothermal dilute solutions.

In the absence of chemical reaction, the chromatographic column is characterized by :

- the residence time distribution of the mobile phase $E_o(t_o)$ and its Laplace transform (transfer function) $G_o(s)$.

- the dynamics of the interaction at the elementary level and its transfer function $M(s) = \bar{n}'/\bar{n}$. We have shown (réf. [28] sec.5) that when the stationary phase is made up of porous particles or a stack of retention layers in series :

$$M(s) \neq \frac{K'}{1 + t_m s} \qquad (2-1)$$

whereas when it consists of a set of retention sites acting in parallel (without external resistance) :

$$M(s) = K' \sum_j \frac{\omega_j}{1 + s\tau_j} \text{ or } K' \int_0^\infty \frac{f(\tau)d\tau}{1 + \tau s} \qquad (2-2)$$

where t_m and τ_j are characteristic mass transfer time constants.

Then, we have established a powerful theorem : the transfer function of the column is obtained by the substitution :

$$G(s) = G_o \left[s + s M(s) \right] \qquad (2-3)$$

We propose to generalize this theorem to the case of first order chemical reactions occuring either in the mobile or in the stationary phase.

2.1. First-order irreversible reaction

Let us consider the reaction A $\longrightarrow$ products and examine the behaviour of species A in the column.

Reaction in the mobile phase only. In order to demonstrate theorem (2-3), the column was decomposed into a bundle of small plug flow reactors, each with its own residence time (see ref. [28] sec. 6). Adding the reaction term $k_m c$, the mass-balance is now written for an elementary plug flow stream as :

$$u \frac{\partial c}{\partial z} + \frac{\partial c}{\partial t} + k_m c + \frac{1 - \varepsilon}{\varepsilon} \frac{\partial c'_m}{\partial t} = 0 \qquad (2-4)$$

Applying the Laplace transform :

$$u \frac{d\bar{c}}{dz} + \bar{c} \left[s + k_m + s \frac{1 - \varepsilon}{\varepsilon} L(s) \right] = 0 \qquad (2-5)$$

the elementary transfer function is then :

$$g(s,t_o) = \exp \left[-(s + k_m)t_o - s t_o M(s) \right] \qquad (2-6)$$

Associating the plug flow streams in parallel according to the mobile phase RTD

$$G(s) = \int_0^\infty g(s,t_o) \, E_o(t_o) \, dt_o \qquad (2-7)$$

Recalling that

$$G(s') = \int_0^\infty \exp(-s't_o) E_o(t_o) \, dt_o \qquad (2-8)$$

One finally obtains :

$$G(s) = G_o \left[s + k_m + s \, M(s) \right] \qquad (2-9)$$

which generalizes (2-3). The s M(s) term is unchanged and the reaction rate constant in the mobile phase k_m has simply been added to s in the first term.

For instance, the transfer function of the MCE model (see eq. (7-2) in réf. [28] sec. 7) becomes :

$$\left. \begin{array}{l} G(s) = \left[1 + \dfrac{St_o}{J} \right]^{-J} \\[3ex] \text{with } S = s + k_m + \dfrac{K's}{1 + t_m s} \end{array} \right\} \qquad (2-10)$$

Reaction in the stationary phase. Let us first treat the case where the dynamics of the interaction is described by one single transfer time t_m (homogeneous porous particle or retention layers in series). Following the same method as above, the mass balance in an elementary plug flow reactor is now written, taking into account the reaction term $k_s c'_m$

$$u \frac{\partial c}{\partial z} + \frac{\partial c}{\partial t} + \frac{1-\varepsilon}{\varepsilon} \left(\frac{\partial c'_m}{\partial t} + k_s \, c'_m \right) = 0 \qquad (2-11)$$

In the Laplace domain :

$$u \frac{d\bar{c}}{dz} + \bar{c} \left[s + \frac{1-\varepsilon}{\varepsilon} (s + k_s) \frac{\overline{c'_m}}{\bar{c}} \right] = 0 \qquad (2-12)$$

The mass balance in the stationary phase is :

$$\frac{\partial c'_m}{\partial t} = \frac{\alpha}{t_m} \left(c - \frac{c'_m}{\alpha} \right) - k_s c'_m \qquad (2\text{-}13)$$

From which the transfer function $\bar{c}'_m/\bar{c}$ results :

$$\frac{\bar{c}'_m}{\bar{c}} = \frac{\alpha}{1 + t_m(s + k_s)} = L(s + k_s) \qquad (2\text{-}14)$$

Setting $M(s) = \dfrac{1 - \varepsilon}{\varepsilon} L(s)$, (2-12) is written

$$u \frac{d\bar{c}}{dz} + \bar{c} \left[s + (s + k_s) M (s + k_s) \right] \qquad (2\text{-}15)$$

The derivation is easily continued along the same lines, and one finally obtains an alternative form to (2-9), generalizing (2-3) :

$$G(s) = G_o \left[s + (s + k_s) M (s + k_s) \right] \qquad (2\text{-}16)$$

The modification here concerns the interaction term $s\,M(s)$. For instance, the transfer function of the MCE model with a chemical reaction in the stationary phase is :

$$G(s) = \left[1 + \frac{St_o}{J} \right]^{-J} \qquad (2\text{-}17)$$

with $\quad S = s + \dfrac{K'(s + k_s)}{1 + t_m(s + k_s)}$

Obviously, when chemical decomposition of the solute occurs simultaneously in both phases, the transfer function is obtained by superposition as

$$G(s) = G_o \left[s + k_m + (s + k_s) M (s + k_s) \right] \qquad (2\text{-}18)$$

Let us now examine the case of retention sites in parallel and assume that chemical reactions may occur on each site (j) with a particular rate constant k_{sj}.

Following the derivation of ref. [28] sec. 5.3., the mass balance for the j-population in the stationary phase is written, in presence of chemical reaction :

$$V_2 \frac{dc'_j}{dt} = k_j S_j \left(c - \frac{c'_j}{\alpha_j} \right) - k_{sj} V_2 c'_j \qquad (2\text{-}19)$$

or, in reduced form, setting $\tau_j = \dfrac{\alpha_j V_2}{k_j S_j}$

$$\tau_j \frac{dc'_j}{dt} + c'_j (1 + k_{sj}\tau_j) = \alpha_j c \qquad (2\text{-}20)$$

and, applying the Laplace transformation

$$\frac{\bar{c}'_j}{\bar{c}} = \frac{\alpha_j}{1 + (s + k_{sj})\tau_j} \qquad (2\text{-}21)$$

Recalling that $c'_m = \sum_j c'_j$, the mass balance in an elementary plug flow reactor is, as above

$$u \frac{\partial c}{\partial z} + \frac{\partial c}{\partial t} + \frac{1 - \varepsilon}{\varepsilon} \left(\frac{\partial c'_m}{\partial t} + \sum_j k_{sj} c'_j \right) = 0 \qquad (2\text{-}22)$$

or, in the Laplace domain

$$u \frac{d\bar{c}}{dz} + \bar{c} \left[s + \frac{1 - \varepsilon}{\varepsilon} \sum_j \frac{\alpha_j (s + k_{sj})}{1 + (s + k_{sj})\tau_j} \right] = 0 \qquad (2\text{-}23)$$

Leading to the final theorem (where $\omega_j = \alpha_j/\alpha$) :

$$\boxed{\begin{array}{l} G(s) = G_o(S) \\[2mm] S = s + K' \sum_j \dfrac{\omega_j (s + k_{sj})}{1 + (s + k_{sj})\tau_j} \end{array}} \qquad (2\text{-}24)$$

If $k_{sj} = k_s$ is the same on each site, (2-24) is obviously equivalent to (2-16).

Calculation of the moments of the impulse response : conversion, mean retention time and variance of a single peak of reactant.   Now that we have derived expressions for the transfer function in various case, it is an easy task to obtain the moments in applying Van der Laan's theorem (see ref. [28] sec. 2). The column's transfer function is expressed as $G(s) = G_o(S)$ where $G_o(s)$ is the transfer function of the mobile phase alone (unretained solutes) and S a function of s, M(s) and the rate constants which has been established above.

The relative amount of reactant **A** remaining at the column's exit is simply :

$$\frac{\int_0^\infty c(t,L)\,dt}{\int_0^\infty c(t,o)\,dt} = \mu_o = G(o) = G(S_o) \tag{2-25}$$

where S_o is the expression for S   with s = 0.

In particular, for an ideal column in which axial dispersion is negligible

$$G_o(s) = \exp(-\,st_o) \text{ and thus } G(s) = \exp(-\,St_o)$$

Then $\mu_o = \exp(-\,S_o t_o)$ \hfill (2-26)

Keeping this assumption of plug flow in the mobile phase, the mean and variance are easily obtained :

$$m_1 = -\left(\frac{\partial G}{\partial s}\right)_o = -\left(\frac{\partial G_o}{\partial S}\right)_o \cdot \left(\frac{\partial S}{\partial s}\right)_o = t_o \exp(-\,S_o t_o)\left(\frac{\partial S}{\partial s}\right)_o$$

the mean retention time is then

$$t_R = \mu_1 = m_1/\mu_o = t_o\left(\frac{\partial S}{\partial s}\right)_o \tag{2-27}$$

In the same way, the second-order moment is :

$$m_2 = \left(\frac{\partial^2 G_0}{\partial S^2}\right)_0 \left(\frac{\partial S}{\partial s}\right)_0^2 + \left(\frac{\partial G_0}{\partial S}\right)\left(\frac{\partial^2 S}{\partial s^2}\right)_0$$

and $\quad \mu_2 = \dfrac{m_2}{\mu_0} = t_0^2 \left(\dfrac{\partial S}{\partial s}\right)_0^2 - t_0 \left(\dfrac{\partial^2 S}{\partial s^2}\right)_0 \qquad\qquad$ (2-28)

We pass on to the variance $\sigma^2 = \mu_2 - \mu_1^2$

$$\sigma^2 = - t_0 \left(\frac{\partial^2 S}{\partial s^2}\right)_0 \qquad\qquad (2\text{-}29)$$

This calculation can be continued to obtain higher order central moments, involving the successive derivatives of S.

Table II below gives the results of these calculations corresponding to the various expressions for S $(2 - 10)$, $(2 - 17)$, $(2 - 24)$:

TABLE II : MOMENTS OF AN IMPULSE RESPONSE

expression of S	Reaction in the mobile phase (2 - 10)	Reaction in the stationary phase, One single transfer time (2 - 17)	Reaction in the stationary phase, Sites in parallel (2 - 24)
μ_0	$\exp\left[-k_m t_0\right]$	$\exp\left[-\dfrac{k_s K' t_0}{1 + k_s t_m}\right]$	$\exp\left[-K' t_0 \sum_j \dfrac{\omega_j k_{sj}}{1 + k_{sj}\tau_j}\right]$
$t_R = \mu_1$	$t_0(1 + K')$	$t_0\left[1 + \dfrac{K'}{(1 + k_s t_m)^2}\right]$	$t_0\left[1 + K' \sum_j \dfrac{\omega_j}{(1 + k_{sj}\tau_j)^2}\right]$
$\sigma^2 = \mu'_2$	$2 K' t_0 t_m$	$\dfrac{2 K' t_0 t_m}{(1 + k_s t_m)^3}$	$2 K' t_0 \sum_j \dfrac{\omega_j \tau_j}{(1 + k_{sj}\tau_j)^3}$
$\dfrac{\sigma^2}{t_R^2}$	$\dfrac{2 K'}{(1 + K')^2} \cdot \dfrac{t_m}{t_0}$	$2 K' \dfrac{t_m}{t_0} \cdot \dfrac{1 + k_s t_m}{\left[(1 + k_s t_m)^2 + K'\right]^2}$	
$\dfrac{\mu'_3}{\mu_1^3}$	$6 K' \dfrac{t_m^2}{t_0^2}$	$6 K' \dfrac{t_m^2}{t_0^2} \dfrac{(1 + k_s t_m)^2}{\left[(1 + k_s t_m)^2 + K'\right]^3}$	

A few observations can be made with regards to table II :

For a first-order reaction in the mobile phase, the overall conversion of a single pulse $(1 - \mu_o)$ is the same as in a plug flow reactor without retention. The chromatographic peak is unshifted (μ_1) and undistorted (σ^2) : it is simply reduced homothetically in the ratio μ_o, as referred to a peak without reaction.

On the contrary, for a first-order reaction in the stationary phase, the surface and position of the peak are both affected by mass-transfer and chemical reaction. A characteristic parameter is the Damköhler number $\delta = k_s t_m$. If δ is small $(\delta << 1)$, the conversion is close to $1 - \exp \left[- k_s K' t_o \right]$, which is that of a plug-flow reactor with the mean residence time equal to the corrected retention time $K' t_o$. The peak is then unshifted and undistorted. Conversely, if δ is large $(\delta >> 1)$, the reaction becomes mass-transfer controlled : the conversion is $1 - \exp \left[- K' t_o / t_m \right] = 1 - \exp(-NTU)$ which is that of a normal mass exchanger. The stationary phase acts then as a mere sink. The peak is shifted toward the region of unretained solutes. For intermediate values of δ, the chemical reaction causes a "chemical shift" of the mean retention time

$$\frac{\Delta t_R}{t_{Ro}} = \frac{K'}{1 + K'} \cdot \frac{2\delta + \delta^2}{(1 + \delta)^2} \tag{2-30}$$

where t_{Ro} refers to the absence of reaction. We see also that chemical reaction reduces the peak broadening.

2.2. First order reversible isomerization

Let us now consider the case of a reversible isomerization $A \underset{k_2}{\overset{k_1}{\rightleftharpoons}} B$ in which A and B may both be simultaneously retained by the stationary phase. The reactions take place either in the mobile or in the stationary phase. We will treat these two cases successively, under the assumption of negligible axial dispersion.

Reaction in the mobile phase. The mass balances of A and B are the standard forms :

(A)
$$\begin{cases} u \dfrac{\partial C_{Am}}{\partial z} + \dfrac{\partial C_{Am}}{\partial t} + k_A \dfrac{a}{\varepsilon} \left(C_{Am} - \dfrac{C_{AS}}{\alpha_A} \right) + k_{1m} C_{Am} - k_{2m} C_{Bm} = 0 \\[4mm] \dfrac{\partial C_{AS}}{\partial t} + \dfrac{k_A a}{1-\varepsilon} (C_{AS} - C_{Am}) = 0 \end{cases}$$

$$\tag{2-31}$$

$$
(B) \begin{cases} u\,\dfrac{\partial C_{Bm}}{\partial z} + \dfrac{\partial C_{Bm}}{\partial t} + k_B\,\dfrac{a}{\varepsilon}\left(C_{Bm} - \dfrac{C_{Bs}}{\alpha_B}\right) + k_{2m}C_{Bm} - k_{1m}C_{Am} = 0 \\[4mm] \dfrac{\partial C_{Bs}}{\partial t} + \dfrac{k_B a}{1-\varepsilon}\left(\dfrac{C_{Bs}}{\alpha_B} - C_{Bm}\right) = 0 \end{cases} \qquad (2\text{-}32)
$$

Boundary conditions : A is assumed to be injected as a Dirac pulse at $z = 0$. C_0 is the concentration which would be observed if the amount of A injected was uniformly dissolved in the mobile phase present in the column.

$$
z = 0 \qquad C_{Am} = C_o\,t_o\,\delta(t) \qquad , \qquad C_{Bm} = 0
$$

$$
t = 0 \qquad C_{Am} = C_{Bm} = 0
$$

Reduction of variables :

$$
y_A = C_{Am}/C_o \;,\; x_A = C_{As}/\alpha_A C_o
$$

$$
M_A = K'_A = \alpha_A\,\dfrac{1-\varepsilon}{\varepsilon} \qquad,\qquad N_A = \dfrac{k_A\,a\,t_o}{\varepsilon} = M_A\,\dfrac{t_o}{t_{mA}}
$$

where $t_{mA} = \dfrac{\alpha_A(1-\varepsilon)}{k_A a}$. Analogous variables are defined for B.

$$
x = z/L,\; t_o = L/u,\; \theta = t/t_o
$$

$$
K_{1m} = k_{1m}t_o,\; K_{2m} = k_{2m}t_o
$$

Then, the mass balances are written, in reduced form :

$$
(2\text{-}33) \begin{cases} \dfrac{\partial y_A}{\partial x} + \dfrac{\partial y_A}{\partial \theta} + N_A\,(y_A - x_A) + K_{1m}y_A - K_{2m}y_B = 0 \\[4mm] \dfrac{\partial x_A}{\partial \theta} + \dfrac{N_A}{M_A}\,(x_A - y_A) = 0 \end{cases}
$$

$$
(2\text{-}34) \begin{cases} \dfrac{\partial y_B}{\partial x} + \dfrac{\partial y_B}{\partial \theta} + N_B\,(y_B - x_B) + K_{2m}y_B - K_{1m}y_A = 0 \\[4mm] \dfrac{\partial x_B}{\partial \theta} + \dfrac{N_B}{M_B}\,(x_B - y_B) = 0 \end{cases}
$$

$$x = 0 \qquad Y_A = \delta(\theta), \quad x_A = Y_B = x_B = 0$$

$$\theta = 0 \qquad x_A = Y_A = x_B = Y_B = 0$$

Applying the Laplace transformation to both systems, the elimination of x_A and x_B yields (capital letters hold for the transforms)

$$(2\text{-}35) \begin{cases} \dfrac{dY_A}{dx} + Y_A \left(s + N_A + K_{1m} - \dfrac{N_A^2}{M_A s + N_A} \right) - K_{2m} Y_B = 0 \\[4mm] \dfrac{dY_B}{dx} + Y_B \left(s + N_B + K_{2m} - \dfrac{N_B^2}{M_B s + N_B} \right) - K_{1m} Y_A = 0 \end{cases}$$

$$x = 0 \qquad Y_A = 1, \qquad Y_B = 0$$

Straightforward, but tedious calculations finally lead to the solution of this second order linear differential system. The solution is :

$$(2\text{-}36) \begin{cases} Y_A = \dfrac{(w + f_B - f_A) \exp(r_1 x) - (w + f_A - f_B) \exp(r_2 x)}{2w} \\[4mm] Y_B = \dfrac{K_{1m}}{w} \left[\exp(r_1 x) - \exp(r_2 x) \right] \end{cases}$$

The parameters have the following significance :

$$(2\text{-}37) \begin{cases} f_A = s + N_A + K_{1m} - \dfrac{N_A^2}{sM_A + N_A} \\[4mm] f_B = s + N_B + K_{2m} - \dfrac{N_B^2}{sM_B + N_B} \\[4mm] w = \left[(f_A - f_B)^2 + 4 K_{1m} K_{2m} \right]^{1/2} \\[4mm] 2r_1 = - f_A - f_B + w, \quad 2r_2 = - f_A - f_B - w \end{cases}$$

The peak of A and B at the column exit are obtained in setting $x = 1$. They can be simulated numerically by inversion of the Laplace transforms Y_A and Y_B as usual.

Calculation of the moments - The moments are derived by means of Van der Laan's theorem in deriving Y_A and Y_B with respect to s.

The results are as follows : starting from a unit-impulse of pure A at the column inlet, the surfaces of the peaks of A and B at the column outlet are

$$\mu_{Ao} = 1 - \lambda + \lambda e \tag{2-38}$$

$$\mu_{Bo} = \lambda(1 - e) \tag{2-39}$$

$$\text{where} \quad \lambda = \frac{K_{1m}}{K_{1m} + K_{2m}} = \frac{K_m}{1 + K_m} \tag{2-40}$$

K_m being the equilibrium constant of equilibrium A = B.

e is a kinetic parameter :

$$e = \exp\left[- K_{1m} - K_{2m}\right] \tag{2-41}$$

It is interesting to notice that the conversion is the same as that of a plug flow reactor without retention in the stationary phase. Moreover, expressions (2-38) and (2-39) do not depend on the linearity of the partition isotherm, neither on the mass transfer kinetics, as can be shown below : let Φ_A be the transfer flux of A between the two phases. The mass balance is written :

$$(2\text{-}42)\begin{cases} u\,\dfrac{\partial C_{Am}}{\partial z} + \dfrac{\partial C_{Am}}{\partial t} + \dfrac{1}{\varepsilon}\,\Phi_A + k_{1m}\,C_{Am} - k_{2m}\,C_{Bm} = 0 \\[2ex] \dfrac{\partial C_{As}}{dt} - \dfrac{1}{1-\varepsilon}\,\Phi_A \qquad\qquad\qquad\qquad = 0 \end{cases}$$

Let us integrate these equations with respect to t between t = 0 and t = ∞, and set

$$m_{Ao} = \int_0^\infty C_{Am}\,dt \quad, \quad m_{Bo} = \int_0^\infty C_{Bm}\,dt \tag{2-43}$$

$$\int_0^\infty \frac{\partial C_{AS}}{\partial t}\,dt = C_{AS}\Big|_{t=0}^{t=\infty} = 0, \text{ thence} \int_0^\infty \Phi_A\,dt = 0$$

We thus obtain equations relative to the zeroth-moments which are independent of Φ_A . In reduced form :

$$(2\text{-}44) \begin{cases} \dfrac{d\mu_{Ao}}{dx} + K_{1m}\,\mu_{Ao} - K_{2m}\,\mu_{Bo} = 0 \\[2ex] \dfrac{d\mu_{Bo}}{dx} - K_{1m}\,\mu_{Ao} + K_{2m}\,\mu_{Bo} = 0 \end{cases}$$

$$x = 0 \qquad \mu_{Ao} = 1 \quad ; \quad \mu_{Bo} = 0$$

The resolution of (2-44) obviously yields the same result as (2-38) and (2-39).

From (2-36), the first-order moments may be obtained in a similar way. In the real time domain, one obtains :

$$\mu_{A1}/t_o = 1 + M_A + (M_B - M_A)\,\frac{\lambda(1-\lambda)}{1-\lambda+\lambda e}\left[1 + e - \frac{2(1-e)}{K_{1m} + K_{2m}}\right] \qquad (2\text{-}45)$$

$$\mu_{B1}/t_o = 1 + M_A + \frac{M_B - M_A}{1 - e}\left[\lambda - (1 - \lambda)e + \frac{(1 - 2\lambda)(1 - e)}{K_{1m} + K_{2m}}\right] \qquad (2\text{-}46)$$

Here again, the result <u>does not depend on mass transfer kinetics</u> but it assumes that the <u>partition isotherm is linear</u>. This can be shown in starting from (2-42), multiplying both equations by t and integrating from t = 0 to t = ∞. Equation in μ_{A1} and μ_{B1} are obtained provided linearity is assumed.

It is interesting to see what happens when the reactions become infinitely rapid (K_{1m}, $K_{2m} \to \infty$), but $K_m = K_{1m}/K_{2m}$ remains finite. Then, the two peaks are no longer separate, they coalesce to a single peak having an intermediate position between peaks of pure A and pure B :

$$\mu_{A1} \to \mu_{B1} \to \mu_1 = t_o\left[1 + M_A + \lambda(M_B - M_A)\right] = t_o(1 + M) \qquad (2\text{-}47)$$

with $M = M_A + \lambda(M_B - M_A)$

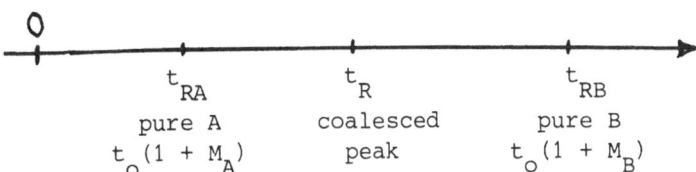

$$t_R = t_{RA} + \lambda(t_{RB} - t_{RA}), \quad \lambda = \frac{K_m}{1 + K_m} \tag{2-48}$$

The calculation of the variances is really tedious. Although we have not taken it to its conclusion, it is likely that in the case of coalesced peaks ($K_{1m} = K_{2m} = \infty$) with $N_A = N_B = N$, the result is

$$\frac{\sigma^2}{t_o^2} = \frac{2M^2}{N} \tag{2-49}$$

<u>Reaction in the stationary phase</u> - The method of reasoning is exactly the same as above. The basic mass balances are

(A)
(2-50)
$$\begin{cases} u \dfrac{\partial C_{Am}}{\partial z} + \dfrac{\partial C_{Am}}{\partial t} + k_A \dfrac{a}{\varepsilon}\left(C_{Am} - \dfrac{C_{AS}}{\alpha_A}\right) = 0 \\[3mm] \dfrac{\partial C_{As}}{\partial t} + k_A \dfrac{a}{1-\varepsilon}\left(\dfrac{C_{As}}{\alpha_A} - C_{Am}\right) + k_{1s}C_{As} - k_{2s}C_{Bs} = 0 \end{cases}$$

(B)
(2-51)
$$\begin{cases} u \dfrac{\partial C_{Bm}}{\partial z} + \dfrac{\partial C_{Bm}}{\partial t} + k_B \dfrac{a}{\varepsilon}\left(C_{Bm} - \dfrac{C_{Bm}}{\alpha_B}\right) = 0 \\[3mm] \dfrac{\partial C_{Bs}}{\partial t} + k_B \dfrac{a}{1-\varepsilon}\left(\dfrac{C_{Bs}}{\alpha_B} - C_{Bm}\right) + k_{2s}C_{Bs} - k_{1s}C_{As} = 0 \end{cases}$$

The rate constants k_{1s} and k_{2s} are now relative to the stationary phase.

To save space, we will not give the reduced equations in detail, these are derived in a similar way as above, the reduced rate constants being simply replaced by $K_{1s} = k_{1s} t_o$ and $K_{2s} = K_{2s} t_o$. The Laplace transform of A and B peaks resulting from a pulse injection of A are now given by :

(2-52)
$$\begin{cases} Y_A = \dfrac{(v + b_B - a_A)\exp(r_1 x) + (v - b_B + a_A)\exp(r_2 x)}{2v} \\[3mm] Y_B = \dfrac{b_A}{v}\left[\exp(r_2 x) - \exp(r_1 x)\right] \end{cases}$$

The parameters have the following significance :

$$
\begin{cases}
a_A = s + \dfrac{N_A}{d}\left[(s + K_{1s})\left(s + k_{2s} + \dfrac{N_B}{M_B}\right) - K_{1s}K_{2s}\right] \\[4mm]
a_B = -\dfrac{N_A N_B K_{2s}}{M_A d} \\[4mm]
b_B = s + \dfrac{N_B}{d}\left[(s + K_{2s})\left(s + K_{1s} + \dfrac{N_A}{M_A}\right) - K_{1s}K_{2s}\right] \\[4mm]
b_A = -\dfrac{N_A N_B K_{1s}}{M_B d} \\[4mm]
d = \left(s + K_{1s} + \dfrac{N_A}{M_A}\right)\left(s + K_{2s} + \dfrac{N_B}{M_B}\right) - K_{1s} K_{2s} \\[4mm]
2r_1 = -(a_A + b_B) + v, \quad 2r_2 = -(a_A + b_B) - v \\[4mm]
v = \left[(a_A - b_B)^2 + 4\, a_B b_A\right]^{1/2}
\end{cases} \tag{2-53}
$$

Calculation of the moments - Before giving the expressions for the moments, it may be noticed that a correspondence exists between parameters for the mobile phase and the stationary phase

$$
A_m \underset{k_{2m}}{\overset{k_{1m}}{\rightleftarrows}} B_m
$$

A reversible reaction in the stationary phase can be "seen" as if it occured in the mobile phase with the following correspondence :

$$
A_s \underset{k_{2s}}{\overset{k_{1s}}{\rightleftarrows}} B_s
$$

$$
(2\text{-}54)\begin{cases}
k_{1m} = \alpha_A \dfrac{1-\varepsilon}{\varepsilon} k_{1s} = M_A k_{1s}, \quad K_{1m} = M_A K_{1s} \\[4mm]
k_{2m} = \alpha_B \dfrac{1-\varepsilon}{\varepsilon} k_{2s} = M_B k_{2s}, \quad K_{2m} = M_B K_{2s} \\[4mm]
K_m = \dfrac{k_{1m}}{k_{2m}} = K_s \dfrac{M_A}{M_B}
\end{cases}
$$

This enables us to compare the expression relative to both situations.

Now, the results for the zeroth order moments (surfaces of peaks) are :

$$\mu_{Ao} = 1 - \lambda + \lambda e' \qquad (2\text{-}55)$$

$$\mu_{Bo} = \lambda(1 - e') \qquad (2\text{-}56)$$

where $\lambda = \dfrac{K_m}{1 + K_m}$ has the same meaning as above. But e' is different from e :

$$e' = \exp\left[-\frac{K_{1m} + K_{2m}}{1 + \dfrac{K_{1m}}{N_A} + \dfrac{K_{2m}}{N_B}}\right] \qquad (2\text{-}57)$$

$$\frac{K_{1m}}{N_A} = k_{1S}\, t_{mA} = \delta_{1A} \quad \text{and} \quad \frac{K_{2m}}{N_B} = k_{2S}\, t_{mB} = \delta_{2B}$$

are Damköhler numbers characterizing mass transfer of A and B compared to chemical reactions.

We see thus that mass transfer interferes with conversion, as in the case of an irreversible reaction.

We have not performed the calculation for the first order moments μ_{A1} and μ_{A2} in the general case, because the mathematical expressions are too complicated. Two special cases may be dealt with more easily : that of instantaneous chemical reaction, and that of instantaneous mass-transfer.

Let us first assume that K_{1m}, $K_{2m} \to \infty$, $K_m = K_{1m}/K_{2m}$ remaining finite. Moreover, the number of mass transfer units are supposed to be the same for A and B : $N_A = N_B = N$. The center of gravity of peaks A and B at the column outlet are :

$$\mu_{A1}/t_o = 1 + \frac{M_A + K_m M_B}{(1+K_m)(1+K_m e^{-N})} = 1 + \frac{M}{1 + K_m e^{-N}} \qquad (2\text{-}58)$$

$$\mu_{B1}/t_o = 1 + \frac{M_A + K_m M_B}{(1 + K_m)(1 - e^{-N})} = 1 + \frac{M}{1 - e^{-N}} \qquad (2\text{-}59)$$

M has already been defined as $M_A + \lambda (M_B - M_A)$. The mass transfer resistance prevents the two peaks from coalescing into a single peak with the mean t_o $(1 + M)$. The corresponding values of the variances are :

$$\frac{\sigma_A^2}{t_o^2} = \frac{M^2}{1 + K_m e^{-N}} \left(\frac{2}{N} + \frac{K_m e^{-N}}{1 + K_m e^{-N}} \right) \tag{2-60}$$

$$\frac{\sigma_B^2}{t_o^2} = \frac{M^2}{1 - e^{-N}} \left(\frac{2}{N} - \frac{e^{-N}}{1 - e^{-N}} \right) \tag{2-61}$$

If we now consider the sum of the two peaks A + B as whole, this double peak is located at the mean retention time

$$\mu_1 = t_o (1 + M) \tag{2-62}$$

The variance of the double peak is

$$\frac{\sigma^2}{t_o^2} = \frac{2M^2}{N} \tag{2-63}$$

which is exactly that of a single peak with the mean capacity factor M. This result shows that there is <u>no chemical broadening</u> of the peak, even if μ_{A1} and μ_{B1} are so close to each other that A and B are no longer distinguishable.

The second special case is that of instantaneous mass transfer. For a reaction with a finite rate, the expressions of μ_{A1} and μ_{B1} are found to be exactly the same as those for a reaction in the mobile phase i.e. (2-45) and (2-46). This can be understood because the compositions in the two phases are in partition equilibrium at any time.

If we assume in addition that the reactions are very fast, then the two peaks are coalesced into a unique peak at the common value of μ_1

$$\mu_{A1} = \mu_{B1} = \mu_1 = t_o (1 + M) = t_o \left[1 + M_A + \lambda (M_B - M_A) \right] \tag{2-64}$$

and the variance is obviously the same as in (2-63) : $\sigma^2/t_o^2 = 2M^2/N$ This problem was studied for the first time by Klinkenberg [29].

Examples of simulations - Chromatograms of the system A $\xrightarrow[k_{2s}]{k_{1s}}$ B have been simulated by numerical inversion (Fast Fourier Transform algorithm, see ref. [28]) of the Laplace Transforms (2-52) in the case of a reaction in the stationary phase. The parameters have the following values :

Capacity factors : $M_A = 5$, $M_B = 2$

Number of mass transfer units : $N_A = N_B = N = 10^4$

This corresponds, for species A, to a number of theoretical plates :

$$NTP = \left(\frac{1 + M_A}{M_A}\right)^2 \cdot \frac{N_A}{2} = 7200$$

The simulation consists in studying the shape and deformation of peaks as a function of kinetic parameters K_{1s} and K_{2s} (which are noted K_1 and K_2 on the figures).

Fig. 1 represents the case of an irreversible reaction A $\xrightarrow{k_{1s}}$ B ($k_{2s} = 0$), the column being fed with a pulse of pure A. The A peak is decreased in surface without deformation whereas B produced by the reaction is eluted continuously between retention times t_{RA} and t_{RB} of pure species. Such a simulation had already been performed by Kallen and Heilbronner [30] by the method of staged cells in series (Martin and Synge model), while Kocirik has published the expressions for the moments [31].

Figures 2A to 2H represent the case of a reversible reaction where the equilibrium constant is kept at the constant value

$$K = K_{1s}/K_{2s} = K_1/K_2 = 2$$

(corresponding to $K_m = K_{1s} M_A/K_{2s} M_B = 5$)

K_{1s} is varied between 0.2 and 48. The column is fed with pure A. As K_{1s} is small, B is eluted in the shape of a flat plateau between t_{RA} and t_{RB}. A and B are partially separated. This result is interesting because it would be possible to stop elution just before the rise of A in order to collect the product B. As K_{1s} is increased, the initial A peak disappears into a broader zone largely superimposed on that of B. When K_{1s} is large (fig. 2-H), A and B are eluted as a single peak located between t_{RA} and t_{RB}, in the equilibrium proportions.

Figures 3A to 3H relate to the same rate constants as in figures 2 but the column is here fed with a pulse of a mixture A/B already at thermodynamic equilibrium. As K_{1s} is increased, the

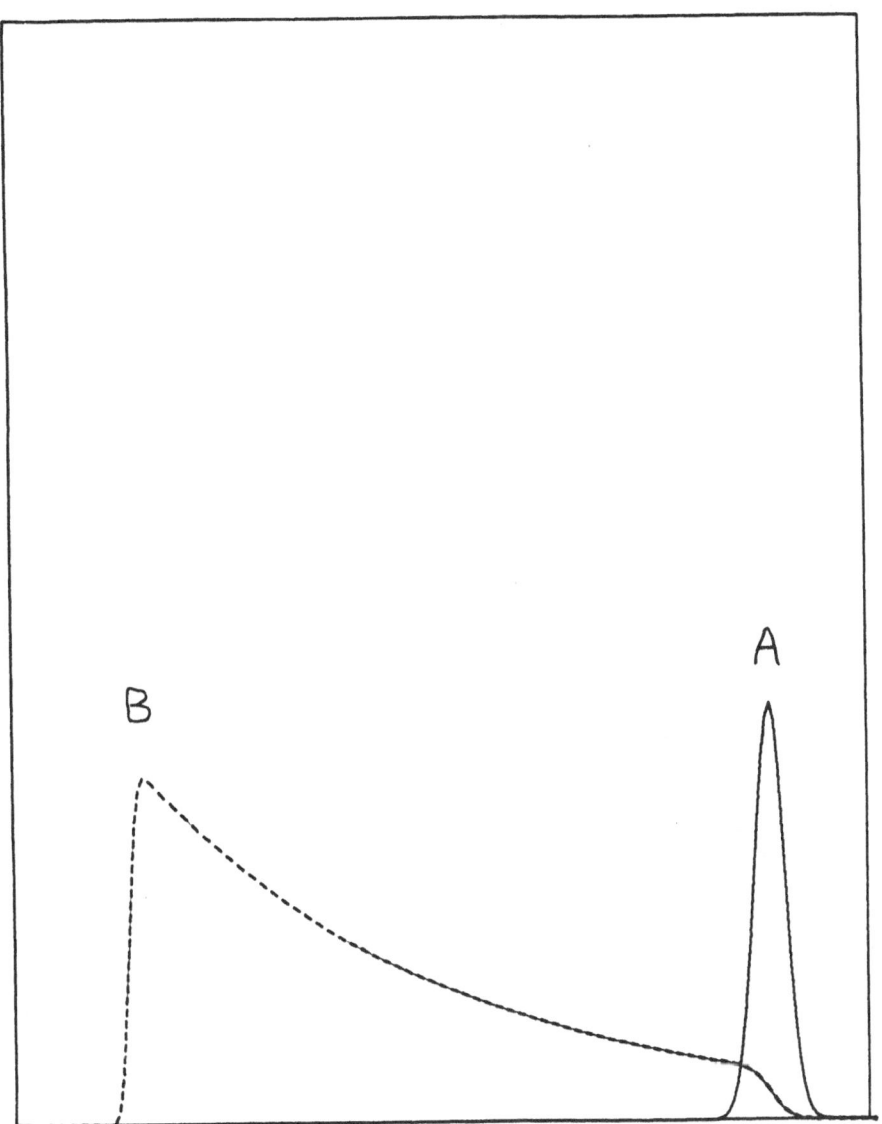

K1= 0·4 K2= 0·0

FIG-1

562

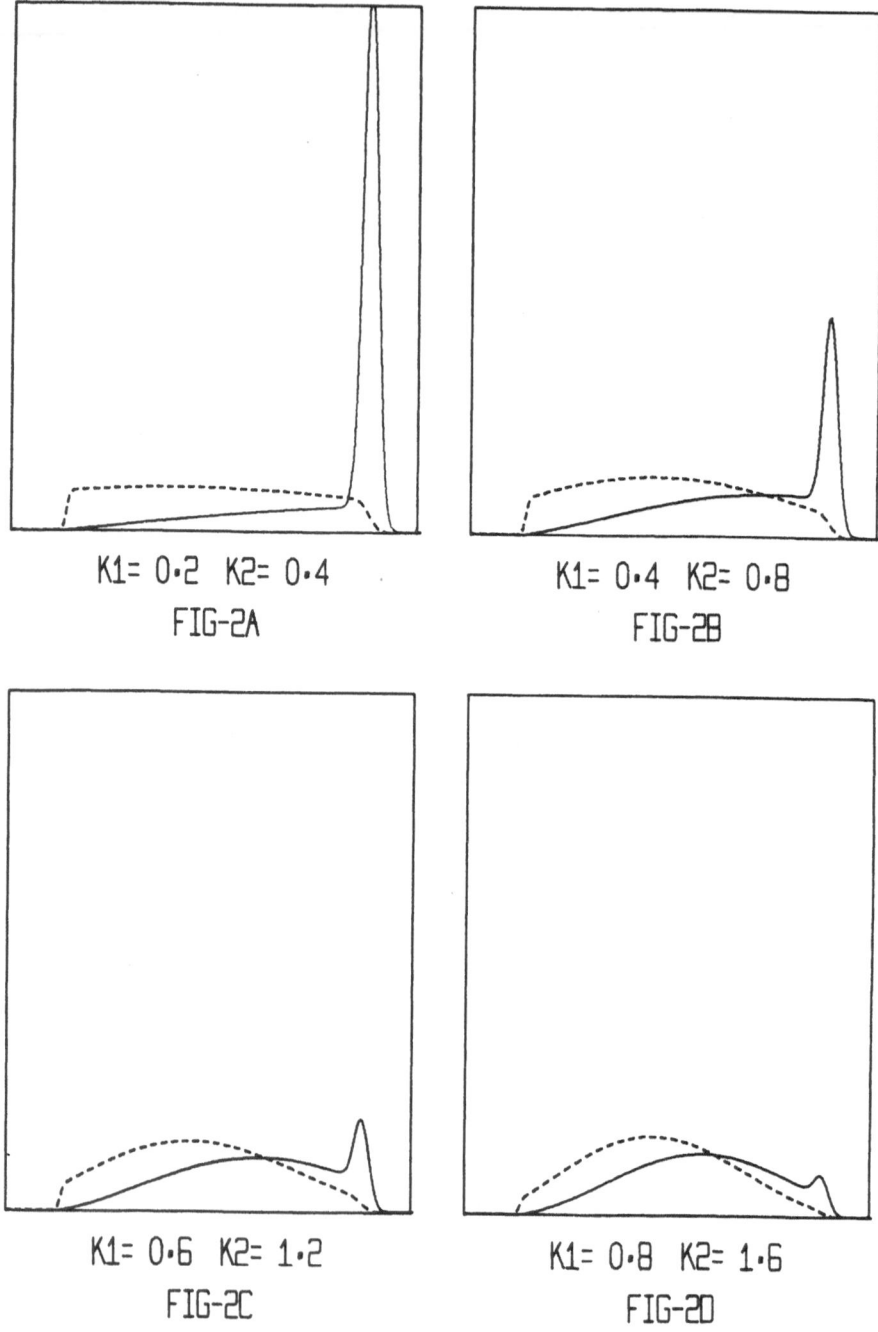

K1= 0·2 K2= 0·4
FIG-2A

K1= 0·4 K2= 0·8
FIG-2B

K1= 0·6 K2= 1·2
FIG-2C

K1= 0·8 K2= 1·6
FIG-2D

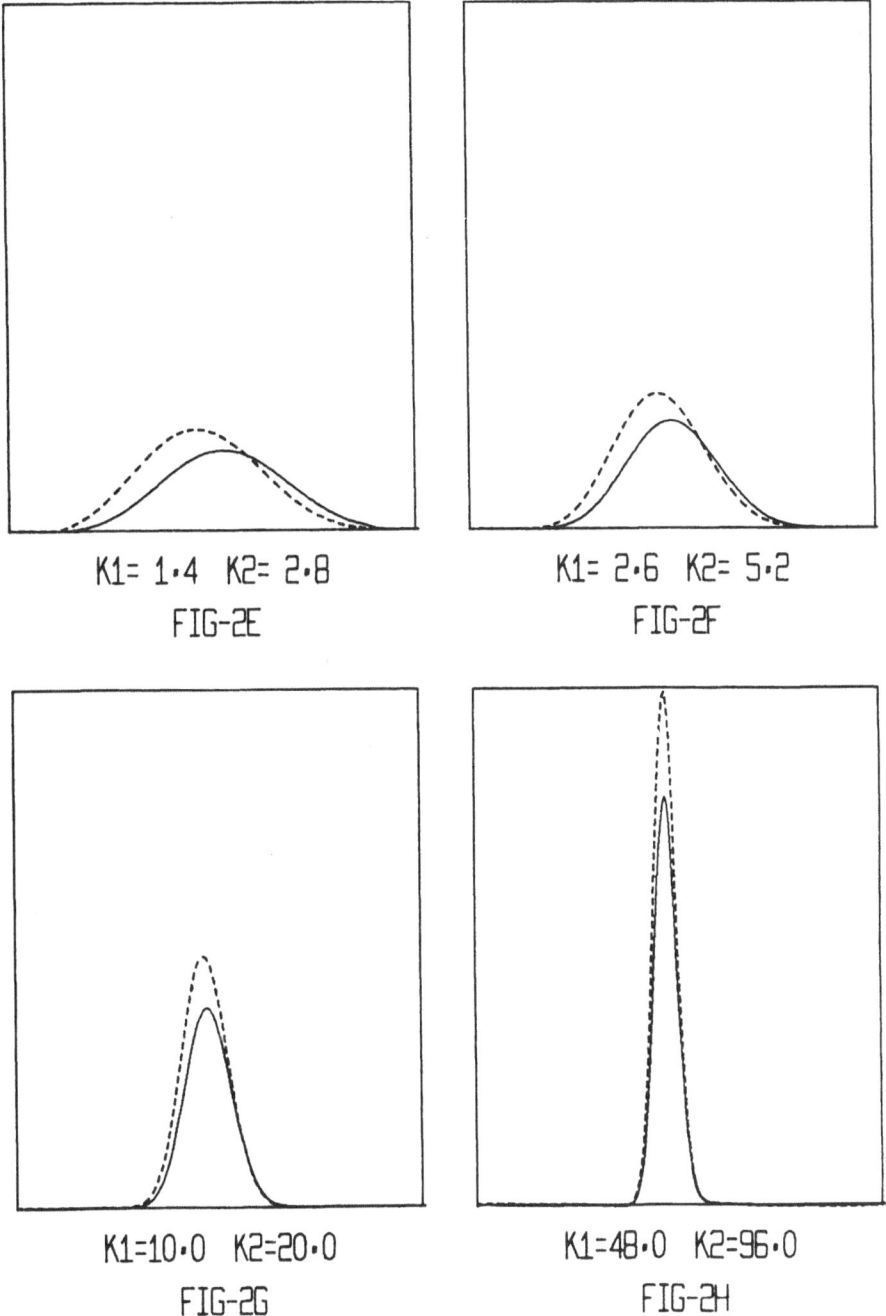

K1= 1·4 K2= 2·8
FIG-2E

K1= 2·6 K2= 5·2
FIG-2F

K1=10·0 K2=20·0
FIG-2G

K1=48·0 K2=96·0
FIG-2H

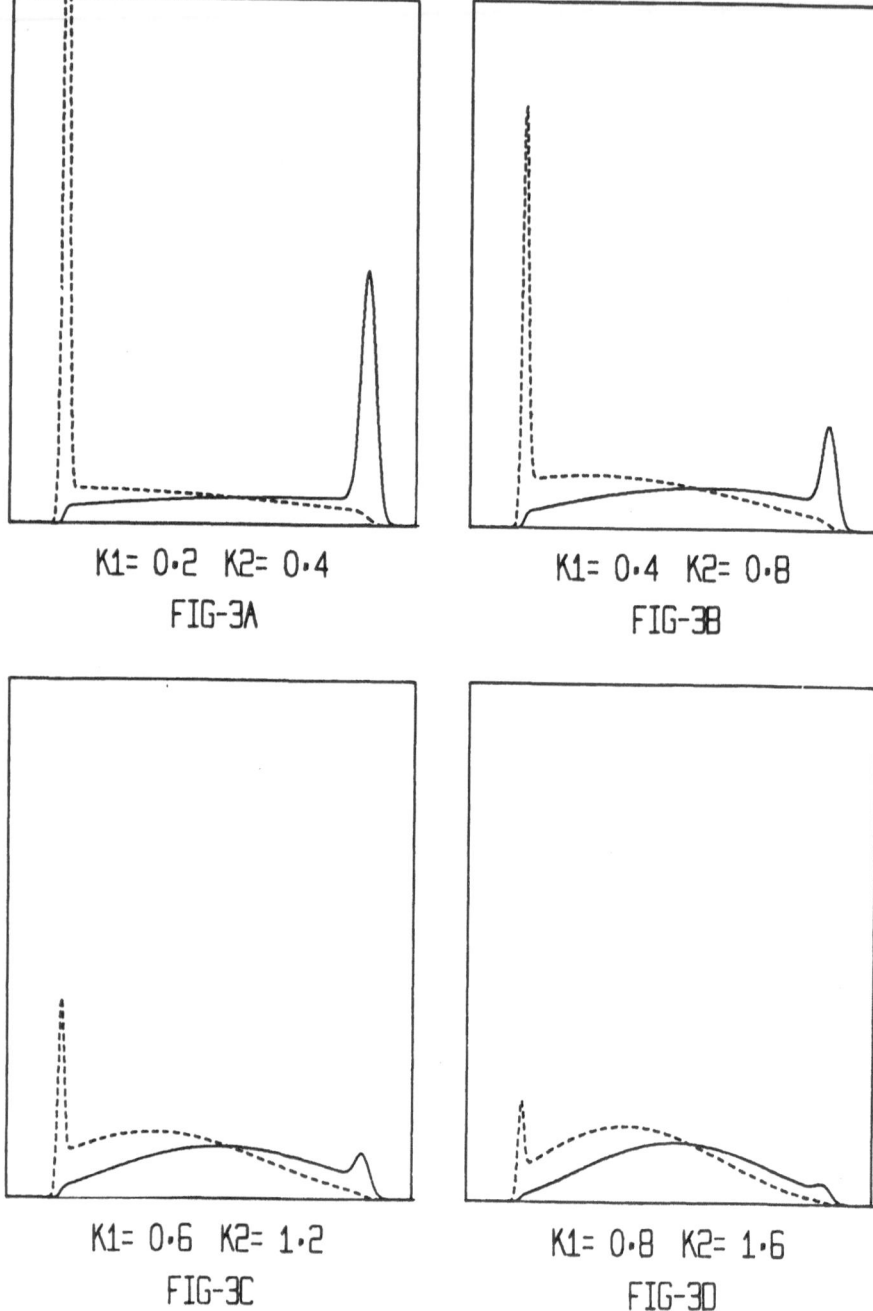

K1= 0·2 K2= 0·4

FIG-3A

K1= 0·4 K2= 0·8

FIG-3B

K1= 0·6 K2= 1·2

FIG-3C

K1= 0·8 K2= 1·6

FIG-3D

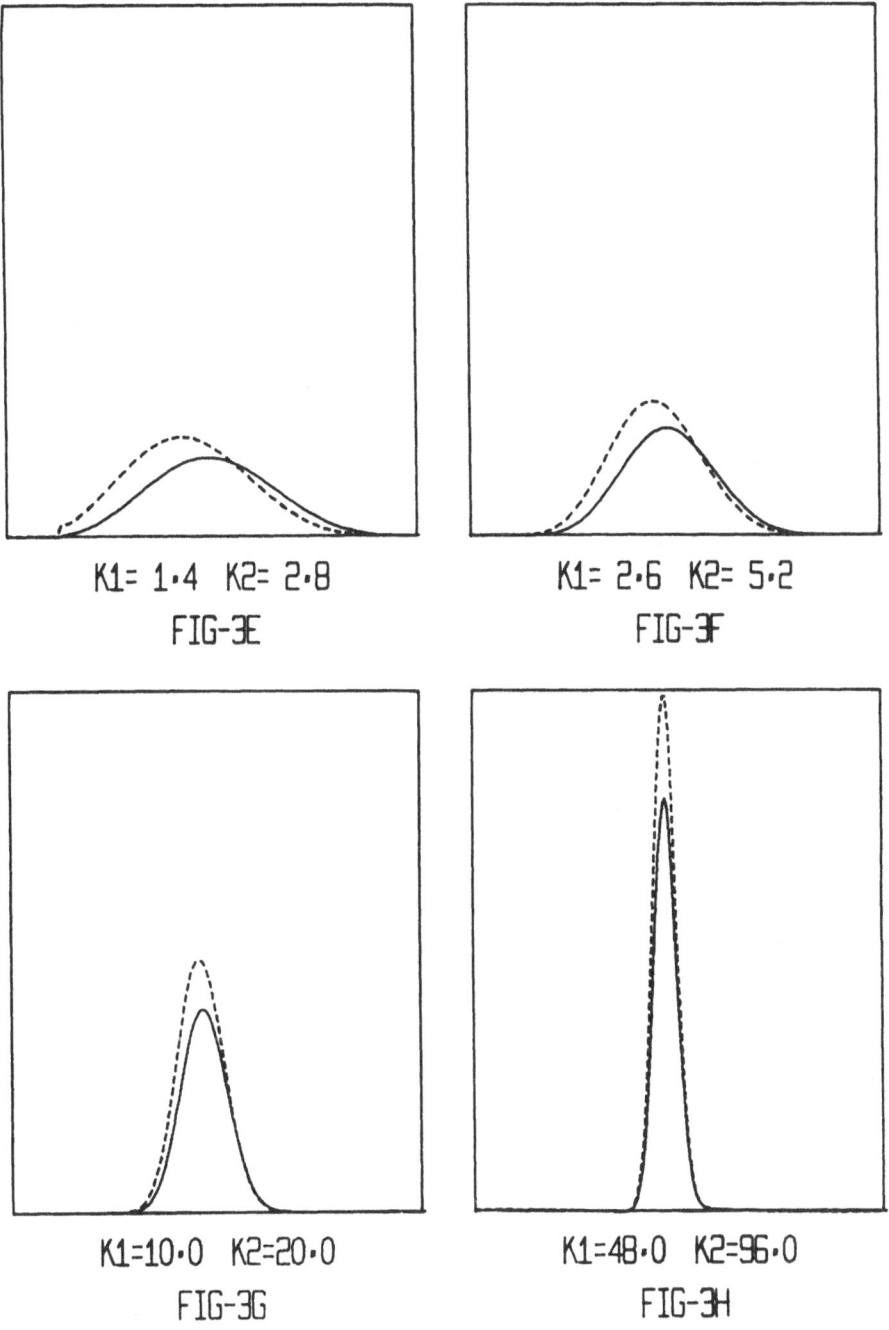

K1= 1·4   K2= 2·8

FIG-3E

K1= 2·6   K2= 5·2

FIG-3F

K1=10·0   K2=20·0

FIG-3G

K1=48·0   K2=96·0

FIG-3H

species first tend to be separated (fig. 3A to 3C), then they react again to return to the equilibrated mixture in a single peak (fig. 3D to 3H). Notice that as $K_{1s} > 1$ in this example, the shape of the chromatogram is almost the same, whatever the composition of the inlet pulse (compare fig. 2E-3E, 2F-3F, 2G-3G, 2H-3H).

Such a behaviour has been experimentally observed by Yurchak [32] in the isomerization of Syn/anti-acetaldoxim.

Comparative yield in pulsed runs and at steady state - Now the question arises : given the reversible first-order reaction $A \underset{k_2}{\overset{k_1}{\rightleftarrows}} B$ limited by thermodynamic equilibrium, is it possible to obtain a better yield of B (possibly higher than that of equilibrium) by running the reactor in the chromatographic pulsed mode rather than running it at steady state in the same conditions?

The answer to this question is no, owing to the properties of linear dynamic systems.

Let $C_A(z,t)$ be a transient concentration signal of A in the column. If the injected amount is finite, it is given by

$$\int_0^\infty C_A(z,t)\ dt = \overline{C_A}(z,o) \qquad (2-65)$$

where $\overline{C_A}(z,s) = \int_0^\infty C_A(z,t)\ e^{-st}\ dt \qquad (2-66)$

is the Laplace transform of C_A. The relative amount of A remaining at the column outlet is thus

$$\frac{\int_0^\infty C_A(L,t)dt}{\int_0^\infty C_A(0,t)dt} = \frac{\overline{C_A}(L,0)}{\overline{C_A}(0,0)} = Y_A(L,0) \qquad (2-67)$$

where $Y_A(z,s)$ is the transfer function of species A in the column.

Now, let us feed the column with a positive step injection $C_A(0)$, the concentration being kept at this value of the inlet. The Laplace transform of this injection is

$$\overline{C_A}(0,s) = C_A(0)/s \qquad (2-68)$$

And at the outlet :

$$\overline{C_A}(L,s) = Y_A(L,s) \cdot \overline{C_A}(0,s) \tag{2-69}$$

After a very long time, the outlet concentration goes to a constant value $C_A(L)$ which is given by the final value theorem

$$C_A(L) = \lim_{s \to 0} \left[s \ \overline{C_A}(L,s) \right] \tag{2-70}$$

Taking (2-68) and (2-69) into account, one finds

$$C_A(L) = Y_A(L,o) \cdot (C_A(o) \tag{2-71}$$

Thus, the relative amount of A remaining at the column outlet in steady state processing is

$$\frac{C_A(L)}{C_A(o)} = Y_A(L,o) \tag{2-72}$$

which is exactly the same as (2-67) in transient processing.

Thus, no quantitative yield improvement over steady state is to be expected in using a pulsed mode in the case of linear processes. The situation is different for non linear processes, as will be shown in sections 3 and 4 below.

3. NON LINEAR PROCESSES : A SIMPLIFIED MODEL FOR THE CHROMATOGRA-PHIC REACTOR

The linear methods of section 2 are no longer applicable if non-linear kinetics are involved in the reaction system. This is the case of the reversible reactions which have been historically carried out for the first time in chromatographic reactors, and which are of the kind

$$A \underset{\longleftarrow}{\longrightarrow} B + \nu C \tag{3-1}$$

Without reducing the generality of the results, Schweich et al. [20] have investigated the simple reaction in greater detail:

$$A_1 \underset{\longleftarrow}{\longrightarrow} A_2 + A_3 \tag{3-2}$$

The performance of the reactor can be measured by the improvement in yield in pulsed runs with respect to steady state runs with the same feed.

3.1. "Rigorous" model

Chromatograms may be simulated using the Martin and Synge model of J reactors in series (see ref. [28] sec. 3), and adding the reaction terms in the mass balance equations. In order to simplify the analysis, the following assumptions are made in a first step :

- infinite forward and backward reaction rates : equilibrium is attained at any point :

$$C_1 = K_t \, C_2 \, C_3 \tag{3-3}$$

- instantaneous sorption rates

- dilute incompressible and isothermal medium

- linear partition isotherms (capacity factors M_1, M_2, M_3)

- reactant A_1 is eluted between products A_2 and A_3 :

$$M_2 < M_1 < M_3$$

These conditions are very similar to those of Langer and Matsen (see sec. 1), except for axial dispersion, which is accounted for by the number J of mixing cells in series.

A_1 is injected periodically into the column as rectangular pulses defined by three parameters : θ, duration of the feed pulse, C_0 reactant concentration of A_1 in the pulse before chemical equilibrium is reached, and T, period of pulsations. As the reactor is not in steady-state conditions, the yield is defined as the ratio of the amount of product A_2 or A_3 which flows out of the reactor during a period, to the amount of reactant fed during this same period. In order to compare the yield in pulsed runs to that in steady state, the average feed composition is kept constant : $\theta C_0/T$ = constant.

The derivation of mass balances in each reactor of the cascade is straightforward, and can be found in reference [20]. For instance, the mass balance on component A_i in cell number k is written :

$$Q \, C_{i,k-1} + v_i r \, \frac{V}{J} = Q \, C_{i,k} + (1+M_i) \frac{V}{J} \frac{d \, C_{i,k}}{dt} \qquad k = 1,\ldots J \tag{3-4}$$

Figure 4 shows examples of chromatograms obtained in integrating the differential mass-balance equations with the following values of the parameters : $M_1 = 1$, $M_2 = 0$, $M_3 = 3$, $J = 100$, $K_t = 2500$ l/mol.

569

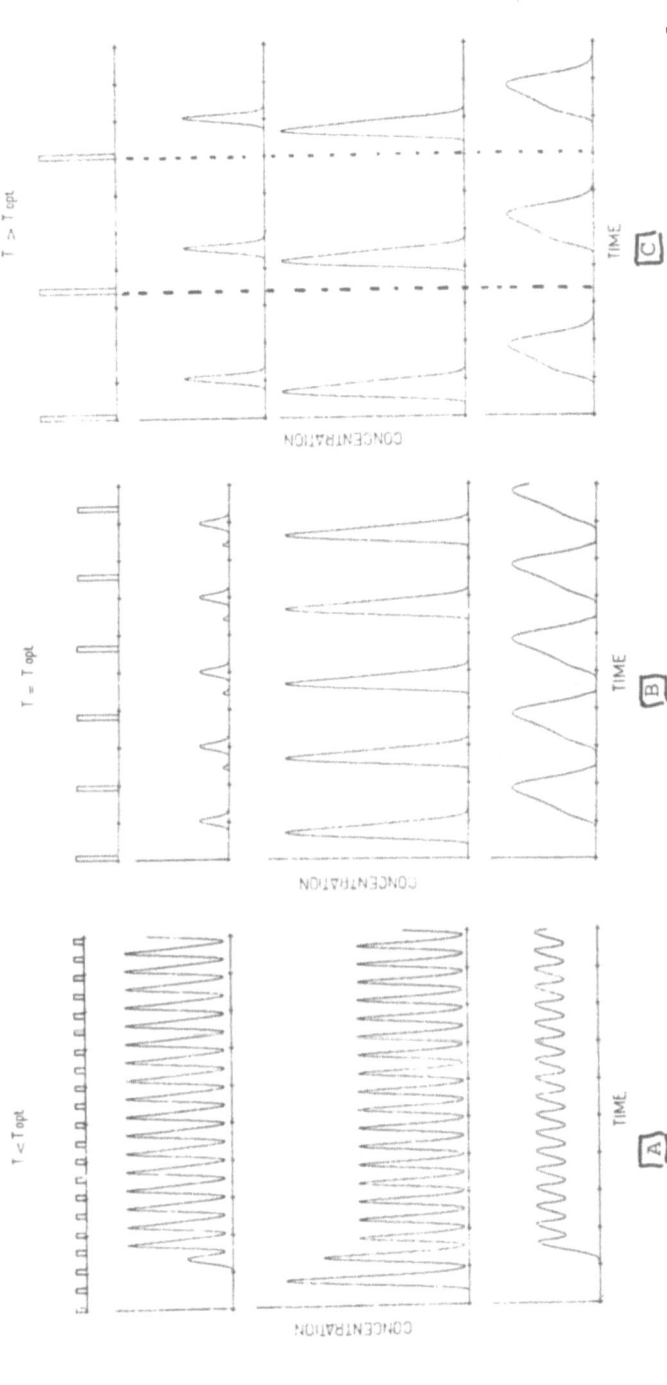

Figure 4. Simulated chromatograms. From top to bottom : reactant feed, reactant, product A_2, product A_3 chromatograms. $\boxed{A}$ Short period pulsed run at low feed concentration ($T<T_{opt}$). Note the interferences between successive pulses which prevent zero concentration from being reached (simulation parameters same as in figure 5, $\theta=0,25$). $\boxed{B}$ Optimal pulse period ($T=T_{opt}$). Note the duplication of the reactant peaks. The rounded peak is the remaining part of the feed pulse. The sharp peak is recovered reactant due to the reaction occurring between products of successive pulses. $\boxed{C}$ Low pulse frequency at high feed concentration ($T>T_{opt}$). Note the independence of successive pulse chromatograms as shown by the dotted vertical lines.

The corresponding yields are plotted against the period T on figure 5.

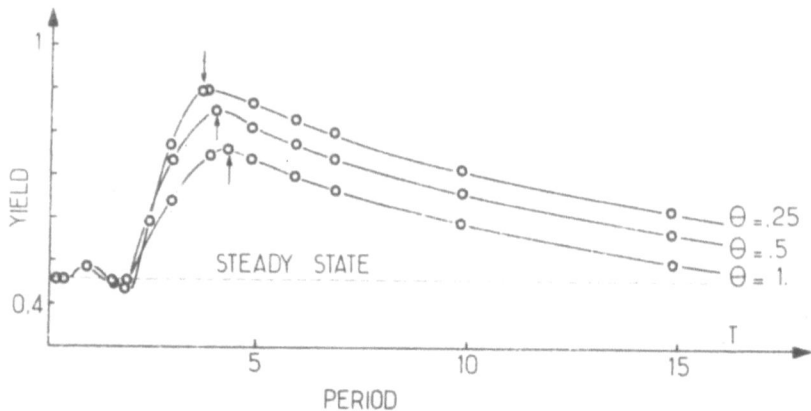

Figure 5. Yield improvement with respect to steady state as function of pulse period. The shorter the pulse, the larger the yield improvement. The arrows point the approximate position of the maximum calculated with our formula ($M_1=1$, $M_2=0$, $M_3=3$, $J=100$, $K_t = 2500$ L/mol, $\theta C_0/T=10^{-3}$ mol/L).

As pointed out by Chieh et al. [18], there is an optimum pulse period T_{opt} leading to a maximum yield improvement. The shorter the pulse, the greater the yield improvement. The chromatograms show that elution curves of consecutive pulses are independent of each other provided that the period is greater than T_{opt}. They begin to overlap when $T = T_{opt}$. Matsen et al [23] found the same qualitative results by experiments. The first simplifying rule can be derived from these remarks : provided that $T \gg T_{opt}$, a superposition principle can be applied for obtaining the yield in repetitive pulse experiments from the behavior of the chromatographic reactor in single-pulse experiments. Repetitive pulse chromatograms are no more than single-pulse chromatograms which have been suitably translated.

The time required to elute all the components from a single pulse is of great interest. It is estimated as follows. Three contributions are shown in Figure 6 namely : (1) the difference between shortest and largest retention times without any chemical reaction : $t_{R3} - t_{R2} = t_0(M_3 - M_2)$; (2) the duration of the pulse and (3) the broadening of the first and last peaks, which are measured by the standard deviation $\sigma_i = [t_0(1 + M_i)]/\sqrt{J}$. This expression is

$$\sigma_2 + \sigma_3 = \frac{t_0}{\sqrt{J}} (2 + M_2 + M_3) \tag{3.5}$$

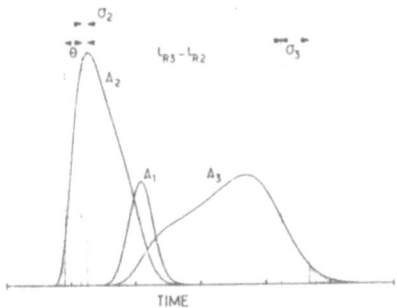

Figure 6. Estimate of the optimum period : during optimum runs overall chromatograms are in contact with one another, T_{opt} is estimated by means of the time required to elute the chromatogram resulting from a single pulse.

It is supposed that reaction does not contribute to the broadening of the reactant peak. This assumption is true provided that product peaks are on either side of the reactant peak.

We are now ready to pass on the property of this elution duration. The optimum period is such that a reconverted peak of reactant begins to appear in the chromatogram as shown in Figure 4B. This goal is achieved when the pulse period is equal to the elution duration of the chromatogram from a single-pulse experiment which allows the second simplifying rule to be laid down : the optimum T_{opt} depends on the chromatographic characteristics of the components without any chemical reaction. A good approximation is

$$T_{opt} = \theta + t_o (M_3 - M_2) + \frac{t_o}{\sqrt{2}} (2 + M_2 + M_3) \qquad (3-6)$$

In Figure 5, the small arrows point to the approximate optimum period. The good agreement between the approximation and the true maximum proves the validity of the assumptions. The consequence of these two simplifying rules is that the main features of the optimal repetitive pulsed runs may be determined from those of single-pulse experiments.

Before dealing with this new problem, it is of interest to examine the decrease of the improvement in yield when the period is sufficiently large. It is not an intrinsic property of the chromatographic reactor, but merely a consequence of our choice of the performance index. As the time average feed composition is fixed, an increase in the period results in an increase in the concentration of each feed pulse. Thermodynamic equilibrium displacement rules show that a concentration increase is not favorable to dehydrogenation reactions. If the part of the curves of yield improvement which increases with the period is a consequence of the chromatographic separation, the decreasing part arises from the unfavorable effect of the thermodynamics. This remark has been pointed out by Gore [16] as a limitation to the use of the chromatographic reactor.

3.2. Simplified model

The two simplifying rules allow us to state the problem as follows : how can the behavior of the chromatographic reactor in single pulse experiments be predicted ?

The rigorous model answers this question, but the mathematical simulations are rather complicated and time consuming. Moreover, in order to describe the physical situation, pressure gradient and nonlinear isotherms must be taken into account.

As we are interested in overall characteristics, such as yield, the shape of the chromatogram is not important. To answer our question we shall set up a new model where concentration gradients will be neglected. Let us recall that Hattori [10] showed that the shape of the feed pulse is of little importance. We shall only consider the small slice of fluid containing the reactant peak. This slice can be considered as a small semi-batch reactor moving down the column, the volume of which increases due to peak broadening. Products and reactant differ in this flow velocities so that the products A_2 and A_3 escape from this reactor (Figure 7).

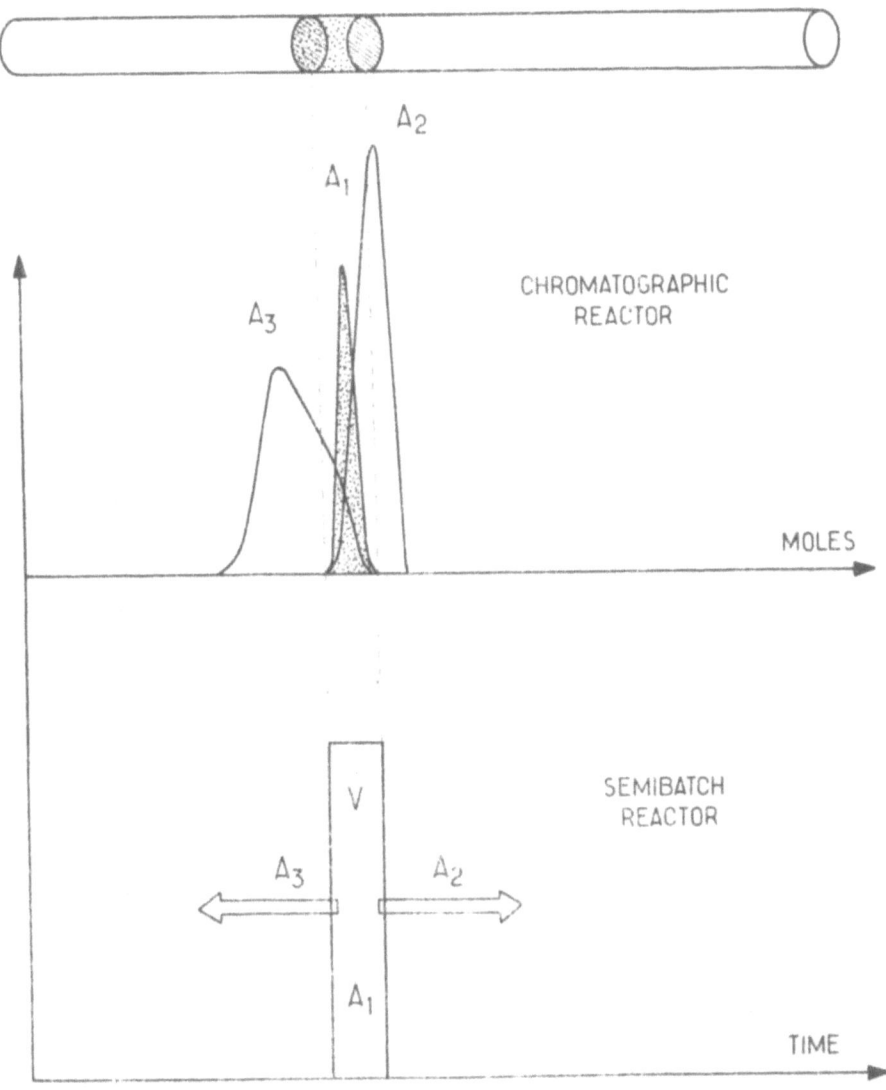

Figure 7. Principle of the simplified model. Only the small slice containing the reactant is considered. It behaves like a semibatch reactor out of which reaction products flow.

Let us consider the equilibrated reaction $A_1 \rightleftarrows A_2 + A_3$. The simplified approach asummes that the reactant is in a batch reactor. The mass balance equation for A_1 is thus :

$$r V = (1 + M_1) \frac{dn_1}{dt} \qquad (3-7)$$

Products A_2 and A_3 have escaping fluxes which account for chromatographic separation :

$$r \, V = (1 + M_i) \, \frac{dn_i}{dt} + F_i \qquad i = 2,3 \qquad\qquad (3\text{-}8)$$

The escaping flow-rate F_i can be expressed as the product of the overall mole fraction of A_i, i.e. $n_i(1 + M_i)/n_T$ times the overall flow rate carrying A_i with respect to A_1, i.e.

$$\frac{F_T}{1 + M_i} - \frac{F_T}{1 + M_1} \qquad\qquad (3\text{-}9)$$

Hence, the mass balance equation for A_2 or A_3 is :

$$r \, V = (1 + M_i) \, \frac{dn_i}{dt} + \frac{n_i}{n_T} \, F_T \, \frac{\left| M_i - M_1 \right|}{1 + M_1} \qquad i = 2,3 \qquad (3\text{-}10)$$

Absolute values are introduced to get the right sign in the equations. The volume of the reacting slice V and the whole quantity of matter n_T depend on elution time t and are related to the amount of reactant injected and to the broadening parameter J. For the sake of simplicity we will neglect the effect of the broadening parameter (complete calculations are exposed in [20]). Let C_o and θ be the concentration and the duration of the reactant pulse. If Q is the volumetric flow rate then

$$n_T = f \, \frac{\theta C_o}{1 + M_1} \, Q \qquad \text{and} \quad V = f \, \frac{\theta}{1 + M_1} \, Q \qquad\qquad (3\text{-}11)$$

f is a shape factor which accounts for the neglected concentration profiles in the slice. f is of the order of 1 and would be equal to 1 if the chromatogram of the reactant were a rectangular pulse which did not get out of shape. The expressions (3-11) are slightly more complicated if dispersion processes are accounted for via J. The set of three equations (3-8), (3-10), whatever J, is integrated with the following boundary conditions : t = 0, $n_1 = n_o$, $n_2 = n_3 = 0$. Integration is performed up to $t = t_{R1} = t_o(1 + M_1)$, the composition of the semi batch reactor at this time being the composition observed at the outlet of the chromatographic reactor.

The simplified model is still valid with a more complex reaction scheme such as $A + B \rightarrow R$, $R + B \rightarrow S$. R being the valuable product. Assuming A is fed continuously at concentration C_{AO} and B as a pulse at concentration C_{BO}, the mass balance equations are :

$$(3\text{-}12) \begin{cases} - k_1 C_A C_B V = Q \dfrac{|M_A - M_B|}{1 + M_B} (C_A - C_{AO}) + (1 + M_A) V \dfrac{dC_A}{dt} \\[3ex] - k_1 C_A C_B V - k_2 C_R C_B V = (1 + M_B) V \dfrac{dC_B}{dt} \\[3ex] k_1 C_A C_B V - k_2 C_R C_B V = Q \dfrac{|M_R - M_B|}{1 + M_B} C_R + (1 + M_R) V \dfrac{dC_R}{dt} \end{cases}$$

It is assumed that no broadening effects take place. We may expect that the separation of R and B will give a better yield of R in pulsed runs than in steady state runs. Figure 8 shows the results of the integration as the yield of the intermediate R versus the converted fraction of reactant B. The ratio of the mean concentration of A to B during the elution duration of the reaction chromatogram is kept constant and equal to 1. This constraint allows us to draw the dotted curve measuring the yield in a steady state reactor operated with the same mean concentrations of reactants. In fact, no improvement is observed because the concentration of B in each pulse is so high (with respect to steady state) that the chromatographic separation is not sufficient to separate R before it reacts again with B.

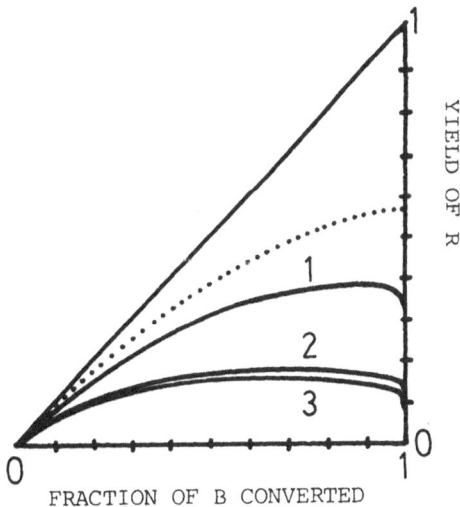

Figure 8. Yield of intermediate product R as a function of the fraction of converted B. The shape factor is 1. The ratio of the mean concentration of reactant A to B during the elution duration of the reaction chromatogram is kept constant and equal to 1. $M_A = 6$, $M_B = 0$, $M_R = 1$, t_o variable.

$$k_1 = 20 \ min^{-1}.\ell.mole^{-1}, \quad k_2 = 7.5 \ min^{-1}.\ell.mole^{-1}$$

curve no.	1	2	3
C_{AO} (mole/ℓ)	8.2×10^{-4}	8.2×10^{-4}	8.2×10^{-4}
C_{BO} (mole/ℓ)	4.92×10^{-3}	5×10^{-2}	4.93×10^{-1}
θ (min)	1	0.1	0.01

The elution duration of the reaction chromatogram is about 61 minutes. The dotted line represents the yield obtained in a steady state reactor operated with the same mean concentration.

The third example showing the flexibility of the simplified model is the isomérisation $A \longrightarrow B$ inhibited by B. Assuming the reaction rate is

$$(3-13) \quad \begin{cases} r = \dfrac{kC_A}{1 + KC_B} \ , \text{ the mass balance equations are :} \\[2ex] -\dfrac{kC_A}{1 + KC_B} \ V = (1 + M_A) \ V \ \dfrac{dC_A}{dt} \\[2ex] -\dfrac{kC_A}{1 + KC_B} \ V = Q \ \dfrac{|M_B - M_A|}{1 + M_A} \ C_B + (1 + M_B)V \ \dfrac{dC_B}{dt} \end{cases}$$

In this case we may expect that the separation of A and B will decrease the inhibition of the reaction rate. Figure 9 shows the extent of reaction in the chromatographic reactor as a function of the extent observed in a steady state reactor operated with the time average concentration of reactant during the elution duration of the reaction chromatogram. Once more, it is seen that the chromatographic effect is not sufficient to give a better yield in pulsed runs than in steady state runs.

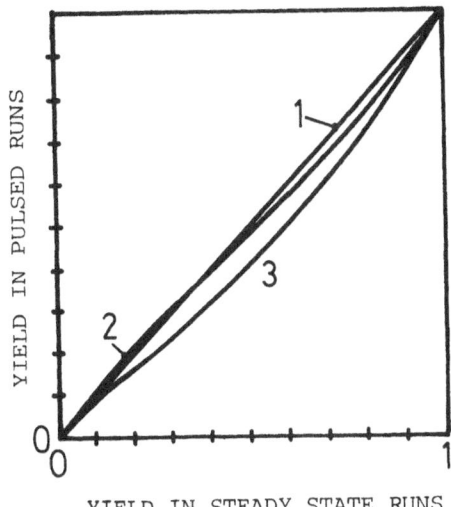

YIELD IN PULSED RUNS

YIELD IN STEADY STATE RUNS

Figure 9. Extent of reaction in pulsed runs as a function of the extent observed in a steady state reactor operated with the time average concentration of reactant during the elution duration of the reaction chromatogram. The reaction is supposed to be inhibited by the product B. The shape factor is 1

$M_A = 0$, $M_B = 1$, $k = 2 \text{ min}^{-1}$, $\theta = 0.25 \text{ } t_o$, t_o variable

curve	1	2	3
C_{Ao} (mole/l)	1.	1.	10.
K (1/mole)	0.	19.2	192.

Inhibition increases from 1 to 3.

3.3. Comparison between the simplified model and the rigorous model, example $A_1 \rightleftharpoons A_2 + A_3$

In the previous section we have shown that the simplified model is a useful and easy to use tool for investigating the behaviour of a chromatographic reactor. Nevertheless the simplified model is revealed to be really efficient when compared with the rigorous model and analytical solutions when they are known. Rigorous and simplified approaches can be compared via the computer time requirements to get the solution. On the first example $A_1 \rightleftharpoons A_2 + A_3$ it is easily seen that the time required to compute the solution of equations (3-4), whatever the integration method, must increase with J which is the number of equations. A precise analysis

(20) of the problem shows that the time increment which ensures the convergence and the stability of the integration decreases with J so that the computer time required to solve the rigorous model increases as J^2. On the other hand the time requirement of the simplified model is revealed to be independant of J. As an example, data points of Figure 11 below require several hours of computer time with the rigorous model whereas about 15 seconds only are required with the simplified model on a CII Mitra 15 computer. From the point of view of computer core requirement the simplified model is not limited whereas J cannot exceed 1000 in a 32 K words memory with the rigorous model when slow reaction rate is assumed. This limit would be even lower if more complex reaction schemes were studied. Another way of improving the efficiency of the model is to measure the sensitivity of the yield prediction to the unknown shape factor f. This factor was adjusted to fit the yields calculated from the rigorous model for various M_i, J, θ and thermodynamic equilibrium constant. The result is that f lies between .40 and .58 and seems to be independant of the pulse duration θ. Morever if f is varied within the limits of $\pm$ 10% the predicted yields vary within the limits of $\pm$ 2%. It seems that the critical parameters affecting the value of f may be the capacity factors M_i, J and the stoichiometry of the reaction. Special atten tion was paid to the case where no dispersion processes take place, fast reaction rate is assumed and only one product escapes from the reacting zone ($M_1 = M_2 \neq M_3$) because the analytical solution of both simplified and rigorous model are known. The main results when unfavourable thermodynamic is assumed (large K_t) are summarized in table III. The best value of the shape factor f is found to be .41 which is in good agreement with the previous values. Nonetheless it is important to note that the simplified model fails to predict that the reaction can be forced to completion with a finite residence time as it can be **seen** from the expression relative to the rigorous model in table III.

TABLE III : TIME FOR CONVERSION X ACCORDING TO DIFFERENT MODELS.

Reaction $A_1 \rightleftarrows A_2 + A_3$

$M_1 = M_2 \neq M_3$

No dispersion processes

Fast reaction rate

Large thermodynamic equilibrium constant K_t

TABLE III (CONTINUED)

Simplified model (1/2 hour of a mathematicians time)

$$t_o (1 + M_1) = f \, K_t \, C_{AO} \, \theta \, \frac{1 + M_1}{|M_3 - M_1|} \left[- \text{Log} \, (1 - X) - X \right]$$

Rigorous model (about 20 hours of a mathematicians time)

$$t_o (1 + M_1) = K_t \, C_{AO} \, \theta \, \frac{1 + M_1}{|M_3 - M_1|} \left[\sqrt{1 - X} - 1 \right]^2$$

---

4. EXAMPLES OF EXPERIMENTAL RESULTS

We report here a few examples of experimental work performed in our laboratory to illustrate the theoretical concepts mentioned above.

4.1. Dehydrogenation of cyclohexane

In order to test the accuracy of the simplified mode, single pulse experiments of cyclohexane dehydrogenation into benzene and hydrogen were performed by Schweich [20] on a Pt/Al_2O_3 catalyst. Increasing amounts of cyclohexane were injected into helium as the carrier gas. Figure 10 shows typical chromatograms.

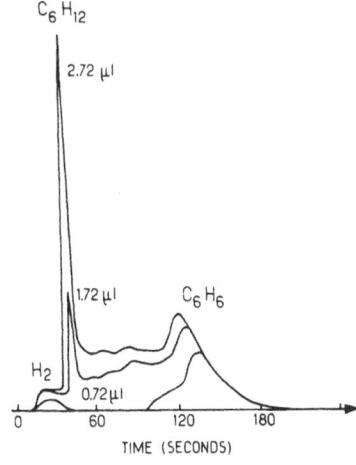

Figure 10. Cyclohexane dehydrogenation performed at 186°C. Behavior

of the overall chromatogram as a function of the amount of reactant injected. No cyclohexane was detected in the 0.72 μL pulse, thus showing improvement above equilibrium.

When the injected amount of reactant is small enough, total conversion is achieved. The species involved in the reaction have non linear adsorption isotherms on this particular catalyst. In order to perform simulations with the simplified model, chromatographic data (capacity factor, variance) were recorded in a non-catalytic column with the pure components as a function of the volume of injection. The model was adapted to the stoichiometry $A_1 \rightleftharpoons A_2 + 3\ A_3$ with instantaneous equilibrium in presence of a pressure gradient. Figure 11 shows typical results. The dots represent the extent of reaction as a function of the amount of C_6H_{12} injected. The shape factor f was optimized to fit experimental data. The curve on figure 11 was drawn with f = 0.84. As can be seen, the agreement is quite satisfactory.

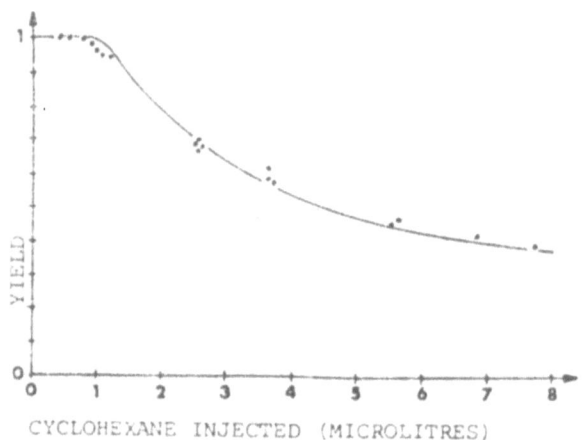

Figure 11. Cyclohexane dehydrogenation performed at 186°C in a one meter length of 1/8 inch stainless steel tubing packed with 2·05 g of a platinum on alumina catalyst which acts as the chromatographic adsorbant. Inlet pressure : 2·36 Bars, outlet : 1·00 Bar. The dots are experimental yield as a function of the amount of cyclohexane injected (μℓ). The curve is the prediction of the simplified model with a shape factor f = 0.84.

4.2. Dehydrogenation of isobutane

A necessary condition for a chromatographic reactor to exist is that both chromatographic separation and chemical reaction occur simultaneously in the column. The classical demonstration of feasability of chromatographic reactors has been made by Roginskii et al. [8] with the dehydrogenation of butane into butene on a catalytic packing at more than 400°C. It is doubtful if the hydrocarbons are retained by adsorption to any extent on the solid at so high a temperature. Indeed, Schweich [14] was unable to find any solid support achieving a chromatographic separation at this temperature. In order to clear up this point, experiments were performed with isobutane on a mixture of catalyst (Cr_2O_3/Al_2O_3) and molecular sieve 13 X at 455°C. After determining the rate of the reaction at steady state, which showed inhibition by the product (isobutene), it was impossible to observe a yield in isobutene larger in pulsed regime than at steady state, under the same average feed conditions, whatever the pulse duration and the period. This negative result seems to indicate that the improvement observed by Roginskii et al. must be ascribed to a cause other than the "chromatographic reactor effect".

Such improvements may be due to the effect of feed cycling on the kinetic behaviour of the catalyst (non linear Langmuir-Hinshelwood kinetics - adsorption/desorption steps - transient fouling) and have been observed by other authors [34] [35] [36] in conditions excluding any chromatographic separation.

4.3. Acid catalyzed esterification of alcohols by carboxylic acids

It is sometimes impossible to meet simultaneously the two requirements of 1) good catalytic activity 2) good retention properties of the stationary phase in a single substance. The solution is then to tailor the stationary phase by using an appropriate mixture of species exhibiting the two functions.

An example of this operation has been given by Sardin et al. [37] who performed the esterification of ethanol by acetic acid in the liquid phase in a chromatographic reactor. The reaction was catalyzed by a strong acid ion-exchange resin in H^+ form.

The reaction obeys the scheme

$$A + B \rightleftharpoons R + S$$

where A = acetic acid, B = ethanol, R = ethylacetate and S = water.

The requirements are :

1) A and B must travel at the same speed in the column (no separation of the reactants)

2) R must be separated from S, and from A + B

3) The rate of reaction must be high enough to promote the conversion of reactants during their transit in the column

4) The spread of the elementary pattern resulting from a square pulse injection of A + B must be as small as possible in order to maximize the productivity of the column. In particular, the period of the pulses is conditioned by the spread of these elementary patterns as the optimum is achieved when these are adjacent at the column outlet (see sec. 3 above).

The best compromise between these (sometimes) contradictory requirements was obtained as follows : the stationary phase was made of a mixture of ion exchange resin (7.5 % in volume), acting as a catalyst, and alumina acting as a sorbent. Both solids exhibit specific retention properties with respect to the reacting species. As it is difficult to completely mix the two solids before filling the column, these were packed in successive layers. The "striped column" obtained is this way contained 16 alternate zones of resin (0.85 cm) and alumina (10.5 cm) and this arrangement was shown to provide as good results as an homogeneous packing.

The composition of the solvent offered a second way of tailoring the capacity factors. A mixture of heptane and dioxanne (50/50) was found to be suitable.

Under these conditions, ethylacetate is eluted first as an unretained solute whereas acetic acid and ethanol have almost the same retention time (about 1.5 t_o, depending on the concentration). Water is strongly retained and flows out of the column as a weak background which constitutes a limitation to total conversion.

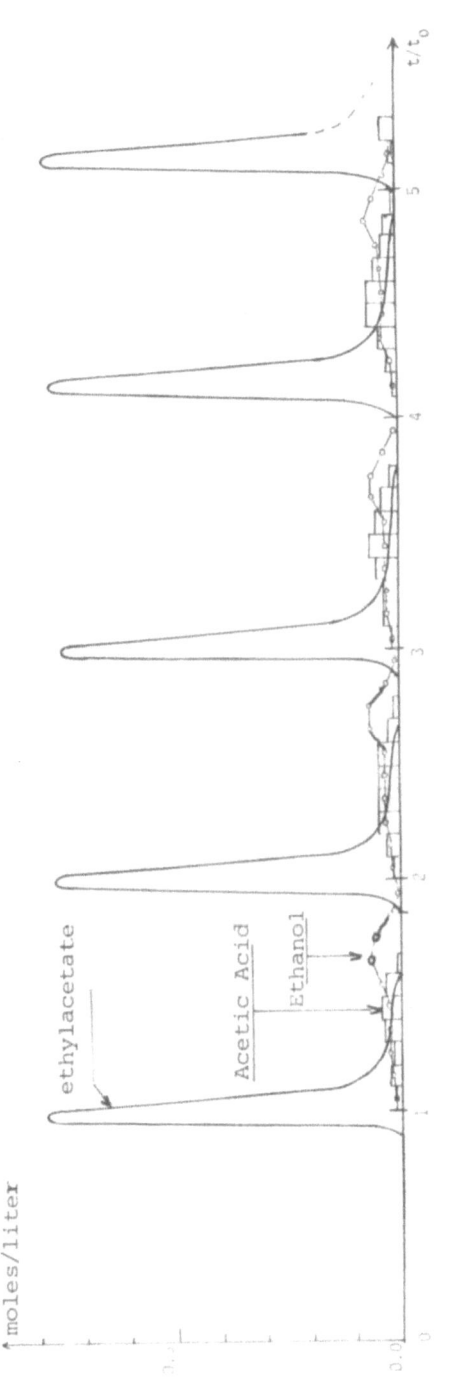

Figure 12.

Column : diameter 7 mm, length 144 cm

Stationary phase : activated acid alumina 92.5 %

                    catalyst : ion exchange resin in H^+ form (AH-10 Bayer) 7.5 %

                    particle diameter : 75 μm

Solvent : dioxanne/heptane 50/50

Injection : acid + alcohol at equilibrium 8.6 M

Flowrate : 1 cm^3/min

Space time : t_o = 37.5 min

Pulse duration θ = 0.013 t_o

Period : T = t_o

Figure 12 shows a typical chromatogram obtained in the pulsed regime after optimization of the conditions. A conversion of 0.85 is achieved, well above the steady state conversion (0.66) and the product (ethylacetate) is obtained almost pure and separated from the reactants. The simplified model has been successfully applied to account for the observed conversion. The shape factor in this case was found equal to $f = 1.1$

4.4. Productivity versus purity

If the objective is to utilize the reactant at best to obtain a high degree of conversion X_R and a concentrated and almost pure product, this requirement is contradictory with the achievement of a high productivity. The ratio of the rate of production of ester R in pulsed regime to that in steady state is

$$\frac{(X_R)_{pulsed}}{(X_R)_{steady\ state}} \cdot \frac{\theta}{T} \qquad (4-1)$$

where θ is the duration of feed pulses and T the period of the injections. In the case of the experiments described above, the gain in conversion $(X_R)_{pulsed}/(X_R)_{steady\ state}$ is moderate, about 1.25 whereas the space utilization of the column θ/T is low, about 0.1. Thus, important values of the gain in conversion are to be reached in order to compensate for the poor space utilization, and to overcome the steady state hourly yield.

This is no longer true when the yield is referred to conversion of reactants. In this case, any gain in conversion is valuable : the chromatographic reactor here shows its use as a tool for chemical synthesis. Not only the yield of conversion of reactants is improved, but **the** desired product is obtained in a concentrated form and separated from the reactants and unwanted products in a solvent from which it is easily separable.

The injection of reactants in the form of concentrated pulses has another kinetic advantage, as the rate of reaction is enhanced by the concentration increase for reaction orders higher than 1 [38].

The purity of the product (relative amount of desired product in the collected output fraction) depends on the experimental conditions. It tends to decrease as the injected quantity of reactants is increased. In the example of production of ethylacetate reported above, the purity may easily be higher than 90 %.

5. CONCLUSION

In conclusion, the chromatographic reactor is a device allowing products of single reversible reactions to be obtained, with improved conversion yields and under a concentrated and/or separated form.

The design of a chromatographic reactor requires the control and the independent adjustment of retention and kinetic parameters of the reacting species. These constraints seem to be more easily reconciled in the liquid phase than in the gas phase. This problem may be partially overcome by using a mixed stationary phase with distinct chromatographic and catalytic properties. We have clearly pointed out such a possibility in the case of the reversible esterefication of an alcohol by an acid - where the ester was obtained almost pure.

In the field of gas-solid reactions, it seems that the simultaneous obtention of a good chromatographic separation and a good chemical reactivity is difficult to achieve (especially with respect to temperature. Some of our experimental results lead us to question the examples given in support of the patent claims for catalytic chromatographic reactors. The observed improvements might be attributable to the cyclic operation of the catalyst, but this point deserves further investigation.

Besides, it has been shown that it was possible to achieve total conversion in the case of a reversible reaction by using a column of finite length fed with a pulse injection of reactants.

A new and simplified model, based on a Lagrangian approach has been developped. This model allows potential users of the chromatographic reactors to evaluate the possibilities and to predict performance with a minimum of measurable physicochemical data without beeing discouraged by too complex calculations.

However, besides its analytical application, the chromatographic reactor is a relatively sophisticated technique better adapted to the production of fine chemical than to high tonnage reactions.

REFERENCES

1. S.H. Langer and J.E. Patton. "Chemical reactor applications of the gas chromatographic column" in "New developments in gas chromatography" edited by H. Purnell, Wiley-New York N.Y., 1974, pp. 293-373
2. S.H. Langer, J.Y. Yurchak and J.E. Patton, Ind. and Engng.

Chem., $\underline{61}$, 10 (1969)

3.   S.Z. Roginskii, M.I. Yanovskii and G.A. Gaziev, Dokl. Akad. Nauk SSSR, $\underline{140}$, 1185 (1961)

4.   E.M. Magee, Canadian Patent, 631, 882 (1961)

5.   J.A. Dinwiddie and W.A. Morgan, U.S. Patent, 2, 976, 132 (1961)

6.   S.Z. Roginskii, R.A. Zimin and M.I. Yanovskii, Dokl. Akad. Nauk. SSSR, $\underline{164}$, 144 (1965)

7.   S.Z. Roginskii, E.I. Semenenko and M.I. Yanovskii, Dokl. Akad. Nauk SSSR, $\underline{153}$, 383 (1963)

8.   S.Z. Roginskii, M.I. Yanovskii and G.A. Gaziev, Kinetika i kataliz, $\underline{3}$, 529 (1962)

9.   E.I. Semenenko, S.Z. Roginskii and M.I. Yanovskii, Kinetika i kataliz. $\underline{5}$, 490 (1964)

10. Y. Murakami, T. Hattori and T. Hattori
     Journal of Catalysis, $\underline{10}$, 123 (1968)

11. C.S.G. Phillips, Private communication

12. B.D. Unger and R.G. Rinker, Ind. Eng. Chem. Fundam., $\underline{15}$, 225 (1976)

13. P.Antonucci and N. Giordano, J. of Chromatography, $\underline{150}$, 309 (1978)

14. D. Schweich, M. Sardin and J. Villermaux, Compte rendu de fin d'études, Contrat DGRST 76-7-0267

15. E.M.Magee, Ind. Eng. Chem. Fundam. $\underline{2}$, 32, (1963)

16. F.E.Gore, Ind. Eng. Chem. Proc. Des. Develop., $\underline{6}$, 10 (1967)

17. T.Hattori and Y.J. Murakami, J. of Catalysis, $\underline{10}$, 114 (1968)

18. C. Chieh and L.C. Tsang, Ind. Eng; Chem., $\underline{10}$, $\overline{47}$ (1971)

19. R.G. Wetherold, E.H. Wissler and K.B. Bischoff, Adv. Chem. Series, Chemical Reaction Engineering (Evanston), $\underline{133}$, 181 (1974)

20. D.Schweich and J. Villermaux, Ind. Eng. Chem. Fundam., $\underline{17}$, 1 (1978)

21. G.A. Gaziev, S.Z. Roginskii and M.I. Yanovskii, USSR Patent 149, 398 (1962). Chem. Abstr. $\underline{58}$, 5082 (1963)

22. J.A. Dinwiddie, U.S. Patent, 3, 243, 472 (1966)

23. J.M. Matsen, J.W. Harding and E.M. Magee, J. Phys. Chem., $\underline{69}$, 522 (1965)

24. S. Viswanathan, "Chromatographic reactors", Ph. D. Thesis, 1973. University of Minnesota

25. S. Viswanathan and R. Aris, Advances in Chemistry Series, Chemical Reaction Engineering (Evanston), $\underline{133}$, 191 (1974)

26. K. Takeuchi and Y. Uraguchi, J. Chem. Eng. Japan, $\underline{9}$, 246,(1976)

27. K. Takeuchi and Y. Uraguchi, J. Chem. Eng. Japan, $\underline{9}$, 164,(1976)

28. J. Villermaux, "Theory of linear chromatography", Percolation Processes, Theory and Applications, NATO-ASI, 1978

29. A.Klinkenberg, Chem. Eng. Sci., $\underline{15}$, 255 (1961)

30. J. Kallen and E. Heilbronner, Helv. Chim. Acta, $\underline{43}$, 489 (1960)

31. M. Kocirik, J. Chromatogr. $\underline{30}$, 459 (1967)

32. J.Y. Yurchak, M.S. Thesis, University of Wisconsin, Madison Wis. (1966)

33. D. Schweich and J. Villermaux, Paper A-142 Meeting : "Contribution of computers to the development of chemical engineering and Industrial Chemistry", Paris 7-10 March 1978

34. A.S. Al-Taie and L.S. Kershenbaum - Chemical Reaction Engineering - Houston, 1978 - ACS Symposium series N° 65, p. 512

35. M.R. Bilimoria and J.E. Bailey, Chemical Reaction Engineering Houston 1978, ACS Symposium Series N° 65, p. 526

36. E.I. Leupold and I. Renken, Chem. Ing. Techn., $\underline{49}$, 667 (1977)

37. M. Sardin and J. Villermaux, to be published

38. G.E. Klinzing, Chem. Eng. Sci., $\underline{32}$, 1536, (1977)

NOMENCLATURE

a	specific interfacial area	m^2/m^3
a_A, a_B	parameters eq (2-52)	
b_A, b_B	parameters eq (2-52)	
c	(various subscripts and superscripts - see text) concentration	mol/m^3 or mol/kg
d	parameter eq (2-52)	
e	kinetic parameter eq (2-41)	
e'	kinetic parameter eq (2-56)	
$E_o(t_o)$	Residence time distribution of the mobile phase	s^{-1}
f	shape factor	
f_A, f_B	parameters eq (2-37)	
$f(\tau)$	transfer time distribution	s^{-1}
F_i	molar flowrate of species i	mole/s
$g(s, t_o)$	elementary transfer function	
$G_o(s)$	transfer function (mobile phase)	
$G(s)$	transfer function (column)	
J	number of cells in series	
k_m	rate constant in the mobile phase	s^{-1}
k_s, k_{sj}	rate constant in the stationary phase	s^{-1}
$k_1, k_2, k_{1m}, k_{2m}, k_{1S}, k_{2S}$	rate constants (reversible reaction)	s^{-1}
k_A	mass transfer coefficient	$m\ s^{-1}$
$K_{1m} = k_{1m}t_o$ $K_{2m} = k_{2m}t_o$	dimensionless rate constants	
$K_m = k_{1m}/k_{2m}$	equilibrium constant	
K_t	equilibrium constant	$1/mol$
K'	capacity factor	
$L(s)$	transfer function	
L	column length	m
m_k	unnormalized moment of kth order about the origin	
M, M_A, M_B	capacity factors = K', K'_A, K'_B	
$M(s)$	transfer function	
n, n', n_i, n_T	mole numbers	mol
N, NTU	number of transfer units	
Q	volumetric flowrate	m^3/s

r	rate of reaction	
r_1, r_2	parameters eq (2-37)-(2-52)	
s	Laplace variable	s^{-1}
S_j	interfacial surface area	m^2
S	generalized Laplace variable	s^{-1}
t	time	s
t_m	transfer time	s
t_R	retention time	s
t_o	residence time of unretained solutes	s
T	period of pulsation	s
u	intersticial velocity	m/s
v	parameter eq (2-37)	
V, V_2	volume	m^3
w	parameter eq (2-52)	
$x = z/L$	reduced abscissa	
x_A	reduced concentration	
X_A	Laplace transform of the latter	
X_R	yield in product R	
Y_A	reduced concentration	
Y_A	Laplace transform of the latter	
z	abscissa	m

α, α_j	partition coefficient	
$\delta = k_s t_m$	Damköhler number (mass-transfer)	
$\varepsilon =$	interparticle porosity	
Φ_A	transfer flux	mol/s
θ	reduced time	
θ	duration of a square pulse	s
λ	parameter eq (2-40)	
μ_k	normalized moment of kth order about the origin	
μ'_k	normalized moment of kth order about the mean (central moment)	
ν	stoichiometric coefficient	
σ^2	variance	
τ, τ_j	transfer time	
ω_j	fraction of site occupancy	

Acknowledgement : Mathematical treatments of models on the computer were carried out with the help of Bernard ANTOINE. His valuable contribution to this work is gratefully acknowledged.